Extinctions in Near Time

Causes, Contexts, and Consequences

ADVANCES IN VERTEBRATE PALEOBIOLOGY

Series Editors

Ross D. E. MacPhee
American Museum of Natural History, New York, New York

Hans-Dieter Sues
Royal Ontario Museum, Toronto, Ontario, Canada

THE EMERGENCE OF WHALES
Evolutionary Patterns in the Origin of Cetacea
Edited by J. G. M. Thewissen

EXTINCTIONS IN NEAR TIME
Causes, Contexts, and Consequences
Edited by Ross D. E. MacPhee

Extinctions in Near Time

Causes, Contexts, and Consequences

Edited by

Ross D. E. MacPhee

American Museum of Natural History
New York, New York

Kluwer Academic / Plenum Publishers
New York, Boston, Dordrecht, London, Moscow

Cover illustration: Megafaunal losses in "near time": shaded species survive; the rest are extinct (see Chapter 11, art by Patricia J. Wynne).

ISBN 0-306-46092-0

233 Spring Street, New York, N.Y. 10013

10 9 8 7 6 5 4 3 2 1

A C.I.P. record for this book is available from the Library of Congress.

Printed in the United States of America

Ab actu ad posse valet illatio

It is possible to infer the future from the past

"Humans and Other Catastrophes: A New Look at Extinction and the Extinction Process," a symposium held April 17, 1997, in the Hall of Ocean Life, American Museum of Natural History, New York City.

Contributors

Josep Antoni Alcover Institut Mediterrani d'Estudis Avançats, 07071 Ciutat de Mallorca, Balears, Spain

John Alroy Department of Paleobiology, Smithsonian Institution, Washington, DC 20560; *Present address:* National Center for Ecological Analysis and Synthesis, University of California, Santa Barbara, California 93101

Pere Bover Institut Mediterrani d'Estudis Avançats, 07071 Ciutat de Mallorca, Balears, Spain

David A. Burney Department of Biological Sciences, Fordham University, Bronx, New York 10458

B. Sunday Eiselt Department of Anthropology, University of Michigan, Ann Arbor, Michigan 48109

Niles Eldredge Department of Invertebrates, American Museum of Natural History, New York, New York 10024

T. F. Flannery Mammal Section, Australian Museum, Sydney, New South Wales 2000, Australia

Clare Flemming Department of Mammalogy, American Museum of Natural History, New York, New York 10024

Ian J. Harrison Department of Ichthyology, American Museum of Natural History, New York, New York 10024

Gary Haynes Anthropology Department, University of Nevada, Reno, Reno, Nevada 89557

Richard N. Holdaway Palaecol Research, Christchurch, New Zealand

R. D. E. MacPhee Department of Mammalogy, American Museum of Natural History, New York, New York 10024

Paul S. Martin The Desert Laboratory, Department of Geosciences, University of Arizona, Tucson, Arizona 85701

Donald A. McFarlane Joint Science Department, The Claremont Colleges, Claremont, California 91711

Norman Owen-Smith Center for African Ecology and Department of Zoology, University of the Witwatersrand, Wits 2050, South Africa

R. G. Roberts School of Earth Sciences, La Trobe University, Melbourne, Victoria 3083, Australia

Bartomeu Seguí Departament de Ciències de la Terra, Universitat de les Illes Balears, 07071 Ciutat de Mallorca, Balears, Spain

David W. Steadman Florida Museum of Natural History, University of Florida, Gainesville, Florida 32611-7800

Melanie L. J. Stiassny Department of Ichthyology, American Museum of Natural History, New York, New York 10024

Anthony John Stuart Norwich Castle Museum, Norwich NR1 3JU, United Kingdom

Preface

"Near time"—an interval that spans the last 100,000 years or so of earth history—qualifies as a remarkable period for many reasons. From an anthropocentric point of view, the outstanding feature of near time is the fact that the evolution, cultural diversification, and global spread of *Homo sapiens* have all occurred within it. From a wider biological perspective, however, the hallmark of near time is better conceived of as being one of enduring, repeated loss. The point is important. Despite the sense of uniqueness implicit in phrases like "the biodiversity crisis," meant to convey the notion that the present bout of extinctions is by far the worst endured in recent times, substantial losses have occurred throughout near time. In the majority of cases, these losses occurred when, and only when, people began to expand across areas that had never before experienced their presence. Although the explanation for these correlations in time and space may seem obvious, it is one thing to rhetorically observe that there is a connection between humans and recent extinctions, and quite another to demonstrate it scientifically. How should this be done?

Traditionally, the study of past extinctions has fallen largely to researchers steeped in such disciplines as paleontology, systematics, and paleoecology. The evaluation of future losses, by contrast, has lain almost exclusively within the domain of conservation biologists. Now, more than ever, there is opportunity for overlap and sharing of information. Such an opportunity to collect these threads together and reweave them in a larger framework came in April 1997, when an international conference on extinction was convened at the American Museum of Natural History in New York City. Over a two-day period, conference participants and their audiences had the opportunity to hear the latest word on issues ranging from how Quaternary extinctions were caused, to how to maintain biodiversity in New York City urban landscapes. The symposium held on the first day, entitled "Humans and Other Catastrophes: A New Look at Extinction and the Extinction Process," presented by a panel of leading theorists, became a delightfully contentious examination of cause-and-effect processes. Papers delivered at this symposium form the core of the present volume. Although some participants were unable to provide a written version of their presentations in time for inclusion in this book, the salient points of each can be found in an issue-oriented white paper available from the Center for Biodiversity and Conservation (CBC) of the American Museum of Natural History.*

My first thanks must go to my co-convenor Francesca Grifo, director of the CBC, who arranged funding for the conference as a whole and acted as convenor of the second day's

***From Catastrophe to Conservation: Perspectives on Extinction. A White Paper from the 1997 Spring Symposium on Extinction at the American Museum of Natural History.* Center for Biodiversity and Conservation, AMNH, New York, 1999.

proceedings ("Preventing Extinctions: Advances in Biodiversity Conservation"). I am also very grateful to the following individuals, all of whom are associated with the AMNH: My fellow curators Joel Cracraft, Michael Novacek, and David Thomas, who acted as moderators; our exceptional volunteer Susan Zetkis, whose organizational talents got things moving at an early stage; CBC office manager Margaret Law, who made sure that everyone was well taken care of; studio photographer Craig Chesek, who documented the proceedings (see frontispiece); and my assistant and coauthor Clare Flemming, who helped with all phases of this book's gestation. There were many others whose help was vital to the success of the meeting. By mentioning some I fear I might omit others: Thus, a blanket thanks to all. However, I would be remiss if I did not especially mention the heroic work of the staff of the CBC and the Museum's Audio-Visual Department. Finally, I thank my friend and editor at Kluwer Academic/Plenum Publishers, Michael Hennelly, for the patience and good humor he unfailingly exercised in bringing this project to a conclusion.

R. D. E. MacPhee
New York

Contents

Extinctions in Near Time

Causes, Contexts, and Consequences

CHAPTER 1

Cretaceous Meteor Showers, the Human Ecological "Niche," and the Sixth Extinction

NILES ELDREDGE

1. Introduction

Mass extinctions are real events that have had a profound effect on the course of biological evolutionary history. In retrospect, it is astonishing that it took a full century and a half since the publication of Cuvier's (1812) monograph detailing successive episodes of extinction and consequent biotic rebounds—rebounds that modern science attributes to evolution, but that Cuvier understood as separate acts of Creation—for the importance of extinction to come to the fore.

Modern understanding of mass extinctions of the geological past began with the seminal mid-twentieth-century work of invertebrate paleontologist Norman D. Newell (e.g., 1967), who aptly termed them *crises* in the history of life. But Newell stood virtually alone in his focus on mass extinctions (especially the Permo-Triassic, his specialty) until three more-or-less independent factors coincided finally to make extinction a central focus of paleobiological research in the 1980s and 1990s.

First among these, perhaps, was the famous "Alvarez hypothesis" (e.g., Alvarez *et al.*, 1980) that posited the mass extinction at the end of the Mesozoic (the Cretaceous–Tertiary, or "K–T" boundary) to have been triggered by a collision between the earth and one or more "bolides"—extraterrestrial objects such as asteroids or (currently preferred) comets. With the famed iridium layer providing the smoking gun, and the eventual identification of an impact crater off the Yucatan peninsula as helpful corroborative evidence, the Alvarez hypothesis captured the fancy of the public (itself triggering a dinosaur cover story in *Time* magazine) and spurred a great deal of scientific research as well. The other four earlier global mass extinction boundaries (i.e., Upper Ordovician, Upper Devonian, Permo-Triassic boundary [end of the Paleozoic], and Upper Triassic) were soon scoured for iridium and other evidence of extraterrestrial impacts—with somewhat ambiguous results.

NILES ELDREDGE • Department of Invertebrates, American Museum of Natural History, New York, New York 10024.

Extinctions in Near Time, edited by MacPhee. Kluwer Academic/Plenum Publishers, New York, 1999.

Best of all, the Alvarez hypothesis made profound ecological sense: At its heart lay the notion that gases and particulate matter released into the upper atmosphere would likely have occluded solar radiation severely, lowering global temperatures and greatly inhibiting photosynthesis—the base of the food chain in all terrestrial and aquatic ecosystems, save the vent faunas of the marine abyss. Finally, gone were the theories that focused on the possible causes for the demise of particular taxa—such as selenium poisoning of dinosaurs and other such selective speculations.

But extinction has attracted recent scientific attention for at least two reasons other than Alvarez's work on the K–T boundary. The current biodiversity crisis is often called the *sixth extinction.* Though ecologists and systematists debate how many species currently exist on earth (e.g., May [1990] estimates some 10–13 million, whereas T. Erwin [1982] estimates there are some 30 million species of insects in the tropics alone), most agree with Wilson (e.g., 1993) that rates of habitat destruction imply a loss of some 30,000 species per year. If so—and I number among those who consider Wilson's estimate in the right ballpark—then we are currently dealing with a rate of species extinction far in excess of most estimates of "background" extinction—the so-called "normal" expected rate of extinction of species, often given as ca. 1 species/million years (e.g., Jablonski, 1986).*

Biologists have rightly wondered if mass extinctions of the geological past have light to shed on the spasm of intensified extinction currently gripping earth's biota (e.g., Eldredge, 1991). The implications of past mass extinctions for understanding the current biodiversity crisis appear to be twofold: (1) We see that global mass extinctions have indeed occurred in the past, thus defusing whatever lingering doubts that current events are at least potentially as devastating as they appear headed to be, and (2) as reviewed in somewhat greater detail in the next section, all such events, whatever their initial cause, involve ecosystem disruption, collapse, and even downright destruction.

The third reasons why extinction has commanded much biological attention in the past two decades is purely intellectual: The phenomenon of extinction, especially patterns of cross-genealogical (ecosystemic) extinction of varying spatiotemporal magnitudes, has recently been shown to play a critical role in the evolutionary history of life. Specifically, some paleontologists (myself included; Eldredge, 1995a, 1999) have come to the conclusion that little evolution is likely to occur *without* prior disruption of stable ecosystems and the consequent loss of significant numbers of preexisting species. Specifically, adaptive change is highly correlated with true speciation (counterintuitive, but empirically the case; see Eldredge, 1995a, for review), whereas speciation itself is highly correlated with post-extinction recovery phases. It further seems empirically clear that the more global and devastating an extinction episode, *the more species and consequently the higher ranked the taxa tend to be that suffer extinction.* As a corollary, the higher ranked the taxa that become extinct, the higher ranked the taxa that subsequently evolve—or simply proliferate—tend to be (see especially Eldredge, 1999).†

*A caveat here: as discussed briefly in the following paragraphs, *most* extinctions seem to be concentrated in relatively brief temporal intervals—with little or no extinctions occurring within long periods of relative ecosystem stability. Such periods may last up to 5–7 million years in marine ecosystems (e.g., Brett and Baird, 1995). Thus, the entire idea of "background" versus episodes of intensified extinction may be specious.

†Although attention is most generally focused on the great global mass extinctions, the fossil record is literally littered with thousands of more local, less taxically encompassing events. Reference to

Thus, extinction—originally imagined by Darwin a mere side effect of the evolution of progressively more efficient or "highly evolved" forms, and thus uninteresting in an evolutionary sense—is now seen in some circles at least as the sine qua non trigger of evolution. Not least, such models (such as my "sloshing bucket" concept [Eldredge, 1999]—a meld of empirical pattern and hierarchy theory [see below]) offer an explicit link between the evolutionary process and the entities and processes of the physical world—the world of "matter in motion." As such, the environmental determinism at the core of this style of evolutionary theory presents a stark contrast to the reductive, internalist genic determinism of ultradarwinian evolution theory—as most clearly exemplified by the notion of "selfish genes" (Dawkins, 1976).

2. What Causes Extinction?

As already mentioned, habitat alteration and ecosystem disruption lie at the heart of all well-studied examples of cross-genealogical extinction—from the five major mass extinctions of the geological past, the myriad lesser-degree events, up to and including the present-day "sixth" extinction.

The Alvarez hypothesis has stood the test of time—argument and scrutiny—quite well. Its only rival for late Cretaceous extinctions appears to be the counterthesis that extraordinary volcanism may be responsible for both the iridium anomaly *and* the same sort of upper atmosphere pollution imagined in the Alvarez scenario. The Deccan traps of India—the world's greatest sequence of lava flows—have recently been redated, placing the bulk of the flows at the end of the Cretaceous. Though geologists tend to attribute such enormous outpourings of lava to plate tectonics processes, it has been suggested that the flows themselves might have been triggered by the cometary impact on the other side of the world (i.e., Yucatan)—forming a hybrid model.

Though some evidence of impact has been found at the end-Triassic boundary, earlier claims for an iridium layer at the end of the Permian (i.e., for China) have not been verified. In general, extraterrestrial impact appears not to provide a general model for global mass extinction—nor especially for the myriad lesser biotic turnover events that regularly (if not truly periodically) occur on a regional basis.

Often it has proven difficult to pin down the exact physical causes of such events. This is especially the case for the Permo-Triassic event—according to Raup (1986) the most dev-

---

any detailed chart of geological (Phanerozoic) time bears this out: The Paleozoic–Mesozoic and Mesozoic–Cenozoic boundaries, naturally enough, comprise two of the five major global extinction events. But the subdivisions of the geological systems are all based on faunal and floral "turnover events"—albeit of lesser magnitude.

An example of postextinction rebound episodes where evolution of replacement taxa approximates the rank of recently extinct taxa (i.e., as determined historically by standard taxonomic practice) are the rugose corals (order Rugosa), replaced by order Scleractinia after the Permo-Triassic extinction; a classic example of proliferation of one higher taxon after the demise of another are the mammals, present alongside the ecologically dominant and far more diverse dinosaurs and collateral aquatic and aerial reptilian kin since the Upper Triassic—but only equally diverse and ecologically dominant beginning within the Paleocene, after final extinction of the dinosaurs at the K–T boundary.

astating such event to have occurred so far, at least in the 4.6-billion-year history of the Metazoa. Perhaps as many as 96% of all species extant at the end of the Permian perished in the event—leaving a scant 4% of the existing genetic variation on which the evolutionary process acted to produce new species—the basis for newly reconstructed ecosystems. The latest explanatory models for the Permo-Triassic event (e.g., D. H. Erwin, 1993, 1996; Knoll *et al.,* 1996), rejecting earlier notions of global climate cooling (cf. Stanley, 1987), are complex, involving anoxic marine turnovers, regressions, and climatic fluctuations.

Many extinction events do appear to reflect global climate change—especially relatively abrupt episodes of global cooling, on the order of 10 to 15°C in a few hundreds or thousands of years. For example, pulses of global cooling still appear to be the best hypothesis accounting for the end-Ordovician and Upper Devonian events—as well as many more lesser-magnitude events, such as the well-known "turnover" episode documented by Vrba (e.g., 1985) in African ecosystems 2.8–2.5 million years ago (Ma). On the other hand, the closely comparable repeated patterns of "coordinated stasis" (documented by Brett and Baird, 1995) for the marine Paleozoic of the Appalachian realm involve sea level changes and, at times, environmental disruptions (e.g., changes in sedimentation regime) occasioned by plate collisions. For a more complete—if now somewhat dated—review of causal theories and other details pertaining to mass extinctions of the geological past, see Eldredge (1991).

With respect to the current biodiversity crisis, there is little disagreement that it is humans who are, in effect, mimicking the end-Cretaceous bolide (Wilson, 1993; Eldredge, 1995b). Three sets of causes are generally identified as the culprits underlying the current mounting loss of species—and all are side effects of human activity. These three, in probably order of importance, are (1) direct habitat destruction, (2) overexploitation, and (3) introduced species.

Habitat degradation and destruction is the side effect of clearing of lands, initially for agriculture, later for urban centers, roads, and the like. It is also caused by air and water pollution. There is no real evidence to suggest—nor any reasons even to suspect—that preagricultural people engaged in sufficient habitat alteration to lead directly to the demise of other species. Indeed, it is only after some 10,000 years of steady but logarithmically intensifying growth of the total human population that habitat alteration has emerged as the prime cause of species loss.

Overexploitation is overharvesting of biological resources—perhaps most prominently fisheries and forests utilized for lumber and firewood. Such utilization of natural biological resources is, of course, a holdover from the days when all humans were hunter-gatherers. Unlike habitat transformation, overexploitation to the point of driving other species extinct apparently long predates the invention of agriculture. Indeed, the major theme of the chapters in this very volume are devoted to examining the pros and cons of "Pleistocene overkill"—to my mind, a compelling account of the systematic disruption of local ecosystems by invading *Homo sapiens* as we spread around the globe during the past 100,000 years. That the African megafauna—and to some extent that of Asia—has survived while elsewhere the megafauna disappeared just as humans showed up seems to me compelling, if admittedly circumstantial, evidence that (1) hominid hunting was well established in our African birthplace—and in Asian regions where early humans had long since spread: The hunted biota was, in some meaningful sense, in behavioral equilibrium with early hominid hunters. But (2) as *Homo sapiens* arrived where no earlier hominids had been, the fauna was as naive as the indigenous populations (largely birds and reptiles) of the present-day Galápagos Islands.

One alternative to the overkill hypothesis (see MacPhee and Marx, 1997) corresponds exactly to the third general cause of the present biodiversity crisis: introduced species. There is ample evidence of the ecologic havoc culminating in outright extinction of endemic species through the introduction of alien species. For example, the biota of Hawaii has been severely depleted in historic times—in part through direct habitat alteration, but perhaps in greatest measure through the introduction of alien species. MacPhee and Marx (1997) argued, in effect, that this is nothing new—that as humans spread around the globe, it was not so much their (overhunting) proclivities as their microbial and other companions that played hob with the local ecosystems they invaded. That other, smaller and arguably nonprey species (such as passerine birds of the La Brea fauna) also became extinct at about the same time that the larger mammalian species succumbed as humans arrived indeed suggests that the human disruption of ecosystems transcended simple overexploitation of an unwary food resource.

Yet now it is the direct destruction of the habitats—hence ecosystems—that is the main cause of species loss. And though it is simple enough to point to a global population of 6 billion humans that, coupled with modern technology, are effecting all of this destruction, the question of why humans have come to mimic Cretaceous bolides is far from answered. What follows is a brief sketch arguing that there has been a trend apparent in the human archaeological/paleontological record for at least the past 2.5 million years toward the increasing dominance of culture over purely biological (adaptive phenotypic) means of "making a living"—of identifying the basic human hunter-gathering ecological "niche." But to understand how the dawn of agriculture some 10,000 years ago so radically transformed this niche—indeed, humans essentially abandoned the niche in the traditional sense—I take a brief detour through the twin system of ecological and evolutionary hierarchies. Hierarchy theory reveals what is so radically different about humans, ecologically speaking. It puts us in a position to understand how humans have become the Cretaceous bolides that we are.

3. The Rise of Culture and the Hunter-Gatherer Niche

I take it as given that *Homo sapiens* nests neatly and progressively with the great apes, catarrhines, other primates, and so forth in the great cladogram of life's history. In other words, we have evolved along with the rest of life's present and past diversity. It therefore follows that, as the null hypothesis, the genetic and ecological structure of hominid species is the same in basic outline (though not of course in specific detail) as that of all other species of Metazoa—and arguably of all sexually reproducing eukaryotes.

This is to say that all species consist of minimally one, but virtually invariably many more than one, local population integrated into the matter–energy flow of their local economic system ("ecosystem")—of which each such population ("avatars" sensu Damuth, 1985) forms a component part.* The role of each such avatar in its local ecosystem is what

*Sewall Wright (e.g., 1931, 1932), ironically enough, appears to be the founder of this vision of the internal structure of species. I say ironically merely because, as a mathematically inclined theorist

ecologists traditionally mean by the term *niche*. As a corollary, it is important to see that it is not individual organisms—nor is it entire species (see below for more on this)—that have such "niches." Rather, a "niche" is a population-level construct: the functional, interactive role played by entire populations of conspecifics entirely within the context of local ecosystems.

Hunter-gathering bands of *Homo sapiens* (see Gowdy, 1998, for a handy compendium) appear to me to conform very well to this vision of local populations of conspecifics acting in a concerted economic fashion within the confines of their local natural economy. Fashion has too often dictated disdain for such supposedly "primitive" groups as the San of the Kalahari, the BaAka of the Congo Basin, or the Yanomamo of the Amazon–Orinoco, who, along with the handful of other hunter-gathering peoples, today barely cling to existence, in most instances with only vestiges of their traditional cultures intact. Yet enough has been documented of the essential modes of life of various hunter-gathering bands to make it quite clear that, in their often complex culturally mediated behaviors,[†] such bands are indeed human avatars whose existence depends solely on their culture-dependent ability mutually to wrest a living from their physical and biotic surroundings. It is essential also to recognize that the size of such bands is literally determined by the "carrying capacity" of that local ecosystem: the interface of the actual productivity of the system and the ability of the local band to exploit that productivity.

With some modern human cultures (however severely distressed) still displaying clear signs of living within the context of local ecosystems, it is an obvious and fair presumption that all preagricultural *Homo sapiens* were also hunter-gatherers. This of course implies that all preagricultural *Homo sapiens* were localized in quasi-independent, resource-limited bands—bands that were playing concerted economic roles in their respective local ecosystems. In other words, such bands had "niches." We can also presume, again based on well-documented ethnographic examples of contemporary hunter-gatherers, that culture (i.e., complex, language-mediated systems of learned behavior) forms the very basis of such modern and immediately preagricultural human "niches."

The question that immediately arises is whether we can point to a time in hominid evolutionary/ecological history when culture manifestly was of lesser importance. Although tool use and other aspects of "culture" are increasingly attributed to, e.g., great apes, the

---

and analyst of experimental breeding data, Wright appears never to have spent a professional day in his life examining the biology of species in the wild. Nonetheless, Dobzhansky (1937)—who is justly famed for his penchant of matching experimental results with precisely such observations in the wild—adopted Wright's vision of species structure completely and virtually verbatim.

As an aside, I note that the tendency of some contemporary population geneticists (e.g., Coyne and Charlesworth, 1997) to treat species as if they are simply "large populations" does violence to these empirically well-established percepts on the internal structure of species, as well as to Wright's (1931, 1932) and many subsequent biologists' conclusions: The mutational and selectional histories of each of these local populations, integrated as they are in slightly to greatly differing local economic systems, can hardly be expected to be identical specieswide. This has led me (Eldredge, 1995a, 1999) to conclude that stasis, rather than protracted directional evolutionary change, should be the expected norm throughout the large measure of a species's history.

[†]For example, San hunters utilize poisons from beetle larvae and plants—poisons that enable the San to hunt without delivering an otherwise lethal blow from the physical wound itself, yet poisons that themselves do not endanger those who eat the flesh. This sophisticated technology has profound ramifications for the entire structure of meat consumption in San bands.

physical and behavioral adaptations that underlie resource utilization—thus the local population's niche—of nonhuman species is quintessentially and purely biological. Lions, in other words, typically hunt in prides; but though their hunting behavior is social, it emphatically is not "cultural" in any meaningful sense of the term.

Clearly, at some point in hominid history, culture began to take precedence over purely adaptive biological approaches to simply living life. And I think we can, in a general way, at least, document that change over the past 3 million years or so. I'll mention just three points in hominid history that seem to me to record precisely just this transition from purely biological to culture-mediated approaches to living within the context of local ecosystems.

3.1. Ecosystem and Species Turnover 2.5 Ma

I have already alluded to Vrba's (1985) and others' documentation of a major biotic turnover event in eastern and southern African ecosystems. Wet-woodland plant and animal species drop out of the record, for the most part rather abruptly at 2.5 Ma—and are replaced in the next succeeding layers by species on the whole adapted to cooler, drier conditions. The cause appears to be a drop of some 10–15°C in mean annual temperature, apparently beginning 2.8 Ma. According to Vrba and others, though the cause was progressive and cumulative, the effect came as a rather abrupt ecosystem makeover—as evidenced especially in the spread of open-savannah conditions, meaning drier grasslands at the expense of wetter woodlands.

Vrba has adduced a "turnover pulse" model—with distinct ecological and evolutionary components—using this event in particular as a case study. In brief, the ecological component involves migration out of such disturbed areas by species not adapted to the new conditions; symmetrically, species that are so adapted will migrate into the area. In this ecological component, the abrupt change recorded at the stratigraphic boundary is purely ecobiogeographic, i.e., *not* evolutionary. Though, for example, the eurytopic impalas occur on both sides of the 2.5-Ma datum, woodland antelope species give way to grassland-adapted species.

On the other hand, true extinction also occurs when species cannot continue to recognize familiar habitat elsewhere. And, significantly, ecosystemic disruption can also lead to true speciation. After all (as pointed out by Vrba, 1985), habitat fragmentation (e.g., fragmentation of woodlands as the result of the spread of grasslands) is at least in prospect a prime source of geographic isolation.

Hominid fossils, of course, are found along the length of the East African Rift Valley System—as well as in caves in southern Africa (which recently have yielded strong evidence of a cooler, drier climate replacing warmer wetter conditions some 2.5 Ma). Though the Taung child may be as young as 2.3 Ma, the stratigraphic occurrence of *Australopithecus africanus* appears to end right around 2.5 Ma. This species is known from the breccia deposits of southern Africa.

In both eastern and southern Africa, species of both the robust australopithecine lineage *and* gracile hominids (*H. rudolfensis* and/or *H. habilis*) appear 2.5 Ma. So, for that matter, do the earliest known, determinedly worked stone tools. The gracile species have expanded braincases (i.e., over the condition of earlier gracile australopithecines) and are the main candidates for authorship of those crude early tools.

From these salient facts—especially because they fit in with the more general pattern of biotic turnover—several conclusions are suggested: (1) Ecological control over distribution, plus the extinction and evolution of hominid species of this general vintage, followed exactly the same "rules" governing the appearance and disappearance of all other mammalian species of that time. (2) Development of material culture was closely associated with—and arguably coincident in its development with—the evolution of particular hominid species. In general, (3) early hominid species were living in the context of small bands within local ecosystems—with niches not conspicuously informed by complex cultural systems (though the tools imply a point of departure; this last conclusion becomes more apparent after consideration of points 2 and 3 following).

3.2. The Evolution of *Homo ergaster*

The cooling pulse of some 2.8–2.5 Ma produced no as-yet-detected signs of glaciation in the Northern Hemisphere. In contrast, 1.65 Ma marked the onset of the first of the four major pulses of Northern Hemisphere continental glaciation. With it came another cooling pulse, and once again Africa became cooler and drier.

Right about then, *Homo ergaster* (the African forerunner to, and originally classified as, *Homo erectus*) appeared. Though the oldest fossils attributable to this species are some 1.8 million years old, the dates are sufficiently intriguingly close enough to suggest that this may well have been the very last time a hominid species evolved in concert with an episode of global climate change.

The nearly complete skeleton of an individual of this species—the 1.6-Ma "Turkana boy"—was some 5′4″ at the time of his death, but is estimated to have had a potential adult height of ca. 6′ (Tattersall, 1995). Though the cranial capacity of Turkana boy was not great, in general the *ergaster/erectus* lineage shows an expanded average of nearly 1000 ml, an average, moreover, that according to Rightmire (1981) was destined to remain virtually constant statistically throughout the combined 1.3-Ma range of *H. ergaster* and its Asian offshoot, *H. erectus.*

The tool kit of *ergaster/erectus* was likewise expanded—now including more sophisticatedly wrought hand axes and cleavers. And, though earlier claims that late Chinese populations of *H. erectus* had fire have been disputed, the evidence that early *H. ergaster* in Africa did indeed have the use of controlled fire is compelling (Tattersall, 1995). However, though the increased complexity of early *H. ergaster* material culture is noteworthy, its evolutionary appearance still seems to have been linked to a climatic event—more in keeping with earlier patterns than what was to follow:

3.3. The First Diaspora

The onset of the second major glacial advance 0.9 Ma is correlated with an ecological event unparalleled, not just in prior hominid history, but to my knowledge in the annals of the history of life in general. In response to yet another global cooling event, instead of becoming extinct, or perhaps undergoing yet another speciation event, and rather than huddling along the equator for whatever residual tropical warmth there might have been, parts of the *ergaster/erectus* species lineage actually moved north into the very teeth of the glacial advance.*

Though examples of changing distribution of individual taxa abound in the fossil record—examples ranging from biogeographic changes evident in Devonian seas (e.g., Eldredge, 1971) to the well-documented and analyzed instances of Pleistocene range expansions and contractions in conjunction with glacial/interglacial intervals—none appears to me to suggest actual *expansion* of niche. Rather, all fit comfortably under the category of "habitat tracking"—meaning simply movement to occupy habitats that are "recognizable," i.e., acceptably close to the ecosystemic conditions to which the organisms in question are already adapted (see, e.g., Eldredge, 1995a,b, for reviews).

Not so in the case of *ergaster/erectus* expanding into Eurasia during the second glacial. Here, apparently, the range of tolerable climatic conditions, not to mention comestible plants and animals, was increased. The species as a whole did not leave Africa; rather, parts of that species became able to live in the starkly different conditions afforded by Pleistocene Eurasia, while the rest of the species remained in far more familiar surroundings.

No conspicuous morphological evolutionary change accompanies this dramatic expansion in range and environment. Rather, it must have been culture—fire, clothing, possession of successful hunting strategies, and countless other cultural nuances—that allowed this truly remarkable event, this unprecedented *expansion,* of the ancestral *ergaster* niche to occur.

And thus, in a very general way, I think we can understand how culture came to inform, and eventually to dominate, the human ecological "niche." Bear in mind, though, that as important as culture becomes in hunter-gathering approaches to life within the local ecosystem, the ecological situation remains precisely that: life within the context—and confines—of the local ecosystem. And so it remained, right up to and including the very sophisticated cultural approaches to living documented by anthropologists among the remnant contemporary bands of hunter-gatherers.

Culture evidently enabled hominids to control their ecological life sufficiently such that each new successive episode of global climate change no longer elicited the knee-jerk pattern of habitat tracking cum extinction cum speciation—the patterns that stamp the entire history of life, including early phases of hominid evolution. For hominids, speciation and extinction, in particular, had become decoupled from physical environmental events by the mid-Pleistocene. Soon thereafter, changes in material culture were to become decoupled from speciation events as well (Eldredge and Tattersall, 1982).

But, no matter how pervasive the impact of culture on human ecology became, up to the point around 10,000 years ago when people started to become farmers, humans were still living inside ecosystems. They were just, through culture, getting better at it. To set the analytic stage a bit more clearly for understanding that the (first) Agricultural Revolution

*Recent finds and redating of previously collected specimens suggest that *Homo ergaster,* and perhaps even species of *Australopithecus,* had arrived in parts of Eurasia well before 0.9 Ma. However, the evidence is scarce, and the dates in no case fully confirmed. If confirmed, the evidence still suggests that whatever migrations into Eurasia had occurred prior to 0.9 Ma, they were nothing compared with the volume of hominids that began leaving their tools in abundance all over Eurasia beginning 0.9 Ma.

was not yet another example of niche expansion—but rather a radical change in the billion-year-old history of the occupation of ecological niches by populations of sexually reproducing organisms—I turn now to a brief summary of the hierarchical structure of large-scale biotic systems.

4. The Economic and Genealogical Hierarchies: A Thumbnail Sketch

Organisms do two—and only two—general sorts of things: They differentiate, grow, and maintain the soma, and they reproduce. Both classes of activity require energy. Otherwise, the relationship is asymmetrical: Organisms need a soma simply to "be," and they need a soma to reproduce (as the "vehicles" of Dawkins, 1982). But the converse is not true: The soma does not "require" organismic reproduction.* Reproduction is required only to make more organisms, i.e., descendant somas. Put another way, of the standard list of organismic physiological functions, all (e.g., respiration, digestion, elimination) are required for an organism's survival *except* reproduction.

All organisms (save a very few whose lifetime nourishment ends with the egg, or some other preadult stage of development) must obtain energy and nutrients—to build and conserve the soma, and to reproduce. The business of obtaining that energy and those nutrients is the organism's economic life. Organisms within a single species (defined as that collectivity of sexually reproducing organisms sharing what Paterson [e.g., 1985] has called the specific mate recognition system (SMRS)—or simply, reproductive adaptations) are similar in morphology, physiology, and behavior as a result of their common descent, their shared genomes. With similar economic proclivities, conspecifics in the same local area find themselves exploiting the same resource base—and thereby variously competing and cooperating for said resources. It is this collective pressure by these individual avatars (local populations of conspecifics in an explicitly economic context) on the local resource base that conforms to the traditional notion of ecological "niche," as already mentioned.

Thus, avatars are parts of economic systems—systems composed of other avatars plus the physical environment. Ecosystems include everything living in a given place: all avatars of fungi, plants, and animals, plus the populations/clones of bacteria and eukaryotic microbes, with their complex reproduction (which often does, but may not, include gametic or other form of genic exchange).

Boundaries of ecosystems are almost of necessity difficult to recognize and specify. Energy flows across the boundaries of even the most clear-cut of such systems—as when an osprey, otherwise living in trees on the margin of a waterway, takes a fish from its aqueous environment. And that is the point, really: Although I have argued at length (e.g., Eldredge, 1985, 1986; Eldredge and Grene, 1992) that ecosystems are individuated systems, however fuzzily bounded, it is at the same time true that local ecosystems are connected geographically into progressively larger systems, up to and including the global economic system that some know as "Gaia."

*The soma is maintained by ongoing mitotic "reproduction" of its constituent cells. I have in mind here the contrast between the economic and reproductive activities of *entire* organisms.

Thus, by sheer dint of the necessary procurement of energy and nutrients simply to *be,* organisms find themselves as parts of local groups of conspecifics, which themselves are parts of economic systems of matter–energy transfer (from the physical environment to organisms; through complex food/energy webs among avatars; and ultimately back to the physical world); these local systems in turn are linked by the generically same matter–energy transfer processes into progressively larger-scale, regional systems, forming a hierarchy of nested systems within systems of matter–energy transfer. By sheer dint of being alive and trying to stay alive, organisms find themselves as basal units in a hierarchy of matter–energy transfer systems. As such, economically speaking, every organism is a citizen of the world—albeit linked up through its local avatar/local ecosystem/regional biota—up to Gaia.

But organisms, at least potentially, or in some cases at least vicariously, also reproduce. For sexually reproducing organisms, this means finding a mate—and mates are generally selected from a sample of possibilities, the eligible (by age, sex, and so forth) of those conspecifics living in close enough proximity given the vagility and other biological factors pertinent to those organisms. Local populations of conspecifics in an explicitly reproductive sense are traditionally called *demes.* The composition of local demes may converge to near identity with the corresponding avatar—or the two might be quite different. For example, in many mammalian species (e.g., African elephants), breeding bulls typically spend the bulk of the year living apart from matriarchal herds of females and young—while older immature males may form their own clusters. The avataral structure is thus very different from the deme—which in any given year involves only the prime males and females.

Demes, too, are parts of large-scale systems. But in this context of reproduction, that larger-scale system is the species—already defined as the largest collectivity of sexually reproducing organisms sharing reproductive adaptations (i.e., anatomies, physiologies, and behaviors allowing them to reproduce should the opportunity arise).

But that is not all—for species are known to speciate: Species occasionally divide into two or more daughter species. This is, of course, in itself a form of reproduction. Should changes in the SMRS occur (or should sufficient genetic change pertaining to somatic adaptations occur, sufficient to prevent reproduction—the latter a less likely event) in isolation, true speciation is said to have occurred.

Thus, species are themselves parts of larger-scale systems—in this instance, skeins of ancestral—descendant species known as *monophyletic taxa.* Smaller such related clusters are genera; larger ones are families, orders, classes, phyla. Though unnamed, there is at least in principle a category of "all life" (i.e., above the "kingdom" level) that would include absolutely every "species" (i.e., including prokaryotic "bacteria"), and all eukaryotes that have ever graced the planet.

Thus, moment-by-moment matter–energy transfer involves organisms in a nested hierarchy of progressively larger economically interactive systems; each such system is made of components whose members are engaging in matter–energy transfer, be it organisms within avatars; avatars within ecosystems; local ecosystems within regional ones; or regional systems within Gaia.

At the same time, occasional reproduction links sexually reproducing organisms together into local demes; demes themselves split, merge, and become extinct, the sum of them providing the structure and internal genetic dynamics of the entirety known as the "species"; and species, too, "reproduce"—producing more species, making them parts of hierarchically arrayed monophyletic taxa. It is the interactions among components of these two hierarchies that govern the evolutionary process.*

The purpose of this rapid, yet abstract, description of the twin biotic hierarchical systems is twofold: we can see at a glance that (1) *absolutely all (sexually reproducing) organisms belong to avatars that are parts of local ecosystems* and (2) *species are not parts of economic systems.* There are no large-scale economic biotic systems in which species play concerted roles analogous to avatars within local ecosystems—except in the trivial case where species are reduced to a single avatar/deme.† Though we had encountered the first proposition in general terms in conjunction with the foregoing discussion, the structure of the economic hierarchy clarifies why that proposition must be so. As to proposition No. 2—i.e., that species are parts of genealogical, not economic, systems—the implications are many. For evolutionary theory, it explains why species do not have niches, nor do higher taxa occupy "adaptive zones," and other similar constructs in the literature. More importantly, in the present context proposition No. 2 has much to tell us about how different we postagricultural *Homo sapiens* truly are from any other species that has ever been present on earth. Indeed, though I believe that absolutely all species adhere absolutely to both propositions, it is at once stunning and undeniable that we, *Homo sapiens,* have violated them both.

5. Agriculture, *Homo sapiens,* and the Human Ecological Niche

Violation of proposition No. 1—that all organisms are parts of avatars playing concerted economic roles in local ecosystems—came with the invention of controlled planting and animal husbandry. Though conventionally seen as an "expansion" of the human ecological niche, the invention of agriculture was much more like a declaration of independence from—or even war on—local ecosystems. Nor is such imagery unduly extreme: Substitution of monocultures (or at most a few crop varieties) for the normally occurring plant species of the local ecosystems at the very least perturbs such local systems—and carried out extensively amounts to the literal physical destruction of local terrestrial systems. Prior to agriculture, there had been no concept of "weeds"—which, at least originally, were simply nonutilizable local plant species competing for growing space with planted crops.

Agriculture—the human culturally mediated control of food production—removes local bands of people from the local ecosystem. With agriculture, people are no longer governed by their abilities to wrest energy sources from the productivity of their immediate surroundings. Indeed, as agricultural technology and transportation have become more sophisticated, the agricultural activities carried out in one place have increasingly come to sustain peoples living elsewhere—a process that began with division of labor (and social stratification) that immediately accompanied early agricultural societies.

*See especially Eldredge and Salthe (1984), Eldredge (1985, 1986, 1989, 1999) and Eldredge and Grene (1992) for deeper discussion of the nature and structure of both hierarchical systems and how they interact in the evolutionary process.

†See references in preceding footnote for a more detailed consideration of why species are purely genealogical, and not economic, entities.

Agriculture enabled settled existence,* and with that came the rise of towns, cities, and nation-states.

Anthropologists have long pointed out that agriculture is no panacea to the exigencies of economic existence. Famine stalks the land as droughts, floods, and other natural disasters strike. Social upheaval also imperils food production. But on the whole, and in addition to the finer aspects of "civilization" (e.g., invention of writing, science, technology, and the arts), agriculture has been enormously successful—as measured in traditional evolutionary ecological terms: The human population has mushroomed from some 5 *million* only 10,000 years ago to 6 *billion* at the millennium.†

Darwin and Wallace were probably the first to see the biological implications of the avataral structure of species: Though, of course, they did not express themselves explicitly in such terms, resource limitation lies at the very heart of natural selection. Darwin (1859) said it best when, taking a fanciful look at elephant population growth, he noted that the world was hardly "standing room only" (his words) in elephants—which long since would have been the case had there been no limits to population expansion even for such slow-breeding animals as elephants.

The stunning growth of the human population after the invention of agriculture can mean only one thing, namely, that the primordial limits to such growth were demolished. We did not simply get better at wresting a living from ecosystems: We actually stopped doing that (for the most part, and vestiges of hunter-gathering to the contrary notwithstanding) in favor of agriculture. No longer limited by the productivity of local ecosystems, the Malthusian lid on population size no longer obtained. With the Malthusian lid thus popped off, the human population simply took off—with an accelerated doubling time that threatens to see 12–14 billion of us by the middle of the next century.

That's why there are, suddenly, so many of us—and how it has come to be that the initially modest transformation of the landscape (and concomitant pollution of the air and aqueous ecosystems, plus the juggernaut of [sub]urbanization) has grown to such dimensions that we have begun to destroy so many ecosystems—and to lose, as a direct consequence, so many species that we are beginning to talk of the "sixth extinction."

Yet, even though the human population continues to double at ever-diminishing intervals, most demographers have traditionally seen an eventual leveling off (see Cohen, 1995)—and why this prediction will doubtless prove true involves our second violation of the laws of ecological nature:

While the human population has expanded over the past 10,000 years, and while technology, especially communication and travel, remained short of the present-day capabilities developed especially in the twentieth century, humans were free to think of themselves as somehow outside of—or even "above"—the rest of nature. Indeed, given our first violation of ecological nature, this was an accurate and even appropriate view. The vast majority of us simply do not live inside ecosystems, and haven't for the past 10,000 years.

But the advances in communication and travel in the twentieth century have changed the situation yet again. Talk of the global economy is not idle. Kennedy (1993) reported

*Slash-and-burn agriculture, of course, being a partial exception to this generalization.

†Cohen (1995) reported that estimates of the earth's human population 10,000 years ago range from a low of 1,000,000 to a high of 10,000,000—hence my choice of the average figure.

that, as of the mid-1990s, humans were already exchanging in excess of $1 trillion in goods and services *daily.*

Hierarchy theory tells us that species per se are not economic entities and do not play concerted roles in specifiable ecosystems. The eastern gray squirrels of New York's Central Park rely on a different range of plant species, and predated on by different species, than is the case for their conspecific avatars in South Carolina and Ontario. Only modern *Homo sapiens* interact economically with far-flung groups of conspecifics—meaning people are doing business with each other literally around the globe, and all societies are connected in some fashion (and still to varying degrees) with absolutely all other postagricultural societies. And this is the second wholly unique aspect of human ecological behavior—absolutely unprecedented in the 3.5-billion-year history of life: We are the first species to have become an economic entity—to exert a cohesive economic force as an entity—in addition to being a shared genome, a reproductive, genealogical entity.

But if *Homo sapiens* may fairly be said to have emerged as a cohesive economic entity, what economic system is it a part of? Clearly it's the global system—Gaia itself. Unconsciously, demographers have grasped this point all along: The ultimate limit to human growth will be the limits to agricultural productivity and other aspects of "natural" resources afforded by our planet itself. These limits very definitely also include the diminution (if not downright loss) of ecosystem services (e.g., such things as the carbon and nitrogen cycles, oxygen production, clean air and water) already well under way.*

Thus, hierarchy theory clarifies the human ecological condition—preagriculturally, postagriculturally, and now as postindustrialized, high-technology interconnected organisms. To the long, traditional list of uniquely human attributes—i.e., upright posture, opposable thumbs, expanded forebrains, possession of culture that is rooted in language—we can add two critical ecological percepts: (1) freedom from local ecosystems, but (2) economic role-playing membership in the global ecosystem.

It is these two factors that have caused humans to resemble Cretaceous comets in our devastating impact on the global ecosystem. Amelioration of our status as vectors of the sixth extinction will only come when we collectively realize—and exert the corresponding political will—that we are, in the end, neither outside nor above the economy of nature, but have, rather, simply, if dramatically, redefined our role in the natural economy of this planet.

References

Alvarez, L. W., Alvarez, W., Asaro, F., and Michel, H. V. 1980. Extraterrestrial cause for the Cretaceous–Tertiary extinction. *Science* **208**:1095–1108.

Brett, C. E., and Baird, G. 1995. Coordinated stasis and evolutionary ecology of Silurian to Middle Devonian fau-

*At this writing, the President of the United States had just announced a firmer stance on the subject of greenhouse gas emissions ("global warming") than his prior statements indicated. His toughened stance (though of course decried by some environmentalists as too little too late, and by some industrialists as harmful to economic growth) included the explicit acknowledgment that such issues are inherently global. The President's statement came a few days after a similar strong stance articulated by the Prime Minister of Great Britain—who urged all members of Parliament to read the report of his science adviser on the subject. His science adviser is Robert M. May, who has already contributed so much to pure and applied ecology, and whose work has been cited in this text.

nas in the Appalachian Basin, in: R. Anstey and D. H. Erwin (eds.), *Speciation in the Fossil Record,* pp. 285–315. Columbia University Press, New York.

Cohen, J. E. 1995. *How Many People Can the Earth Support?* Norton, New York.

Coyne, J. A., and Charlesworth, B. 1997. [Letter]. *Science* **276**:339–341.

Cuvier, G. 1812. Discours sur les Revolutions de la Surface du Globe. Paris.

Damuth, J. 1985. Selection among "species": a formulation in terms of natural functional units. *Evolution* **39**:1132–1146.

Darwin, C. 1859. *On the Origin of Species.* John Murray, London.

Dawkins, R. 1976. *The Selfish Gene.* Oxford University Press, London.

Dawkins, R. 1982. *The Extended Phenotype: The Gene as the Unit of Selection.* Freeman, San Francisco.

Dobzhansky, T. 1937. Genetics and the Origin of Species. Columbia, New York.

Eldredge, N. 1971. The allopatric model and phylogeny in Paleozoic invertebrates. *Evolution* **25**:156–167.

Eldredge, N. 1985. *Unfinished Synthesis: Biological Hierarchies and Modern Evolutionary Thought.* Oxford University Press, London.

Eldredge, N. 1986. Information, economics and evolution. *Annu. Rev. Ecol. Syst.* **17**:351–369.

Eldredge, N. 1989. *Macroevolutionary Dynamics: Species, Niches and Adaptive Peaks.* McGraw–Hill, New York.

Eldredge, N. 1991. *The Miner's Canary: Unraveling the Mysteries of Extinction.* Prentice–Hall, Englewood Cliffs, NJ.

Eldredge, N. 1999. *The Pattern of Evolution.* Freeman, New York.

Eldredge, N. 1995a. *Reinventing Darwin: The Great Debate at the High Table of Evolutionary Theory.* Wiley, New York.

Eldredge, N. 1995b. *Dominion.* Holt, New York.

Eldredge, N., and Grene, M. 1992. *Interactions: The Biological Context of Social Systems.* Columbia University Press, New York.

Eldredge, N., and Salthe, S. N. 1984. Hierarchy and evolution. *Oxford Surv. Evol. Biol.* **1**:182–206.

Eldredge, N., and Tattersall, I. 1982. *The Myths of Human Evolution.* Columbia University Press, New York.

Erwin, D. H. 1993. *The Great Paleozoic Crisis: Life and Death in the Permian.* Columbia University Press, New York.

Erwin, D. H. 1996. The mother of mass extinctions. *Sci. Am.* **July**:72–78.

Erwin, T. 1982. Tropical forests: their richness in Coleoptera and other arthropod species. *Coleopt. Bull.* **36**:74–75.

Gowdy, J. M. (ed.). 1998. *Limited Wants, Unlimited Means.* Island Press, Covelo, Calif.

Jablonski, D. 1986. Background and mass extinctions: the alteration of macroevolutionary regimes. *Science* **231**:129–133.

Kennedy, P. 1993. *Preparing for the Twenty-first Century.* Random House, New York.

Knoll, A. H., Bambach, R. K., Canfield, D. E., and Grotzinger, J. P. 1996. Comparative earth history and late Permian mass extinction. *Science* **273**:452–457.

MacPhee, R. D. E., and Marx, P. A. 1997. The 40,000-year plague: humans, hyperdisease, and first-contact extinctions, in: S. M. Goodman and B. D. Patterson (eds.), *Natural Change and Human Impact in Madagascar,* pp. 169–217. Smithsonian Institution Press, Washington, D.C.

May, R. M. 1990. How many species? *Philos. Trans. R. Soc. London Ser. B* **330**:293–304.

Newell, N. D. 1967. Revolutions in the history of life. *Geol. Soc. Am. Spec. Pap.* **89**:63–91.

Paterson, H. E. H. 1985. The recognition concept of species, in: E. S. Vrba (ed.), *Species and Speciation. Transvaal Mus. Monogr.* **4**:21–29.

Raup, D. M. 1986. Biological extinction in earth history. *Science* **231**:1528–1533.

Rightmire, P. 1981. Patterns in the evolution of *Homo erectus. Paleobiology* **7**:241–246.

Stanley, S. M. 1987. *Extinction.* Scientific American Books, New York.

Tattersall, I. 1995. *The Fossil Trail: How We Know What We Think We Know About Human Evolution.* Oxford University Press, London.

Vrba, E. S. 1985. Environment and evolution: alternative causes of the temporal distribution of evolutionary events. *S. Afr. J. Sci.* **81**:229–236.

Wilson, E. O. 1993. *The Diversity of Life.* Harvard University Press, Cambridge, Mass.

Wright, S. 1931. Evolution in Mendelian populations. *Genetics* **16**:97–159.

Wright, S. 1932. The roles of mutation, inbreeding, crossbreeding, and selection in evolution. *Proc. Sixth Int. Congr. Genet.* **1**:356–366.

CHAPTER 2

Prehistoric Extinctions on Islands and Continents

PAUL S. MARTIN and DAVID W. STEADMAN

1. Overview of a Prehistoric Catastrophe

Geological extinction of a continental megafauna of Holarctic mammoths, American ground sloths, and Australian diprotodonts, to name a few mammalian examples, rivals pulsing ice sheets and fluctuating sea levels in being a hallmark of the Quaternary. To these more familiar examples of late Quaternary extinction (LQE), younger fossils recently recovered from oceanic islands, including bird and land snail taxa from the Pacific and various endemic terrestrial vertebrates from the Caribbean, add many thousands of species and endemic populations to the extinction list. Apart from loss of a few pinnipeds and sirenians (large coastal or estuarine mammals), the LQE was strictly a terrestrial accident. Unlike the case on islands, on the continents virtually all small vertebrates (other than commensals or parasites of the large mammals) escaped extinction. Looking toward the continents from deep-water islands promises to aid in our understanding of what happened, when it did, and what forced the change.

LQE swept away half of the 167 genera of large mammals (>44 kg) found on the continents (Martin, 1990). Most or all disappeared within the last 40,000 radiocarbon years, an interval we name "Libby time," after Nobel laureate Willard Libby. Uncalibrated radiocarbon dates in this interval will be designated "yrbp" and calendar years, "yrBP." Calibration of radiocarbon measurements has revealed departures of as much as 20% with dates from 7000 to 10,000 calendar years too young by 1000–1500 years and those in excess of 13,000 years too young by roughly 2000 years or more (Hughen *et al.*, 1998).

Unlike the loss of large reptiles, ammonites, and numerous other marine taxa that disappeared in a mass extinction 65 million years ago (Ma) at the end of the Cretaceous, along with much earlier mass extinctions, the late Quaternary losses involve no radical loss of marine life and no known or hypothetical cosmic intruder. Although LQE is characterized by selective losses of large vertebrates (including "charismatic megafauna") and thousands of

---

PAUL S. MARTIN • The Desert Laboratory, Department of Geosciences, University of Arizona, Tucson, Arizona 85701. DAVID W. STEADMAN • Florida Museum of Natural History, University of Florida, Gainesville, Florida 32611-7800.

Extinctions in Near Time, edited by MacPhee. Kluwer Academic/Plenum Publishers, New York, 1999.

insular species or populations, the character of the losses is quite unlike the five much older mass extinctions that are typical of earlier parts of the Phanerozoic. The strong possibility that at least some of the mass extinctions were triggered by an asteroid hit is not an issue in the case of LQE. The youngest extraterrestrial accident severe enough to attain the threshold of impacts believed to have had global consequences occurred at the end of the Pliocene, 2.15 Ma, when the Eltanin asteroid struck the South Pacific at roughly 57°S and 91°W. Its diameter is estimated at 1–4 km, large enough to generate tsunamis 25–50 m in height in the open ocean, running up on continental shelves to 10 to 25 times that height (Gersonde *et al.*, 1997). Any low-lying islands in the region at that time should have been purged of most of their terrestrial biota. Tsunamis forcing extinctions are not known in the Late Quaternary.

The geochronology of LQE differs from that of the mass extinctions in having a finer resolution. Radiocarbon dates demonstrate stepwise extinction pulses accompanying human spread across the globe (Figs. 1–3). The primary competing explanations for LQE are (1) human impacts, especially at first contact (FC), and (2) global or regional climatic change (see various models explored by Burney, 1993; MacPhee and Marx, 1997; Martin and Klein, 1984; Putshkov, 1997; Stuart, 1991). Although many models commonly combine aspects of both cultural and climatic forcing, MacPhee and Marx (1997:209) note that from a testing standpoint "this is a relatively weak approach epistemologically; . . . an admission of defeat rather than a breakthrough in understanding." They take the high ground and mount a strong case for what appears to be a long shot, the relatively neglected model of highly lethal disease as the driving force in FC extinctions.

The most severe case of mammalian LQE is found in the Americas. Around 11,000 years BC, near the end of the last glaciation, North and South American lost most of their terrestrial megafauna, defined here as animals over 44 kg (about 100 pounds) in adult body weight. North America lost 34 genera of large mammals (Table 1) and 3 genera of small mammals (Table 2). South America lost 51 genera, all large (Table 3). Subtracting 11 shared extinctions, the hemisphere lost 74 large genera along with three mammalian orders, the Litopterna, Notoungulata, and Proboscidea. The synchronous extinction of three mammalian orders on one continent is extraordinary. Prior to ordinal extinction in the Lujanian (the South American equivalent of the Rancholabrean in North America), the Litopterna and

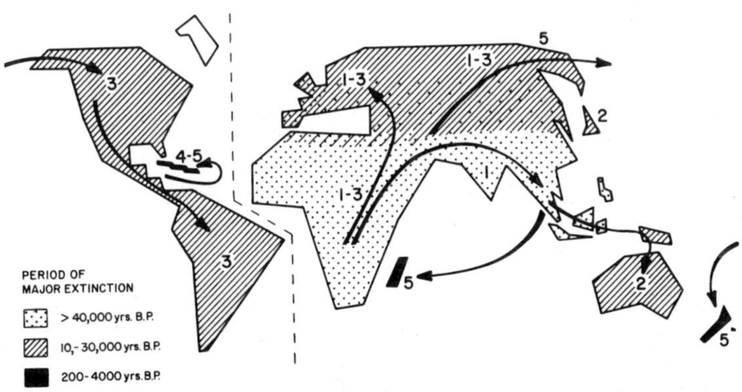

FIGURE 1. Global sequence of Late Quaternary extinction and prehistoric human dispersal (after Martin, 1967).

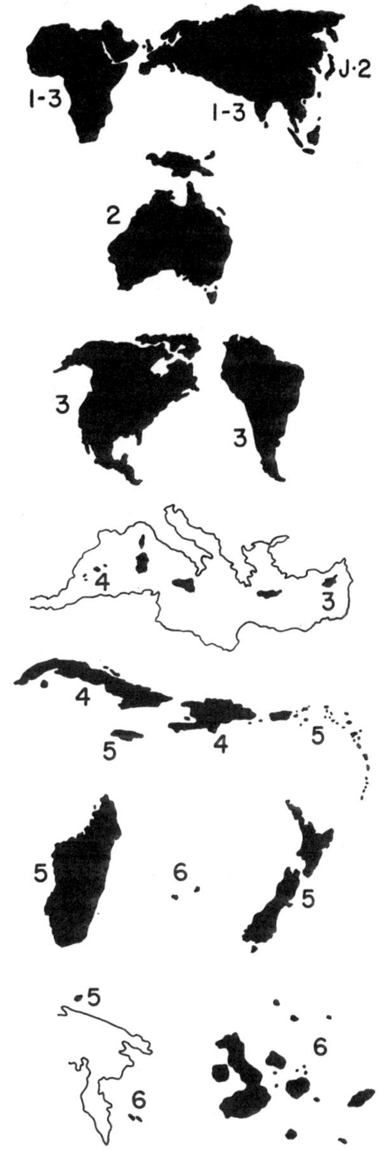

FIGURE 2. Continents and islands in time-transgressive sequence of extinction (after MacPhee and Marx, 1997). *Afroeurasia* (1–3): Sequential extinctions of large mammals over the past 100,000 years including Nauman's elephant and giant deer in Japan (J-2) ca. 30,000 years ago (Fig. 3).

Meganesia (2): Humans arrive 30,000–40,000 yrbp; major extinction follows; chronology of both events is disputed.

Americas (3): Well-dated evidence for human arrival and for megafaunal extinctions center on 11,000 yrbp (13,000 yrBP).

Mediterranean Islands: Epipaleolithic arrival and dwarf hippo and dwarf elephant extinction on Cyprus (3) 10,500 yrbp; mid-Holocene colonization and extinction of goat antelope on the Balearic Islands (4); FC and LQE chronology elsewhere is uncertain.

Antilles: Humans arrive in Cuba and Hispaniola in early or mid-Holocene (4) and in Jamaica and the Lesser Antilles in the Late Holocene (5). Few radiocarbon dates on extinctions.

Madagascar (5): Humans arrive ca. 2000 yrbp; major episode of extinction terminates ca. AD 1500. *Mascarenes,* east of Madagascar (6): Humans arrive AD 1600, extinctions of dodo, solitaire, other birds, and giant tortoises in next 300 years. *New Zealand* (5): Humans colonize ca. 800 yrbp; the Polynesian rat perhaps earlier. Moas extinct by ca. AD 1500.

Wrangel Island (5): Last dwarf woolly mammoths dated 4000 yrbp, 1000 years older than oldest cultural material. *Commander Islands* (6): Humans arrive AD 1741; Steller's sea cow extinct within 30 years. *Galápagos Islands* (6): Humans arrive AD 1535; extinction rates increase over background by two orders of magnitude.

Notoungulata had declined to only 2 genera each (Marshall and Cifelli, 1990). In contrast, the order Proboscidea was thriving with 3 families and 5 genera, as many as had existed at any time in the late Tertiary. Prior to their extinction the bones and teeth of various species of Proboscidea characterize late Cenozoic fossil deposits throughout the New World.

At least 20,000 years before LQE in America, Australia lost all but one of 16 genera of terrestrial vertebrates exceeding 44 kg. The small mammal fossil record of the late Quaternary in Australia is rich enough to discount sample bias. All of the animals involved in LQE exceeded 10 kg in body weight (Murray, 1991:1154–1155).

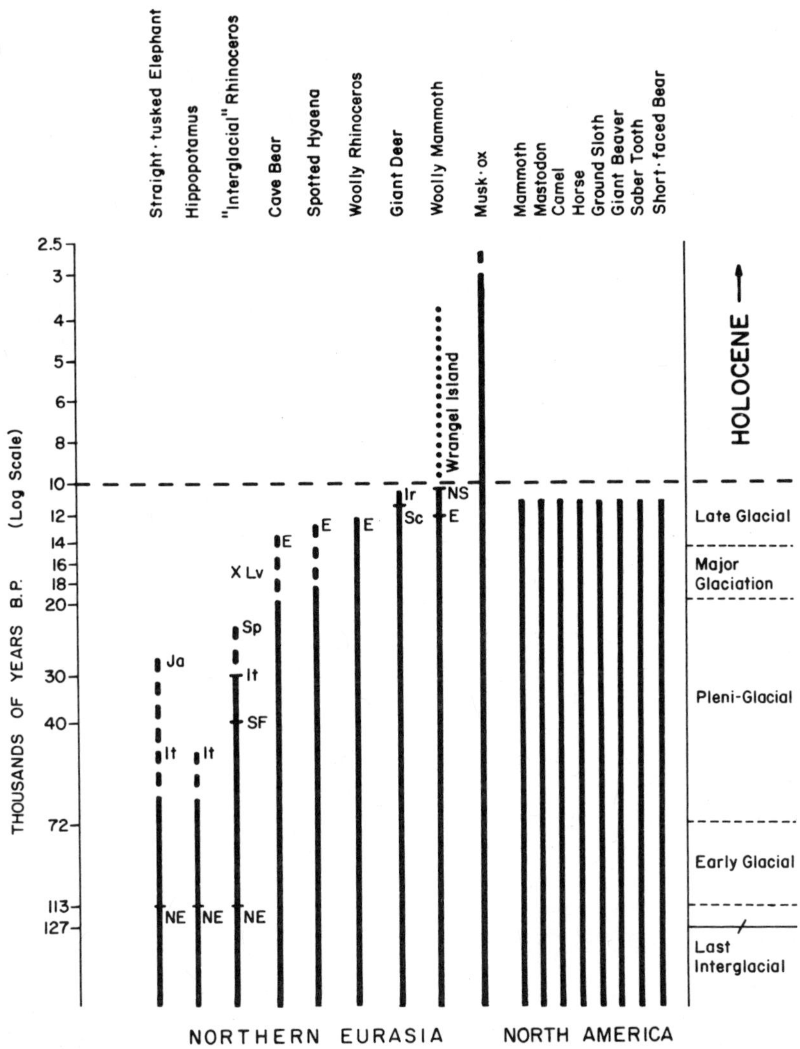

FIGURE 3. Extinction or extirpation of large mammals in northern Eurasia and some from North America (after Martin and Stuart, 1995, and Stuart, 1991). Loss of the former was sequential both between and within species over at least 50,000 yrbp. NE, northern and central Europe; It, Italy, Ja, Japan; SF, southern France; Sp, Spain; Lv, Levant; E, England; Sc, Scandinavia; Ir, Ireland; NS, north-central Siberia. Straight-tusked elephant, *Elephas (Palaeoloxodon) antiquus;* hippopotamus, *Hippopotamus amphibius;* "interglacial" rhinoceros, *Dicerorhinus* spp.; cave bear, *Ursus spelaeus;* spotted hyena, *Crocuta crocuta;* woolly rhinoceros, *Coelodonta antiquitatis;* giant deer, *Megaloceros giganteus;* woolly mammoth, *Mammuthus primigenius;* muskox, *Ovibos moschatus.* North American extinctions: mammoth, *Mammuthus primigenius* and *M. columbi;* mastodon, *Mammut americanum;* camel, *Camelops hesternus;* horse, *Equus* spp.; ground sloth, *Nothrotheriops shastensis;* saber-tooth, *Smilodon fatalis;* and short-faced bear, *Arctodus simus.*

Table 1. Large (>44 kg) Plio-Pleistocene Terrestrial Mammals of North America North of Mexico

	Blancan					IRV			RLB		H
	1	2	3	4	5	E	M	L	E	L	
Stage duration, Ma	0.5	0.5	0.5	0.5	0.2	0.9	0.4	0.2	0.2	0.1	0.01
Xenarthra											
+*Glyptotherium*				X	X	X	X	X	X	X	
**Dasypus*				X	X	X	X	X	X	X	
+*Holmesina*				X	X	X	X	X	X	X	
+*Pachyarmatherium*								X			
+*Eremotherium*						X	X	X	X	X	
+*Nothrotheriops*						X	X	X	X	X	
+*Megalonyx*		X	X	X	X	X	X	X	X	X	
+*Paramylodon*			X	X	X	X	X	X	X	X	
Carnivora											
+*Borophagus*	X	X	X	X	X	X					
**Canis*					X	X	X	X	X	X	X
+*Protocyon*						X					
Ursus			X	X	X	X	X	X	X	X	X
[+]*Tremarctos*			X	X	X	X	X	X	X	X	
+*Arctodus*							X	X	X	X	
+*Chasmaporthetes*		X	X	X	X	X					
+*Meganteron*		X	X	X	X						
+*Smilodon*						X	X	X	X	X	
+*Ischyrosmilus*	X										
+*Homotherium*						X	X	X	X	X	
+*Dinofelis*			X	X	X						
**Panthera*							X	X	X	X	X
+*Miracinonyx*					X	X	X	X	X	X	
Puma						X	X	X	X	X	X
Rodentia											
+*Procastoroides*		X	X	X							
+*Castoroides*							X	X	X	X	
+*Neochoerus*						X	X	X	X	X	
[+]*Hydrochoerus*								X	X	X	
Proboscidea											
+*Mammut*		X	X	X	X	X	X	X	X	X	
+*Stegomastodon*			X	X	X	X					
+*Rhynocotherium*		X	X	X	X						
+*Cuvieronius*				X	X	X	X	X	X	X	
+*Mammuthus*						X	X	X	X	X	
Sirenia											
+*Hydrodamalis*	X	X	X	X	X	X	X	X	X	X	
Trichechus	X	X	X	X	X	X	X	X	X	X	X
Perissodactyla											
+*Cormohipparion*	X	X	X								
+*Nannipus*	X	X	X	X	X						
+*Plesippus*	X	X	X	X	X						
[+]*Equus*			X	X	X	X	X	X	X	X	
[+]*Tapirus*						X	X	X	X	X	
Artiodactyla											
+*Megalotylopus*	X	X	X	X							
+*Blancocamelus*					X						

(continued)

Table 1. *(Continued)*

	Blancan					IRV			RLB		H
	1	2	3	4	5	E	M	L	E	L	
Stage duration, Ma	0.5	0.5	0.5	0.5	0.2	0.9	0.4	0.2	0.2	0.1	0.01
+*Titanotylopus*		X	X	X	X	X					
+*Camelops*					X	X	X	X	X	X	
+*Hemiauchenia*		X	X	X	X	X	X	X	X	X	
+*Palaeolama*								X	X	X	
+*Mylohyus*				X	X	X	X	X	X	X	
+*Platygonus*		X	X	X	X	X	X	X	X	X	
+*Bretzia*			X							X	
+*Odocoileus*	X	X	X	X	X	X	X	X	X	X	X
+*Torontoceras*										X	
+*Navahoceros*										X	
Rangifer					X	X	X	X	X	X	X
**Alces*										X	X
+*Cervalces*										X	
Cervus					X	X	X	X	X	X	X
+*Tetrameryx*						X	X	X	X	X	
+*Hayoceros*								X			
+*Stockoceros*								X	X	X	
Antilocapra										X	X
[+]*Saiga*										X	
Oreamnos										X	X
Ovis										X	X
+*Euceratherium*						X	X	X	X	X	
+*Soergelia*						X	X	X			
+*Bootherium*								X	X	X	
+*Praeovibos*								X			
Ovibos										X	X
Bison									X	X	X
+*Platycerabos*									X		
Primates											
Homo										X	X
Originations	—	9	6	5	6	12	3	7	2	10	0
Extinctions	1	0	2	2	6	5	0	4	1	34	0
Total genera	9	17	24	27	31	37	35	42	40	49	15

[a]Abbreviations: IRV, Irvingtonian; RLB, Rancholabrean; H, Holocene; E, early; M, middle; L, late; Ma, million years. Symbols: +, exinct genus; ∗, extinct species; brackets indicate generic survival on other continents. Living genera including extinct species of large size that vanished with late RLB extinction include *Dasypus belli* (giant armadillo), *Canis dirus* (dire wolf), *Panthera leo atrox* (American lion), *Alces latifrons* (broad-fronted moose), and *Oreamnos harringtoni* (Harrington's mountain goat). All but *D. belli* >44 kg.

Age estimates in millions of years (Ma): Blancan 1, 4.0–3.5 Ma; Blancan 2, 3.5–3.0 Ma; Blancan 3, 3.0–2.5 Ma; Blancan 4, 2.5–2.0 Ma; Blancan 5, 2.0–1.8 Ma (Blancan–Irvingtonian Boundary); Early Irvingtonian, 1.8–0.9 Ma; Middle Irvingtonian, 0.9–0.5 Ma; Late Irvingtonian, 0.5–0.3 Ma; Early Rancholabrean, 0.3–0.1 Ma; Late Rancholabrean, 0.1–0.01 Ma; Holocene, last 10,000 years.

There may not be specimen records for all cells within the temporal range of a genus as plotted here. Checklist and age estimates are courtesy of Elaine Anderson, 1996, Denver, Colorado, revised from Kurtén and Anderson (1980). Mckenna and Bell (1997) provide phylogenetic relationships and may differ in details; for example, they list *Bretzia* from the late Pleistocene. *Ovis* and *Oreamnos* are now recorded from the Irvingtonian (Anderson, personal communication, 1998).

In the late Holocene, at least 8000 years after extinctions in America, Madagascar lost half of its 41 species of lemurs. The 21 extant lemurs are small, with body weights of 0.05–10 kg, the largest being *Indri.* The extinct ones were large, weighing 10–200 kg (Godfrey *et al.,* 1997:231; Simons, 1997:142). Although the fossil sample from Madagascar may be biased in favor of larger animals, "all mammals having body sizes >12 kg are extinct except for the bush pig . . . probably a human import" (MacPhee and Marx, 1997:187).

Table 2. Small (<44 kg) Extinct Plio-Pleistocene Terrestrial Mammals of North America North of Mexico[a]

	Blancan					IRV			RLB		H
	1	2	3	4	5	E	M	L	E	L	
Stage duration, Ma	0.5	0.5	0.5	0.5	0.2	0.9	0.4	0.2	0.2	0.1	0.01
Didelphimorphia											
living	0	0	0	1	1	1	1	1	1	1	1
Insectivora											
+*Beckiasorex*			X								
+*Paracryptotis*	X	X	X								
+*Planisorex*			X	X							
living	2	4	5	6	6	6	6	8	8	8	8
Xenarthra											
[+]*Myrmecophaga*						X	X	X			
living	0	0	0	0	0	0	0	0	0	0	1
Carnivora											
+*Tisisthenes*							X				
+*Ferinestrix*		X									
+*Trigonictis*		X	X	X	X	X					
+*Sminthosinus*			X								
+*Canimartes*					X						
+*Satherium*		X	X								
+*Buisnictis*	X	X	X								
+*Brachyprotoma*						X	X	X	X	X	
+*Osmotherium*								X			
+*Brachyopsigale*		X	X								
[+]*Cuon*										X	
+*Parailurus*			X								
living	2	8	10	10	10	11	14	15	15	17	18
Rodentia											
+*Paenemarmota*	X	X	X	X	X						
+*Cryptopterus*				X							
+*Pliogeomys*			X								
+*Nerterogeomys*		X									
+*Prodipodomys*	X	X	X	X							
+*Dipoides*	X	X	X	X							
+*Bensonomys*		X	X	X							
+*Symmetrodontomys*	X										
+*Ogmodontomys*	X	X	X	X							
+*Pliophenacomys*		X	X	X							
+*Nebraskomys*			X								
+*Atopomys*							X				
+*Mimomys*					X	X	X				
+*Pliolemmus*			X								
+*Allophaomys*								X			
+*Lasiopodomys*					X	X					
+*Terricola*								X			
+*Repomys*							X				
+*Guildayomys*								X			
+*Hibbardomys*					X						
+*Proneofiber*								X			
+*Pliopotamys*							X	X			

(*continued*)

Table 2. (*Continued*)

	Blancan					IRV			RLB		H
	1	2	3	4	5	E	M	L	E	L	
Stage duration, Ma	0.5	0.5	0.5	0.5	0.2	0.9	0.4	0.2	0.2	0.1	0.01
+*Predicrostonyx*					X						
+*Plioctomys*				X	X						
+*Javazapus*								X			
+*Tyrannomys*								X			
living	4	9	12	15	19	19	20	34	35	42	42
Lagomorpha											
+*Hypolagus*	X	X	X	X	X						
+*Pewelagus*				X	X						
+*Notolagus*		X	X								
+*Pratilepus*			X								
+*Aluralagus*					X						
+*Alilepus*		X	X	X	X						
+*Aztlanolagus*					X	X	X	X	X	X	
+*Nekrolagus*		X	X	X	X						
living	0	0	0	1	2	3	4	4	4	4	4
Artiodactyla											
+*Ceratomeryx*		X	X								
+*Capromeryx*			X	X	X	X	X	X	X	X	
living	0	0	0	0	0	0	0	0	0	0	1
Originations	—	23	14	10	12	4	9	24	1	10	3
Extinctions	0	2	13	7	11	2	4	9	0	3	0
Total genera	16	39	51	48	53	46	53	73	65	75	75

[a]For age estimates, abbreviations, symbols, and so forth, see footnote to Table 1. Brackets indicate survival elsewhere (*Myrmecophaga* in tropical America, *Cuon* in Eurasia). Numbers of living genera are included in tabulating originations and in totals. Checklist and age estimates are modified from Kurtén and Anderson (1980), courtesy of Elaine Anderson, 1996, Denver Museum of Natural History, Denver, Colorado.

Considerable recent fieldwork has not added many associations of artifacts with Madagascar's extinct lemurs, hippos, giant flightless birds, and giant tortoises. Despite the temporal association of a dramatic extinction spasm with FC, the proximal causes of extinction in Madagascar are unknown. In contrast, cultural associations (if not kill sites) are much more common in New Zealand where Anderson (1997) argues that early colonists took an optimizing approach, mining first the easily accessible and protein-rich seam of big animals (moas) before turning to less profitable resources. We apply the same economic optimization model to all cases of FC extinction in lands of human colonization. In each case, whether on islands or continents, the first animals lost appear to be high-ranked taxa, the largest, tastiest, slowest, most accessible (least cryptic), or most inviting from the perspective of the human invaders.

On continents of human origin the LQE spasm is mild and diffuse. In Europe and northern Asia, megafaunal extinctions started earlier than in North America and ended later (Fig. 3; Stuart, 1991). Straight-tusked elephants, temperate hippos, and temperate rhinos disappeared from Italy over 60,000 years ago (Stuart, 1991), perhaps as early as the Middle Pleistocene in the case of elephants in the Levant (Tchernov and Shoshani, 1996), and in either case long before proboscidean extinctions in Japan, America, and Siberia. Some European extinctions in temperate latitudes were regional, thus properly extirpations, such

Table 3. Large (>44 kg) Extinct Mammals of the Lujanian in South America[a]

Stratum 1	Stratum 2
Order Xenarthra	Order Rodentia
Family Dasypodidae	Family Hydrochoeridae
Eutatus	*Neochoerus*
Holmesina	Family Octodontidae
Pampatherium	*Dicolpomys*
Propraopus	Stratum 3
Family Glyptodontidae	Order Carnivora
Chlamydotherium	Family Canidae
Doedicurus	*Theriodictis*
Glyptodon	Family Felidae
Hoplophorus	*Smilodon*
Lomaphorus	Family Ursidae
Neosclerocalyptus	*Arctodus*
Neothoracophorus	Order Proboscidae
Panochthus	Family Gomphotheriidea
Parapanochthus	*Cuvieronius*
Plaxhaplous	*Haplomastodon*
Sclerocalyptus	*Notiomastodon*
Family Megalonychidae	*Stegomastodon*
Nothropus	Order Perissodactyla
Nothrotherium	Family Equidae
Ocnopus	*Equus*
Valgipes	*Hippidion*
Family Megatheriidae	*Onohippidium*
Eremotherium	Order Artiodactyla
Megatherium	Family Camelidae
Paramegatherium	*Eulamaops*
Family Mylodontidae	*Hemiauchenia*
Glossotherium	*Palaeolama*
Lestodon	Family Cervidae
Mylodon	*Agalmaceros*
Scelidodon	*Charitoceros*
Scelidotherium	*Morenelaphus*
Order Litopterna	*Paraceros*
Family Macraucheniidae	Family Tayassuidae
Macrauchenia	*Platygonus*
Windhausenia	
Order Notoungulata	
Family Toxodontidae	
Mixotoxodon	
Toxodon	

[a]After Marshall (1990) and Marshall and Cifelli (1990).

as interglacial populations of *Hippopotamus* and *Dicerorhinus* which survive within tropical Asia and Africa.

Two populations of large mammals, muskoxen on the Siberian coast and a population of dwarf woolly mammoths on Wrangel Island in the Arctic Ocean, almost escaped the fate of their species elsewhere in Eurasia. They lingered into the late Holocene, the time of the Pharaohs, to die out shortly before extinctions began on Madagascar. Although the cause of dwarf mammoth extinction on Wrangel Island continues to be debated (Lister and Sher,

1995; Martin and Stuart, 1995; Sher, 1997; Vartanyan *et al.*, 1992), there is overwhelming evidence that numerous Pacific island extinctions that occurred around the same time and under similar circumstances were anthropogenic (see following discussion). The late survival of mammoths and muskoxen reflects their good fortune in occupying a corner of the Arctic where they were safe from human hunters until the late Holocene, long after human economies dominated temperate parts of Eurasia.

The pattern of extinctions within the last 40,000 years so changed the face of the earth that historic observations and assumptions about what is natural are misguided. Until LQE the large mammals of North and South America equaled or exceeded in size those of Africa (Stuart, 1991). The world's islands had many more endemic species than historic biogeographers imagined, some even more bizarre than the dodo. Australia harbored a megafauna of many, not just a few species of large marsupials, the largest weighing a metric ton and perhaps ambushed by a reptile 6 m in length, much larger than a Komodo dragon (see Flannery and Roberts, this volume). If the extinctions happened during times of human colonization, as appears to be the case, what was the nature of the process?

2. Manifold Effects of Human Arrival

During FC North American extinctions of megafauna occurred rapidly (Martin, 1990; Stuart, 1991), leaving few kill sites or processing sites. To accommodate these features, Mosimann and Martin (1975) modeled sudden "overkill" or "blitzkrieg" (lightning war), a wave of extinctions sweeping across a continent in a few centuries. In the Pacific heartland, rapid extinctions of megapodes and other attractive and vulnerable species such as parrots, large pigeons, and flightless rails suggest an insular blitz, in this case sweeping across Polynesian islands (Steadman, 1997). Although likely to be dramatic at FC, insular extinctions continued at a lower pace up to the present.

Unlike New Zealand (Anderson, 1989) or some islands in the Polynesian heartland (Steadman, 1995), in America, Madagascar, and on islands in the Mediterranean, an unambiguous association of artifacts or other cultural manifestations with extinct animals is rarely found. Although Hawaiian extinctions of flightless ibises, rails, and moa-nalo (large gooselike ducks, some with a tortoiselike beak), as well as numerous volant species, apparently coincide with the arrival of Polynesians over 1000 years ago, archaeological associations are few (Storrs Olson, personal communication).

On islands the ecological upsets accompanying human arrival include (1) human predation on a naive fauna; (2) the introduction of rats and domestic animals as well as their diseases; and (3) clearing of forest for cultivation or grazing. Perhaps for these reasons animals smaller in size than those lost on the continents became extinct. Throughout the world the side effects or "secret weapons" that are part of the syndrome of human impact include the following:

- *Aliens.* Ground nesting birds on islands lacking mammalian carnivores, are vulnerable to predation by introduced pigs, dogs, and rats (Diamond, 1984; Steadman, 1997). Very soon after FC, synanthropic continental land snails replaced endemic land snails on such Pacific Islands as Tikopia (Christensen and Kirch, 1981) and Ofu in Samoa (Hunt and Kirch, 1997).

- *Habitat destruction.* First detected in cores from Lakeba (Fiji) by Hope and Hughes (in Brookfield *et al.*, 1979), high rates of charcoal and sediment influx along with a fern spore spike and other evidence of disturbance in the pollen profile accompany clearing for cultivation in many islands, for example, in Puerto Rico (Burney *et al.*, 1994), the Cook Islands (Ellison, 1994), and New Caledonia (Stevenson and Dodson, 1995). In the process forest biota suffers (Steadman, 1995).
- *Hyperdisease.* Besides humans the potential disease vectors include domesticates, commensals, and synanthropics. According to MacPhee and Marx (1997:185), "blitzkreig is essentially descriptive of the epizootiology of a disease of unusual lethality." They cite rapid die-out of native species of *Rattus* on Christmas Island, south of Java, as an example. Another example is avian malaria in Hawaiian forest birds (Van Riper *et al.*, 1996).
- *Keystone cascade.* Brush invasion detrimental to grazing ungulates will result following loss of keystone megaherbivores such as proboscideans, which are themselves especially vulnerable to human predation (Owen-Smith, 1988; Mithen, 1993, 1997; Putshkov, 1997).
- *Predation overload.* Doomed but still potent predators such as North American lions, sabertooths, short-faced bears, and dire wolves interacted with human hunters at FC, albeit briefly, to create a predator pit that may have vastly amplified human predation pressure, engulfing the large herbivores (Janzen, 1983).
- *Surplus killing.* FC humans may have killed easy prey in excess, contrary to what might be expected of a prudent predator. For example, Mandan hunters sought to kill all of the buffalo in a herd, believing that any escapees would warn others (Henry, 1806, in Coues, 1897:336). Surplus killing need not be maladaptive. Kruuk (1972) argued that although rare, in foxes and hyenas "excessive" killing enables predators to return to a known larder at a later time. Saturated predators are not inhibited from further catching and killing of vulnerable prey. There are ethnographic examples of surplus killing in foraging societies (Kristan Hawkes, personal communication).

3. Island Faunas: Testimony of Screen Washing

In the last two decades the sieve screens of archaeologists have provided rich new data for the computer screens of modelers. Although zoogeographers have long known of severe extinction on oceanic islands in the West Indies, they did not bother to look for fossils elsewhere. Before the 1970s, few anticipated the remarkable wealth of fossil data to be dredged from prehistoric deposits in the remote Pacific. The discoveries began with the description of a new genus of flightless ibis, *Apteribis,* and a new genus of large flightless duck, *Thambetochen* (Olson and Wetmore, 1976), followed by documentation that within the last 2000 years on any one Hawaiian island, over half of the native land birds had disappeared (James and Olson, 1991; Olson and James, 1991).

Subsequent finds in the eastern Pacific included flightless rails, pigeons, parrots, doves, and other extinct species (Steadman, 1995). To the east of their present range, new species of mound builders or megapodes were dug up in Tonga, Samoa, and Fiji. In New Caledonia, Balouet and Olson (1989) reported bones of extinct birds including a new genus of giant flightless megapode, *Sylviornis.* The remarkable horned turtle, *Meiolania,* survived into the late Holocene in New Caledonia, much longer than in its Australian homeland. Its last records are associated with cultural remains radiocarbon dated at around 1500 yrbp (Gaffney *et al.*, 1984). Balof Cave on New Ireland, Papua New Guinea, has yielded bones of the world's largest known extinct flightless rail (Steadman *et al.*, 1999). Not all change was negative. The Bismarcks are remarkable for their record of prehistorically introduced mammals such as the cuscus (*Phalanger orientalis*), pademelon (*Thylogale brunii*), various murid rodents (Flannery *et al.*, 1988; Flannery, 1994), and other mammals that may or may not have been part of the package of introduced horticultural components (Allen, 1997:36).

Although only a few species of pelagic seabirds suffered extinction, large breeding colonies no longer occur on New Zealand, the Society Islands, the Marquesas, Mangaia (Cooks), the Hawaiian islands, Easter Island, and many other archipelagos in the Remote Pacific. The seabird colonies vanished soon after human arrival. In the process the reduction in numbers of shearwaters and petrels dimmed an avian beacon guiding prehistoric voyagers to uninhabited islands in the first place (Steadman, 1997). Pelagic seabirds endure on uninhabited islands in colder waters, or in warm waters on rat-resistant karst limestone outcrops, or on unscalable rat-free pinnacles ("islets") adjoining densely settled larger islands.

An anthropogenic agency can be held in limbo or ruled out when oceanic islands yield evidence of extinctions that predate possible human involvement. For example, the extinct Miocene *Hipparion* faunas found in continental sediments predate Crete's isolation and long predate any possibility of human activity (van der Made, 1996). In the Seychelles of the Indian Ocean, repeated flooding by rise in sea level obliterated colonies of giant tortoises (*Aldabrachelys*), which recolonized during marine regression, presumably from Madagascar (Taylor *et al.*, 1979).

Very early human involvement in extinction has been proposed for the loss of giant tortoises and pigmy elephants east of Wallace's line on Timor, Flores, and perhaps Sumba, thought to coincide with arrival of *Homo erectus* 750,000 years ago (Gibbons, 1998; Morwood *et al.*, 1998; Van den Bergh *et al.*, 1996). Neanderthals may also be involved in the mid-Pleistocene loss of dwarf elephants, dwarf hippos, and giant rodents on Crete (Spaan, 1996). These cases of early or mid-Pleistocene insular extinctions and human colonizations suggest a much longer and more dynamic record of island turnovers than envisioned by Martin (1984). They may be expected on oceanic islands within warm waters close to the Eurasian continent and need not be limited to anatomically and behaviorally modern members of *Homo sapiens*. For example, *Homo erectus* may have survived until 30,000 years ago in Java (Swisher *et al.*, 1996). Nevertheless, most island extinctions of endemic faunas known to date happened within Libby time. On islands or island arcs ("stepping stones") more than roughly 100 km from adjacent continents, the numerous extinctions accompanying FC are not likely to be older than the Holocene.

Only the most remote deep-water islands, not linked by stepping stones to any adjacent land mass, escaped prehistoric discovery. Examples include the Atlantic islands of the Azores and Ascension; the Galápagos; Lord Howe in the Tasman Sea; Christmas Island 360 km south of Java; and the Mascarenes, east of Madagascar. Historic accounts, such as Darwin's on the Galápagos, reveal an avian tameness unknown in more predatory environments.

Beyond the notable exception of dwarf woolly mammoth on Wrangel Island, we know of no populations of large mammals that managed to survive for a time after mainland extinctions by finding a refuge on shelf (land bridge) islands adjoining the continents. Presumably rising sea levels did not bring shelf islands into existence until after extinctions had swept the continents. Oceanic islands in warm waters could serve as refugia for dwarf populations of large terrestrial mammals if they were not too remote to be colonized by chance arrivals of good swimmers such as proboscideans, hippo, giant deer, or ground sloths, and yet far enough from the continents to escape rapid detection and simultaneous invasion by humans.

A powerful argument for human agency in driving the extinctions is the delay seen between continental and adjacent insular extinctions, the latter always lagging behind the former. For example, extinctions of Madagascar's giant lemurs postdate by over 1 million years extinctions of giant baboons in Africa. New Zealand's moas lasted 30,000 years longer than Australia's extinct giant bird, the mihirung (*Genyornis*). Finally, in many cases if not all (MacPhee and Marx, 1997:180–183; MacPhee *et al.,* 1989), extinctions on the Greater and Lesser Antilles followed those on the American mainland.

In the Pacific, many new records of island bird extinctions and extirpations have been reported in recent years (James and Olson, 1991; Meredith, 1991; Olson and James, 1991; Steadman, 1989, 1995, 1997). On any remote Pacific island with a respectable fossil record, one can expect to find at least two to three times the number of land bird taxa before contact than after. The human colonization of remote Oceania occurred in the late Holocene. Prehistoric human explorers missed only the Galápagos and a very few other out-of-the-way places as they surged east out of the Solomons, island-hopping thousands of kilometers through the Polynesian heartland to reach Hawaii far to the north, Easter Island over 7500 km to the east, and New Zealand to the south (Irwin, 1992). Between 3500 and 1000 years ago, this "Pacific express" left a window of extinctions in its wake. In all of Oceania (Micronesia, Melanesia, Polynesia), there are about 800 islands over 1 square km in area. Steadman (1995:1127) estimates the loss of 10 species or populations from each, a total loss of 8000 species or indigenous populations of land birds (including an estimated 2000 endemic species of flightless rails).

Very few tropical islands and archipelagos in the Pacific remained undiscovered by the sixteenth century. Some, such as Henderson, Pitcairn, and Norfolk, were inhabited and abandoned prehistorically. The fossil record on such "islands of doom" strongly suggests resource depletion as the cause of abandonment (Steadman and Olson, 1985; Schubel and Steadman, 1989). At historic contact in 1722, Easter Island was on the threshold of economic collapse with no wood left to build seaworthy canoes suitable for fishing (Diamond, 1995; Van Tilburg, 1994). Prehistoric deforestation began at least 1200 years ago and was essentially over some 300 years before European contact (Bahn and Flenley, 1992). Large colonies of pelagic seabirds disappeared along with endemic land birds (Steadman *et al.,* 1994; Steadman, 1995).

Within the first few hundred years of colonization, archaeological sites reveal major loss of taxa of land birds and of colonies of pelagic seabirds. The harvest of parrots, pigeons, doves, flightless rails, and seabird rookeries made a significant contribution to the food supply of colonists only in the early years. In Tonga, hunting of pigeons (*Ducula pacifica*) became the prerogative of chiefs after most avian resources had dwindled (Burley, 1996). The fact that "birds are widely regarded in Oceania as omens, signs, spirit helpers and spirit messengers" (Van Tilburg, 1994:55) did not save thousands of species and disjunct populations from postcontact extinction.

Oceanic islands provide many interesting variations of the extinction scenario. On low featureless islands such as Lifuka in Tonga, bones of the extinct megapodes and pigeons occur only in the earliest deposits of the pioneering Lapita culture (Steadman *et al.,* in press). Succeeding Polynesian cultural layers postdate most bird extinction. On raised islands with variable lithologies such as Mangaia in the Cooks, rich volcanic soils ideal for cultivation adjoin rugged forested outcrops of limestone (*makatea*) unsuitable for cultivation. The *maka-*

tea habitat buffered certain species of birds from extinction for more than 1000 years after human arrival, much longer than would have been possible on low islands like Lifuka.

For these and many other losses there were no literate witnesses. No Magellan, Melville, or Mundy arrived in time to tell of the flightless moa-nalo on Hawaii, the giant flightless megapode *Sylviornis* of New Caledonia (30 kg in weight; Balouet and Olson, 1989), or the legendary immense elephant bird or *roc* of Marco Polo, which some think was *Aepyornis* of Madagascar, whose egg exceeded in size that of any other bird. The literary burden of symbolizing extinction was placed on the "Bird of Disgust," the dodo of Mauritius, anatomically a flightless pigeon. Both extinction and its symbol once seemed remote, freakish, and bizarre, the inevitable fate of the totally unfit. In no way does the detailed inventory of historic extinctions and dynamic models derived from modern distributions alone prepare the biogeographer for the magnitude of prehistoric extinction unearthed in the last two decades, especially from Pacific Islands.

Nevertheless, a sense of what the FC experience was like can be gained from historic accounts. Sailors exploring Mauritius in 1598 in the days of the dodo knew they were on an uninhabited island when "they caught a multitude of birds, which remained still when they were approached, not attempting to fly away, so that the sailors went on and caught them in their hands" (Barnwell, 1948:80). Other parties found a rich avifauna of pigeons, parrots, crows, hawks, falcons, flightless rails, and an immense flightless pigeon they named the dodo. A captured gray parrot, made to call out, attracted hundreds more of its kind, soon killed with sticks (van der Hagen, 1607, in Barnwell, 1948). Remarkable tameness of land birds and other animals remains an attribute of the Galápagos, undiscovered until the sixteenth century, not widely colonized until the nineteenth century, and with some inhospitable islands still uninhabited. The Indian Ocean islands near Madagascar, including Reunion, Mauritius, the Comoros, and the Seychelles, once supported living populations of large land tortoises, presumably rafted in from Madagascar, until historic FC. Remoteness and official government protection has helped to spare giant tortoises on Aldabra and on the Galápagos from extinction.

4. Islands as Controls

Islands provide opportunities to test two common extinction models, climatic change and cultural change. On occasion these models can be decoupled. Although any island first contacted in the Holocene will serve if it yields older (pre-Holocene) faunas, Diamond (1984) believed that the best opportunities are offered by New Zealand and Madagascar. Discovery of glacial or late-glacial age extinctions on these islands coinciding with global climatic change such as the Allerod–Upper Dryas switch would be striking support for climatic models. Conversely, it is hard to see how climatic models could be rescued if no generic extinctions of megafauna can be discovered on these and other oceanic islands prior to the well-known if not well-understood loss of moas in New Zealand and giant lemurs in Madagascar, both occurring during the period of human colonization.

4.1. The Galápagos

Among archipelagos of the tropical Pacific, the Galápagos are unique in having been uninhabited before the historic period (Steadman *et al.*, 1991). For this reason the Galápagos serve as a valuable control, as informative to extinction biologists as they are to evolutionary biologists.

Natural deposits in lava tubes on Santa Cruz, Floreana, and other islands have yielded an abundance of fossils. In a search for evidence of prehistoric extinction, Steadman *et al.* (1991) recovered and identified 500,000 bones from species of rodents, bats, Galápagos finches, mockingbirds, other land birds, land lizards, tortoises, and snakes. Despite the size of the sample, no obvious cases of generic extinction turned up, with one glaring exception. Among the mammal bones were skulls and jaws of a giant (muskrat-sized) rodent, *Megaoryzomys,* related (despite its name) not to the rice-rats (*Oryzomys, Nesoryzomys*) found in the Galápagos, but to an Andean genus *Thomasomys* (Steadman and Ray, 1982). *Megaoryzomys* has never been found alive.

Although the bones from the dry caves did not appear unusually old, that alone was no guarantee of youthfulness. Caves in arid regions are well known for the quality and fresh appearance of bones that are many thousands of years old, including those of extinct animals. As well as bones, dry caves in Arizona and Utah yielded well-preserved perishable remains, dung, hair, claws, horn sheaths, and dried tissues of ground sloths, extinct mountain goats, and even Columbian mammoths. Their age equals or exceeds 10,000 yrbp (Agenbroad and Mead, 1996; Martin, 1987, 1990; Mead and Agenbroad, 1992). Did *Megaoryzomys* die out along with Arizona's ground sloths and other extinct megafauna, long before any human presence in the Galápagos? Such a finding would strengthen the case for noncultural models of prehistoric extinction.

However, accelerator radiocarbon dates (sensu Stafford *et al.,* 1991) on 11 samples of fossil bone of *Megaoryzomys* were found to span the last 4000 years only (Steadman *et al.,* 1991). The results eliminated both the possibility that the giant rats from the Galápagos disappeared at the same time as Arizona's ground sloths and the possibility of an extinction in advance of FC. When calibrated by tree rings, the three youngest dates for giant rats intersect historic time, which began in the Galápagos with the arrival of the Bishop of Panama in 1535. Historic records familiar to us, including Darwin's account of the biota of the Galápagos in *Voyage of the Beagle,* do not hint at the existence of a giant rat. Only the radiocarbon dates link giant rat extinction to the centuries of human presence. Whatever its fate, the extinction of *Megaoryzomys,* along with all other extinctions on islands in the eastern Pacific of which we are aware, is too young to fit any Younger Dryas or other late-glacial model of terminal Pleistocene extinction.

The Galápagos are famous for the intensity of El Niños, warm wet periods interrupting the more normal cold currents and aridity. Fossil evidence relating island extinctions to El Niño Southern Oscillation (ENSO) phenomena would be of great interest. The most severe El Niño recorded in historic time took place in 1982–1983, accompanied by a crash in seabird populations, which have recovered (Glynn, 1990). In the Galápagos, Steadman *et al.* (1991) estimated prehistoric background extinctions at only 0–3 populations over the last 4000 to 8000 years, at least two orders of magnitude below the rate of extinction in the five centuries following human contact. The record suggests that the endemic faunas of islands are highly selected for resistance to extinction by natural causes such as ENSO driven climatic changes. Although ENSO might have operated as a natural selection mechanism in some species of Darwin's finches (Grant and Grant, 1989), we have no evidence that it forced extinctions in the Galápagos or elsewhere.

4.2. New Zealand

Thus far we have considered insular extinctions that are too young to be explained by events of the last glaciation. New Zealand has a fossil record of moas and other extinct fauna that can be traced to and beyond the Upper Dryas into the full glacial, locally known as the Otiran glaciation. Climatic change in New Zealand involved essentially the same glacial-age changes seen elsewhere, a 5°C drop in mean annual temperature, expanded alpine glaciers, extensive loess deposition in the lee (east) of the glaciers, and vegetation change restricting forests to the northern part of the North Island (Anderson and McGlone, 1992). As the climate warmed, rising sea level reduced land mass. In the Holocene, the North Island saw severe volcanic eruptions and tephra deposits yielding cobalt-deficient soils. All of these changes are potential ingredients of a climate-driven extinction model favored by those Putshkov (1997) calls "climatists."

Since the advent of radiocarbon dating in the 1950s, it has been increasingly apparent that the extinction of New Zealand's 11 species of moas, all over 20 kg in weight, coincided with Polynesian arrival within the last 1000 years. Besides moas, the losses included geese, ducks, hawks, an eagle, rails, coots, adzebills, snipe, an owlet nightjar, a crow, two New Zealand wrens, lizards, frogs, and even large invertebrates. Moa extinctions are attributed to human hunting, whereas the loss of smaller species may involve introduced dogs, the Polynesian rat, and other side effects of human colonization. From AMS dates on rat bone gelatin, Holdaway (1996, in press) has proposed that colonization of New Zealand by Polynesian rats (as stowaways on boats whose crews did colonize New Zealand themselves) preceded large-scale human colonization by 1000 years. Significantly before human presence or overt involvement, Holdaway (in press) anticipates extinctions of certain native species vulnerable to the new predator. If the chronology proves robust, the circumstances should afford a unique opportunity to evaluate impact of *Rattus exulans* independent of all other biotic change brought by humans (see Holdaway, this volume). Although a novel development in the inventory of Pacific Island extinctions, the pre-moa extinctions of small species forced by Polynesian rats are still within the model of cultural causality.

Were there even earlier extinctions in Libby time, 10,000 years or more in advance of those of the late Holocene? If global extinctions were driven at least in part by the Bolling–Upper Dryas shift, as some have argued in the case of North American extinctions, they might be expected in New Zealand at a time when any coincidental human impacts are out of the question.

Recent exploration of caves as well as open sites and judicious application of radiocarbon dating has enriched the record of Otiran (last glacial and late-glacial) fossil faunas of New Zealand (Worthy and Holdaway, 1993, 1994; Worthy, in press). In the Otiran, shrinkage of forest, especially in the South Island, and expansion of grassland were accompanied by changes in range of forest and open country species of moas (Worthy, in press). However, the range changes were unaccompanied by any Otiran moa extinction, or that of any other species. Furthermore, extinctions do not accompany the rise in sea level or the deposit of tephras on the North Island (Anderson and McGlone, 1992).

In the absence of any biotic extinctions predating human (and Pacific rat) arrival, we conclude that, as in the case of the Galápagos, the winnowing effect of earlier Quarternary climatic switches buffered the New Zealand fauna against Otiran climatic change. Presumably the world's oceanic islands had been purged of climatically sensitive species throughout the Quaternary. On the other hand, FC constituted a planetary accident so unusual that, like the impact of an asteroid, it could not be anticipated by natural selection or any other evolutionary process.

4.3. 'Eua and Oahu

The Galápagos and New Zealand are not the only examples of oceanic islands as controls. From bones of 21 species of birds above and below a dripstone layer in Ana Tu Cave on 'Eua (Kingdom of Tonga) dated by uranium–thorium to 60,000–79,000 yrbp, Steadman (1993) found no clear indication of prehuman extinctions. In Tonga most species were lost following FC around 3000 years ago. On Oahu in the Hawaiian archipelago, James (1987, 1995) found no evidence of prehuman extinctions in a rich fauna of both flightless and volant species screened from lacustrine sediments over 120,000 years in age. If, as there is every reason to expect, the fossil record from oceanic islands is much richer in glacial-age or older faunas than the few that are known at present, many more tests of this sort can be attempted. For example, although Holocene extinctions appear to be the rule, the possibility of late-glacial or much earlier extinction in the West Indies, as advanced by MacPhee *et al.* (1989) for the giant rodents *Clidomys* and *Amblyrhiza,* is still open.

5. Continental Extinctions

Without abandoning our island-centered view, which means comparing oceanic island extinctions with those on the continents, we turn to the latter. We begin with the continents of human colonization, North and South America and Australia. We end with Eurasia and Africa, the continents of human origins. Only Antarctica escaped LQE.

5.1. North America

North American ground sloths, mastodons, mammoths, camels, horses, and other taxa repeatedly dated by radiocarbon (Martin, 1987, 1990; Stuart, 1991) disappeared around 11,000 yrbp. The event immediately predates or coincides with the Younger Dryas cold snap, a reversal to glacial conditions after 3000 years of interglacial warmth known to pollen stratigraphers in western Europe as the Bolling–Allerod. Throughout North America, megafaunal extinctions coincided with both Clovis colonization and with the end of the Allerod or the start of the Younger Dryas.

In North America there are less than two dozen well-dated carefully documented sites with Clovis points and other stone tools associated with extinct animals, mostly mammoth and bison. Most Clovis sites are found in flood plains or in buried stream deposits and most are in the West (Taylor *et al.,* 1996). Sizable portions of North America lack any well-dated association of Clovis artifacts with extinct fauna. Purported associations have not always withstood stratigraphic and geochronological scrutiny (Steadman *et al.,* 1997).

Because the alleged sites lack associated stone tools, archaeologists have been uncertain of how to interpret Michigan mastodon assemblages, including claims for a sunken meat cache (summarized in Fisher, 1996). Even if these provisional man–mastodon associations are accepted, the total number of kill or processing sites linking extinct animals with artifacts seems small. As the Clovis–mammoth sites are of the same age, the only model involving human forcing that can accommodate field evidence is one of a spontaneous or cataclysmic extinction (Mosimann and Martin, 1975). Their model is based on rapid population growth and spread of experienced hunters encountering naive prey.

The blitz, or "bow wave" model as it has been called, accommodates a tight chronology, with no more than a few hundred years of overlap between human colonization and megafaunal extinction within the United States and as little as a decade of overlap of extinct fauna and first colonists in any one region. The model loses credibility if, for example, mastodon hunting in Michigan persisted for 1000 years after FC as proposed by Fisher (1996), or if humans in Chile lived in houses covered with skins of gomphotheres over 1000 years before proboscideans became extinct in North America (Dillehay, 1989). Monte Verde is one of three claims of pre-Clovis occupation of the New World that, according to Adavasio and Pedler (1997), imply that the Clovis Curtain has fallen and that archaeological sites older than Clovis have attained legitimacy. Whether or not the "Clovis Curtain" has fallen remains to be seen. We probe this possibility further in our treatment of Australia. Meanwhile, the Clovis coincidence, of first human presence concordant with the rapid elimination of large animals in the Americas, fuels overkill models.

Not all extinct genera to be accounted for are well dated by radiocarbon. Fossils of some, such as the American cheetah, *Miracinonyx,* have yet to yield dates contemporaneous with mammoth, mastodon, camel, horse, and many other species last recorded at 12,000–10,000 yrbp (Stuart, 1991). Proponents of a gradual extinction take this as evidence that some genera vanished before the rest (Grayson, 1991). Proponents of catastrophic extinction note that in the case of rare taxa, the chance of recovering a terminal date that represents the true date of extinction is small (the Signor–Lipps effect). Further sampling should increase (and in the last two decades has increased) the number of concurrent terminal dates, an interpretive argument familiar to those engaged in the debate on the Cretaceous–Tertiary event and dinosaur extinction (Ward, 1997).

Although the purpose of our review is to compare islands with continents in radiocarbon time, the record of older extinctions of megafauna in the Americas over the last 4.0 Ma, since the latest Miocene, deserves inspection. From estimates of body weights of extinct mammals from Kurtén and Anderson (1980), the genera are divided into those larger and those smaller than 44 kg. Of the 70 large genera known in North America in the last 4 Ma, a total of 55 became extinct, 34 in the Late Rancholabrean (Table 1, Fig. 4). Of 126 genera of small mammals recorded in North America over the last 4 Ma, a total of 52 became extinct, 3 in the late Rancholabrean (Table 2, Fig. 5). The 3 small genera that disappeared, an antilocaprid (*Capromeryx*), a skunk (*Brachyprotoma*), and a rabbit (*Aztlanolagus*), are large enough to imagine that possibly they too were victims of FC trauma. A total of 15 genera of large mammals and 75 genera of small mammals survived. Clearly there was no extinction pulse among the small mammals (Fig. 5) to match the LQE (late Rancholabrean) pulse that obliterated two-thirds of the genera of large animals (Fig. 4).

Despite our change in scale in Fig. 4, the LQE spike for large mammals shown there should be even more extreme, a matter of just a few thousand or possibly just a few hundred years in the Late Rancholabrean. The increase in large mammal extinction rate far exceeds anything earlier in the Plio-Pleistocene.

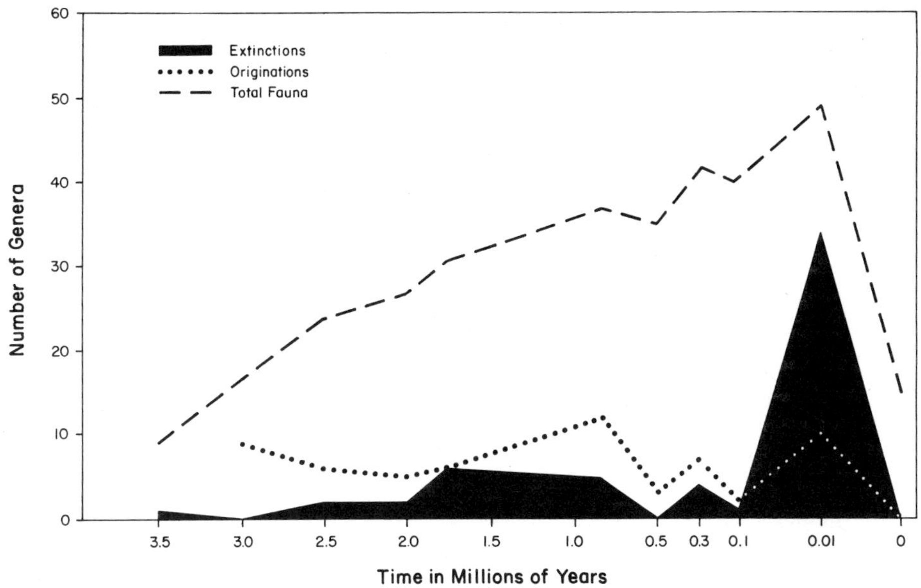

FIGURE 4. North American large mammal originations (dotted line), extinctions (solid line), and total fauna (dashed line) since the Miocene from data shown in Table 1. Note change of scale in the last 0.5 Ma.

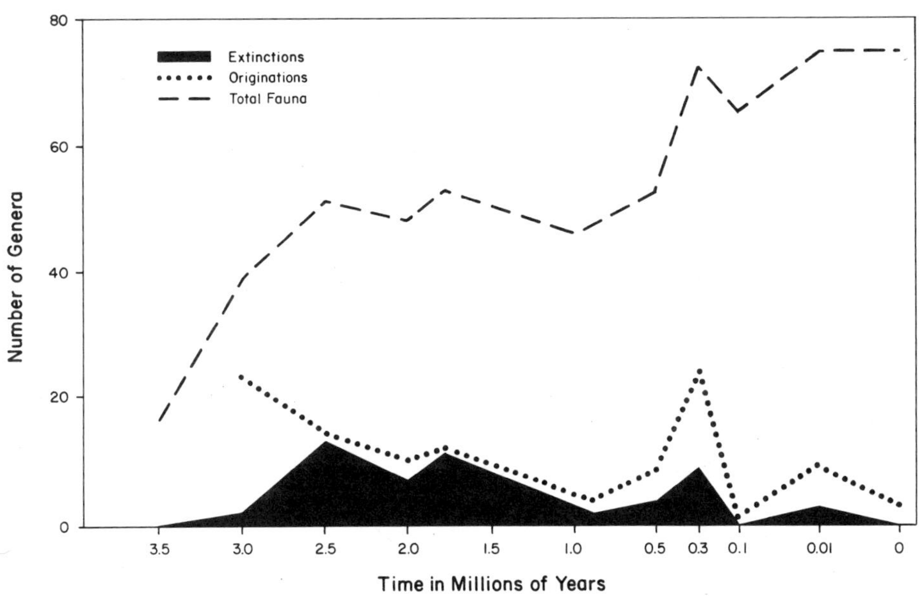

FIGURE 5. North American small mammal originations (dotted line), extinctions (solid line), and total fauna (dashed line) since the Miocene from data shown in Table 2. Note change of scale in the last 0.5 Ma.

Webb (1984) noted the loss of a great many mammals of both large and small size around the Hemphillian–Blancan boundary, over 4 Ma, with almost as many of large size as in the late Quaternary. Although geochronologically elusive, events of the late Hemphillian (including the Messinian salinity crisis) invite further study. At issue is an important and yet unexplained extinction event that differs from LQE by including many medium-sized carnivores and smaller mammals. It may not have been catastrophic.

Since Kurtén and Anderson's (1980) work, there have been few changes in the inventory of large mammals listed in Table 1. Among the extinct megafauna of the Rancholabrean, the bonnet-headed muskox *Symbos* is now considered to be the male of *Bootherium,* an older name that takes priority. According to Elaine Anderson (personal communication), the supposed cervid *Sangamona* is apparently a mixture of bones of other genera of deer. Its status is dubious. In the case of the yak (*Bos poephagus*), long listed in Alaskan Pleistocene faunas, radiocarbon dating has supported the suspicions of Guthrie (1990:50–51) that the alleged records were actually based on discolored cow bones from historic garbage dumps intruding into Pleistocene gravels rich in extinct animals.

Overall, the record of origins and extinctions (arrivals and departures) of large mammals has been stable since Kurtén and Anderson's (1980) assessment. There has been more change, especially the addition of extinct genera, to the record of small mammals. Nevertheless, the results of changes in the checklist of both large and small mammal extinctions (Tables 1 and 2; Alroy, this volume) do not affect our conclusion that after 4 million years of quiescence, a severe, unbalanced extinction event struck two-thirds of North American large mammals, including coextinction of some of their parasites, commensals, and scavengers. As noted, this happened at or close to the time of FC (Stuart, 1991).

Not all of the 34 extinct genera were equally important. Some arrived in North America quite late, within roughly 100,000 years of the time of their extinction, such as *Cervalces.* Some are rare (*Torontoceros,* recently proposed on few specimens) or penetrated North American only in Alaska and northwestern Canada (*Saiga*). Three other genera (*Equus, Tapirus,* and *Tremarctos*) also survive outside of North American in either Asia or South America or both and thus are not extinct globally. Although most extinct genera are monotypic, *Equus* and *Mammuthus* harbor multiple biological species.

With the exception of deer (*Odocoileus*), the common fossils found in Quaternary deposits preceding LQE are not the bones or teeth of large mammals that survive the extinctions but those that didn't, especially mammoth, mastodon, camelids, equids, giant peccary, and extinct taxa of bison (Graham and Lundelius, 1994). One might imagine that if LQE were a natural catastrophe, the ranges of the survivors (15 species listed in Table 1) would have expanded rapidly. In the absence of competitors, they would have moved into habitats and niches massively vacated by extinctions of other large herbivores. With only eight glacial-age records (Graham and Lundelius, 1994:442), *Cervus* (elk, wapiti) is a possible example. Relative to *Odocoileus,* the postextinction numbers of *Cervus* soared, suggesting competitive release.

On the other hand, after LQE the range of bison shrank (McDonald, 1981) along with that of jaguar (Graham and Lundelius, 1994:377). If some lethal climate conditions at the end of the last glaciation eliminated their North American relatives, one might expect that with early Holocene warming the surviving populations of the jaguar, the spectacled bear (*Tremarctos*), and the tapir would expand north out of the tropics to recover lost range in lands purged of former competitors. The fossil record of the Holocene indicates no such expansion.

Given the apparent opportunity to exploit niches recently emptied by multiple extinctions of megafauna, one might expect the survivors to increase in size. Instead, postglacial bison, antelope, mountain sheep, wolves, and wolverine all are typically smaller than their glacial-age ancestors (Gilbert and Martin, 1984:144). Although proponents of climatic extinction models regard Holocene size reduction as the outcome of climatic stress, the same data should reflect human predation on high-ranked prey (Fisher, 1996; G. Haynes, 1991:272). Evidently there is a possibility of distinguishing among these competing explanations by examination of rates of growth revealed in tusk rings of proboscideans. Preliminary results suggest that prior to extinction, Michigan mastodons were not stressed by dietary deficiency (Fisher, 1996).

As we have noted, whereas Clovis points are widespread in North America, there are fewer than two dozen Clovis sites intimately associated with extinct fauna, mainly mammoth, that constitute robust field evidence for human involvement with extinct fauna. Some archaeologists believe this is much too slender a record of human predation to explain the extinctions and to indict Clovis hunters as the ultimate agent of extinction. If their spear points were elegant and their experience in hunting megafauna was extraordinary, Clovis hunters did not have rifles. "But the Clovis hunter is still a leading candidate to fill the role of the unique and irresistible agency behind extinction, at least until evidence for more abrupt and much more disastrous climatic change everywhere at the end of the Pleistocene is found" (G. Haynes, 1991:317, see Haynes and Eiselt, this volume).

5.2. South America

The Quaternary of South America includes the Ensenadan and Lujanian land mammal ages (LMA) dated at 1.5 to 0.5 Ma and 0.5 to 0.01 Ma, respectively. The preceding Uquian LMA from 2.5 to 1.5 Ma is of great interest as it immediately follows the intercontinental suturing of North and South America. The Uquian shows the effects of the famous Great American Faunal Interchange, a natural experiment in which two continents exchanged long-isolated terrestrial faunas across a new land bridge.

In addition, the land bridge blocked marine circulation through a seaway between the Pacific and the Atlantic (Gulf of Mexico). Although marine extinctions are tangential to our focus on terrestrial events, those of the late Pliocene and early Quaternary in waters around the land bridge merit at least passing notice. Loss of the seaway is held accountable for loss of enrichment of coastal waters in western Florida by loss of upwelling. This in turn eliminated marine invertebrates, pelagic seabirds, and marine mammals including an endemic walrus, a cold-water animal hardly to be expected so far south (Allmon *et al.,* 1996). Whatever its cause (a cold climate initiating the Quaternary ice age is another possibility), biotic extinctions in marine waters in the Pliocene and early Quaternary did not continue into the late Quaternary. Thus, the marine turnovers were out of step with megafaunal extinction on land, adding to the argument that LQE had unique properties unrelated to global climatic change.

The land bridge connecting the continents at the end of the Pliocene brought Northern Hemisphere carnivores, gomphotheres, artiodactyls, and perissodactyls into contact with

South American endemic orders, the xenarthrans, notoungulates, and litopterns. Long before intercontinental suturing, megalonychid ground sloths found their way into the Greater Antilles and into North America.

Marshall and Cifelli (1990) followed Simpson (1980) in distinguishing three principal strata of land mammals based on their time of first appearance in South America. Stratum 1 includes taxa present in South America through most of the Cenozoic. Stratum 2 includes primates and caviomorph rodents dispersing across an oceanic barrier from some other land mass (not necessarily North America) to reach South America in the Oligocene or early Miocene. Although they play a very minor role in Lujanian extinctions on the continent (Table 3), Stratum 2 taxa contributed to the endemic fauna of the West Indies and suffered Holocene or earlier extinctions there.

Stratum 3 includes North American families that reached South America with or (in a few cases) without help of the land bridge after its final emergence about 2.5 Ma. They include shrews, rabbits, gophers, pocket mice, squirrels, canids, felids, mustelids, bears, gomphotheres, tapirs, equids, camelids, deer, and peccary, preceded by procyonids and cricetine rodents. Although small-mammal extinctions occurred in the Uquian (Marshall and Cifelli, 1990), there are no appreciable extinctions of the megafauna in either Stratum 1 or Stratum 3 faunas until FC in the late Lujanian (see Table 3), over 2 million years after the land bridge opened (Marshall, 1981; Marshall and Cifelli, 1990). The main paleobiological conclusion to draw from the marvelous mix of animals that ended South America's "splendid isolation" is that before the interchange neither continent was saturated in large mammals.

Just as in North America, to survive LQE in South America it paid to be relatively small, cryptic in habit, fleet, and fecund. It also helped to favor dense cover, or, conversely, wide open spaces of the high dry inter-Andean plateau not easily exploited by human societies. South American capybara, tapir, two camelids, various deer, and the spectacled bear possess at least some of the above attributes. Although too small to qualify as megafauna as here defined, the ultimate cryptophiles are tree sloths in the branches of tropical forest. In Madagascar, large-bodied, suspensory lemurs, convergent on tree sloths, suffered FC extinction (Simons, 1997).

Geochronological information on large xenarthran extinction is of high priority for those interested in first arrivals of people in the New World. There is every reason to suspect that ground sloths and glyptodonts would have been very vulnerable to human hunters. Defensible radiocarbon dates of fossil glyptodonts have yet to appear and will be difficult to obtain, given the poor preservation of bone collagen in the tropical sediments where fossil glyptodonts are often found.

On the other hand, on the basis of 15 determinations of dung and perishable tissue, *Mylodon* from a cave in southern Chile has been dated to 10,600 yrbp (Markgraf, 1985). These dates along with further excavation at Cueva del Mylodon (Borrero *et al.,* 1991) cast doubt on earlier claims that ground sloth extinction postdated human arrival by thousands of years (Saxon, 1979).

South America lost over 50 genera of large mammals, more than any other continent (Table 3). If climatic change contributed to forcing South American extinctions, it had to operate across 60 degrees of latitude, spanning a vast variety of vegetation types as we judge them from the modern setting. The types include tropical forests (tall, dense, and wet to short, open, and dry); a variety of savannas; Andean grasslands both wet (paramo) and dry (puna); and both coastal and Andean deserts. Extinction had to overwhelm large herbivores in the most durable habitat of all, the seasonally inundated and highly productive riparian flood plains of the lowlands and altiplano.

In their review of the South American Cenozoic fossil mammals, Marshall and Cifelli (1990) found no earlier megafaunal extinction event as massive as that seen in the late Lujanian (Table 3). They reported that "of the 153 genera recorded in Lujanian faunas, 56 (37%) are now extinct and 35 belong to [extinct families]." To this Marshall (1990) added: "These extinctions were differential and involved primarily mammals of large body size. . . . [B]ased on carbon-14 dates, the majority of these extinctions appear to have occurred between 21,000 and 10,000 kyr. . . . [T]he timing of this event is consistent with the blitzkrieg overkill hypothesis and the arrival of man in the Neotropics at the end of Lujanian time."

5.3. Australia

The Australian environment has limited potential for sustaining human economies, especially in comparison with the New World (Flannery, 1994). Nevertheless, it has an older prehistory. Australian sites have repeatedly yielded geochemical dates indicating Pleistocene habitation to and beyond 30,000 years ago (various chapters in Allen and O'Connell, 1995; Davidson, 1997; O'Connell and Allen, 1998; Webb and Rindos, 1997). In contrast, in the New World archaeologists have seen that a variety of sites claimed to be 13,000 to 22,000 years or older fail to be verified or replicated, despite the assertions of their proponents. Some of the claims rejected or discounted by subsequent findings include Calico Hills, Laguna Beach, Old Crow, Pikimachai Cave, Shaguinda, Taber child, and Tule Springs. One infamous case, Sandia Cave, New Mexico, the source of supposed Sandia Cave points and alleged evidence of pre-Clovis habitation, is now regarded by archaeologists familiar with the site and its claimant as a deliberate (if inexplicable) hoax (Preston, 1995).

Even if we accept the three pre-Clovis sites in America recently endorsed by Adavasio and Pedler (1997), namely, Meadowcroft in Pennsylvania; Blue Fish Caves, Yukon, Canada; and Monte Verde, Chile, the record for the New World is miniscule. There is no rich harvest of hundreds of pre-Clovis (pre-12,000 yrbp) radiocarbon dates on dozens of sites to support claims of glacial-age people in the New World that can match or come close to that of Australia. For example, from Papua New Guinea and the Bismarck archipelago through Australia to Tasmania, Davidson (1997) mapped 68 localities that are over 12,000 yrbp with no dispute regarding evidence for human presence up to 40,000 years ago (O'Connell and Allen, 1998).

If we accept the pre-Clovis claims of Adavasio and Pedler, the result, when compared with Australia, is ecologically weird. Australia is and has long been much more arid than either North or South America. Precipitation in Australia is highly variable (longer droughts, wetter wets). As stockmen are well aware, range resources boom and bust. Australian soils are notably poor in nitrogen and phosphorus. Aridity, high variance in annual precipitation, and edaphic impoverishment severely limit carrying capacity over the vast, sparsely settled interior (Flannery, 1994). In the Holocene (as today), Australia supported a much smaller human population, one much less visible archaeologically than that of the conterminous

United States which is comparable in area (8 million km^2). Furthermore, the number of professionals and amateurs surveying archaeological sites is much larger in the New World than in Australia. Thus, in terms of both carrying capacity and potential archaeological visibility, one might expect that if people occupied what is now the United States thousands of years before Clovis, the number of authentic American sites prior to 12,000 years in age should be considerably more, not far less than the 68 now listed by Davidson (1997) from Australia. Despite the burst of acclaim for Monte Verde recently unleashed by pre-Clovis partisans, there is very good reason to call for independent verification of it and any other pre-Clovis claim in the New World. In the meanwhile, it is certainly fair to maintain considerable skepticism about the entire enterprise.

Despite driving back the time of human colonization to about 40,000 years, if not beyond (O'Connell and Allen, 1998), Australian prehistorians have had much less success at detecting last occurrence of Australia's extinct fauna, the sort of evidence needed to determine rate and possible cause of extinction. This methodological problem is quite understandable if, as appears to be the case, megafaunal extinctions in Australia bore a close relationship to FC and thus were at or beyond the limits of the radiocarbon dating technique. One of a variety of problems is the fact that a very small amount of modern carbon, rich in carbon-14, can seriously contaminate a 35,000-year-old sample yielding a barely detectable level of radioactivity. Another is the fact that the hazards of time averaging (redeposition and palimpsest problems) increase with age. Perhaps the main problem is scarcity of hyperarid caves suitable for preservation not only of bone collagen but also of perishable organic remains of extinct mammals and birds such as dung, keratinous nails or horns, tissue, feathers, and hair, as in the arid southwestern United States (Martin, 1990).

As in America, LQE in Australia targets large animals. Flannery and Roberts (this volume) report that around the time of FC, Sahul lost about 28 genera and 55 species of large vertebrates (body weights exceeding 10 kg). Australia is a land of human colonization and, as we have emphasized, the relationship between FC and LQE is critical to any model of overkill. The pattern of extinction in Australia resembles that in America and Madagascar. In all three land masses, kill sites or processing sites are few, in dispute, or absent. In America and Madagascar, and perhaps in Australia as well, the fossil record is in accord with a rapid wipeout, not necessarily by overhunting alone, of those vulnerable species likely to have been of maximum interest to the first colonists, usually the largest. Except for a few possible cases of coextinction, small mammals (up to 10 kg) were not affected.

Despite its problems, the record of Australian extinctions yields important leads. For example, if eight extinct species of the genus *Macropus* were in fact thriving in Australia until about 15,000 years ago (Murray, 1991:1112), at least 15,000 years after the arrival of modern humans, the scarcity of cultural associations is inexplicable. In Tasmania the record of LQE is rich, including a variety of extinct giant species of marsupials. One might expect that some of the 21 dated Pleistocene sites in caves or rock shelters in the Franklin River region of southwestern Tasmania ranging in age from 10,000 to at least 25,000 yrbp might yield bones of the extinct species. To date they have not. The sites have yielded hundreds of thousands of animal bones, all of living species and mainly of Bennett's wallaby, *Macropus rufogriseus* (Allen, 1997).

This negative evidence seems sufficient to question the survival of extinct megafauna as late as 15,000 or even 25,000 years ago. More to the point, the record as it stands strongly suggests a pattern similar to that seen in North America and Madagascar, that is, rapid extinctions at the time of first contact with very little if any field evidence of hunting, killing, or processing of the remains. The alternative in Australia would mean at least 15,000 years of coexistence of extinct fauna with prehistoric people, without a robust archaeological association. Such a lengthy overlap between human colonization and last occurrence of potentially preferred prey would eliminate any thought of rapid extinction ("blitzkrieg"). Is the geochronological support for such a lengthy coexistence robust? Dates on the cultural record appear to be secure, although those on time of extinction are not. Baynes (1995) and White and Flannery (1995) have raised questions about post-30,000-year dates reported for extinct Australian megafauna in the past.

A well-excavated stratigraphic association between extinct fauna and cultural material from 1.0 to 1.7 m depth is reported from Cuddie Springs, central New South Wales (Dodson *et al.*, 1993; Furby, 1995; Field and Boles, 1998). The archaeological record includes flaked and ground stone artifacts, ochre, copious quantities of charcoal and human-modified bone. Eight radiocarbon dates between 1.0 and 1.6 m fall in a narrow range from 29,000 to 34,000 years suggesting time averaging or stratigraphic smearing. As is true in many other Australian records of extinct megafauna, the bones themselves are unsuitable for direct dating and the possibility of a secondary association cannot be rejected.

Although worked bone is reported at Cuddie Springs, there is no evidence of killing or processing, simply the conjunction of artifacts and extinct fauna. With the exception of the thick eggshells of the mihirung (*Genyornis*), Australian deposits rarely yield fossils of extinct animals suitable for direct dating. On the other hand, the rich record of stratified deposits in Australian cave and dune sites offers the opportunity to build a case for absence. If they were still alive 30,000 years ago or less, it is hard to believe that the extinct mihirung, giant macropods, diprotodonts, and the others would not have been found in archaeological sites.

The record from Cuddie Springs, the Tasmanian caves, and new radiocarbon dates of *Genyornis* eggshell (Miller *et al.*, 1999) support an earlier interpretation (Martin, 1984) based on similar evidence that viewed full glacial-age droughts (23,000 years ago or later) as too young to have played a part in initiating Australian extinctions. The case is by no means closed and some Australian archaeologists believe they have or will have sites a good deal older than 35,000 years. Beyond the ever-popular search for the oldest archaeological sites, we urge a relentless search for the youngest extinct faunas. The opportunity both to date and to recover environmental information from stable isotopes from samples of *Genyornis* eggshell is especially promising. Miller and colleagues (1999) claim extinction of *Genyornis* 45,000–55,000 yrbp in several parts of its wide range and in more than one climatic province.

As in the New World, the scarcity or lack of kill sites or processing sites, viewed by some archaeologists on both continents as fatal to the case for anthropogenic extinctions, is a prediction of the bow wave model (Mosimann and Martin, 1975). An Australian modification based on the assumption of limited carrying capacity of extinct fauna, as seems entirely reasonable on environmental grounds (see above), overrides the objection that the colonization process would have to be slow because the Australian environment was so much less favorable for human foragers than their tropical source area. Webb and Rindos (1997) have structured a colonization model that is inherently fast. Perhaps some of the potential food plants adopted in later times were first recognized as desirable by observations made on the feeding habits of diprotodon, *Genyornis,* and other doomed megafauna during the coexistence, however brief, between people and large animals.

5.4. Eurasia and Some of Its Islands

Compared with the catastrophic loss of Quaternary megafauna in the New World, Madagascar, and arguably in Australia, the losses in Eurasia were relatively minor and they occurred gradually (Fig. 3; Martin and Stuart, 1995; Stuart, 1991). The extinct fauna can be divided into "warm" or interstadial faunas with straight-tusked elephants (*Elephas [Palaeoloxodon] antiquus*), temperate rhinoceros (*Dicerorhinus*), hippo (*Hippopotamus amphibius*), and cave bear (*Ursus spelaeus*), and "cold" or stadial ones with woolly rhinoceros (*Coelodonta antiquitatis*), woolly mammoth (*Mammuthus primigenius*), and muskox (*Ovibos moschatus*). We are unaware of any climatic models of extinction that embrace both the protracted pattern of megafaunal loss in Eurasia (first of warm and then of cold faunas), as well as the universal, catastrophic late-glacial extinctions of megafauna in the New World (Fig. 3).

The intriguing differences in the pattern of Old and New World extinction can be well illustrated in the case of proboscideans. Climatic changes alone may be responsible for gomphothere (*Anancus*) extinction along with the loss of many temperate forest trees from western Europe in the early Quaternary. The loss of straight-tusked elephants (*Elephas antiquus*) is another matter. Extinct throughout their European range by 60,000 years ago (Stuart, 1991), *Elephas (Palaeoloxodon)* persisted in Japan until 30,000 years ago or less. "This extinction phenomenon [which included giant deer, as in Europe] seems to have been caused at least partly by human hunting activities" (Kawamura, 1991). The Holocene survivors were not larger than sika deer (*Cervus*), wild boar (*Sus*), or black bear (*Selenarctos*). Dwarf relatives of *Elephas (Palaeoloxodon)* survived as late or later on Mediterranean islands, occurring in the Akrotíri *Aetokremnos* hippo bone cave, 10,500 years ago, an epipaleolithic site on Cyprus (Simmons, 1991). Just as Wrangel Island served as a refugium for the last of the boreal woolly mammoths, extinct throughout the rest of their range, Japan and Cyprus sheltered the last populations of temperate Eurasian *Elephas.*

In all likelihood, woolly mammoths survived longer than straight-tuskers not simply because they were better adapted to an extreme subarctic environment, which they were, but because the subarctic was penetrated later by humans (Martin, 1982). Woolly mammoths attained their maximum range *before the height of the last glaciation,* a curious occurrence if climate controlled their range. Only before 20,000 years ago did they reach the shores of the Mediterranean and Black Seas (Sher, 1997; Soffer, 1993; Stuart, 1991) and their southern limit in China, the 45th parallel (Liu and Li, 1984). By 12,000 yrbp, woolly mammoth had vanished from much of northern Europe. In the next millennium they suffered further reduction in range. They survived on the Arctic coast until 9500–10,000 yrbp with a population of dwarf mammoths (up to 1.8 m in shoulder height) on Wrangel Island in the Arctic Ocean until 4000 years ago (Vartanyan *et al.,* 1992). If climatic change alone drove the retreat of Eurasian Proboscidea, one might expect similar results in North America.

In contrast, in 50 kyr prior to their extinction around 11 kyr, North American Proboscidea suffered no loss of taxa, no clear-cut reduction in range, and no apparent decline in numbers (see maps in Graham and Lundelius, 1994). Had North American Proboscidea followed the Old World pattern, the temperate species of midlatitudes such as the Columbian mammoth would have predeceased high-latitude woolly mammoths by tens of thousands of years whereas the order Proboscidea would endure in South America, represented by one or two living species.

Because kill sites or processing sites are not known, and because mammoth bones used in construction might have been picked up out of carcasses from animals that died of natural causes, Soffer (1993) and others have warned against assuming that apparent associations constitute evidence of hunting. At least some of the late Paleolithic mammoth ivory used for artistic purposes could have come from fossil deposits.

Nevertheless, both ivory carvings and cave paintings by stone-age artists reveal an intimate knowledge of the anatomy and behavior of mammoths and many other species. Low rates of increase of modern megaherbivores mean that modern elephants, rhinoceros, and hippo are highly vulnerable to human predation (Owen-Smith, 1988). Without governmental protection, all would be extinct. The gradual decline of Eurasian mammoths could readily result from a minimal amount of human hunting or interference, especially if younger age classes were removed selectively (Budyko, 1967; Mithen, 1993, 1997). Some archaeologists have felt that the lack of extensive field evidence of Paleolithic mammoth hunting, compared with abundant evidence of human predation on *Rangifer, Bison,* and *Cervus,* is sufficient to discount human involvement in the process of mammoth extinction. Here as elsewhere the assumption that ample field evidence in the form of kill sites would accompany extinction by overkill has misled skeptics.

Because the Wrangel Island dwarf mammoths are not found associated with archaeological remains, Vartanyan *et al.* (1992) discount overkill. The same argument is used in other cases including *Megaloceros,* the giant deer or "Irish elk" in Ireland (Mitchell and Parkes, 1949; Barnosky, 1986); Pleistocene extinct *Hippopotamus,* deer, and elephants in Crete (Reese, 1996); and dwarf mammoths whose bones are numerous on the Channel Islands off the California coast. Although archaeological evidence of human activity is evident in younger strata postdating the extinctions, in these cases no acceptable associations with the extinct fauna are known. At first glimpse these findings look damaging to the FC model of human-triggered extinctions.

Similar arguments on Cyprus subsided when an epipaleolithic association of abundant extinct hippo (*Phanuoris*) and dwarf elephant bones with thumbnail scrapers and extralocal marine shells were found at Akrotíri *Aetokremnos,* a small cave above a sea cliff (Simmons, 1991). The weighted average of 26 dates on shell, charcoal, and burnt bone is 10,464±25 yrbp, indicating an occupation in the 11th millennium bp, 1500 years earlier than the earliest previously known sites on Cyprus.

Although the nature of the association and its meaning in modeling insular faunal extinctions in the Mediterranean is disputed by Bunimovitz and Barkai (1996), the record at Akrotíri *Aetokremnos* suggests a Paleolithic "commando raid" in which stone-age hunters or foragers overran an island, slaughtered preferred prey (dwarf hippo, dwarf elephant, and endemic deer), did not find sufficient resources for a sustainable economy, and soon left with no other evidence of their passage, beyond the remarkably rich contents of one cave. Is it possible that Crete, Ireland, Wrangel Island, and the Channel Islands of California also lost megafauna to commando raids that left scant evidence, or at least none discovered to date? Given the scarcity of convincing kill or processing sites on continents, the rarity of such features on off-shore islands is not surprising. As we have seen, only in the remote Pacific such as on the low limestone islands in Tonga are extinct animals (bones of megapodes) abundantly associated with artifacts of the first colonists. In other cases (Hawaii, Madagascar) such associations are rare or unknown.

5.5. Africa and Madagascar

Africa's intriguing record of Quaternary extinctions, long overshadowed by African mammalian evolution, remains to be explained (see Burney, 1996, for a recent attempt). In Africa the extinction of carnivores, large primates, and a variety of suids accompanied rapid evolution of bovids around 2 million years ago. Whatever caused the turnovers (climatic change at the start of the Quaternary is a popular model), the event is not matched by similar turnovers of large mammals at the start of the Quaternary in North America (see Table 1). An even more interesting comparison is that with Madagascar.

Some six species of bovids disappeared from South Africa as late as the early Holocene, although it was much later in the Holocene when Madagascar lost half of its lemurs, all of them larger than the largest living lemur, *Indri,* and all estimated to exceed 10 kg in body weight. Also in the late Holocene, Madagascar lost two hippos, the highly endemic bibymalagasy (*Plesiorycteropus*), a large (bobcat size) species of carnivore (*Cryptoprocta spelea*), an extinct rodent (*Hypogeomys*), two genera of giant flightless birds, and a giant tortoise (MacPhee and Marx, 1997). Although most if not all of these extinctions in Madagascar coincide with human colonization, they have not been found in well-defined kill sites, adding support to MacPhee and Marx's model of hyperdisease. Yet as we have seen in the Americas and Australia, lands of human colonization, the association of prehistoric cultural material with remains of extinct faunas is remarkably scarce. We have attributed lack of associations in Australia to FC extinction near the limit of radiocarbon dating over 30,000 years ago, with ample opportunity for smearing and time averaging of the associations in a continent where aridity limited population size of potential prey and human colonists alike.

Such an explanation is less plausible in Madagascar where extinctions occurred in the last millennium before the historic period. If aridification contributed, it was evidently a potential problem only in the Southwest. Furthermore, the model of a blitz, a sweep of the vulnerable megafauna, seems unlikely if the start-up of colonization took over 1000 years. Dewar (1997) suggested that the impact of wild cattle rather than human hunting forced the extinctions.

All of this may be resolved eventually by enrichment of the stratigraphic and geochronological record, including a major effort at building a radiocarbon chronology of the extinct Malagasy megafauna. Recent dates of 13,000 and 26,00 yrbp on *Megaladapis* (Simons, 1997) are the first step in testing the remote possibility that there were glacial-age extinctions in Libby time long before FC, a test that (as we have seen) has failed in New Zealand, Tonga, and Hawaii.

Perhaps the most interesting comparison with Africa is the case of giant tortoises. Giant tortoises can be regarded as extremely vulnerable to human foraging. They survived historically only on remote warm-water oceanic islands that escaped prehistoric colonization, such as the Mascarenes, Seychelles, and Galápagos. In the late Holocene in Madagascar, giant tortoises vanish with the giant lemurs. In Africa, they disappear much earlier, perhaps over 2 million years ago. In our view, that event marks the beginning of extinctions forced by hominid ancestors.

Africa has been held up to modelers of overkill as a conundrum. If people are such effective exterminators, how did the African megafauna survive? However, the argument cuts two ways. If late Pleistocene climatic change played a major role in triggering megafaunal extinctions elsewhere, how did so many large animals manage to survive in Africa and Asia, continents no more immune to Pleistocene climatic change than other corners of the globe?

The standard reply to the conundrum is that in Africa and Asia there was sufficient time to evolve a relatively benign predator–prey relationship, with human activity suppressed by an extra load of human parasites and diseases (Martin, 1984), and by intertribal warfare. In any case, the extinction record in both Africa and Asia is so different in chronology and intensity from that found elsewhere on the planet (Figs. 1–3) that important differences in causality may be postulated. The record of gradual human evolution and radiation confined to the Afro-Asian land mass, ending in the late Quaternary with human colonization on other land masses, fits the model of a time transgressive overkill.

6. Didn't Climatic Change Make a Difference?

Repeated, severe, and rapid climatic change is a hallmark of the Quaternary (the last 1.7 million years in some definitions, 2.4 million in others). Climatic forcing has proven highly effective in accounting for so many features of the ice ages that it is naturally considered first whenever some feature of Quaternary biogeography requires an explanation. Surely climatic change is involved in LQE.

Nevertheless, the popular answer, "the climate did it," is unsatisfactory in the case of LQE, for two reasons. The first is that, unlike major climatic pulses that are synchronous globally, radiocarbon and other methods of geochemical dating have shown that the geographical sequence of late Quaternary biotic extinctions is sequential. Mass extinction of megafauna struck first at least 30,000 years ago in Meganesia (Australasia) and Japan; then in the late-glacial it swept North and South America; in the late-glacial or early Holocene it reached nearshore (epicontinental) oceanic islands, such as Cuba and Cyprus; in the late Holocene shortly before historic explorations extinctions burned through hundreds of more remote oceanic islands of the Pacific and Indian Oceans, up to the size of Madagascar and New Zealand. Finally, in the last 500 years, the historic epoch of "discovery" saw anthropogenic extinction of the dodo in Mauritius, Steller's sea cow in the Commander Islands, a giant rat in the Galápagos, and 100 other mammals mainly on oceanic islands (MacPhee and Flemming, this volume). With or without any help from natural environment or climatic changes, these and all other extinctions in Libby time accompany the arrival of people (Figs. 1–3).

The second is that proxy climatic data from deep-sea cores and ice cores feature repeated switching from glacial to interglacial or interstadial conditions. Between the last (Eem) interglacial 110,000 years ago and the Younger Dryas 99,000 years later, the Greenland ice core (GRIP) reveals 24 interstadial climatic pulses or switches from cold to warm to cold (Dansgaard *et al.*, 1993). Six of the pulses had previously been detected in European pollen stratigraphy. Thus, the Younger Dryas switch is by no means unique. For this reason it is hard to imagine that one final switch, from the warm Bolling–Allerod to the cold Younger Dryas, could wreak such havoc, including mammoth extinction on two continents (see McFarlane, this volume; Putshkov, 1997).

However, regional climatic models remain popular. They are used to explain LQE in North America, for example in terminating the "mammoth steppe" of Beringia (Guthrie, 1990; Sher, 1997; for contra see Zimov *et al.,* 1995); in a "Clovis drought" coinciding with megafaunal extinction in western North America (C. V. Haynes, 1991); and in the disharmony model of Graham and Lundelius (1984) with large herbivore extinctions driven by a change from more equable interglacial and glacial climates to less equable (supposedly more extreme) climate in the Holocene.

Whatever their regional merit, these models do not generalize well. Woolly mammoth extinction in the Beringian "mammoth steppe" driven by alleged habitat loss 11,000–16,000 years ago came at a time when expanding steppes and outwash plains in the newly deglaciated parts of northern North America should have exposed mineral-rich soils ideal for grasses and sedges to support expanding populations if not of woolly mammoths then certainly of Columbian mammoths and their associated megafauna. If a "Clovis drought" 11,000 radiocarbon years ago helped force extinctions of mammoths and tapirs along the Colorado River drainage of Arizona, would drought also drive extinction of mammoths, mastodonts, and tapirs along rivers and deltas in much wetter habitats to the southeast, such as the riparian gallery forests and flood plains from Texas to Florida? And if increased continentality (cooler winters, warmer summers) caused problems for the ice-age megafauna in the midwest of the United States 11,000 years ago (which would be curious in itself given the apparent success of Asian and African megafauna in continental climates), what will explain massive megafaunal extinctions along the Amazon and elsewhere in tropical America?

It is no surprise that climatic changes coinciding with LQE have been detected as abundant proxy climatic data from the Late Quaternary show that "the times are always changing," a conclusion reached by Dickinson (1995) regarding the Holocene that applies with even greater force to the ice ages. Presumably any species surviving into the Late Quaternary had already survived numerous glacial–interglacial cycles as well as innumerable stadial switches within a cycle. We are left to wonder why Late Quaternary climates suddenly turned lethal when earlier ones were not.

7. Discussion and Conclusions

Like other Quaternary phenomena, the record of late Quaternary extinction is especially revealing when viewed globally. The LQE pattern involving large terrestrial vertebrates in North America, South America, and Australia differs fundamentally from that in Africa and Eurasia (Martin, 1967, 1990). In the Americas and probably also in Australia, the losses of large mammals are massive and catastrophic. In Africa and Eurasia, LQE involves a smaller portion of the large mammal fauna and the losses occurred much more gradually.

Both the Americas and Afro-Asia include highlands and lowlands, temperate and tropical climates, forests and grasslands, wetlands and deserts. Both sets of continents underwent changes in climate and in plant communities during numerous glacial-to-interglacial transitions throughout the Quaternary as well as near its end when the extinctions occurred.

Apart from the difference in intensity of extinctions, the one distinguishing feature separating North America, South America, and Australia from Africa and Eurasia is the sudden appearance of *Homo sapiens* emigrating from the latter to the former.

In addition, the "human bolide" struck hard on oceanic islands. Island faunas were innocent of human impacts; the point cannot be overemphasized. Animals accustomed to human predation are much more difficult to hunt or capture than are conspecific populations or closely related species that are naive (Steadman, 1997). Such tameness develops rapidly in unhunted parks or reserves. The first encounter between innocent prey and highly skilled, sophisticated humans may well have involved considerable surplus killing, transgressing the regulated harvest of a "prudent" predator. In their capacity to amplify human impact, North America, South America, and Australia were "islands" at FC. In Africa and Eurasia where hominids had evolved alongside other vertebrates, the patterns of behavior and community resilience were quite different.

A geologically rapid and biotically monumental extinction event will follow human establishment on a virgin land mass. Although various factors can fine-tune the rapidity and intensity of extinctions, a first tier of highly vulnerable species succumb to predation or other ecological traumas to which they cannot adjust in time to escape extinction. On continents, the first-tier losses are biased strongly toward large mammals and their commensals. On remote oceanic islands, the first tier is dominated by seabirds, large flightless land birds, and, when present, giant tortoises and large lizards. If Polynesian rats accompany human colonization, small ground-nesting or flightless species may be lost independent of direct human predation.

In the Americas and on certain tiny oceanic islands, such as Lifuka, the first tier (most large mammals in America, most birds on Lifuka) was lost rapidly in time scales that defy resolution by current methods of geochronology. After first-tier losses, the survivors (the second tier, for various reasons more extinction resistant) came to dominate the faunal landscape. On some islands such as Mangaia and possibly in Australia and Madagascar, loss of the first tier was not as rapid. A blend of physical, biotic, and cultural factors could have retarded first-tier extinctions, with cultural factors perhaps being the most important. During the first millennium of human influence, Mangaia and Madagascar may not have been settled permanently or extensively. The early zooarchaeology of Mangaia suggests a "fishing camp" with little or no agriculture and no introduction of alien fauna (Kirch and Ellison, 1994; Steadman, 1995).

Whereas archaeological sites rich in bones of extinct large mammals are few, and some of those (e.g., the mammoth-bone huts of the Ukraine) have been discounted as secondary (Soffer, 1993), there are over a dozen North American Clovis sites incorporating mammoth. Those unfamiliar with fossils and the fossil record sometimes imagine that finding terminal populations of species on the threshold of extinction is easy, because large numbers of bones can be expected. This intuitive notion is dead wrong. Finding the ultimate population of a species on the edge of extinction is a major geochronological challenge, not to mention difficulties in determining directly from field evidence the proximal cause of death.

In the South Pacific, large numbers of megapode bones have recently been found in 3000-year-old Lapita pottery sites, the signature of an explosive first contact by humans on small (1–10 km diameter) low islands in the Kingdom of Tonga (Steadman *et al.,* in press; Burley *et al.,* submitted). Bones of extinct pigeons, parrots, and flightless rails also appear regularly in prehistoric middens on Pacific islands (Steadman, 1995). Thus, at least in parts of the Pacific, there are ample archaeological associations to suggest the probable cause, if not the details, of the extinction process. On Madagascar and the continents of colonization, intimate associations between bones of doomed species and evidence of human invaders are exceptional.

What fuels the overkill model is (1) size and ecology of the species affected and (2) chronology. The losses strongly suggest human pursuit of preferred (high ranked) prey. Around the time of human arrival, extinctions of large (>44 kg) genera occur rapidly and massively on continents of human colonization (Figs. 1 and 2). On continents of human origins, extinctions were relatively few and those few vanished gradually (Fig. 3). Smaller animals were also hunted and, after extinctions of the large, they dominate the archaeofaunal record. With time as resource harvests intensify, the populations of preferred prey species decline or disappear locally. On islands, smaller species were much more vulnerable and the survivors continue to be at risk. Human activity may deplete island species through side effects of agriculture (habitat change) and by the introduction of aliens including diseases.

Before Libby time, it is possible to associate island extinctions with hominid activity at least 700,000 years ago (Morwood *et al.*, 1998). Within the last 40,000 years, the chronology of extinctions closely follows human colonizations. Extinctions struck first (over 30,000 years ago) with disappearance of giant marsupials in Australia and of Nauman's elephants and giant deer in Japan. Roughly 20,000 years later, continental extinctions erupted in the Americas with the loss of 74 genera of large mammals. Coincidentally, extinctions involving dwarf hippo and dwarf elephants occurred in the Mediterranean islands along with extinction of giant deer in Ireland. Early in the Holocene, dwarf megalonychid sloths vanished from the West Indies, although direct dating of this event is lacking. Until 4000 years ago, the last mammoths on the planet lingered as dwarfs on Wrangel Island in the Arctic Ocean, soon to be followed by the last Eurasian muskoxen. Extinctions of giant lemurs, giant tortoises, and giant birds on Madagascar and the moas on New Zealand were under way within the last millennium or two, apparently preceded in New Zealand by some rat-driven reductions in ground-nesting birds and large invertebrates, the rats a harbinger of Polynesian landings prior to Polynesian colonization (Holdaway, in press). Flightless rail and other land bird extinctions swept from west to east through hundreds of Pacific islands beginning in the eastern Solomons 3500 years ago and ending on Easter Island less than 1000 years ago.

In some cases such as Wrangel Island, Ireland, and Crete, extinctions appear to predate human contact, although they may not if the archaeological record turns out to be older or the extinction event younger than the present chronology indicates (Signor–Lipps effect). There is no doubt about anthropogenic forcing of extinctions in historic time (see MacPhee and Flemming, this volume; Harrison and Stiassny, this volume).

What happens when climatic change and human FC are decoupled? Diamond (1984) illustrates the point by what we call the "Silver Blaze" test after the A. C. Doyle detective story. The incomparable detective Sherlock Holmes considered the case of a stolen racehorse. A neglected clue, a watchdog that did not bark in the night, provided the solution. In the case of island extinctions, Diamond asks what happened before people arrived, at a time when glacial-age climates changed? We now have a few such cases, especially well documented in New Zealand. There a rich glacial-age fossil record coincides with the time when climatic change is often held accountable at least in part for extinctions in Ireland, the Mediterranean islands, and North America. Contrary to predictions of the climate model, there is no record of prehuman extinction in Libby time in New Zealand, despite evidence of significant change in distribution of glaciers, loess deposits, grassland–forest habitats and, not least, in associated range changes in moas (Worthy, in press). Neither moas nor any other species known from the glacial-age fauna of New Zealand became extinct until late in the Holocene, after human arrival. Glacial-age climatic and other environmental change left no imprint on the 70,000 yrbp or older avifaunas obtained on Oahu (Hawaiian Islands) and 'Eua (Kingdom of Tonga).

Given the failure of climatic extinction models to yield robust examples of extinctions independent of possible human influence, we find it useful to regard all extinctions of mammals, birds, and other terrestrial vertebrates in Libby time (including coextinctions of carnivores, parasites, and commensals) as fundamentally anthropogenic. FC extinctions invariably accompany human arrival in new lands (Burney, 1993; MacPhee and Marx, 1997; Martin, 1990; Steadman, 1997). In few if any cases would the colonizers realize the consequences of their actions.

Given the variable conditions of preservation of prehistoric faunas and artifacts, it is too much to expect that each local region of the earth will inevitably yield the kind of fossil data needed to provide unequivocal associations of humans and extinct species. Although extinction chronologies of all species and geographic regions offer major opportunities for discovery, the global picture is clear; it is more revealing than the sum of its parts. LQE are not in step with proxy paleoclimatic data. They fall in step with human colonization (Figs. 1–3) in a "dreadful syncopation" (MacPhee and Marx, 1997). Extinctions of the last 40 millennia represent the lethal overture to the sixth great mass extinction, currently under way.

8. Summary

Close comparison of islands and continents during Libby Time (the last 40,000 years) indicates sequential extinctions of prehistoric terrestrial mammals and birds. On oceanic islands in the Pacific 50%–90% of the native land birds, along with other terrestrial species vanished in the last 3000 years. Among the continents North and South America and Australia lost over two-thirds of their large mammals and other megafauna >44 kg in body weight. In the New World a unique extinction spasm at the end of the Pleistocene obliterated 74 genera, 10 families, and three orders of large mammals. In North America the sum of all generic extinctions of megafauna in the previous four million years is less than the number coinciding with spread of the Clovis Culture and with rapid climatic change of the late-glacial (15,000–10,000 yrbp). On remote oceanic islands extinction spasms of land birds postdate the late-glacial, tracking prehistoric human first contact (FC) and settlement in the Holocene. In a few crucial cases oceanic islands that experienced severe FC extinction late in the Holocene yield much older fossil records that predate and known or expected human arrivals. Examples include the Galapagos with fossil faunas extending through the Holocene and the islands of New Zealand, 'Eua (Kingdom of Tonga), and O'ahu (Hawaiian Islands). The latter three examples are critically important. They yield fossil faunas equivalent in age to rapid climatic changes (switching) in the late Quaternary. If climate changes forced extinction independent of the time of human arrivals, these records should reveal it. Evidently, in the absence of humans, very few extinctions of endemic insular species are known, at least in the late Quaternary. While much more fossil data on island faunas is needed, the results to date suggest that by the late Quaternary the global biota already had been thoroughly purged of taxa susceptible to ice-age climatic or environmental changes. To force terrestrial extinctions in Libby time required an impact transcending the range of environmental shocks typical of the Quaternary. The unique environmental crisis that triggers the "dreadful syncopation" of sequential extinctions is arrival of anatomically and behaviorally modern *Homo sapiens.* Features of the fossil record indicate that the anthropogenic extinction process involved more than human predation alone. Nevertheless, the repeated coincidence throughout the globe of human contact and colonization soon followed by an extinction pulse demonstrates cause and effect.

Acknowledgments

We thank Ross McPhee for the opportunity to participate in the AMNH conference on prehistoric extinctions and for crucial editorial aid. We benefited by collegial assistance from the following: L. Agenbroad, E. Anderson, J. Betancourt, D. Burney, I. Davidson, J. Field (Furby), K. Hawkes, G. Haynes, V. Haynes, R. Holdaway, H. James, C. Kay, R. Kimber, R. Klein, G. Miller, S. Mithen, M. Morbeck, S. Olson, M. O'Rourke, N. Owen-Smith, J. Quade, D. Reese, C. Schrire, P. Sondaar, J. Speth, T. Stafford, M. Stiner, A. Stuart, and Y. Tomida.

References

Adavasio, J. M., and Pedler, D. R. 1997. Monte Verde and the antiquity of humankind in the Americas. *Antiquity* **71**:573–580.

Agenbroad, L. D., and Mead, J. I. 1996. Distribution and palaeoecology of central and western American *Mammuthus,* in: J. Shoshani and P. Tassy (eds.), *The Proboscidea: Evolution and Palaeoecology of Elephants and their Relatives,* pp. 280–288. Oxford University Press, London.

Allen, J. 1997. The impact of Pleistocene hunters and gatherers on the ecosystems of Australia and Melanesia: in tune with nature?, in: P. V. Kirch and T. L. Hunt (eds.), *Historical Ecology in the Pacific Islands: Prehistoric Environmental and Landscape Change,* pp. 22–38. Yale University Press, New Haven.

Allen, J. and O'Connell, J. F. (eds.). 1995. Transitions: Pleistocene to Holocene in Australia & Papua New Guinea. *Antiquity* **69** (Special No. 265).

Allmon, W. D., Emslie, S. D., Jones, D. S., and Morgan, G. S. 1996. Late Neogene oceanographic change along Florida's West Coast: evidence and mechanisms. *J. Geol.* **104**:143–162.

Anderson, A. 1989. *Prodigious Birds: Moas and Moa-Hunting in Prehistoric New Zealand.* Cambridge University Press, London.

Anderson, A. 1997. Prehistoric Polynesian impact on the New Zealand environment: Te Whenua Hou, in: P. V. Kirch and T. L. Hunt (eds.), *Historical Ecology in the Pacific Islands: Prehistoric Environmental and Landscape Change,* pp. 271–283. Yale University Press, New Haven.

Anderson, A., and McGlone, M. 1992. Living on the edge: Prehistoric land and people in New Zealand, in: J. Dodson (ed.), *The Naive Lands: Prehistory and Environmental Change in Australia and the Southwest Pacific,* pp. 199–241. Longman Cheshire Pty. Limited, Melbourne.

Bahn, P., and Flenley, J. 1992. *Easter Island: Earth Island.* Thames & Hudson, London.

Balouet, J., and Olson, S. L. 1989. Fossil birds from late Quaternary deposits in New Caledonia. *Smithson. Contrib. Zool.* **469**:1–38.

Barnosky, A. D. 1986. Big game extinction caused by late Pleistocene climatic change: Irish elk (*Megaloceros giganteus*) in Ireland. *Quat. Res.* **25**:128–135.

Barnwell, P. J. 1948. *Visits and Despatches* [sic] (Mauritius, 1598–1948). The Standard Printing Establishment, Port Louis, Mauritius.

Baynes, A. 1995. The question "Why did the Australian Pleistocene megafauna become extinct" is currently unanswerable because there are no reliable dates for any of the species. Abstract of a paper presented at the Taphonomy Symposium, held at the Australian National University, Canberra, 28–29 April 1995.

Borrero, L. A., Lanata, J. L., and Cardenas, P. 1991. Reestudiando cuevas: Nuevas excavaciones en Ultima Esperanza, Magallanes. *Ann. Inst. Patagonia* (Punta Arenas, Chile) **20**:101–110.

Brookfield, H. C., Latham, M., Brookfield, M., Salvat, B., McLean, R. F., Bedford, R. D., Hughes, P. J., and Hope, G. S. 1979. Lakeba: environmental change, population dynamics and resource use, in: The UNESCO/UNFPA Population and Environment Project in the Eastern Islands of Fiji. *Man and the Biosphere (MAB) Programme, Project 7: Ecology and Rational Use of Island Ecosystems,* pp. 93–110. Canberra, Australia.

Budyko, M. I. 1967. On the causes of the extinction of some animals at the end of the Pleistocene. *Sov. Geogr. Rev. Transl.* **8**:783–793.

Bunimovitz, S., and Barkai, R. 1996. Ancient bones and modern myths: ninth millennium BC *Hippopotamus* hunters at Akrotiri *Aetokremnos,* Cyprus? *J. Mediterr. Archaeol.* **9**:85–96 (responses and bibliography on pp. 96–126).

Burley, D. V. 1996. Sport, status, and field monuments in the Polynesian chiefdom of Tonga: the pigeon snaring mounds of northern Ha'apai. *J. Field Archaeol.* **23**:421–435.

Burley, D. V., Nelson, E., and Shutler, R. Submitted. A radiocarbon chronology for the eastern Lapita frontier: new dates from the Ha-apai Islands, Central Tonga. *Archaeol. Oceania.*

Burney, D. A. 1993. Recent animal extinctions: recipes for disaster. *Am. Sci.* **81**:530–541.

Burney, D. A. 1996. Paleoecology of humans and their ancestors, in: T. R. McClanahan and T. P. Young (eds.), *East African Ecosystems and Their Conservation,* pp. 19–36. Oxford University Press, London.

Burney, D. A., Burney, L. P., and MacPhee, R. D. E. 1994. Holocene charcoal stratigraphy from Laguna Tortuguero, Puerto Rico, and the timing of human arrival on the island. *J. Archaeol. Sci.* **21**:273–281.

Christensen, C. C., and Kirch, P. V. 1981. Non-marine mollusks from archaeological sites in Tikopia, southeastern Solomon Islands. *Pac. Sci.* **35**:75–88.

Cosgrove, R. 1995. Late Pleistocene behavioural variation and time trends: the case from Tasmania. *Archaeol. Oceania* **30**:83–104.

Coues, E. (ed.). 1897. The Manuscript Journals of Alexander Henry and David Thompson: Exploration and Adventure among the Indians on the Red, Saskatchewan, Missouri and Columbia Rivers. Francis P. Harper, New York.

Dansgaard, W., Johnsen, S. J., Clausen, H. B., Dahl-Jensen, D., Gundestrup, N. S., Hammer, C. U., Hvidberg, C. S., Steffensen, J. P., Sveinbjörnsdottir, A. E., Jouzel, J., and Bond, G. 1993. Evidence for general instability of past climate from a 250-kyr ice-core record. *Nature* **364**:218–220.

Davidson, I. 1997. "Pervye liudi, stavshie australiitsami" (First people becoming Australian), in: A. A. Velichko and O. Soffer (eds.), Chelovek saseliaet planeru zemli. Global' noe rasselenie gominid. Institute of Geography, Russian Academy of Sciences, Moscow.

Dewar, R. E. 1997. Were people responsible for the extinction of Madagascar's subfossils, and how will we ever know?, in: S. Goodman and B. Patterson (eds.), *Natural Change and Human Impact in Madagascar,* pp. 364–377. Smithsonian Institution Press, Washington, D.C.

Diamond, J. 1984. Historic extinctions: a Rosetta stone for understanding prehistoric extinctions, in: P. S. Martin and R. G. Klein (eds.), *Quaternary Extinctions: A Prehistoric Revolution,* pp. 824–862. University of Arizona Press, Tucson.

Diamond, J. 1995. Easter's end. *Discover* **16**(8):63–68.

Dickinson, W. R. 1995. The times are always changing: the Holocene saga. *Geol. Soc. Am. Bull.* **107**:1–7.

Dillehay, T. D. (ed.). 1989. *Monte Verde—A late Pleistocene site in Chile, Vol. 1. Paleoenvironmental and site context.* Smithsonian Institution Press, Washington, D.C.

Dodson, J. R., Fullagar, R., Furby, J., Jones, R., and Prosser, I. 1993. Humans and megafauna in a late Pleistocene environment from Cuddie Springs, north western New South Wales. *Archaeol. Oceania* **28**:94–99.

Ellison, J. 1994. Paleo-lake and swamp stratigraphic records of Holocene vegetation and sea-level changes, Mangaia, Cook Islands. *Pac. Sci.* **48**:1–15.

Field, J. H., and Boles, W. E. 1998. *Genyornis newtoni* and *Dromaius novaehollandiae* at 30,000 b.p. in central northern New South Wales. *Alcheringa* **22**:177–188.

Fisher, D. C. 1996. Extinction of proboscideans in North America, in: J. Shoshani and P. Tassy (eds.), *The Proboscidea: Evolution and Palaeoecology of Elephants and Their Relatives,* pp. 296–315. Oxford University Press, London.

Flannery, T. F. 1994. *The Future Eaters.* Reed Books, Melbourne.

Flannery, T. F., Kirch, P. V., and Spriggs, M. 1988. Holocene mammal faunas from archaeological sites in island Melanesia. *Archaeol. Oceania* **23**:89–94.

Furby, J. H. 1995. Megafauna under the microscope: archaeology and palaeoenvironment at Cuddie Springs. Ph.D. thesis, University of New South Wales, Sydney. 373 pp.

Gaffney, E. S., Balouet, J. C., and De Broin, F. 1984. New occurrences of extinct meiolaniid turtles in New Caledonia. *Am. Mus. Novit.* **2800**:1–6.

Gersonde, R., Kyte, F. T., Bleil, U., Diekmann, B., Flores, J. A., Gohl, K., Grahl, G., Hagen, R., Kuhn, G., Sierro, F. J., Volker, D., Abelmann, A., and Bostwick, J. A. 1997. Geological record and reconstruction of the late Pliocene impact of the Eltanin asteroid in the Southern Ocean. *Nature* **390**:357–363.

Gibbons, A. 1998. Ancient island tools suggest *Homo erectus* was a seafarer. *Science* **279**:1635–1637.

Gilbert, B. M., and Martin, L. D. 1984. Late Pleistocene fossils of Natural Trap Cave, Wyoming, and the climatic model of extinction, in: P. S. Martin and R. G. Klein (eds.), *Quaternary Extinctions: A Prehistoric Revolution,* pp. 138–147. University of Arizona Press, Tucson.

Glynn, P. W. (ed.). 1990. *Global Ecological Consequences of the 1982–1983 El Niño-Southern Oscillation.* Elsevier Oceanography Series Volume 52, New York.

Godfrey, L. R., Jungers, W. L., Reed, K. E., Simons, E. L., and Chatrath, P. S. 1997. Subfossil lemurs, in: S. Goodman and B. Patterson (eds.), *Natural Change and Human Impact in Madagascar,* pp. 218–256. Smithsonian Institution Press, Washington, D.C.

Graham, R. W., and Lundelius, E. L. 1984. Coevolutionary disequilibrium and Pleistocene extinctions, in: P. S. Martin and R. G. Klein (eds.), *Quaternary Extinctions: A Prehistoric Revolution,* pp. 223–249. University of Arizona Press, Tucson.

Graham, R. W., and Lundelius, E. L. 1994. *Faunmap: A Database Documenting Late Quaternary Distributions of Mammal Species in the United States.* Springfield: Illinois State Museum Scientific Papers Volume 30, No. 1 and 2.

Grant, B. R., and Grant, P. R. 1989. *Evolutionary Dynamics of a Natural Population: The Large Cactus Finch of the Galapagos.* University of Chicago Press, Chicago.

Grayson, D. K. 1991. Late Pleistocene mammalian extinctions in North America: taxonomy, chronology, and explanations. *J. World Prehist.* **5**:193–232.

Guthrie, R. D. 1990. *Frozen Fauna of the Mammoth Steppe: The Story of Blue Babe.* University of Chicago Press, Chicago.

Haynes, C. V. 1991. Geoarchaeological and paleohydrological evidence for a Clovis-Age drought in North America and its bearing on extinction. *Quat. Res.* **36**:438–450.

Haynes, G. 1991. *Mammoths, Mastodonts, and Elephants: Biology, Behavior, and the Fossil Record.* Cambridge University Press, London.

Holdaway, R. N. 1996. Arrival rats of New Zealand. *Nature* **384**:225–226.

Holdaway, R. N. In press. A spatio-temporal model for the invasion of the New Zealand archipelago by the Pacific rat, *Rattus exulans. J. R. Soc. N. Z.*

Hughen, K. A., Overpeck, J. T., Lehman, S. J., Kashgarian, M., Southon, J., Peterson, L. C., Alley, R., and Sigman, D. M. 1998. Deglacial changes in ocean circulation from an extended radiocarbon calibration. *Nature* **391**:65–68.

Hunt, T. L., and Kirch, P. V. 1997. The historical ecology of Ofu Island, American Samoa, 3000 YR BP to the present, in: P. V. Kirch and T. L. Hunt (eds.), *Historical Ecology in the Pacific Islands: Prehistoric Environmental and Landscape Change,* pp. 105–123. Yale University Press, New Haven.

Irwin, G. 1992. *The Prehistoric Exploration and Colonization of the Pacific.* Cambridge University Press, London.

James, H. F. 1987. A late Pleistocene avifauna from the island of Oahu, Hawaiian Islands. *Doc. Lab. Geol. Lyon* **99**:221–230.

James, H. F. 1995. Prehistoric extinctions and ecological changes on oceanic islands, in: P. M. Vitousek, L. L. Loope, and H. Andersen (eds.), *Islands: Biological Diversity and Ecosystem Function,* pp. 87–102. Springer-Verlag, Berlin.

James, H. F., and Olson, S. L. 1991. Descriptions of thirty-two species of birds from the Hawaiian Islands: Part II. Passeriformes. *Ornithol. Monogr.* **46**:1–88.

Janzen, D. 1983. The Pleistocene hunters had help. *Am. Nat.* **121**:598–599.

Kawamura, Y. 1991. Quaternary mammalian faunas in the Japanese Islands. *Quaternary Research* (Tokyo) **30**:213–220.

Kirch, P. V., and Ellison, J. 1994. Paleoenvironmental evidence for human colonization of remote Oceanic islands. *Antiquity* **68**:310–321.

Kruuk, H. 1972. Surplus killing by carnivores. *J. Zool.* **166**:233–244.

Kurtén, B., and Anderson, E. 1980. *Pleistocene Mammals in North America.* Columbia University Press, New York.

Lister, A. M., and Sher, A. V. 1995. Ice cores and mammoth extinction. *Nature* **378**:23–24.

Liu, T., and Li, X. 1984. Mammoths in China, in: P. S. Martin and R. G. Klein (eds.), *Quaternary Extinctions: A Prehistoric Revolution,* pp. 517–527. University of Arizona Press, Tucson.

MacPhee, R. D. E., and Marx, P. A. 1997. The 40,000-year plague: humans, hyperdisease and first-contact extinctions, in: S. Goodman and B. Patterson (eds.), *Natural Change and Human Impact in Madagascar,* pp. 169–217. Smithsonian Institution Press, Washington, D.C.

MacPhee, R. D. E., Ford, D. C., and McFarlane, D. A. 1989. Pre-Wisconsinan mammals from Jamaica and models of late Quaternary extinction in the Greater Antilles. *Quat. Res.* **31**:94–106.

Markgraf, V. 1985. Late Pleistocene faunal extinctions in southern Patagonia. *Science* **228**:1110–1112.

Marshall, L. G. 1981. The Great American Interchange: An invasion induced crisis for South American mammals, in: M. Nitecki (ed.), *Biotic Crises in Geological and Evolutionary Time,* pp. 133–229. Academic Press, New York.

Marshall, L. G. 1990. The evolution and paleoecological context of mammalian faunas in South America during the Pleistocene. *L'Anthropologie* No. 2.

Marshall, L. G., and Cifelli, R. L. 1990. Analysis of changing diversity patterns in Cenozoic land mammal age faunas, South America. *Paleovertebrata* **19**:169–210.

Martin, P. S. 1967. Prehistoric overkill: The global model, in: P. S. Martin and H. E. Wright (eds.), *Pleistocene Extinctions: The Search for a Cause,* pp. 75–120. Yale University Press, New Haven.

Martin, P. S. 1982. The pattern and meaning of Holarctic mammoth extinction, in: D. M. Hopkins, J. V. Mathews, C. E. Schweger, and S. B. Young (eds.), *Paleoecology of Beringia,* pp. 399–408. Academic Press, New York.

Martin, P. S. 1984. Prehistoric overkill: the global model, in: P. S. Martin and R. G. Klein (eds.), *Quaternary Extinctions: A Prehistoric Revolution,* pp. 354–403. University of Arizona Press, Tucson.

Martin, P. S. 1987. The promise of TAMS 14-C dating, in: H. E. Gove, A. E. Litherland, and D. Elmore (eds.), *Accelerator Mass Spectrometry,* pp. 179–186. North-Holland, Amsterdam.

Martin, P. S. 1990. 40,000 years of extinctions on the "planet of doom." *Palaeogeogr. Palaeoclimatol. Palaeoecol.* (Global and Planetary Change Section) **82**:187–201.

Martin, P. S., and Klein, R. G. (eds.). 1984. *Quaternary Extinctions: A Prehistoric Revolution.* University of Arizona Press, Tucson.

Martin, P. S., and Stuart, A. J. 1995. Mammoth extinction: two continents and Wrangel Island. *Radiocarbon* **37**:7–10.

McDonald, J. N. 1981. *North American Bison: Their Classification and Evolution.* University of California Press, Berkeley.

McKenna, M. C., and Bell, S. K. 1997. *Classification of Mammals above the Species Level.* Columbia University Press, New York.

Mead, J. I., and Agenbroad, L. D. 1992. Isotopic dating of Pleistocene dung deposits from the Colorado Plateau, Arizona and Utah. *Radiocarbon* **34**:1–19.

Meredith, C. 1991. Vertebrate fossil faunas from islands in Australasia and the Southwest Pacific, in: P. Vickers-Rich, J. M. Monaghan, R. F. Baird, and T. H. Rich (eds.), *Vertebrate Paleontology of Australia,* pp. 1346–1382. Pioneer Design Studio Pty. Ltd., Lilydale, Victoria, Australia.

Miller, G. H., Magee, J. W., Johnson, B. J., Fogel, M. L., Spooner, N. A., McCulloch, M. T., and Ayliffe, L. K. 1999. Pleistocene extinction of *Genyornis newtoni:* Human impact on Australian megafauna. *Science* **283**:205–208.

Mitchell, G. F., and Parkes, H. M. 1949. The giant deer in Ireland. *Proc. R. Ir. Acad.* **52**(B):291–314.

Mithen, S. 1993. Simulating mammoth hunting and extinction: implications for the Late Pleistocene of the Central Russian Plain, in: G. L. Peterkin, H. M. Bricker, and P. Mellars (eds.), *Hunting and Animal Exploitation in the Later Palaeolithic and Mesolithic of Eurasia,* pp. 163–178. Archaeological Papers of the American Anthropological Association No. 4.

Mithen, S. 1997. Simulating mammoth hunting and extinctions: implications for North America, in: S. van der Leeuw and J. McGlade (eds.), *Time, Process and Structured Transformation in Archaeology,* pp. 176–215. Routledge, London.

Morwood, M. J., O'Sullivan, P. B., Aziz, F., and Raya, A. 1998. Fission-track age of stone tools and fossils on the east Indonesian island of Flores. *Nature* **392**:173–176.

Mosimann, J. E., and Martin, P. S. 1975. Simulating overkill by paleoindians. *Am. Sci.* **63**:304–313.

Murray, P. 1991. The Pleistocene megafauna of Australia, in: P. Vickers-Rich, J. M. Monaghan, R. F. Baird, and T. H. Rich (eds.), *Vertebrate Paleontology of Australia,* pp. 1071–1164. Pioneer Design Studio Pty. Ltd., Lilydale, Victoria, Australia.

O'Connell, J. F., and Allen, C. 1998. When did humans first arrive in Greater Australia and why is it important to know? *Evol. Anthropol.* **6**:132–146.

Olson, S. L., and James, H. F. 1991. Descriptions of thirty-two new species of birds from the Hawaiian Islands: Part I. Non-Passeriformes. *Ornithol. Monogr.* **45**:1–85.

Olson, S. L., and Wetmore, A. 1976. Preliminary diagnoses of two extraordinary new genera of birds from Pleistocene deposits in the Hawaiian Islands. *Proc. Biol. Soc. Wash.* **89**:247–258.

Owen-Smith, R. N. 1988. *Megaherbivores: The Influence of Very Large Body Size on Ecology.* Cambridge University Press, London.

Preston, D. 1995. The mystery of Sandia Cave. *The New Yorker* **71**(16):66–83.

Putshkov, P. V. 1997. Were the mammoths killed by the warming? (Testing of the climatic versions of Wurm extinctions). *Vest. zool. J. Schmalhausen Inst. Zool.* Kyiv, Ukraine, Suppl. 4:3–81.

Reese, D. S. (ed.). 1996. *Pleistocene and Holocene Fauna of Crete and its First Settlers.* Monographs in World Archaeology No. 28. Prehistory Press, Madison, Wisc.

Saxon, E. C. 1979. Natural prehistory: the archaeology of Fuego-Patagonia ecology. *Quaternaria* **12**:329–356.

Schubel, S. E., and Steadman, D. W. 1989. More bird bones from Polynesian archaeological sites on Henderson Island, Pitcairn Group, South Pacific. *Atoll Research Bulletin* No. 325.

Sher, A. V. 1997. Late-Quaternary extinction of large mammals in northern Eurasia: a new look at the Siberian contribution, in: B. Huntley, W. Cramer, A. V. Morgan, H. C. Prentice, and J. R. M. Allen (eds.), *Past and Future Rapid Environmental Changes: The Spatial and Evolutionary Responses of Terrestrial Biota,* pp. 319–339. Springer-Verlag, Berlin.

Simmons, A. H. 1991. Humans, island colonization and Pleistocene extinctions in the Mediterranean: the view from Akrotiri *Aetokremnos,* Cyprus. *Antiquity* **65**:857–869.

Simons, E. 1997. Lemurs: old and new, in: S. Goodman and B. Patterson (eds.), *Natural Change and Human Impact in Madagascar,* pp. 142–166. Smithsonian Institution Press, Washington, D.C.

Simpson, G. G. 1980. *Splendid Isolation: The Curious History of South American Mammals.* Yale University Press, New Haven.

Soffer, O. 1993. Upper Paleolithic adaptations in Central and Eastern Europe and man–mammoth interactions, in: O. Soffer and N. K. Praslov (eds.), *Kostenki to Clovis: Upper Paleolithic–Paleo-Indian Adaptations,* pp. 31–49. Plenum Press, New York.

Spaan, A. 1996. *Hippopotamus creutzburgi:* the case of the Cretan hippopotamus, in: D. S. Reese (ed.), *Pleistocene and Holocene Fauna of Crete and its First Settlers,* Monographs in World Archaeology No. 28, pp. 99–110. Prehistory Press, Madison, Wisc.

Stafford, T. W., Hare, P. E., Currie, L., Jull, A. C. T., and Donahue, D. J. 1991. Accelerator radiocarbon dating at the molecular level. *J. Archaeol. Sci.* **18**:35–72.

Steadman, D. W. 1989. Extinction of birds in eastern Polynesia: a review of the record, and comparisons with other Pacific island groups. *J. Archaeol. Sci.* **16**:177–205.

Steadman, D. W. 1993. Biogeography of Tongan birds before and after human impact. *Proc. Nat. Acad. Sci. USA* **90**:818–822.

Steadman, D. W. 1995. Prehistoric extinctions of Pacific Island birds: biodiversity meets zooarchaeology. *Science* **267**:1123–1131.

Steadman, D. W. 1997. Extinctions of Polynesian birds: reciprocal impacts of birds and people, in: P. V. Kirch and T. L. Hunt (eds.), *Historical Ecology in the Pacific Islands: Prehistoric Environmental and Landscape Change,* pp. 51–79. Yale University Press, New Haven.

Steadman, D. W., and Olson, S. 1985. Bird remains from an archaeological site on Henderson Island, South Pacific: man caused extinctions on an "uninhabited" island. *Proc. Natl. Acad. Sci. USA* **82**:6191–6195.

Steadman, D. W., and Ray, C. E. 1982. The relationship of *Megaoryzomys curioi,* an extinct cricetine rodent (Muroidea: Muridae) from the Galapagos Islands, Ecuador. *Smithson. Contrib. Paleobiol.* **51**, 23 pp.

Steadman, D. W., and Rolett, B. 1996. A chronostratigraphic analysis of landbird extinction on Tahuata, Marquesas Islands. *Archaeol. Sci.* **23**:81–94.

Steadman, D. W., Stafford, T. W., Donahue, D. J., and Jull, A. J. T. 1991. Chronology of Holocene vertebrate extinction in the Galapagos Islands. *Quat. Res.* **36**:26–33.

Steadman, D. W., Vargas, C. P., and Cristino, F. C. 1994. Stratigraphy, chronology, and cultural contest of an early faunal assemblage from Easter Island. *Asian Perspect.* **22**:79–96.

Steadman, D. W., Stafford, T. W., and Funk, R. E. 1997. Nonassociation of Paleoindians with AMS-dated late Pleistocene mammals from the Duchess Quarry Caves, New York. *Quat. Res.* **47**:105–116.

Steadman, D. W., Plourde, A., and Burley, D. V. In press. Prehistoric butchery and consumption of birds in the Kingdom of Tonga, South Pacific. *J. Archaeol. Sci.*

Steadman, D. W., White, P., and Allen, J. 1999. Extinction and biogeography of birds from New Ireland, Papua New Guinea. *Proc. Nat. Acad. Sci. USA* **96**.

Stevenson, J., and Dodson, J. R. 1995. Palaeoenvironmental evidence for human settlement of New Caledonia. *Archaeol. Oceania* **30**:36–41.

Stuart, A. J. 1991. Mammalian extinctions in the Late Pleistocene of northern Eurasia and North America. *Biol. Rev.* **66**:453–562.

Swisher, C. C., Rink, W. J., Anton, S. C., Schwarcz, H. P., Curtis, G. H., Suprijo, A., and Widiasmoro. 1996. Latest *Homo erectus* of Java: potential contemporaneity with *Homo sapiens* in southeast Asia. *Science* **274**: 1870–1874.

Taylor, J. D., Braithwaite, C. J. R., Peake, J. F., and Arnold, E. N. 1979. Terrestrial faunas and habitats of Aldabra during the late Pleistocene. *Philos. Trans. R. Soc. London* **286**:47–66.

Taylor, R. E., Haynes, C. V., and Stuiver, M. 1996. Calibration of the Late Pleistocene radiocarbon time scale: Clovis and Folsom age estimates. *Antiquity* **70**:515–525.

Tchernov, E., and Shoshani, J. 1996. Proboscidean remains in southern Levant, in: J. Shoshani and P. Tassy (eds.), *The Proboscidea: Evolution and Palaeoecology of Elephants and Their Relatives,* pp. 225–233. Oxford University Press, London.

Van den Bergh, G. D., Sondaar, P. Y., De Vos, J., and Aziz, F. 1996. The proboscideans of South-east Asian islands, in: J. Shoshani and P. Tassy (eds.), *The Proboscidea: Evolution and Palaeoecology of Elephants and Their Relatives,* pp. 296–315. Oxford University Press, London.

van der Made, J. 1996. Pre-Pleistocene land mammals from Crete, in: D. S. Reese (ed.), *Pleistocene and Holocene Fauna of Crete and its First Settlers,* Monographs in World Archaeology No. 28, pp. 69–79. Prehistory Press, Madison, Wisc.

Van Riper, C., Van Riper, S. G., Goff, M. L., and Laird, M. 1996. The epizootiology and ecological significance of malaria in Hawaiian land birds. *Ecol. Monogr.* **56**:327–344.

Van Tilburg, J. A. 1994. *Easter Island: Archaeology, Ecology and Culture.* Smithsonian Institution Press, Washington, D.C.

Vartanyan, S. L., Garutt, V. E., and Sher, A. V. 1992. Holocene dwarf mammoths from Wrangel Island in the Siberian Arctic. *Nature* **362**:337–340.

Ward, P. 1997. *The Call of Distant Mammoths: Why the Ice Age Mammals Disappeared.* Copernicus Books, Springer-Verlag, New York.

Webb, R. E., and Rindos, D. J. 1997. The mode and tempo of the initial human colonisation of empty landmasses: Sahul and the Americas compared, in: C. M. Barton and G. A. Clark (eds.), *Rediscovering Darwin: Evolutionary Theory and Archeological Explanation,* pp. 233–250. Archaeological Papers of the American Anthropological Association No. 7.

Webb, S. D. 1984. Ten million years of mammalian extinctions in North America, in: P. S. Martin and R. G. Klein (eds.), *Quaternary Extinctions: A Prehistoric Revolution,* pp. 189–210. University of Arizona Press, Tucson.

White, J. P., and Flannery, T. F. 1995. Late Pleistocene fauna at Spring Creek, Victoria: a re-evaluation. *Aust. Archaeol.* **40**:13–17.

Worthy, T. H. In press. The role of climate change versus human impacts—avian extinction in New Zealand, in: S. L. Olson (ed.), *Smithson. Contrib. Paleobiol.* **88**.

Worthy, T. H., and Holdaway, R. N. 1993. Quaternary fossil faunas from caves in the Punakaiki area, West Coast, South Island, New Zealand. *J. R. Soc. N. Z.* **23**:147–254.

Worthy, T. H., and Holdaway, R. N. 1994. Quaternary fossil faunas from caves in Takaka Valley and on Takaka Hill, northwest Nelson, South Island, New Zealand. *J. R. Soc. N. Z.* **24**:297–391.

Zimov, S. A., Chuprynin, V. I., Oreshko, A. P., Chapin, F. S., Reynolds, J. F., and Chapin, M. C. 1995. Steppe–tundra transition: a herbivore-driven biome shift at the end of the Pleistocene. *Am. Nat.* **146**:765–793.

CHAPTER 3

The Interaction of Humans, Megaherbivores, and Habitats in the Late Pleistocene Extinction Event

NORMAN OWEN-SMITH

1. Introduction

The extinction pulse during the late Pleistocene exhibited the following special features: (1) extinction likelihood was strongly dependent on body size, (2) the severity of the extinctions varied markedly between continents, and (3) extinctions were associated in time with the appearance of humans, and also with a period of climatic and hence habitat change (Owen-Smith, 1987). Causal explanations must be judged by their ability to predict these patterns.

My contribution to this volume will emphasize the ecological processes that potentially led to these species extinctions. Extinctions may result from climatic shift, habitat loss, competition, predation, or disease outbreaks. Competition has not been invoked as a factor in the late Pleistocene (except possibly between human hunters and other predators), and disease panzootics seem unlikely to have been involved (but for a contrary view, see MacPhee and Flemming, this volume). The central issue to be addressed concerns the relative importance of human predation, versus consequences of the climatic shift at the end of the last ice age for habitat conditions (see papers in Martin and Klein, 1984).

My focus is on large mammalian herbivores, as these are both directly responsive to habitat change, and potential prey for human hunters. The time period I cover extends from somewhat after the end of the Sangamonian/Eemian Interglacial into the early Holocene, i.e., from about 100,000 yrbp to perhaps 5000 yrbp.

2. Pattern of the Late Pleistocene Extinctions

Toward the end of the Pleistocene, over half of the species of large mammalian herbivores occupying North America, South America, and Australia became extinct. North-

NORMAN OWEN-SMITH • Center for African Ecology and Department of Zoology, University of the Witwatersrand, Wits 2050, South Africa.

Extinctions in Near Time, edited by MacPhee. Kluwer Academic/Plenum Publishers, New York, 1999.

ern Eurasia (i.e., Europe plus Siberia) lost 37% of its meager fauna of large herbivores, while Africa incurred only a 10% loss (Fig. 1; data for tropical Asia unavailable). The lower size limit of 10 kg that I have used is arguably a more natural subdivision than Martin's (1966, 1967) criterion of 44 kg (see Fig. 17.1 in Owen-Smith, 1988, and Fig. 24 in Stuart, 1991). It encompasses most bovids and macropods, while excluding most rodents and lagomorphs.

Notable distinctions existed in body size distributions among continental Pleistocene herbivore faunas (Fig. 1). Both North and South America exhibited predominantly medium to large species weighing over 100 kg. Africa was and is especially species-rich in the size range of 10–100 kg, while retaining as many medium–large species as formerly in the Americas. Australia formerly had about as many medium–small species as Africa, but very few large species. Eurasia was generally depauperate in species, despite its vast area. Africa's past and present richness lies especially in the radiation of medium-sized bovids, with 16 modern genera represented by almost 50 species in the size range 10–100 kg. North and South America were formerly as species-diverse as Africa in the size range 100–1000 + kg, but lost most of these species at the end of the Pleistocene.

The extinctions took out all megaherbivores (species attaining over 1000 kg in adult body weight), plus over 80% of species weighing 100–1000 kg, throughout the Americas, Eurasia, and Australia. In Africa a pulse of extinctions between the Early and Middle Stone Age, somewhat prior to 100,000 yrbp, involved the disappearance of *Elephas recki, Hippopotamus gorgops, Sivatherium, Hipparion,* the hartebeest *Parmularius,* and several large suids (Klein, 1984a,b). Near the end of the Pleistocene, a giant horse (*Equus capensis*), long-horned buffalo (*Pelorovis antiquus*), giant hartebeest (*Megalotragus priscus*), giant warthog (*Phacochoerus* [*Metridiochoerus*] *compactus*), two medium-sized bovids (*Damaliscus niro* and *Antidorcas bondi*), and an ass (*Equus* c.f. *asinus*) became extinct in southern and eastern Africa (Klein, 1977, 1984a,b; Marean and Gifford-Gonzalez, 1991; Brink, 1994), although *Pelerovis* lingered into the Holocene in the Sahara region. Tropical Asia lost *Stegodon* (a proboscidean) and *Elasmotherium* (a rhinoceros) sometime during the late Pleistocene (Anderson, 1984).

In southern Europe the straight-tusked elephant (*Palaeoloxodon antiquus*) survived until about 40,000 yrbp and the steppe rhinos (*Dicerorhinus hemitoechus* and *D. kirchbergensis*) until about 20,000 yrbp (Stuart, 1991). The woolly rhino (*Coelodonta antiquitatis*) and woolly mammoth (*Mammuthus primigenius*) persisted in Europe until 12,000 yrbp. Mammoths remained in continental Siberia until nearly 10,000 yrbp, and on Wrangel Island until perhaps 4000 yrbp (Vartanyan *et al.,* 1993). *Hippopotamus amphibius* disappeared from Europe around 80,000 yrbp, whereas the giant deer *Megaloceros giganteus* persisted in Ireland until 10,500 yrbp (Stuart, 1991).

In North America, extinctions were concentrated within a narrow time window between 11,500 and 10,500 yrbp (Stuart, 1991). Terminal dates for South America suggest a similar or even slightly earlier timing for extinctions (Markgraf, 1985), but need to be confirmed. Notably only two grazing ungulate species survive today in South America, both very localized: the vicuna (*Vicugna vicugna*) and the pampas deer (*Ozotoceros bezoarticus*). In Australia the extinctions apparently occurred prior to 35,000 yrbp (Flannery and Roberts, this volume). No species weighing over 100 kg survived.

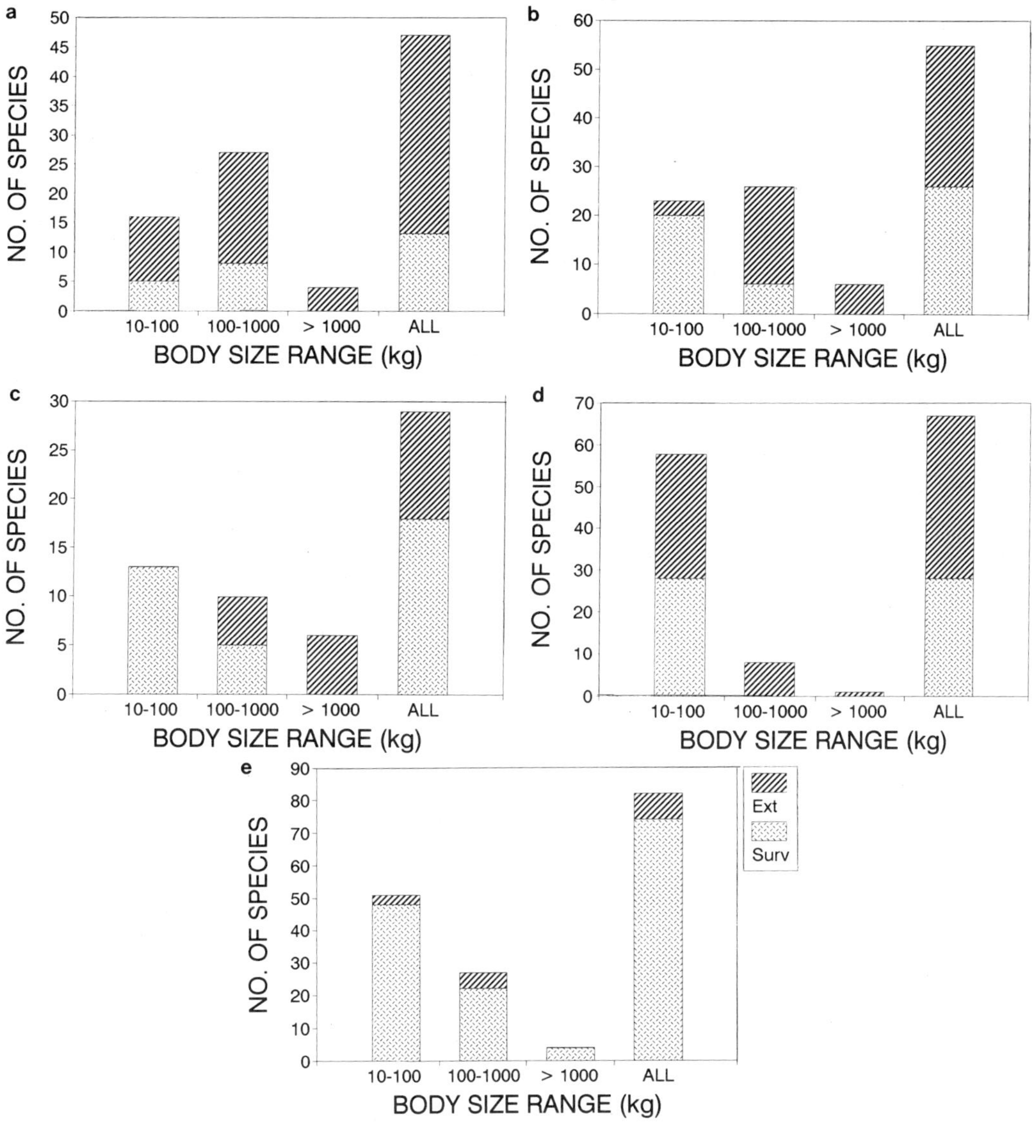

FIGURE 1. (a) North America; (b) South America; (c) Eurasia; (d) Australia; (e) Africa. Body mass distribution of late Pleistocene versus surviving species of large mammalian herbivores in different continents. Eurasia represents Europe plus Siberia. Africa excludes the Mediterranean region. Extinctions tallied from about 100,000 yrbp (i.e., following the Sangamonian/Eemian Interglacial) into the Holocene. For South America, one species per extinct genus was assumed where species-level tallies were unavailable (from Anderson, 1984; Klein, 1984a,b; Macdonald, 1984; Marean and Gifford-Gonzalez, 1991; Murray, 1991; Stuart, 1991).

3. Predation-Induced Extinctions

For predation to drive prey to extinction, three conditions must apply: (1) population losses must exceed recruitment, (2) this imbalance must be maintained even when prey become rare, and (3) no predator-free refuges must exist. Density-dependent recruitment models applied to fisheries highlight the distinction between sustainable and unsustainable harvests. Assuming a simple logistic model, maximum "sustained" yield (MSY) is obtained at a population level equal to half of the equilibrium carrying capacity in the absence of harvests, at a rate equivalent to half of the maximum rate of population growth. But large mammals generally show a right-skewed recruitment curve, so that maximum yield occurs somewhat closer to the equilibrium density than suggested by the logistic model (Fowler, 1981; Fig. 2).

If harvests reduce the resource population below the MSY level, the population collapses, unless the quota is reduced promptly and sufficiently. In natural environments, recruitment success can vary widely over time, so that a quota set initially below the theoretical MSY sooner or later exceeds the sustainable threshold. The history of human resource exploitation shows that, because of uncertainties and economic pressures, an effective response is usually delayed until too late (Ludwig *et al.,* 1993). Accordingly, fisheries have notoriously failed to conserve stocks of fishes and whales. Yet species have not been driven to extinction, because harvesting becomes economically inefficient once the resource gets scarce.

Even in natural predator–prey systems, circumstances may push the prey population into a "predator pit," i.e., a level maintained well below that set by food resources. This may arise from a strongly density-dependent increase in predation rate at low density, sufficient to exceed the resource-dependent recruitment rate (Fig. 3). When prey becomes scarce, predators either switch to other prey species, or decline numerically because of insufficient food. Toward higher prey abundance, individual predators become satiated, and predator populations may reach a density limit set by social factors (e.g., territorial exclusion). A North American example is the moose–wolf interaction, with moose held at a density of under 0.4 per km^2 by the combined impacts of wolves, bears, and human hunters, despite being capable of attaining a food-limited density of over 2 per km^2 (Messier, 1994). In Africa, the contrasting population levels of migratory versus sedentary wildebeest in the Serengeti region of Tanzania may have a similar explanation (Fryxell *et al.,* 1988).

Prey switching will not alleviate predation pressure on favored prey species when these and alternative prey occur together in the same habitat. Accordingly, whether keystone predators enhance or reduce prey diversity depends on the relative abundance of preferred versus alternate prey, and on the capacity of the latter to support the predator population (Begon *et al.,* 1996, Chapter 21). Notably, in the places where wolves depress moose populations, deer or other species provide alternative prey (Messier, 1994). An unpublished South African example of near-extinction of a prey species by a mammalian predator, from the Suikerbosrand Nature Reserve near Johannesburg, is instructive. Following the introduction of cheetah (*Acinonyx jubatus*), blesbok (*Damaliscus dorcas*) declined precipitously to low numbers, at which point the cheetahs started concentrating their hunting on springbok (*Antidorcas marsupialis*). Extinction of the blesbok was avoided by removal of the cheetahs.

A predator-maintained prey density below half of the saturation density of the prey is inherently unstable, unless habitats are sufficiently heterogeneous. Typically the prey species persists in localities where it can effectively evade predation. For example, wilde-

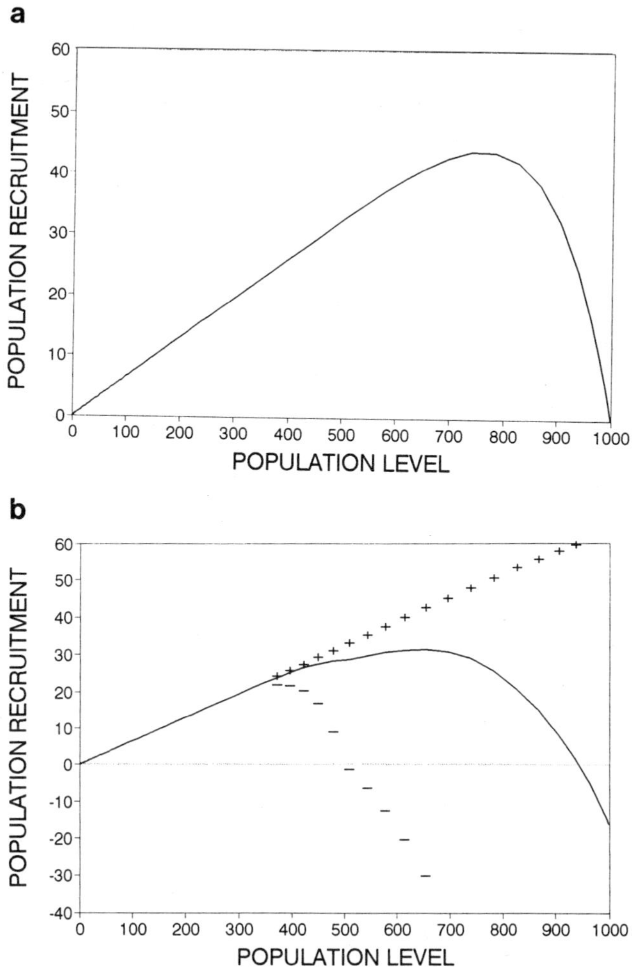

FIGURE 2. Shape of recruitment curve typical for large mammalian herbivores, which also represents potential harvests. (a) Population with a maximum rate of population growth of 6.5% in a constant environment with an equilibrium density ("carrying capacity") of 1000 units (units may represent numbers, density, biomass, or whatever). The maximum sustainable harvest is 44 units from a population level of about 750 units. (b) Population with the same intrinsic growth rate under conditions where the "carrying capacity" varies stochastically between extremes of 0.5 and 1.5 times the mean level of 1000. +, recruitment curve under best conditions (carrying capacity 1500); −, recruitment curve under worst conditions (carrying capacity 500); solid line is the mean recruitment averaged over the range in conditions. Environmental variability reduces the maximum sustainable harvest to on average 33 units. However, to avoid overexploitation in the worst years, the harvest quota would need to be reduced to no more than 22 units, in this specific example.

beest avoid wooded areas providing cover for lions, despite the food resources they offer (Talbot and Talbot, 1963). Deer alleviate predation by wolves by occupying boundary regions between adjacent wolf pack territories (Mech, 1977). Such habitat restrictions underlie the classical distinction between the fundamental and realized niche of a species (Hutchinson, 1957).

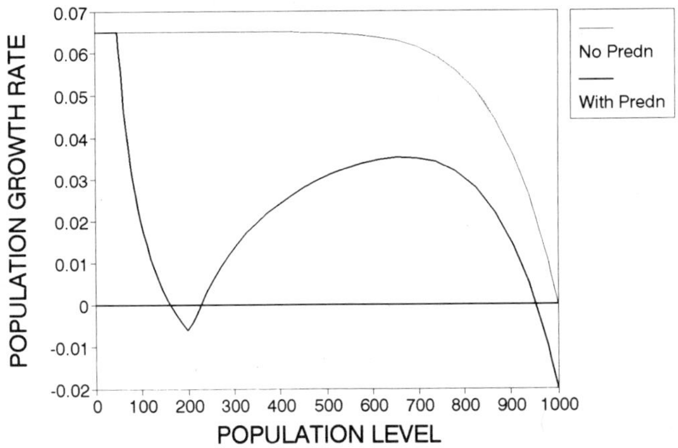

FIGURE 3. Change in population growth rate with increasing population level, indicating a "predator pit" brought about by density-dependent predation, for population parameters as in Fig. 2. Predation level increases more steeply than prey abundance at low prey density, but with predators tending to become satiated toward high prey densities. This generates two alternative equilibrium levels where the net population growth of the prey is zero: (1) a high level of about 950 units with a predation level of about 2% and (2) a low level of about 160 units maintained by the predation of about 10 units per year. The lower population equilibrium is referred to as the predator pit (not the bottoming out of the net recruitment curve, where population growth is negative!).

When populations are depressed to low densities, the "Allee effect," i.e., reduced recruitment rate at low density, may come into operation. The population growth rate of white rhinos in the Hluhluwe–Umfolozi Park in South Africa was evidently lower when the population was small than after high densities had been attained (Fig. 4). Furthermore, white rhinos introduced into other wildlife parks have exhibited longer intercalving intervals than those typical of the dense Umfolozi population (Owen-Smith, 1988). This pattern could be related either to difficulties in finding mates at low densities, or to improvements in food resources brought about by the feeding impacts of the rhinos on vegetation. Notably, the depressed blue whale population has failed to recover, despite cessation of all harvesting. Through "Allee effects," effective population extinguishment may occur well before the last few individuals have been killed.

Refuge habitats ("enemy-free space," Jeffries and Lawton, 1984) may permit prey persistence despite efficient predators. This may also occur dynamically, with prey colonizing vacant habitat patches one step ahead of predators (Huffaker, 1958). Metapopulation extinction (regional extirpation, as opposed to local population extinguishment) occurs only if all suitable habitat is persistently occupied by predators, so that new colonists encounter waiting hunters. This requires that the predator population be maintained by alternate food resources resistant to depletion.

4. Habitat-Induced Extinctions

Here the focus is on habitat components such as food resources and vegetation physiognomy, while recognizing that predators also influence habitat suitability. Spatial varia-

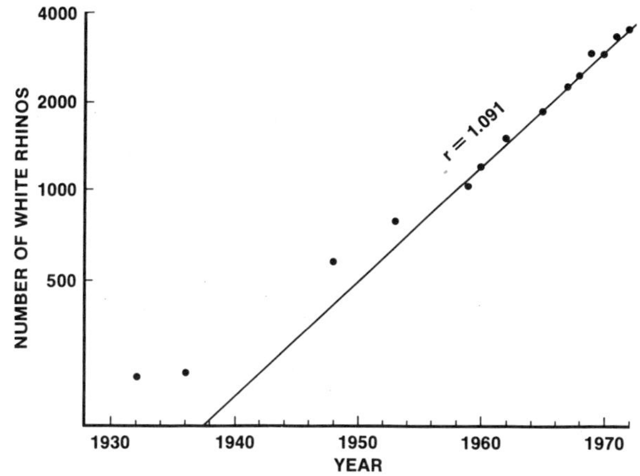

FIGURE 4. Population increase of white rhinos in the Hluhluwe–Umfolozi Park in South Africa, showing indication of an "Allee effect," i.e., reduced population growth rate at low density. The growth rate of the white rhino population increased from about 6.5% per year when the population size was about 200 toward 9.5% per year when the population size exceeded 1000 animals (from Owen-Smith, 1988).

tion in these features underlies the distinction between source and sink habitats, and intervening nonhabitat (Pulliam and Danielson, 1991). Source habitats are places sufficiently favorable to support thriving populations of particular species. Surplus individuals from these localities disperse into surrounding less suitable regions, where population growth would be negative in the absence of continuing immigration. Areas of nonhabitat may be traversed by migrants, but do not support breeding groups.

Environmental fluctuations in precipitation or temperature may cause localities to vary over time in their habitat suitability. A progressive climatic change could bring about species extinctions by (1) transforming formerly source habitat into sink habitat, leading to (2) remaining patches of source habitat becoming isolated, so that (3) local populations become vulnerable to stochastic small-population effects. Isolation retards recolonization following local population extirpation. Unless adequate areas of suitable habitat persist somewhere within the range, species extinction may ultimately result.

The change from glacial to interglacial conditions at the end of the Pleistocene entailed not only a rise in global mean temperatures, but also a widespread increase in precipitation in both high and low latitudes (Ritchie and Cwynar, 1982; Markgraf, 1985; Partridge *et al.*, 1990; deMenocal, 1995). In midlatitudes of North America, vegetation changed from the diverse mosaic of conifers, hardwoods, and grasses that was widespread during the Pleistocene, into the relatively uniform conifer or broadleaf forests and prairie grasslands found today (Guthrie, 1984; Wright, 1984). Such changes were associated in time with a reduction in the ranges occupied by the woolly mammoth and associated grazing ungulates, plus certain low-level browsers, which typified the "mammoth steppe" fauna of North America and Eurasia (Guthrie, 1982; Stuart, 1991). In the tropical and subtropical latitudes of eastern and southern Africa and South America, extensive dry grasslands or open woodlands gave way to expanding forest and moister woodlands around this time (Klein, 1984a; Markgraf, 1985). The general result was a contraction in physiognomically open habitat in favor of closed woodland or forest.

The regional vegetation pattern (biome type) is determined by climatic conditions (temperature and precipitation), whereas local vegetation physiognomy is influenced by topography and soils. The spatial heterogeneity notable within African savannas is largely a result of soil differences determined by underlying geology and downslope sediment movements (catenary patterns). These are in turn a consequence of the high elevation of much of southern and east-central Africa. Toward higher (above 800 mm) and lower (below 400 mm) annual rainfall, soil parent material becomes less influential, because of nutrient and clay leaching, or lack thereof. Superimposed on geophysical factors are the important modifying effects of fire and herbivory on vegetation (Scholes, 1991; Bond, 1997; Owen-Smith, 1997).

Over much of North America and Eurasia, soils are derived from wind-blown loess or former lake and sea beds. Accordingly, disturbance by fire or herbivory may have been relatively more influential in modifying vegetation patterns than in Africa. South America and Australia are mostly low-lying, except for mountain ranges, with soils intensely weathered and especially depauperate in phosphorus and other nutrients (Medina and Silva, 1991; Flannery and Roberts, this volume). Large herbivores concentrate in localities where nutrients accumulate, such as floodplains, as evidenced within regions of nutrient-poor (moist/dystrophic) savanna in Africa today (Huntley, 1982; Scholes and Walker, 1993; Owen-Smith, 1997). The tendency toward large body size in South America's Pleistocene fauna is matched by the pattern typical of the large herbivore guild in regions of nutrient-poor savanna in Africa today (Bell, 1982). The relatively small body size of Australia's mammalian herbivore fauna is perhaps a reflection of widespread aridity. Australian habitats were altered following a drastic increase in burning around 38,000 yrbp following the arrival of humans, with sclerophyllous woodlands almost totally replacing rain forest over extensive regions (Kershaw, 1986).

African elephants at high densities can transform wooded savanna to open savanna, shrubland, or grassland within a few decades (Bourlière, 1965; Laws *et al.,* 1975; Thompson, 1975; Cumming, 1981; Owen-Smith, 1988). Although fire promotes open grassland conditions, elephants contribute significantly by suppressing woody regeneration (Dublin *et al.,* 1990). Within forests, elephants create and maintain open grassy glades, where other ungulate species congregate (Turkalo and Fay, 1995). Heavy grazing pressure by white rhinos and hippos transforms medium–tall grassland into extensive areas where lawnlike grassland predominates (Lock, 1972; Owen-Smith, 1988). As a result, fires are suppressed, promoting invasion by woody plants. The general result of megaherbivore feeding, trampling, and breakage is the creation of a mosaic interspersion of habitat patches characterized by rapidly regenerating relatively nutritious grasses, shrubs, or tree saplings.

However, Africa does not bear witness to the extreme effects that megaherbivores may exert on vegetation and landscapes, except in limited areas, because of historically widespread hunting, and continuing management culling of such species. Nowhere recently in Africa has there been a coexistence of both browsing and grazing megaherbivores at high abundance, except briefly in western Uganda (Laws *et al.,* 1975), eastern Zaire (Bourlière, 1965), and the Luangwa Valley of Zambia (Naylor *et al.,* 1973). The severity of the vegetation transformation associated with the co-occurrence of grazing and browsing proboscideans, ground sloths, rhinos, and other large species through much of the Americas and northern Eurasia would have been more extreme than anything yet documented in Africa. Indeed, Zimov *et al.* (1995) propose that the transformation of the grassy mammoth steppe into the nutrient-deficient shrub tundra found in high northern latitudes today was largely a result of the demise of the grazing megafauna.

5. Interactions

Through attaining a body size exceeding the threshold of 1000 kg, megaherbivores largely escape carnivore predation on adults, at the demographic cost of birth intervals spanning 2 to 4 years (Owen-Smith, 1988, 1989b). Accordingly, maximum population growth rates vary from just over 6% per year for African elephants to 9–10% per year for rhinos and hippos (Owen-Smith, 1988). Harvests exceeding 3–5% per year would be sufficient to drive these populations into decline if sustained. For kudus (*Tragelaphus strepsiceros*), a 180-kg browsing antelope, annual mortality rates ascribed to predation in South Africa's Kruger National Park amounted to 7% per year for prime-aged females, and 12% per year for the overall population, under conditions of high resource abundance, despite the species not being a favored prey of any predator (Owen-Smith, 1990, 1993).

However small the initial founder numbers of humans in the Americas, the population of these hunters will have increased steadily in response to the abundant food resources encountered. With an annual growth rate of 3%, populations would double ten times within 250 years, i.e., 100 founders would end up as 100,000 descendants. Once humans had become sufficiently skilled predators of megaherbivores, it was inevitable that their offtake would eventually rise beyond the sustainable limit for these species having low rates of population increase. Notably, humans are unique in being the only omnivore that is also an effective predator, enabling their populations to be maintained even when prey became scarce. The onslaught by skilled and effective predators, in the form of human hunters, in effect breached the realized niche of megaherbivores.

Within Africa, elephant and rhino species had more time to evolve deterrent responses toward human hunters, until firearms breached this defense. Notably, the relatively inoffensive white rhino persisted precariously, being restricted to the remote southern and northwestern limits of nutrient-rich savanna within historic times. This contrasts with the abundance of this species in Pleistocene fossil deposits at Olduvai Gorge and elsewhere in east-central Africa. Teeth of seemingly modern white rhinos collected from surface deposits in Kenya (Hooijer and Patterson, 1972) suggest that the species survived there into the early Holocene, perhaps being eliminated by early Iron Age hunters. By the time Iron Age people armed with spears colonized southern Africa, they had large herds of livestock, and were no longer dependent on hunting for subsistence.

The prior Stone Age people occupying southern Africa seldom hunted animals above buffalo size (Klein, 1977, 1984b), although a rock painting depicts an elephant being attacked by hamstringing (Woodhouse, 1976). The benefits of killing very large animals are limited in tropical latitudes, because much of the meat would become rotten before it could be consumed by small bands of people. In tropical South America and Australia, this may have been counterbalanced by the great ease of hunting the naive fauna.

Extant megaherbivores are not numerically rare, contrary to common assumptions. White rhinos exhibited a regional density of 3.5 per km^2, with local densities exceeding 5 per km^2, in Umfolozi Game Reserve (Owen-Smith, 1988). African elephants (*Loxodonta africana*) attain regional densities of over 1 per km^2, and sustain local densities up to 3 per km^2 (Owen-Smith, 1988). Such densities are above those generally attained by kudus (*Tragelaphus strepsiceros*), a medium-sized browsing antelope widespread through most of southern Africa.

Coexistence of humans and megafauna for millennia does not preclude human involvement in later extinctions. Effective hunting of very large mammals may require appropriate technology, as well as conditions under which prey populations become vulnerable to extinction. A change in stone tool technology defining the transition from the Early to Middle Stone Age was roughly concurrent with the demise of *Elephas recki* and *Hippopotamus gorgops* in Africa, which may have taken place around the time of climatic change associated with the previous glacial–interglacial transition. The disappearance of *Palaeoloxodon* in southern Europe around 40,000 yrbp followed the appearance in the region of modern humans, who displaced Neanderthal hunters, and was coincident with the beginning of the coldest phase of the last glacial. The ancestors of the Aborigines arriving in Australia perhaps 50,000 yrbp may have lacked prior skills in hunting large open-country mammals, in contrast to the Amerindian ancestors colonizing the Americas toward the end of the last glacial. Accordingly, they may have required more time to become devastating predators.

Ascribing the disappearance of numerous smaller, fleeter, and potentially more rapidly reproducing ungulates solely to human predation is somewhat more problematic. Provided a species remains widely distributed, some populations are likely to persist beyond the range of predators. Habitat changes may have contributed to their demise through range contraction, thereby influencing also the timing of extinctions.

This is where the habitat changes consequent on the elimination of megaherbivores through human predation were probably influential (Owen-Smith, 1987, 1988). In particular, the transformation of the mosaic diversity typical of the Pleistocene into the zonal diversity of modern times (Guthrie, 1984; Graham and Lundelius, 1984) is what would be expected following the removal of the patch-scale effects on vegetation. Climate cannot sustain such small-scale diversity, except in mountainous or hilly terrain.

Without the ecosystem engineering by megaherbivores, grassy glades would revert to closed-canopy forest, while tree and shrub patches within grasslands would become eliminated by hot fires. Loss of local habitat diversity would be detrimental for the survival of medium-sized herbivores with restricted home ranges. Notably, most of the herbivore species becoming extinct in the Americas and Eurasia were open country grazers or browsers, likely to have been sedentary. Even the equids were typically short-limbed (Guthrie, 1984; Anderson, 1984). Surviving species are mainly woodland browsers (e.g., various deer), migratory grazers or low browsers (e.g., bison and reindeer/caribou), or inhabitants of mountainous terrain (various sheep and goats). In Australia, in contrast, extinctions took out most of the browsers, corresponding with the fire-induced change from forest to open woodland and savanna. The low magnitude of the late Pleistocene extinctions in Africa, involving some 10% of the large herbivore fauna, indicates the upper limit to solely climatically related extinctions, although even here human agency cannot be excluded (Klein, 1977).

6. Conclusions

Ecological transformations rarely have single causes. They are subject to contingencies, coincidences of events, interactions between factors, and multiplier effects (Walker, 1993). An ecologically sophisticated explanation must take cognizance of this, in explaining patterns that are richly complex in detail. Habitat changes surely contributed to extinctions of large mammalian herbivores in the late Pleistocene. The climatic shift at the end of the glacial period must have had a strong influence on habitat conditions. But its effects were drastically modified by the elimination of the disturbing influence of megaherbivores on vegetation, especially in the Americas. An increase in fire incidence through human agency may have contributed importantly in Australia. Thus, human presence, in various ways, was the decisive factor transforming a minor extinction pulse into a major extinction cascade.

The processes involved in the late Pleistocene extinction event still operate today. The numbers and distribution of surviving megaherbivores, plus various species of whale, decline progressively through human hunting, today largely for artifacts rather than meat. Populations of these and many other large mammals are isolated within the remnant habitats protected within designated wildlife parks. Global climatic change looms imminent. We may perceive the central involvement of humans in the sixth great extinction episode, but can we do much about it?

References

Anderson, E. 1984. Who's who in the Pleistocene: a mammalian bestiary, in: P. S. Martin and R. G. Klein (eds.), *Quaternary Extinctions,* pp. 40–89. University of Arizona Press, Tucson.

Begon, M., Harper, J. L., and Townsend, C. R. 1996. *Ecology: Individuals, Populations and Communities* 3rd ed. Blackwell Science, Oxford.

Bell, R. H. V. 1982. The effect of soil nutrient availability on community structure in African ecosystems, in: B. J. Huntley and B. H. Walker (eds.), *The Ecology of Tropical Savannas,* pp. 193–216. Springer-Verlag, Berlin.

Bond, W. J. 1997. Fire, in: R. W. Cowling, D. M. Richardson, and S. M. Pierce (eds.), *Vegetation of Southern Africa,* pp. 421–446. Cambridge University Press, London.

Bourlière, F. 1965. Densities and biomasses of some ungulate populations in eastern Congo and Rwanda, with notes on population structure and lion/ungulate ratios. *Zool. Afr.* **1**:199–207.

Brink, J. S. 1994. An ass, *Equus* (*Asinus*) sp., from the late Quaternary mammalian assemblages of Florisbad and Vlakkraal, central southern Africa. *S. Afr. J. Sci.* **90**:497–500.

Cumming, D. H. M. 1981. Elephant and woodlands in Chizaria National Park, Zimbabwe, in: P. A. Jewell, S. Holt, and D. Hart (eds.) *Problems in Management of Locally Abundant Wild Mammals,* pp. 347–349. Academic Press: New York.

deMenocol, P. B. 1995. Plio-Pleistocene African climate. *Science* **270**:53–59.

Dublin, H. T., Sinclair, A. R. E., and McGlade, J. 1990. Elephants and fire as causes of multiple stable states in the Serengeti–Mara woodlands. *J. Anim. Ecol.* **59**:1147–1164.

Fowler, C. W. 1981. Comparative population dynamics in large mammals, in: C. W. Fowler and T. D. Smith (eds.), *Dynamics of Large Mammal Populations,* pp. 437–455. Wiley, New York.

Fryxell, J. M., Greever, J., and Sinclair, A. R. E. 1988. Why are migratory ungulates so abundant? *Am. Nat.* **131**:781–788.

Graham, R. W., and Lundelius, E. L. 1984. Coevolutionary disequilibrium and Pleistocene extinctions, in: P. S. Martin and R. G. Klein (eds.), *Quaternary Extinctions,* pp. 223–249. University of Arizona Press, Tucson.

Guthrie, R. D. 1982. Mammals of the mammoth steppe as paleoenvironmental indicators, in: D. M. Hopkins, C. E. Schweger, and S. B. Young (eds.), *Paleoecology of Beringia,* pp. 307–326. Academic Press, New York.

Guthrie, R. D. 1984. Mosaics, allelochemics and nutrients. An ecological theory of late Pleistocene megafaunal extinctions, in: P. S. Martin and R. G. Klein (eds.), *Quaternary Extinctions,* pp. 259–298. University of Arizona Press, Tucson.

Hooijer, D. A., and Patterson, B. 1972. Rhinoceroses from the Pleistocene of northwestern Kenya. *Bull. Mus. Comp. Zool. Harvard Univ.* **144**:1–26.

Huffaker, C. B. 1958. Experimental studies on predation: dispersion factors and predator–prey oscillations. *Hilgardia* **27**:343–383.

Huntley, B. J. 1982. South African savannas, in: B. J. Huntley and B. H. Walker (eds.), *Ecology of Tropical Savannas,* pp. 101–119. Springer-Verlag, Berlin.

Hutchinson, G. E. 1957. Concluding remarks. *Cold Spring Harbor Symp. Quant. Biol.* **22**:415–427.

Jeffries, M. J., and Lawton, J. H. 1984. Enemy free space and the structure of ecological communities. *Biol. J. Linn. Soc.* **23**:269–286.

Kershaw, A. P. 1986. Climatic change and Aboriginal burning in north-east Australia during the last two glacial/interglacial cycles. *Nature* **322**:47–49.

Klein, R. G. 1977. The ecology of early man in southern Africa. *Science* **197**:115–126.

Klein, R. G. 1984a. Mammalian extinctions and Stone Age people in Africa, in: P. S. Martin and R. G. Klein (eds.), *Quaternary Extinctions: A Prehistoric Revolution,* pp. 553–573. University of Arizona Press, Tucson.

Klein, R. G. 1984b. The large mammals of southern Africa: Late Pliocene to Recent, in: R. G. Klein (ed.), *Southern African Prehistory and Paleoenvironments,* pp. 107–146. Balkema, Rotterdam.

Laws, R. M., Parker, I. S. C., and Johnstone, R. C. B. 1975. *Elephants and Their Habitats. The Ecology of Elephants in North Bunyoro, Uganda.* Clarendon Press, Oxford.

Lock, J. M. 1972. The effect of hippopotamus grazing in grasslands. *J. Ecol.* **60**:445–467.

Ludwig, D., Hilborn, R., and Walters, C. 1993. Uncertainty, resource exploitation and conservation: lessons from history. *Science* **260**:17, 36.

Macdonald, D. (ed.). 1984. *The Encyclopaedia of Mammals,* Volume 2. Allen & Unwin, London.

Marean, C. W., and Gifford-Gonzalez, D. 1991. Late Quaternary extinct ungulates of East Africa and palaeoenvironmental implications. *Nature* **350**:418–420.

Markgraf, V. 1985. Late Pleistocene faunal extinctions in southern Patagonia. *Science* **228**:1110–1112.

Martin, P. S. 1966. Africa and Pleistocene overkill. *Nature* **212**:339–342.

Martin, P. S. 1967. Prehistoric overkill, in: P. S. Martin and H. E. Wright (eds.), *Pleistocene Extinctions: The Search for a Cause,* pp. 75–120. Yale University Press, New Haven.

Martin, P. S., and Klein, R. G. (eds.). 1984. *Quaternary Extinctions: A Prehistoric Revolution.* University of Arizona Press, Tucson.

Mech, L. D. 1977. Wolf-pack buffer zones as prey reservoirs. *Science* **198**:320–321.

Medina, E., and Silva, J. F. 1991. Savannas of northern south America: a steady state regulated by water–fire interactions on a background of low nutrient availability, in: P. A. Werner (ed.), *Savanna Ecology and Management,* pp. 59–70. Blackwell, Oxford.

Messier, F. 1994. Ungulate population models with predation: a case study with the North American moose. *Ecology* **75**:478–488.

Murray, P. 1991. The Pleistocene megafauna of Australia, in: P. Vickers-Rich, J. M. Monaghan, R. F. Baird, and T. H. Rich (eds.), *Vertebrate Palaeontology of Australasia,* pp. 1071–1164. Pioneer Design Studio, Melbourne.

Naylor, J. N., Caughley, G. J., Abel, N. O., and Liberg, O. N. 1973. *Luangwa Valley Conservation and Development Project, Zambia. Game Management and Habitat Manipulation.* FAO, Rome.

Owen-Smith, N. 1987. Pleistocene extinctions: the pivotal role of megaherbivores. *Paleobiology* **13**:351–362.

Owen-Smith, N. 1988. *Megaherbivores: The Influence of Very Large Body Size on Ecology.* Cambridge University Press, London.

Owen-Smith, N. 1989a. Megafaunal extinctions: the conservation message from 11,000 years BP. *Conserv. Biol.* **3**:405–412.

Owen-Smith, N. 1989b. The megaherbivore syndrome: alternative life style or different time frame?, in : M. N. Bruton (ed.), *Alternative Life History Styles of Animals,* pp. 441–458. Kluwer, Dordrecht.

Owen-Smith, N. 1990. Demography of a large herbivore, the greater kudu, in relation to rainfall. *J. Anim. Ecol.* **59**:893–913.

Owen-Smith, N. 1993. Comparative mortality rates of male and female kudus: the costs of sexual size dimorphism. *J. Anim. Ecol.* **62**:428–440.

Owen-Smith, N. 1997. Herbivory, in: R. W. Cowling, D. M. Richardson, and S. M. Pierce (Eds.), *Vegetation of Southern Africa,* pp. 397–420. Cambridge University Press, London.

Partridge, J. C., Avery, D. M., Botha, G. A., Brink, J. S., Deacon, J., Herbert, R. S., Maud, R. R., Scholtz, A., Scott, L., Tolsma, A. S., and Vogel, J. C. 1990. Late Pleistocene and Holocene climatic change in southern Africa. *S. Afr. J. Sci.* **86**:302–306.

Pulliam, H. R., and Danielson, B. J. 1991. Sources, sinks and habitat selection: a landscape perspective on population dynamics. *Am. Nat.* **137**(Suppl.):50–66.

Ritchie, J. C., and Cwynar, L. C. 1982. The late Quaternary vegetation of the north, in: D. M. Hopkins, J. V. Matthews, C. E. Schweger, and S. B. Young (eds.), *Palaeoecology of Beringia,* pp. 113–126. Academic Press, New York.

Scholes, R. J. 1991. The influence of soil fertility on the ecology of southern African dry savannas, in: P. A. Werner (ed.), *Savanna Ecology and Management,* pp. 71–75. Blackwell, Oxford.

Scholes, R. J., and Walker, B. H. 1993. *An African Savanna.* Cambridge University Press, London.

Stuart, A. J. 1991. Mammalian extinctions in the late Pleistocene of northern Eurasia and North America. *Biol. Rev.* **66**:453–562.

Talbot, L. M., and Talbot, M. H. 1963. The wildebeest in western Masailand, East Africa. *Wildl. Monogr.* **12**:1–88.

Thompson, P. J. 1975. The role of elephants, fire and other agents in the decline of a *Brachystegia boehmii* woodland. *J. S. Afr. Wildl. Manage. Assoc.* **5**:11–18.

Turkalo, A., and Fay, J. M. 1995. Studying forest elephants by direct observation: preliminary results from the Dzanga clearing, Central African Republic. *Pachyderm* **20**:45–54.

Vartanyan, S. L., Garutt, V. E., and Sher, A. V. 1993. Holocene dwarf mammoths from Wrangel Island in the Siberian Arctic. *Nature* **362**:337–340.

Walker, B. H. 1993. Rangeland ecology: understanding and managing change. *Ambio* **22**:80–87.

Woodhouse, H. C. 1976. Elephant hunting by hamstringing depicted in the rock paintings of southern Africa. *S. Afr. J. Sci.* **72**:175–177.

Wright, H. E. 1984. Late Quaternary environments of the United States, in: S. C. Porter (ed.), *The Late Pleistocene,* Volume 1. University of Minnesota Press, Minneapolis.

Zimov, S. A., Chuprynin, V. I., Oreshko, A. P., Chapin, F. S., III, Reynolds, J. T., and Chapin, M. C. 1995. Steppe–tundra transition: a herbivore-drive biome shift at the end of the Pleistocene. *Am. Nat.* **146**:765–794.

CHAPTER 4

The Power of Pleistocene Hunter-Gatherers

Forward and Backward Searching for Evidence about Mammoth Extinction

GARY HAYNES and B. SUNDAY EISELT

1. Introduction

The first hunter-gatherers in North America were present at the time mammoths and other large mammals native to the continent rapidly became extinct. These early people may have been at least partly responsible for the extinctions (Martin and Klein, 1984). In this chapter we address some questions about Paleoindian foraging and the possible effect it had on mammoth populations. We also offer suggestions about the nature of the debate that has continued for decades over the cause of the extinction event.

In the first half of the chapter we discuss theoretical and conceptual issues relevant to the extinction debate. In the second half we examine empirical generalizations, archaeological data, and actualistic studies, and we make some suggestions about how both forward and backward searching may lead to one and the same conclusion.

2. Unresolved Issues

2.1. Background to the Problem

The word *Paleoindian* is generally applied by Americanist archaeologists to refer to Pleistocene pioneers in the New World and to Holocene hunters of now-extinct bison species. Fluted-point makers are one kind of Pleistocene Paleoindian, or perhaps several kinds, whose material culture can be found widely throughout much of North America and into Central America (see Bonnichsen and Turnmire, 1991). Clovis is now thought to be the earliest type of fluted point, stratigraphically underlying and predating the Folsom type (Tay-

---

GARY HAYNES • Anthropology Department, University of Nevada, Reno, Reno, Nevada 89557.
B. SUNDAY EISELT • Department of Anthropology, University of Michigan, Ann Arbor, Michigan 48109.
Extinctions in Near Time, edited by MacPhee. Kluwer Academic/Plenum Publishers, New York, 1999.

lor *et al.*, 1996) and a variety of eastern fluted point types such as Cumberland, Gainey, Quad, Suwanee, and numerous others (Mason, 1962).

The subsistence activities of Clovis-point makers is a slippery topic to discuss. The author of one definitive textbook of American prehistory, Gordon Willey (1966), even while helping to create the classic definition for Clovis as a culture, added several qualifiers. According to Wiley (1966:37–38), Clovis fluted points were made by people of a "Big-Game Hunting Tradition," "primarily dependent on killing large game animals." Although cautioning that "their entire livelihood did not depend on the big Pleistocene animals of the hunt," Willey emphasized that "big-game pursuit is the most characteristic and diagnostic feature of the culture."

The existence of mammoth bones bedded with western Clovis points, and occasionally the bones of other large mammals, leads some archaeologists to interpret Clovis culture as centering on big-game hunting. However, eastern Clovis or fluted-point sites very rarely contain associated large-mammal remains. Dincauze and Curran (1983), Meltzer (1988, 1989, 1993, 1996), Dent (1995), Custer (1996), and others have tried to redefine Clovis Paleoindian as an adaptation of "generalists," without any focus on big-game animals at all.

It is worth taking another look at some arguments often used in redefinitions of Clovis subsistence. Here we explicitly try to describe some unique features that may be useful in defining the foraging behavior of Clovis people. What we hope to show is that Clovis fluted-point makers would have hunted proboscideans, killed them, and eaten them, as much as if not more often than they would have eaten blackberries, rabbits, or wild asparagus, and that in fact they were big-game hunters who contributed to the extinction of proboscideans in North America.

2.2. Forward and Backward Searching

The extinction of mammoths, and the dispersal of fluted points in North America, are two penecontemporaneous "events" (in the geologist's sense of the word). There may be a cause–effect relationship between them. If extinction occurred because of Clovis foraging, the situation was one in which a big-game-hunting subsistence system failed badly while at the same time contributing to the failure of an entire ecosystem already under stress from climatic change. The archaeological and paleontological records may provide the needed facts about this important breakdown of late Pleistocene systems.

Analysts of prehistory may structure their inquiry in one of two ways, as would an analyst investigating another kind of broken system, namely a system of computer hardware and software that goes wrong. The search for fault may proceed in a "forward" direction, or in a "backward" direction. Each type of search is carried out in a different way, using different premises and expectations. We do not pretend here to perform error-recovery procedures as they are performed by computer engineers, because fault-finding in computer applications is a complex and much more mathematical enterprise than our evaluations of ideas about mammoth extinction. But we think it is potentially useful to frame our discussion in this way because the forward/backward technique of error recovery encourages a new perspective when comparing competing theoretical and mechanistic explanations for extinction.

When working with computer applications, engineers try to find the causes of failure without jumping to premature conclusions based on partial information. When complicated components of technology break down, the search for the error may be directed forward, or it may proceed backwards. In a forward search, the final outcome(s) is visualized in advance, based on initial conditions and a predicted forward "probability" algorithm. In a backward search, the end effects of error are examined so as to visualize the pathways that could have been directly involved with the failure, working back in time from the moment of failure, and using a backward "probability" algorithm. To rephrase these statements: a forward search is an error-recovery technique that troubleshoots by predicting future outcomes, including possible errors, and traces the effects that all of the various different components could have on each other from the beginning. Backward searching starts with the failed components or the errors and tries to find out what other components were fault-intolerant, thus creating the conditions of failure. Figure 1 illustrates the two different search directions. Note that the different directions of search could actually produce different answers to the question of cause.

We start our search for the cause of extinction by first summarizing the major known facts about mammoth die-out and Clovis foraging. We restate the conclusions that a forward-looking search generally provides, and compare the contradictory conclusions reached through a backward search. Then we look in more detail at the pertinent facts, trying to find possible connections between mammoth extinction and late Pleistocene foraging behavior expected in Clovis culture. We will model some of the features of late Pleistocene Clovis behavior under the specific conditions of the Younger Dryas deglaciation phase that ended the Pleistocene. In other words, we will move forward through models of Clovis foraging behavior to find characteristics that should have contributed to extinction. In the backward part of the search, we examine mammoth bone assemblages, those with and without any artifactual associations, and look for the direct evidence about paleoenvironmental and cultural pressures that affected the mammoths before death.

2.3. The Backward Search and Extinction: How Inevitable Was It?

A backward search begins by examining sites and predicting the characteristics of the people or the events that conceivably could have created them. In this section we seek a few rapid and logical connections between the actual extinction event and previous ecosystem components that may have had the most direct influence on that event.

2.3.1. Statement of Facts

Mammoth bones exist in quantity from the interval spanning 16,000 to 11,000 radiocarbon years before present (rcyrbp). Many are associated with artifacts that have an Upper Paleolithic or Paleoindian character. A high percentage of young mammoths are found in bone-bearing fossil sites. Some bones are carnivore-gnawed, some are trample-broken. Some sites contain bones that were weathered to different extents. Most assemblages consist of greater than 90% mammoth remains, but other taxa are present in all of them. Palynologic, stratigraphic, and geomorphic evidence shows that the sites were created during times of rapid climatic changes.

During the Pleistocene, climates have been characterized as "cold-equable" (Hibbard, 1960; Graham, 1986). Summers were cooler, but winters were no cooler than to-

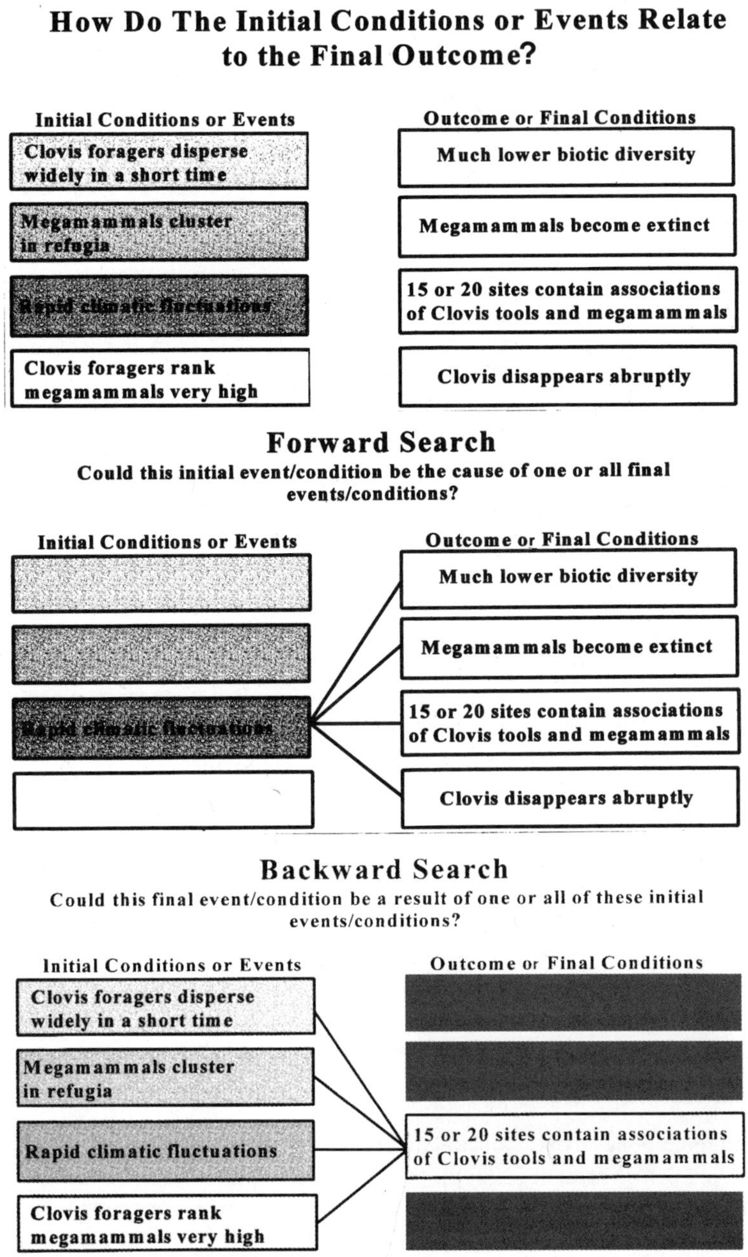

FIGURE 1. In (a), some initial conditions and outcome events are listed; in (b), one initial condition has been selected to determine whether it could have created one or all of the outcomes; in (c), one of the outcomes has been selected to trace any possible relationships with the initial conditions.

day. Glacial conditions and periglacial effects that changed the Arctic jet stream and caused warm Chinook-like winds to descend onto the plains also caused greater biotic heterogeneity and stability, higher species densities, and intermingled biotas on the North American continent. A warming, "continental" climate at the end of the Pleistocene led to the retreat of the glaciers and a decrease in periglacial effects. Seasonality was enhanced through increased solar radiation and the onset of colder Arctic winds. The result was new biotic community patterns, greater biotic homogeneity, unstable biomass, and lowered environmental carrying capacities. Species responded to these changes by dispersing along environmental gradients in different directions, at different rates, and during different times. The most stressful period (in terms of diversity loss) appears to be about 15,000–11,000 rcyrbp.

2.3.2. Conclusion

An acceptable conclusion from this set of data is that natural die-offs occurred in mammoth refuge ranges that reduced mammoth populations to isolated fragments with no hope of recovery, as their ranges were gone by about 10,500 rcyrbp. Advocates of the position include R. Graham (e.g., see Graham and Lundelius, 1984), the late J. Guilday (1984), and D. Grayson (1993).

2.4. Clovis Foraging and a Forward Search

A forward search begins by predicting the characteristics of sites to be expected based on our knowledge of the existing people and environmental events of the late Pleistocene.

2.4.1. Statement of Facts

In North America, Clovis technology contains tools that are well-suited to killing and processing large-mammal carcasses. There are 13 (or more) associations of Clovis tools with mammoths, and this is a very large number given the narrow time range of 10,500 to 11,500 years ago reported for these sites. Clovis dates are similar east to west, north to south, even in sites that do not contain mammoth bones. The earliest evidence of human presence in the overwhelmingly largest part of North and South America is dated to 11,500–8000 rcyrbp, no earlier than Clovis (but see Meltzer *et al.,* 1997, for 12,500-year-old evidence at Monte Verde in southern Chile). In this same time interval, climatic change throughout the Northern Hemisphere reduced diversity, fragmented habitats, and reduced megafauna survival to patches, or refuges.

2.4.2. Conclusion

A logical conclusion from this evidence may be that a wave of rapid human dispersal (or an irruption) would have led to overhunting in those few remaining patches, extinguishing the most vulnerable and largest mammalian taxa. Advocates of this position include C. V. Haynes (1969, 1984), P. Martin (1984), and L. Agenbroad (1984).

2.5. Criticisms of Each Opinion about the Cause of Extinction

The "conclusions" presented above are impressionistic propositions, of course, and each can be effectively contested. For example, it may be argued that mammoths had survived earlier climatic reversals (although these are less understood in nature and in degree), by either migrating, changing behavior, or slowly evolving new features in adaptation, therefore avoiding extinction. Natural die-offs resulting from climatic change alone thus cannot be the only cause of the late Pleistocene extinction, as these kinds of die-offs were never completely fatal to all populations (see Alroy, this volume; Martin and Steadman, this volume). It also may be argued that overhunting by humans is an unacceptable potential cause of extinction because of the difficulties involved in killing so many mammoths, or because of the unlikelihood that Clovis foragers would have continued to seek out mammoths to kill during a time when mammoths must have been getting scarcer and scarcer (see Martin and Steadman, this volume).

What may be most interesting about this debate is that it is still going on, even after decades of adversarial give-and-take. Different but equally qualified experts hold very different ideas about the causes of extinction. This is not an unusual situation in science, although it is difficult for nonscientists to understand how disputes over facts can remain unresolved for so long. Whenever different experts use different methods and prefer different emphases, the results will be different (see Hanks, 1997, for example). The nonexperts wish to adopt the best answer as the "right" one, but if two opinions are endlessly allowed to clash in the science, consensus is impossible.

In the case of mammoth extinction, neither side in the debate can gain ground on the other. The reason simply may be that each side starts the search for cause from a different position. The climate camp begins by modeling the last-mammoth bone sites as the end results of continental processes, producing an explanation that is both parsimonious and accommodative, while the Clovis camp tries to predict overkill based on the foraging characteristics of Clovis culture. Both sides equally emphasize the coincidental nature of the extinction and other changes (in climate and culture). Both sides produce elegant, logical, acceptable conclusions. Extinction was caused by a single external factor.

The computer repairperson, when faced with a set of different diagnoses following the forward and backward searches for faults in a crashed system, may choose to replace or repair more than one component, without ever pinning the full blame on any one. But unless the central error is identified, the system will continue to be in jeopardy. So the repairperson may have to begin searching all over again, and the chances are good that after more searching yet another set of possible faults may be identified (e.g., see MacPhee and Marx, 1997, for their "hyperdisease hypothesis," a potentially additional cause of first contact extinctions among metapopulations brought into contact with foreign species). Initial conditions may or may not link up clearly and directly with the end results; the fault may be impossible to predict when searching forward, or impossible to trace when searching backward. The ideal situation occurs when one and only one component can be seen as defective or doomed, and the component is found by both forward and backward searching. Somewhere the computer repairperson may have an instrument that clearly identifies that one component, and it can be simply and quickly replaced or repaired. But in the case of the extinction searches, we do not possess a single instrument to measure each component equally and fairly. The nature of the evidence about Clovis foraging is not on par with the evidence about late Pleistocene paleoenvironments. To search from one kind of evidence alone, or even in tandem, is to become confused by the inconsistencies, lack of data, theoretical conflicts, and inadequate methodology.

To add a new perspective to the study of extinction, in the next section we attempt to show how the forward and backward search strategies may be adjusted to arrive at the same conclusions about the cause of mammoth extinction.

3. Toward a New Perspective

3.1. Mammoths, Environments, and Foraging at the End of the Pleistocene

In this section we propose that Clovis foragers sought to locate patches containing mammoths or mastodonts by following natural clues that were clear and obvious at the end of the Pleistocene. The end of the Pleistocene is a period of habitat fragmentation and megamammal population isolation. This factor limited proboscidean distribution to nonrandomly situated patches. King and Saunders (1984) demonstrate evidence of increased mastodont insularity caused by changing environmental conditions, a situation in contrast to the earlier Pleistocene paleontological evidence that suggests proboscidean genetic exchange over wide geographic areas (e.g., see Averianov, 1992; Foronova and Zudin, 1999). C. V. Haynes (1991) has argued that a Clovis-age drought affected the southwestern United States, supported somewhat by V. Holliday (1997) who interprets the end of the Pleistocene in the Southern High Plains as increasingly warmer and drier. In the Great Lakes subregion, some evidence can be found that spring discharge was erratic or possibly lacking altogether at certain times within the Clovis interval, for example possibly at the Hiscock site in northeastern New York state. Mastodont and mammoth bones throughout many subregions seem to be clustered in space—at springs, marshes, and salt licks—and in time at the end of the Pleistocene.

Under this set of conditions, megamammal subpopulations may have gone through a period of migration widely in search of food and water, and would have become figuratively tethered to resource oases, or patches that provided better returns than others. Ewer (1968) refers to these areas as "fixed points" where specific activities are carried out, and a network of trails lead to and from the points (see also G. Haynes, 1991:92; Saunders and Daeschler, 1994:23). Fixed points include feeding areas, drinking sites, rubbing trees, and wallows. Here we use the idea of "fixed points" as patches to develop a more theoretical approach concerning Paleoindian foraging behavior.

Mixed vegetational mosaics of the glacial maximum period changed toward broader zonal associations of plants, and diverse botanical communities became very limited in terms of the number and types of taxa growing together. Under these conditions, it is very likely that formerly wide-ranging large mammals must have been subjected to natural pressures that forced them to change their ranges, restrict movements to optimal patches or habitats, and abandon suboptimal ranges. In other words, large mammals sought refuges in shrinking ranges where plant diversity and productivity remained relatively high.

The prey preferred by a theoretically optimizing human predator, which should be the largest (i.e., mammoth and mastodont), thus were nonrandomly distributed in specific patches where water was available along with relatively diverse plant foods. This high diversity of plant forage for megamammals may have been important at the end of the Pleistocene because the size and spacing of stands of optimal feed species were probably changing radically as climate changed. Animals such as mastodonts that preferred to mix browsing in woodlands with feeding on a variety of other plants such as marsh species, shrubs, and herbaceous vegetation, would have found the woodlands closing in on the marshes and open shrublands. Mammoths, which preferred grasses but probably survived at least one season of every year on browse (herbaceous or woody), found their dietary browse elements becoming scarcer, less productive, or located farther away from the grass patches. These megamammal species would have carefully chosen which patches they visited throughout the different parts of the growing seasons so as to optimize their own diets during the end of the Pleistocene interval. Their best choices were more and more limited in size and distribution in North America.

3.2. Is Overkill Impossible?

One example of theoretical conflicts and inconsistencies in the extinction debate has to do with different perceptions among prehistorians about human foraging behavior. In this section we reconsider the argument that mammoth overkill is not predicted by current anthropological, ecological, or economic theories. Overkill may not be predicted, but it clearly is not forbidden by these theories and paradigms.

3.2.1. Optimization Theory and Mammoth-Hunting

In an attempt to understand the evolution of human behavior, a great deal of research has been carried out modeling ecological and social variables of human band-level societies (Binford, 1978). Recent research that relies on abstract modeling of critical variables in human foraging rather than direct analogy to modern populations highlights the utility of mathematical applications as heuristic devices capable of describing human variability and showing their dynamic interaction over time and space. One of the most effective of these applications, evolutionary ecology, is "the study of evolution and adaptive design in ecological context" (Smith and Winterhalder, 1992:3). Given a set of resources with specified characteristics, foraging models propose how these resources will be used. Decision-making is based on an input–output ratio that evaluates a forager's efficiency where individual decision-making is the analytical unit under selection. Because selection acts on feeding behaviors, foraging by animals is not simply a series of fixed behaviors or automatic harvest-responses to resources, nor does foraging behavior depend solely on the order and proportions of food types existing in the environment. Foraging is much more than an attempt to get the most food at the least cost (see Schoener, 1987, for a historical review of misunderstood critiques, oversimplified concepts, and theoretical testing of optimization or foraging theory).

Research in foraging theory has not shied away from examining the types of observed deviations from modeled predictions (e.g., see Alvard, 1995, for why game conservation is not necessarily perceived as optimal by foragers; or Kay, 1994, for aboriginal resource exploitation in North America). Foraging theorists use models and mathematical formulas to understand unique as well as universal decisions and selective processes, and not merely to make programmed predictions about foraging behavior. The models take into consideration a set of probable trade-offs between getting the best food for the time and energy required.

When there is an observed congruence of foraging behavior and the predictions of abstract modeling, it can be said that the theory has explanatory value. Where there are incongruences, it can be said that the model needs to take other perceived costs and benefits into consideration, or to more closely examine the assumptions of the theory and goals of foraging behavior as these relate to other aspects of cultural behavior.

Here we discuss three important components of the body of theory about cost–benefit trade-offs, namely diet breadth, patch choice, and foraging time spent in patches, and we suggest how these conceptualized relationships or theorems may accommodate different predictions concerning Clovis foraging (including the possibility of "overkill" and the so-called "blitzkrieg" model).

3.2.2. Diet Breadth or Prey Selection Models

Diet breadth or prey selection models attempt to predict how potential food items are ranked for inclusion in the diet, and how or why they are selected or not selected when encountered. The graph in Fig. 2 illustrates the trade-offs involved in ranking cost against benefit when prey items are considered for inclusion in the diet. As originally modeled by MacArthur and Pianka (1966), any food item probably will be added to the diet even if it increases handling time, as long as it also decreases search time in compensation.

The highest ranking (1, 2, . . .) of a resource results from a low handling time per calorie value (which takes into consideration an animal's body mass, its ability to return usable food, and the difficulties of killing, butchering, and processing meat) and a low search time per calorie value. Being less common in the environment, very large animals require longer search times than do smaller ones, but while very large animals may be ranked highly they may not necessarily rank the highest. Instead, ranking of one item will vary if search and handling times change—in other words, when an animal taxon becomes easier to track or find and kill, as a result of environmental variables (such as drought that forces some species to crowd into refuges, or the human invention of new and more effective technology).

A forager is faced with decisions about keeping energy intake high per unit of time spent foraging. To optimize, foragers will select combinations of food items that keep the

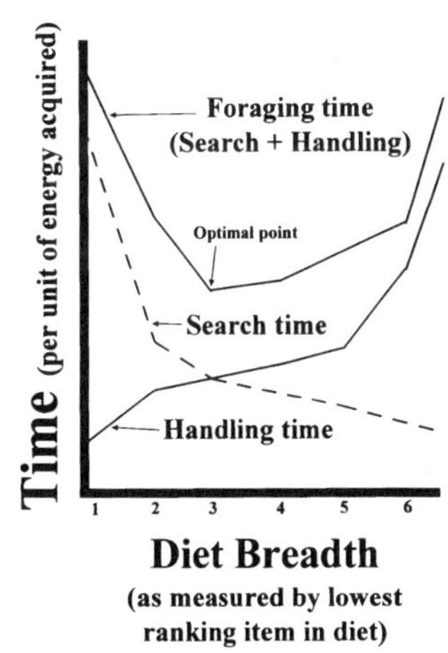

FIGURE 2. Foragers decide to include different items in the diet after evaluating the trade-off between search and handling times. High-ranked foods almost always have greater search expenses, but they also promise greater returns and reduced handling expenses. Redrawn from Bettinger (1991), after MacArthur and Pianka (1966).

intake as high as possible when measured against search time, handling and processing time, and energy yield. Foragers select the range of items from those most preferred to those least preferred, but the preferences are conditioned by the expected difficulties of finding and preparing each food item. If a forager is searching for the highest-ranked prey species and encounters a lower-ranked prey, the decision must be made about whether to stop searching for the top-ranked prey so as to go ahead and procure the lower-ranked one. This is a *contingency* that forces the forager to make a predictably "rational" decision (see Bettinger, 1991:84–87, for further discussion). "The choice to use a given resource is independent of its abundance and depends exclusively on its cost and returns . . . *once it is found*" (italics in original) (Bettinger, 1991:86).

Given this, prey body size is often used as a proxy measure of prey rank where the energetic value of an organism is more or less directly proportional to its weight. In the case of very large animals, ranking can decrease slightly as a result of increased pursuit and processing costs (Griffiths, 1975; Schoener, 1979; Bayham, 1982; see Broughton, 1994:502, for a good review of this concept). Although the issue should be explored further (Smith, 1991:231), these added costs are not seen by us as significant enough to alter our position that mammoths were always taken when encountered in resource patches at the end of the Pleistocene.

In terms of mammoth extinction, what are the implications of these theoretical statements? First, Clovis foragers would not have ignored mammoths or deliberately avoided killing them. Although the ranking of mammoth may have changed depending on its abundance and the abundance of other prey, mammoths should have been taken most times when encountered given their post-encounter return rates, which were relatively high. Second, although it is expectable that other food items may have been included in Clovis diets as mammoths became perceptibly scarcer at the end of the Pleistocene, it is also expectable that mammoths would have been taken if encountered randomly. Furthermore, mammoths confined to resource refugia certainly would have provided a known location for their exploitation, greatly reducing search time and increasing their ranking even when their densities may have been decreasing in the rest of the environment and in relation to other taxa.

Another key concept with important implications in the debate is *specialization,* a word used when foragers place fixed limits on their diet breadth. A forager's *niche* is the role it plays in the environment, which includes the range and ranking of food items selected by the forager. A narrow niche is a specialized one. Keeping the diet breadth model in mind, it should be clear that foragers specialize only insofar as they are able to keep search time, handling time, and energy returns balanced. Foragers do not specialize in low-ranked food items. It is reasonable to predict that if encounter rates for smaller animals were very high, mammoths may not have ended up in the diet of Clovis foragers very often, or at all, because their ranking was relatively low as a result of scarcity. The question then becomes: how much scarcity was required before the largest prey taxon lost its top ranking. Phrased another way, the question is: how many individuals of another and smaller particular prey type (bison, for example) would need to be present and available such that foragers would have turned down a chance to kill a mammoth if one were encountered? Perhaps these questions go to the heart of the term *specialization,* because they are concerned with forager decisions to hunt prey that may be relatively harder to find than others.

Obviously, mammoths were so much larger than other big game that there would be no straightforward one-to-one ratio between exploiting smaller animals and a mammoth; in other words, the payoff of a mammoth once encountered might have been worth several bison standing next to it, so it would still have been profitable to take mammoth even when they were decreasing in the environment.

We cannot state precisely how many bison needed to be present (postencounter) for time spent pursuing mammoths (postencounter) to no longer be profitable. Although mammoth densities were decreasing as a result of environmental changes at the end of the Pleistocene, and this would have contributed to a lower ranking of them by Clovis foragers, we also think that a probable tendency of mammoths to aggregate in refugia would have continued to keep mammoth ranking relatively high. The aggregating would have decreased search time for foragers, because refugia, once discovered—and that would not have been difficult, in the case of proboscideans—would have been known to contain high mammoth densities with a high postencounter rate compared with other large game, regardless of their overall abundance in the environment. Perhaps we can suggest that search time did not figure largely in late Pleistocene situations when mammoth locations were known and predictable. Prior knowledge about prey distribution can clearly affect forager residence times and patch choices (see Giraldeau, 1997).

How is knowledge about this concept relevant in the extinction debate? Mammoth-hunting need never have been a *specialization* that occurred often. However, it could have been wise for foragers to focus their attention on mammoths if the return was perceived as outweighing the time invested in searching for or procuring them versus other food sources.

3.2.3. Patch Choice Modeling

As developed by MacArthur and Pianka (1966), patch choice modeling is used to predict which characteristics of different patches would appeal most to foragers. Not all food patches are identical—some are far away, unevenly distributed, or seasonally depleted. Some are full of low-ranked foods, while others may contain a very few high-ranked prey animals hiding in unexpected spots.

In modeling, patches are defined as spatial areas that contain at least one food resource (although human foragers may define viable patches as sources not only of food, but also of water and a range of other resources such as toolstone). A "patch," however, is not the same thing as a habitat or an environment. Patches are usually measured in terms of their returns, which may be of three types (Kaplan and Hill, 1992) (see Fig. 3). Most terrestrial patches would provide returns as in Graph C. Returns normally taper off as more time is spent in patches, because terrestrial resources tend to be limited in number, density, and distribution. Hunters run out of prey animals, eventually, or find them scarcer and scarcer. However, patches in which the resources are exclusively located may have returns measured as in Graph A or B. In Graph A, the resource never runs out, because of such factors as migration or high reproductive rate, while in Graph B the resource is exhausted abruptly. To harvest optimally, foragers may choose to leapfrog from one such patch to the next.

The word *patchy* refers to an inconsistent or irregular distribution of resources in an environment. Very patchy resources are clumped in space, rarely available, appear only for a short time, and have a small unit size. Nonpatchy resources are dispersed, available frequently and for a long time, and have a large unit size. Mobile resources are patchy, and migratory resources are patchy. Mammoths confined to refugia were a patchy resource, at least in space although not necessarily in time. They were clumped in refugia patches that would have provided returns perhaps unusual when compared with other kinds of patchy resources. Even as foragers exploited the mammoths and other resources in the refugia, mammoth numbers would not have become depleted in a simple way because more animals probably made their way to the refuges over the months and years of climatic change. The returns from this sort of situation may have resembled Graph A in Fig. 3. Foragers would not have experienced the need to travel to other patches as often in this situation. These foragers and the nature of the patches they exploited therefore had an enhanced role in the extinction of mammoths.

Figure 4 shows how foraging theory predicts the time a forager spends in a patch. A forager who must walk from a distant point to reach the patch in mind spends more time foraging in that patch (time = T_y) than one who travels a shorter time (spending T_x in the patch). The forager makes up for the lengthy travel time by staying longer in the patch, until the return rate drops off below the rate other patches on average may provide.

Which are the patches an optimizing forager should choose to visit? If adding a patch to those to be visited can decrease the within-patch foraging time, or decrease the between-patch travel time (at the same time as keeping down the time needed within the patch to find food), foragers will decide to add that patch. The times needed for both searching and foraging are changed by adding or subtracting patches (see the discussion in Bettinger, 1991:87–91). Foragers who tend to spend as much time in search as in pursuit postencounter are not as selective when picking out patches to visit as are foragers who prefer very little search time. That is, foragers who *specialize* in trying to kill prey that are either highly abundant or very predictably located can afford to be specialists in patch use, as well. They can limit the number of patch types visited, or spend far more time within specific patches.

How do theories about forager patch choice relate to the extinction debate? We propose that Clovis foragers spent different lengths of time in different patches, based on (1) two-way travel time, (2) overall returns, (3) food distribution, and (4) returns from higher-ranked, preferred foods in such patches. Some patches, even those containing adequate food plants, may not have been occupied as long as other patches that contained megamammals

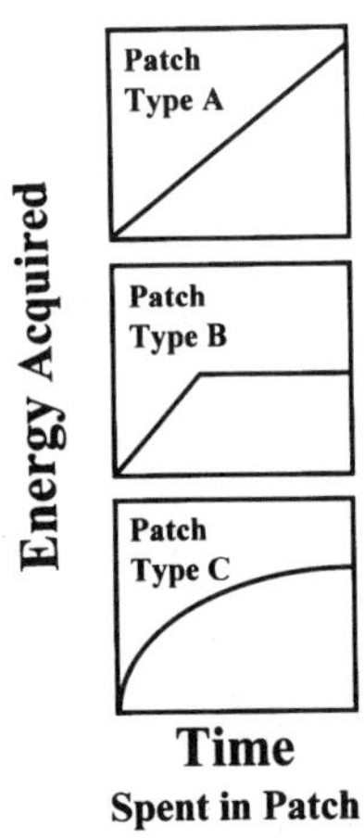

FIGURE 3. Graphs of the different kinds of returns over time in different patch types. Redrawn from Kaplan and Hill (1992).

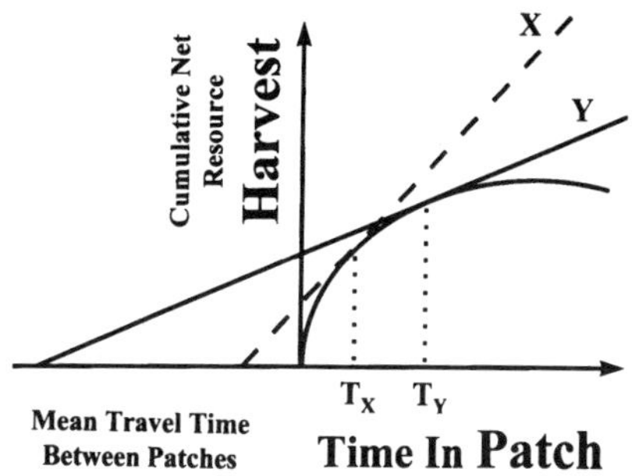

FIGURE 4. Graph predicting the time foragers will spend in a patch that requires different travel times to reach from other patches, and that provides changing returns as more time is spent within it. Redrawn from Charnov (1976), following Bettinger (1991).

(or recent signs of them). Some Clovis sites may be the remains of longer occupations than others, or the result of numerous return visits to the same spots in anticipation of finding proboscideans to procure. Most importantly, as follows from patch choice concepts, it should be possible to understand Clovis prioritization of certain patches. We propose that Clovis foragers preferentially traveled directly to patches containing mammoths or mastodonts—either seasonal clusters of them or localized aggregations in refuges during periods of climatic stress, food scarcity, or other difficult conditions. As we have stated the argument, mammoths or mastodonts were among the most preferred food, and patches that contained them would have been preferentially chosen for food searching. Clovis foragers may have spent more time in proboscidean patches than in patches lacking proboscideans, if the mammoths or mastodonts aggregated seasonally or in response to climatic pressures.

3.2.4. The Marginal Value Theorem

The marginal value theorem as originally articulated by Charnov (1976) and carefully reevaluated by others afterwards (Oaten, 1977; Martin, 1983; see Schoener, 1987, for examples) states that foragers are aware of the energy yield per time in each feeding patch, and they are therefore aware of the optimal time to quit the patch—which is when the capture rate in the patch is equal to the average energy per time obtainable over the entire environment (including travel time, and other variables). In other words, the marginal value theorem attempts to predict the time a forager spends in a patch, based on the time elapsed between the last successful capture and her or his departure from the patch. In richer environments, where patches may be more productive, the "giving up" time may be shorter, but evidence has been found to suggest that decisions about patch abandonment time may vary based on perceptions of prey distribution (Schoener, 1987).

It should be apparent that the marginal value theorem does *not* predict that Clovis foragers would stop killing mammoths when mammoths became scarcer, as critics of overkill have suggested (e.g., see Meltzer, 1993). The theorem actually predicts only that Clovis foragers would have had the choice to leave the patches where mammoths were to be found if food of any sort was perceived to be relatively scarce in those patches, unless other patches were long distances away and the food was believed to be scarce in them as well.

High-ranked food resources need not always retain their high ranking, especially after they become perceptibly scarcer. But in the case of mammoths and mastodonts, because they were so large and aggregated in locatable patches, in which they could be encountered at high rates with low search time, foragers probably continued to prefer hunting them. Mammoths and mastodonts, being the largest land animal contemporary with Clovis foragers, would have provided greater sources of meat and by-products. Further, they would have been available in the same way as any other animal, large or small, requiring one killing action to procure. Modern foragers do not discount the extra effort needed to kill larger mammals, and the extra time needed to process the resource compared with smaller ones, but they are also well aware that a kill is a kill, and the larger the carcass at the end of it, the better the return all around. Hence, large animals may be ranked very highly if not rare, even more highly than relatively more abundant smaller ones, if the overall return from a single large animal outweighs the cumulative return from several smaller animals. The "profitability" of a large animal is high (profitability being the amount of energy harvested per time handling the prey item) if handling is defined as pursuit, kill, and processing costs, and search time is assumed to be shared among prey types (Alvard, 1995:794). An optimizing forager will decide to pursue high-ranked prey whenever it is perceived that the return from that animal will be higher than the return from other types of prey animals that could be sought. In short, the implication of these theoretical statements is that the primary deciding factor in Clovis killing of mammoths or mastodonts would have been the high return rate with the animals, postencounter, and not the expectation that mammoths or mastodonts were big, hazardous, or endangered. Theoretically, then, we conclude that whenever Clovis foragers encountered proboscideans, they tried to kill them.

The paleoenvironmental evidence is strong that at the end of the Pleistocene, megamammal subpopulations were under pressure to congregrate at resource refuges, which would have been situated nonrandomly in North America. Perhaps at any one time thousands of such refugia existed in places where local conditions of hydrology, soil fertility, seasonal weather extremes, plant diversity, and access were favorable for megamammal survival. These places, whose sizes and distribution are only now beginning to be appreciated, were probably the optimal patches for Clovis foragers to choose (see King and Saunders, 1984; some relevant discussion of Pleistocene conditions can be found in Guthrie, 1990). Clovis foragers could find leading prey animals there, in much higher density than outside them. To have the optimal food source so clearly tethered in space was a foraging luxury that may have had other effects on Clovis people; for example, it may have encouraged rapid *in situ* population growth, as well as fostered frequent exploratory or scouting trips far from the patches by parties of Clovis foragers who knew how to retrace their steps back to the optimal patches, if or whenever food resources were found to be too scarce.

Table 1 lists the paleoenvironmental variables that we suggest are relevant for understanding Clovis interactions with proboscideans, and the main implications of the models or theorems from the body of optimization theory mentioned above. In the next section we look for additional empirical support for the proposition that late Pleistocene human foragers did actually kill too many proboscideans.

4. Empirical Evidence for Mammoth Killing by Clovis

In this section we refer briefly to the empirical evidence that proboscidean bone sites created by humans killing the animals are distinguishable from sites created by noncultur-

Table 1. Terminal Pleistocene Conditions Affecting Mammoths and Mastodonts (Left Column), Correlated with Possible Forager Behavior (Right Column), as Predicted from Optimization Theory

Important end-Pleistocene conditions affecting proboscideans	Effects on Clovis foragers (predicted from optimization models)
Proboscideans were present over a wide part of North America until around 11,000 rcyrbp	Proboscideans were the highest-ranked prey animal. (Prey selection/diet breadth model)
In general, habitat fragmentation resulted in increased isolation of proboscidean subpopulations.	Continentwide, proboscidean patchiness was increased, but within-habitat patchiness was greatly decreased for Clovis as proboscideans stayed in or near "refugia." (Patch choice model)
Zonal habitats replaced high-diversity mosaics.	More proboscidean isolation and more stressful regular migrations left signs for Clovis foragers to recognize. The foragers could move directly to the refuge patches. (Patch choice model)
Drought or seasonal extremes of forage availability occurred in some subregions	Clovis foragers spent longer times in certain patches, and had reliable and predictable foraging success, as well as a safety net for out-of-patch explorations. Clovis was "specialized" in refuge patch use. (Marginal value theorem)
Mammoths died off in shrinking refugia; some subpopulations probably died out altogether before Clovis foragers discovered them	As proboscideans became scarcer, Clovis foragers switched patches and/or switched to other prey, specifically the smaller taxa that existed in larger numbers and that were able to survive or recover more quickly from stress periods. Clovis foragers would have continued to kill every proboscidean encountered. (Marginal value theorem)

al processes such as drought-caused starvation. The aim is to demonstrate convincingly that Clovis foragers killed mammoths and did not merely scavenge their carcasses, thus adding support to the theoretical arguments in favor of overkill.

4.1. *Loxodonta* Sites

In the 1980s, Zimbabwe's Department of National Parks and Wild Life Management increased the previously small-scale elephant culling operations in Hwange National Park, northwestern Zimbabwe. The intention was to lower the number of free-ranging elephants in the park, because the ever-growing size of the wild herds was severely and negatively affecting the park's woodlands and natural habitats. The culling resulted in the formation of elephant-bone sites that match or exceed the dimensions of many fossil mammoth or mastodont sites (Fig. 5). These cullsites are the only documented and published mass-kill sites of modern elephants. Also studied at the same time in Zimbabwe were the only known mass die-off sites of elephants (see G. Haynes, 1991), and numerous other sites where individual elephants died natural deaths or were shot (Fig. 6). Sites to which elephant bones had been transported from either die-offs, culls, or individual deaths were compared with the *in situ* sites.

FIGURE 5. Part of a large killsite where 74 elephants were culled, photographed 2 years after its creation.

In both the cullsites and the noncultural sites, there are abundant cultural associations, including dropped artifacts and dugout firepits where elephant meat was cooked. The artifacts in both types of site include bits of thorn-ripped clothing, cigarette filters, matchsticks, newspaper fragments, and pencils or pens, among other classes of objects.

In the killsites (some of which are listed in Table 2), the elephants were butchered en masse by groups of skinners, meat-cutters, and tusk-removers. Shallow basins were always dug out on the cullsite ground surface and wood was kept burning in them for hours, so that water could be boiled for tea and elephant meat could be cooked for snacks. Few to no bones were left in the firepits, but charcoal and ash were abundant. The cullsites contain the bones of elephants of all ontogenic ages, usually, with subadults making up about 50% of the assemblages. Larger adult males are very rare. Most carcasses were fully butchered, which led to immediate disarticulation and scattering of bones. After full butchering and salvaging of meat, skins, and tusks had been completed, animal scavengers fed on the carcass remnants. Herds of elephants also visited many sites, and giraffes and other species often visited to eat bone fragments. Carnivore gnawmarking is present on some bones. Bones in the same sites are mostly in the same weathering stages except for elements that were protected by thick vegetation. Bone representation is not biased by selective element removal except in the case of tusks (all were removed by butchers) and mandibles (all or half of each specimen was removed during butchering).

The noncultural die-off sites (the largest are listed in Table 3) are in upland locales, but lie next to or within depressions that may be filled with pondwater or streams. Wide, thick accumulations of charcoal exist in nearby campsites on the fringes of the die-off depressions, representing leftovers from campfires built by scientists and rangers who visited the

FIGURE 6. Part of a large die-off site where the bones of over 100 elephants are found.

die-off sites. Elephant bones surround the charcoal beds, often as the result of deaths in the sites, or just as often the result of human campers selecting bone elements to use as seats, windbreaks, or protection against lions.

In the noncultural mass deathsites, very little evidence of carnivore utilization can be found (see G. Haynes, 1982, for a very explicit set of carnivore feeding stages and a definition of "carnivore utilization"). Bones in different weathering stages (Behrensmeyer, 1978) are not common and occur mainly when bone burial is gradual or cyclic. The completeness of skeletons (or the bone representation) may not be biased by different agencies of bone subtraction (feeding, trampling, weathering) unless burial is cyclic. Different processes of reworking of bone sites can occur and lead to the disassociation of skeletal parts. These in fact may create biasing of skeletal part representation, as certain elements are fragmented or destroyed.

In terms of mammoth extinction, the implications of these observations (developed at greater length in G. Haynes, 1991, 1999) are (1) modern killsites are different from modern natural die-off sites, although in subtle ways; (2) distinctions can be defined via a multidimensional comparison of bone weathering, bone representation, carnivore utilization, and mortality profiling (Table 4 presents the main differences and similarities); and (3) fossil sites show similar characteristics. It is therefore possible to distinguish kills from die-offs.

The modern cultural and noncultural sites form our "experimental" database. For comparison, we also examined the characteristics of fossil mammoth-bone sites. Most Eurasian fossil mammoth sites and collections forming the basis of the discussion here are described by Soffer (1985, 1993; see also G. Haynes, 1989, 1991). The North American sample is taken from lists described by Saunders (1980) and G. Haynes (1991, 1999).

Table 2. Some Elephant Cullsites in Zimbabwe and the Number of Animals Killed at Each[a]

Elephant mass killsite	Numbers of elephants shot
Dina–Njekwa Road (1984)	74
Past Manga 1 (24 May 1984)	43
Past Manga 1 (29 July 1984)	24
Out of Shapi, Shumba, or Tshabema camps (1984)	21, 28, 18, 29, 16, 16, 32, 36, 40, 7, 36

[a]This small sampling of sites—out of dozens created in the early 1980s—lists only the sites most carefully examined for patterning in the bone accumulations.

4.2. *Mammuthus* Sites

Each individual mammoth site in Eurasia may possess unique features, but collectively they seem to share some important characteristics. The large sites occur in one of two kinds of settings. Some, such as Mezin, Mezhirich, Yudinovo, or Mal'ta, are located in wide river valleys of the Russian plains and Siberia, and contain densely clustered features of mammoth bones used to build shelters or house structures. The other setting is upland land surfaces above and away from floodplains, sometimes overlooking valleys and intersecting rivers. The mammoth bones in upland settings such as the Krakow–Spadzista Street site in Poland may be dump deposits. Both types of sites contain the bones of mammoths of all age groups. The bones may be similar or in different weathering stages, and may or may not bear cutmarks. Some assemblages may contain specimens that were gnawed by carnivores.

Table 5 lists prominent Eurasian (*Mammuthus primigenius*) and North American assemblages reported in the literature (*M. columbi/jeffersonii*). Not all of these have been studied directly, but from our literature review we were able to include general observations about the reported collections from the sites.

A quick comparison of Clovis sites and the Eurasian sample shows some interesting differences. Clovis sites appear to be kills—skeletons bedded with weaponry and butchery tools—and not dwellings, base camps, secondary processing, or dump sites. Eastern Europe's mammoth sites are situated on and within loess on upland slopes or large valley terraces, whereas the Clovis mammoth sites often are in smaller stream valleys, usually within channelfill deposits. Eurasian sites yield bones by the thousands, derived from dozens of individual mammoths, and the bones are often bedded with hundreds of stone artifacts, manufactured using blade technology. The Eurasian sites date to an extended interval, from around the

Table 3. Characterisitics of the Four Largest Sites in Zimbabwe Where Elephants Died en Masse from Noncultural Causes

Name of elephant die-off site	Approximate size of central or main locus	Total MNI (all taxa, all areas, as of 1995)	Number of different animal taxa represented by bones or carcasses
Shabi Shabi	2 areas, 200 × 200 m each	256	14
Nehimba	150 × 300 m	148	11
Lememba	200 × 300 m	50	12
Shakwanki	300 × 300 m	41	10

Table 4. Characteristics of Bone Deposits at Cultural and Noncultural Sites

Bone deposit variable	Characterisitics at kills (cultural)	Characteristics at die-offs (noncultural)
Bone representation	Limited bias at most sites	Limited bias at most sites
Bone weathering	Narrow range of stages	Wider range of stages
Carnivore gnaw-damage	Light to medium	Light or nonexistent
Mortality profile	Catastrophic, unbiased	Biased catastrophic, or attritional

Glacial Maximum to about 12,000 rcyrbp, whereas Clovis mammoth sites date consistently to around 11,000 rcyrbp. The mortality profiles for the Eurasian mammoths are Type A (see G. Haynes, 1991:216–218), a pattern that contains all age groups in proportions that are similar to those in a stable or healthy living population. The cumulative effect of whatever agency that killed the mammoths was nonselective. On the other hand, Clovis mammoth mortality profiles are either Type B or D, both appearing very selective, either as a result of differential natural mortality or of opportunistic serial predation on individual mammoths.

How do these observations relate to the mammoth-extinction debate? We suggest that the Clovis sites are indeed killsites, whereas the Eurasian sites appear to be habitation sites to which bones from kills and natural die-offs were transported. We do not make this suggestion as if it were novel or unexpected; we are here trying to show that not only is it an interesting working hypothesis (as it has been for years), but that when viewed in light of the actualistic studies of modern elephant-bone sites, it also has empirical support now, which was never available before.

The evidence that distinguishes real kills from fortuitously encountered carcasses that were only scavenged is cumulative or compounding—and not merely based on the presence of artifacts bedded with proboscidean skeletons. The variables of weathering stages, bone representation, carnivore utilization, and age profiles from the Clovis-age sites are similar in values to the variables recorded in African elephant killsites, and differ just enough from the values of these variables recorded in noncultural sites to suggest that Clovis sites may be kills, rather than the debris left from only scavenging events. This is a suggestion that may need more testing against the actualistic fit, but it is based on empirical generalizations and theoretical support.

5. Conclusions (Forward and Backward Searches)

We can come to a conclusion now from our forward search, which—as may be recalled—refers to a prediction of Clovis capabilities, based on archaeology, ethnographic analogy, and more analytically based theories about foraging behavior. Clovis foraging could have directly led to mammoth extinction under the conditions existing in North America at the end of the Pleistocene. The first conclusion we reach from comparing and synthesizing the different kinds of data and theory is that Clovis foragers did hunt (and probably also at times scavenged) mammoths and mastodonts. They were very likely highly ranked among preferred prey. These animals were not uniformly or randomly distributed in late Pleistocene environments, but rather were limited to patches of higher diversity. Clo-

Table 5. Mammoth Sites Whose Bone Collections Were Either Directly Examined or Studied in the Available Literature, along with General Observations about Their Characteristics

M. primigenius sites

Central Europe: Vogelherd, Dolni Vestonice, Predmosti, Pavlov, Milovice, Krakow–Spadzistra Street

Russian river valleys and Russian plain: Voronovitsa, Berdyzh, Kirillovskaya, Radomylshl, Mezin, Pushkari, Mezhirich, Timonovka, Yudinovo, Gontsy, Dobranichevka, Borschevo, Kostenki 11 (Anosovka 2), Kostenki 2, Byzovaya

Siberia: Mal'ta, Berelekh

General observations, based on the literature and in a few cases actual study of the collections: There is no clear trend among all of the sites, but the bones in some sites were transported from kills and/or natural deaths, making any potential patterns unclear.

Bone representations	Biased (sometimes very limited bias)
Bone weathering	Mixed stages in some, narrow range of stages in others
Carnivore gnawing	Medium to light
Mortality profile	Catastrophic, or biased catastrophic

M. columbi/jeffersonii sites

Name of site	MNI (mammoth)
Blackwater Draw, NM	13
Colby, WY	7
Dent, CO	14
Domebo, OK	1
Escapule, AZ	1
Lange-Ferguson, SD	2
Lehner, AZ	13
Miami, TX	5
Murray Springs, AZ	2
Naco, AZ	1
Angus, NE	1
Duewall-Newberry, TX	1
Dutton, CO	1+(?)
Lubbock Lake, TX	1+(?)
Rawlins, WY	1

General observations, based on the literature and in a few cases actual study of the collections: Some sites appear quite likely to be kills, although the possibility remains that noncultural deaths also contributed to some or all of the sites.

Bone representations	Limited bias
Bone weathering	Narrow range of stages, but not well reported
Carnivore gnawing	Not well reported—light or nonexistent probably
Mortality profile	Biased catastrophic at larger sites

vis foragers directed themselves to these patches, encountered stressed mammoths or mastodonts, and killed them.

Based on the behavior of modern elephants undergoing environmental stresses in Africa, it seems likely that Clovis foragers searched for their preferred prey by following clearly marked mammoth and mastodont trails to refuge patches (or "fixed points"), if the foragers behaved as predicted from studies of similar (modern) refuge patches (G. Haynes, 1991, 1999). Winterhalder's (1977:567, 1981) work with Cree Indian foraging practices provides some support for this position. Among the Cree, prey mobility increases encounter rate by allowing for easier detection of tracks.

Paleoindian foragers probably stuck to such patches through drought years, and explored away from them whenever possible, perhaps by using game trails and drainage divides as markers and return routes. When all of the mammoths or mastodonts had abandoned the refuges, or when all had been killed, another such refuge could be found rapidly—perhaps requiring a trek of several days to several weeks, which is not beyond the capacity of experienced foragers. Other large mammals most likely also were killed whenever encountered.

When we do a backward search—that is, an examination of the conditions that existed just before mammoths became extinct—we can see that die-offs and kill-offs of mammoths (and probably also mastodonts) were just as likely to have been caused by human hunting alone as by a combination of climate-stress and hunting pressures. The paleoenvironmental conditions at the end of the Pleistocene encouraged mammoths and mastodonts to aggregate in locales where water, food, seasonal weather extremes, and other variables were tolerable (or optimal), and this tethering worked to the advantage of foragers seeking high-ranked prey animals to procure.

References

Agenbroad, L. D. 1984. New World mammoth distribution, in: P. S. Martin and R. G. Klein (eds.), *Quaternary Extinctions: A Prehistoric Revolution,* pp. 90–108. University of Arizona Press, Tucson.

Alvard, M. 1995. Intraspecific prey choice by Amazonian hunters. *Curr. Anthropol.* **36**:789–818.

Averianov, A. O. 1992. Istori nakhodki dvukh skeletov mamonta bliz s. Kamchoe Ust'e, in: I. S. Muraviyov and M. G. Soludukho (eds.), *Volzhskaya fauna pleistotsenovykh mlekopitayushchikh v geologo-mineralogich-eshom musee kazanskogo universiteta,* pp. 54–75. Izdatel'stvo kazanskogo Universiteta, Kazan.

Bayham, F. E. 1982. A diachronic analysis of prehistoric animal exploitation at Ventana Cave. Unpublished Ph.D. dissertation, Department of Anthropology, Arizona State University, Tempe.

Behrensmeyer, A. K. 1978. Taphonomic and ecologic information from bone weathering. *Paleobiology* **4**:150–162.

Bettinger, R. L. 1991. *Hunter-Gatherers: Archaeological and Evolutionary Theory.* Plenum Press, New York.

Binford, L. R. 1978. *Nunamiut Ethnoarchaeology.* Academic Press, New York.

Bonnichsen, R., and Turnmire, K. L. (eds.). 1991. *Clovis: Origins and Adaptations.* Center for the Study of the First Americans, Oregon State University, Corvallis.

Broughton, J. M. 1994. Late Holocene resource intensification in the Sacramento Valley, California: the vertebrate evidence. *J. Archaeol. Sci.* **21**:501–514.

Charnov, E. L. 1976. Optimal foraging: the marginal value theorem. *Theor. Popul. Biol.* **9**;129–136.

Custer, J. F. 1996. Prehistoric cultures of eastern Pennsylvania. *Ethnology Monographs* 12. Pennsylvania Historical and Museum Commission, Harrisburg.

Dent, R. J., Jr. 1995. *Chesapeake Prehistory: Old Traditions, New Directions.* Plenum Press, New York.

Dincauze, D. F., and Curran, M. A. 1983. Paleoindians as generalists: an ecological perspective. Paper presented at the 48th Annual Meeting of the Society for American Archaeology, Pittsburgh.

Ewer, R. F. 1968. *Ethology of Animals.* Logos Press, London.

Foronova, I. V., and Zudin, A. N. 1999. The structure of the lineage *Archidiskidon–Mammuthus* in Eurasia and peculiarities of its evolution, in: G. Baryshnikov, G. Haynes and J. Klimowicz (eds.), *Mammoths and the Mammoth Fauna: Studies of an Extinct Ecosystem. Deinsea,* in press.

Giraldeau, L. A. 1997. The ecology of information use, in: J. R. Krebs and N. B. Davies (eds.), *Behavioural Ecology: An Evolutionary Approach,* 4th ed., pp. 42–68. Blackwell, Oxford.

Graham, R. W. 1986. Plant–animal interactions and Pleistocene extinctions, in: D. K. Elliott (ed.), *Dynamics of Extinction,* pp. 131–154. Wiley: New York.

Graham, R. W., and Lundelius, E. L., Jr. 1984. Coevolutionary disequilibrium and Pleistocene extinctions, in: P. S. Martin and R. G. Klein (eds.), *Quaternary Extinctions: A Prehistoric Revolution,* pp. 223–249. University of Arizona Press, Tucson.

Grayson, D. K. 1993. *The Desert's Past: A Natural Prehistory of the Great Basin.* Smithsonian Institution Press, Washington, D.C.

Griffiths, D. 1975. Prey availability and the food predators. *Ecology* **56**:1209–1214.

Guilday, J. E. 1984. Pleistocene extinction and environmental change: case study of the Appalachians, in: P. S. Martin and R. G. Klein (eds.), *Quaternary Extinctions: A Prehistoric Revolution,* pp. 250–258. University of Arizona Press, Tucson.

Guthrie, R. D. 1990. *Frozen Fauna of the Mammoth Steppe: The Story of Blue Babe.* University of Chicago Press, Chicago.

Hanks, T. C., Jr., 1997. Imperfect science: uncertainty, diversity, and experts. *Eos* **78**:369, 373, 377.

Haynes, C. V., Jr. 1969. The earliest Americans. *Science* **166**:709–715.

Haynes, C. V. 1984. Stratigraphy and late Pleistocene extinction in the United States, in: P. S. Martin and R. G. Klein (eds.), *Quaternary Extinctions: A Prehistoric Revolution,* pp. 345–353. University of Arizona Press, Tucson.

Haynes, C. V., Jr. 1991. Geoarchaeological and paleohydrological evidence for a Clovis-age drought in North America and its bearing on extinction. *Quat. Res.* **35**:435–450.

Haynes, G. 1982. Utilization and skeletal disturbances of North American prey carcasses. *Arctic* **35**:266–281.

Haynes, G. 1989. Late Pleistocene mammoth utilization in northern Eurasia and North America. *Archaeozoologia* **3**:81–108.

Haynes, G. 1991. *Mammoths, Mastodonts, and Elephants: Biology, Behavior, and the Fossil Record.* Cambridge University Press, London.

Haynes, G. 1999. The role of mammoths in rapid Clovis dispersal, in: G. Baryshnikov, G. Haynes and J. Klimowicz (eds.), *Mammoths and the Mammoth Fauna: Studies of an Extinct Ecosystem. Deinsea,* in press.

Hibbard, C. W. 1960. An interpretation of Pliocene and Pleistocene climates in North America. *Annu. Rep. Mich. Acad. Sci. Arts Lett.* **62**:5–30.

Holliday, V. T. 1997. *Paleoindian Geoarchaeology of the Southern High Plains.* University of Texas Press, Austin.

Kay, C. 1994. Aboriginal overkill: the role of Native Americans in structuring western ecosystems. *Hum. Nature* **5**:359–398.

Kaplan, H., and Hill, K. 1992. The evolutionary ecology of food acquisition, in E. Smith and B. Winterhalder (eds.), *Evolutionary Ecology and Human Behavior,* pp. 167–202. Aldine de Gruyter, Hawthorne, New York.

King, J. E., and Saunders, J. J. 1984. Environmental insularity and the extinction of the American mastodont, in: P. S. Martin and R. G. Klein (eds.), *Quaternary Extinctions: A Prehistoric Revolution,* pp. 315–339. University of Arizona Press, Tucson.

MacArthur, R. H., and Pianka, E. R. 1966. On optimal use of a patchy environment. *Am. Nat.* **100**:603–609.

MacPhee, R. D. E., and Marx, P. A. 1997. The 40,000-year plague: humans, hyperdisease, and first-contact extinctions, in: S. M. Goodman and B. D. Patterson (eds.), *Natural Change and Human Impact in Madagascar,* pp. 169–217. Smithsonian Institution Press, Washington, D.C.

Martin, J. F. 1983. Optimal foraging theory: a review of some models and their applications. *Am. Anthropol.* **85**:612–629.

Martin, P. S. 1984. Prehistoric overkill: the global model, in: P. S. Martin and R. G. Klein (eds.), *Quaternary Extinctions: A Prehistoric Revolution,* pp. 354–403. University of Arizona Press, Tucson.

Martin, P. S., and Klein, R. G., (eds.). 1984. *Quaternary Extinctions: A Prehistoric Revolution.* University of Arizona Press, Tucson.

Mason, R. J. 1962. The Paleo-Indian tradition in eastern North America. *Curr. Anthropol.* **3**:227–246.

Meltzer, D. J. 1988. Lake Pleistocene human adaptations in eastern North America. *J. World Prehist.* **2**:1–52.

Meltzer, D. J. 1989. Why don't we know when the first people came to North America? *Am. Antiq.* **54**:471–490.

Meltzer, D. J. 1993. The Pleistocene peopling of the Americas. *Evol. Anthropol.* **1**:57–169.

Meltzer, D. J. 1996. A northern perspective on the peopling of the Americas. *Fundhamentos* Volume I (*Proceedings of the International Meeting on the Peopling of the Americas, Sao Raimundo Nonato, Piaui, Brasil* [1993]), pp. 241–258.

Meltzer, D. J., Grayson, D. K., Ardila, G., Barker, A. W., Dincauze, D. F., Haynes, C. V., Jr., Mena, F., Núñez, L., and Stanford, D. J. 1997. On the Pleistocene antiquity of Monte Verde, southern Chile. *Am. Antiq.* **62**:659–663.

Oaten, A. 1977. Optimal foraging in patches: a case for stochasticity. *Theor. Popul. Biol.* **12**:263–285.

Saunders, J. J. 1980. A model for man–mammoth relationships in Late Pleistocene North America. *Can. J. Anthropol.* **1**:87–98.

Saunders, J. J., and Daeschler, E. B. 1994. Descriptive analyses and taphonomical observations of culturally-modified mammoths excavated at the "Gravel Pit," near Clovis, New Mexico in 1936. *Proc. Acad. Nat. Sci. Philadelphia* **145**:1–28.

Schoener, T. W. 1979. Generality of the size–distance relation in models of optimal feeding. *Am. Nat.* **114**:902–914.

Schoener, T. W. 1987. A brief history of optimal foraging ecology, in: A. C. Kamil, J. R. Krebs, and H. R. Pulliam (eds.), *Foraging Behavior,* pp. 5–67. Plenum Press, New York.

Smith, E. A. 1991. *Inujjuamiut Foraging Strategies.* Aldine de Gruyter, Hawthorne, NY.

Smith, E. A., and Winterhalder, B. (eds.). 1992. *Evolutionary Ecology and Human Behavior.* Aldine de Gruyter, New York.

Soffer, O. 1985. *The Upper Paleolithic of the Central Russian Plain.* Academic Press, Orlando.

Soffer, O. 1993. Upper Paleolithic adaptations in central and eastern Europe and man–mammoth interactions, in: O. Soffer and N. D. Praslov (eds.), *From Kostenki to Clovis: Upper Paleolithic Paleo-Indian Adaptations,* pp. 31–49. Plenum Press, New York.

Taylor, R. E., Haynes, C. V., Jr., and Stuiver, M. 1996. Clovis and Folsom age estimates: stratigraphic context and radiocarbon calibration. *Antiquity* **70**:515–525.

Willey, G. R. 1966. *An Introduction to American Archaeology: Volume One. North and Middle America.* Prentice–Hall, Englewood Cliffs, N.J.

Winterhalder, B. 1977. Foraging strategies and adaptations of the Boreal Forest Cree: an evaluation of theory and models from evolutionary ecology. Unpublished Ph.D. dissertation, Cornell University.

Winterhalder, B. 1981. Foraging strategies in the boreal environment: an analysis of Cree hunting and gathering, in: B. Winterhalder and E. Smith (eds.), *Hunter-Gatherer Foraging Strategies,* pp. 66–98. University of Chicago Press, Chicago.

CHAPTER 5

A Comparison of Methods for the Probabilistic Determination of Vertebrate Extinction Chronologies

DONALD A. McFARLANE

1. Introduction

Much effort has been directed toward determining the extinction dates of late Quaternary megafaunal mammals, usually in the context of testing models of anthropogenic "overkill" or terminal Pleistocene climatic change. These studies have typically relied on searching for the youngest reliable date (e.g., Long and Martin, 1974), or on matching the mode of a sequence of dates to the time of presumed anthropogenic impact (e.g., Martin, 1986). Little attention appears to have been paid to the determination of vertebrate extinctions dates by probabilistic methods, although a number of authors have treated the related problem of determining invertebrate fossil taxon ranges in biostratigraphical analysis (e.g., Hay, 1972; Strauss and Sadler, 1989; Gilinsky and Good, 1991), and Badgley (1990) has addressed the importance of statistical evaluation of the relationship between sample size and apparent extinctions in the Eocene vertebrate record. Here, I use Monte Carlo simulation to compare the performance of two published probabilistic techniques and a third technique developed here that is based on median stratigraphic gap length in estimating the 95% confidence interval relative to a known extinction date. When applied to a sample of the published record of radiocarbon dates, these techniques indicate that the current database is inadequate to constrain the "late Pleistocene" megafaunal extinction event(s) to the narrow chronological window required by the "blitzkrieg" anthropogenic overkill model (but see Alroy, this volume; Martin and Steadman, this volume). Although this result is not disconfirmatory of the blitzkrieg model, it does argue for the necessity of continuing to improve the data base of published radiocarbon dates.

DONALD A. McFARLANE • Joint Science Department, The Claremont Colleges, Claremont, California 91711.
Extinctions in Near Time, edited by MacPhee. Kluwer Academic /Plenum Publishers, New York, 1999.

2. Methods

Consider a sample of N dated fossils, each collected randomly from the total uncollected fossil population. (For the present purpose, it is assumed that the one-sigma errors of these dates are comparable to one another, small compared with the total time span of the database, and can therefore be assumed to cancel out in probabilistic considerations; a list of abbreviations appears in Table 1.) When ordered with respect to increasing age relative to D_{min}, the youngest fossil, the sample yields $N - 1$ time intervals or gaps. Springer and Lilje (1988) have argued that the frequency distribution of such randomly derived interval lengths will be exponential, with $N - 1$ measurable gaps and two unmeasurable gaps (between the youngest date and the true extinction date, and between the oldest date and the stratigraphic "origination" date). If the sampling procedure and stratigraphic distribution of the fossils are homogeneous, then the time interval that one must expect to traverse above the youngest date and below the oldest date so as to include the true termination points is equivalent to the confidence interval of the distribution, given by Springer and Lilje (1988) as

$$x = \ln P(x)/(-m) \tag{1}$$

where P is the specified confidence level for x, the confidence interval, and m is the mean number of gaps per year. Thus, given the maximum time span of an ordered set of N dates, the lower limit of the 95% confidence interval for the extinction date, D_{ext}, can be found as

$$D_{ext} = D_{min} - \{[\ln(0.05)]/ - [(N - 1)/D_{max} - D_{min}]\} \tag{2}$$

where ln(0.05) defines the appropriate proportions of the exponential distribution.

A new approach, also utilizing the exponential distribution, and referred to here as the "median gap" method, can be derived from the fact that the median gap length, i, is equivalent to the "half-life" of the exponential curve. Therefore, one-half of the gaps will be $< i$, three-quarters will be less than $2i$, and so forth. It follows from the proportions of the exponential curve that 93.74% of the gaps will be less than $4i$, and 95% will be less than $4.32i$. Consequently, the 95% confidence interval expected to include the gap between the youngest date and the true extinction date must have its lower limit at

$$D_{min} - 4.32i \tag{3}$$

More recently, Strauss and Sadler (1989) have questioned the validity of the exponential model on the grounds that exponential variables have infinite ranges, whereas gap

Table 1. Abbreviations Used in the Text

D_{min}	Youngest date in the series
D_{ext}	Extinction date for the taxon
D_{max}	Oldest date in the series
D_{hyp}	Hypothetical extinction date to be tested for
N	Number of dated fossils in the series
P	Confidence limit
x	Individual datum (dated fossil)

lengths are constrained within finite limits by their endpoints. These authors propose that a Dirichlet distribution is more appropriate, and derive the 95% confidence interval for the extinction date as

$$D_{min} - a(D_{max} - D_{min}) \tag{4}$$

where a is a parameter that varies with sample size according to the relationship

$$a = (1 - p)^{-1/(n-1)} - 1 \tag{5}$$

where p is the desired confidence, i.e., 0.95.

A simple Monte Carlo simulation can be used to show that the distribution of gap lengths for a realistic time span (40,000 yrbp–10,000 yrbp) and sample size (n = 6) closely approximates the exponential, except at the extremes (Fig. 1). These deficits of very long and very short gaps (i.e., less than the resolution of 1 year, and greater than the time span of 30,000 yrbp) do not significantly influence the median gap length, and confirm the utility of Equations (2) and (3).

The three methods of confidence-interval estimation discussed provide comparable but different values when applied to the same data set. Their performance can be compared using a Monte Carlo simulation in which n random dates are generated within the range D_{max} to D_{min}, where D_{min} is equivalent to the "true" extinction age. Figure 2 illustrates the performance data for a realistic sample size of n = 6.

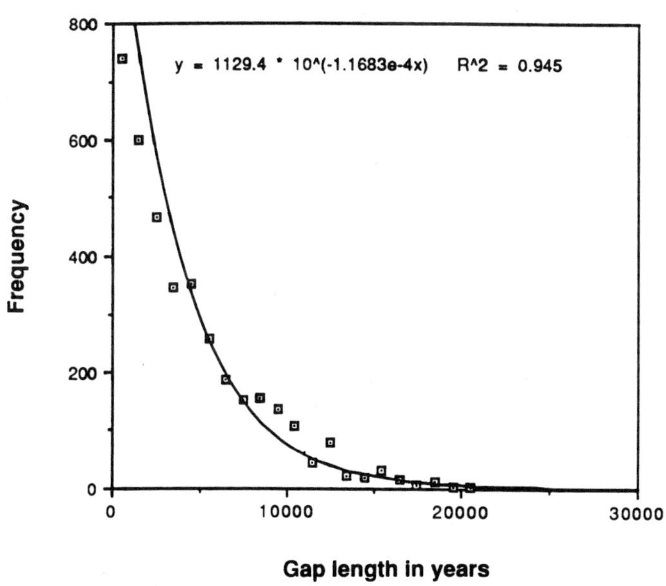

FIGURE 1. Frequency of gap lengths derived from Monte Carlo simulation of six dates drawn at random from a 30,000-year time interval (n = 3750).

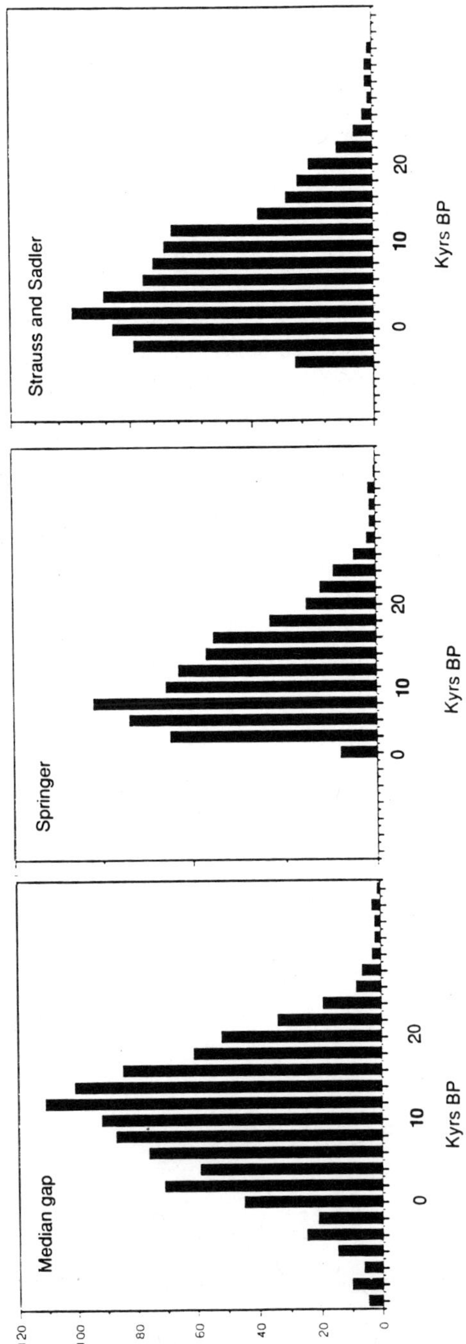

FIGURE 2. Performance of three statistical estimators of extinction age by Monte Carlo simulation.

3. Results

All three techniques provide reasonable estimates of the 95% lower confidence limit. The Springer–Lilje and Strauss–Sadler methods produce distributions of estimates that are asymmetric and right-skewed, whereas the median gap method yields a normal distribution of estimates. For small sample sizes, the Springer–Lilje and median gap methods tend to underestimate the 95% confidence interval, whereas the Strauss–Sadler method overestimates it (Fig. 2).

Long and Martin (1974) have provided a detailed summary of published radiocarbon dates for six ground sloth (*Nothrotheriops shastense*) sites in the American Southwest. Updated with additional dates from Meltzer and Mead (1985) and Beck (1996), this data set provides six "terminal" (i.e., stratigraphically uppermost) dates of range 9840–11,360 yrbp, and a total of 28 finite dates of range 9840–36,200 yrbp. Table 2 provides probabilistic estimates for the extinction date of *N. shastense* in the Southwest using the Springer–Lilje, Strauss–Sadler, and median gap methods.

It will be apparent that whereas the median gap method provides the most conservative estimate using the six terminal dates only, it provides the narrowest confidence interval using the full data set. Because the "true" extinction date for *N. shastense* has been convincingly argued to be greater than 10,000 yrbp (noting the 160-year standard deviation on the 9840 years date; Long and Martin, 1974; Martin and Steadman, this volume), the median gap method provides the most congruent result. These differences result from the fact that the full data set violates the requirement that there be an absence of serial correlation in the gap lengths (i.e., older gaps are longer than younger gaps) presumably because fewer "old" specimens have been dated by researchers interested in finding terminal dates. A correlation coefficient calculated for the first 24 chronologically ordered gaps reveals no serial correlation ($r = -0.128$; $P > 0.5$), and a recomputation of 95% confidence intervals using this culled data set yields a remarkable congruence between the three methods of calculation (Table 2). Similar analyses of the published radiocarbon record of six late Pleistocene north American megafaunal mammals are compared in Table 3, and Eurasian taxa in Table 4.

The median gap method also provides a simple means to test specific hypotheses of extinction chronologies. The probability that a species persisted to a hypothetical date, D_{hyp}, is equivalent to the probability of finding a gap length of $D_{min} - D_{hyp}$, which is given by

$$-\ln(p) = \{(D_{min} - D_{hyp})/I\}/\ln(0.5)$$

Table 2. Probabilistic Extinction Age Estimates (yrbp) for *Nothrotheriops* by Three Methods[a]

	Median gap method	Springer–Lilje method	Strauss–Sadler method
Terminal dates only; $n = 7$	9409	9374	9177
All dates; $n = 29$	9603	7236	7080
Data set culled to remove serial correlation; $n = 25$	9663	9718	9696

[a]Data from Long and Martin (1974) and Meltzer and Mead (1985).

Table 3. Probabilistic Extinction Intervals for Six North American Pleistocene Taxa at the 95% Confidence Level[a]

	Median gap method	Springer–Lilje method	Strauss–Sadler method	95% confidence interval, median gap method
Mammut; n = 6	6809	1587	—	6749–7141
	(6945, n^* = 5)	(7434, n^* = 5)	(5986, n^* = 5)	
Mammuthus; n = 16	10,161	10,134	10,089	9475–10,847
Nothrotheriops; n = 29	9603	7133	6965	9173–10,153
	(9663, n^* = 25)	(9718, n^* = 25)	(9696, n^* = 25)	
Smilodon; n = 40	9186	9258	9184	8647–9725
Equus; n = 5	7291	1292	—	6409–8173
Camelops; n = 6	8210	8303	7539	7524–8896

[a]Dates from Meltzer and Mead (1985) all ranking "8" or "9" on Meltzer and Mead's "2–9" scale and regarded by them as "reliable." Dates in radiocarbon years before present.
*Sample sizes trimmed to remove serial correlation.

where I is the median interval. Thus, the probability that *N. shastense* persisted in the Southwest until 9000 yrbp, assuming a median gap of 110 years (n = 24, culled data set) and a minimum dated age of 9840, is

$$-\ln(p) = (9840/110)/\ln(0.5) = -5.293$$

or $p = 0.005$.

4. Discussion

The Springer–Lilje, Strauss–Sadler, and median gap methods of probabilistic extinction-date determination all provide reliable results for data sets having a high density of dates relative to the age range. The median gap method is the least sensitive to problems of serial correlation and the Strauss–Sadler method performs best in circumstances of low date density relative to the age range.

The North American late Quaternary extinct megafaunal record differs from the northern European record in the scarcity of its archaeological associations. As Martin (1986; Martin and Steadman, this volume) has pointed out, this pattern argues against a gradual depletion of prey species by a slowly increasing population of hunters. Instead, a viable anthropogenic extinction model appears to require a very rapid, intense predatory impact dubbed the "blitzkrieg" model (Mosimann and Martin, 1975; MacPhee and Marx, 1997).

Evidence for the peopling of North America has accumulated rapidly in the last few years. As recently as 1993, a review by Hoffecker *et al.* (1993) continued to affirm the absence of humans (excepting Beringia) before 12,000 yrbp, with no firm dates earlier than 11,200 yrbp. However, a more recent evaluation of the controversial Monte Verde site in Chile (Dillehay, 1997) has pushed that date back to 12,500 yrbp and—by inference—points to an original colonization of North America by 20,000 yrbp or earlier (Meltzer, 1997).

Table 4. Probabilistic Extinction Intervals for Six Eurasian Pleistocene Taxa at the 95% Confidence Level[a]

	Median gap method	Springer–Lilje method	Strauss–Sadler method	95% confidence interval, median gap method
Mammuthus (European Russia); $n = 32$	8052	7778	7677	7542–8562
Mammuthus (Siberia); $n = 55$	8179	4865	4728	8061–8197
Coelodonta (Europe and Siberia); $n = 30$	21,072	"modern"	"modern"	20,487–22,917
Megaloceros (Europe); $n = 10$	9659	9092	8809	9267–10,051
Mammuthus (Wrangel Is.); $n = 6$ (Holocene)	1288	2069	1351	1238–1338
Ovibos (Old World); $n = 16$	523	"modern"	"modern"	2714–3086

[a]Dates from Stuart (1991) and Vartanyan *et al.* (1993) in yrbp.

The intensity of the North American megafaunal extinction has been attributed to the supposed naiveté of the fauna, having evolved in the absence of intelligent bipedal predators. The analogy to naive oceanic island faunas has always been weak—insular faunas evolving in the complete absence of terrestrial predators, such as those of the Galápagos or the Mascarenes, are hardly comparable to North American megafaunal herbivores evolving in synergy with formidable, large-bodied, and intelligent carnivores. If humans entered sub-Beringian North America before 12,000 yrbp, then they had no discernible impact on the Pleistocene megafauna and we must assume that overkill—if it occurred at all—was not initiated until the appearance of Clovis technology around 11,200–10,900 yrbp (Hoffecker *et al.,* 1993). It follows that congruence with the blitzkrieg model requires that megafaunal extinction dates be constrained to dates no later than about 10,500 yrbp, and no earlier than about 11,200 yrbp.

Median-gap analysis of the radiocarbon record of six extinct North American megafaunal mammal taxa for which a series of reliable dates are available (Table 3) reveals that only *Mammuthus* has an extinction date 95% confidence interval that overlaps the blitzkrieg window, taken here to be 10,500–11,500 yrbp. In the extreme case of *Mammut,* the possible extinction age postdates the blitzkrieg window by more than three millennia. In the case of the much debated extinction of the American ground sloth *Nothrotheriops,* the probability of overlap between the 95% extinction confidence interval and the blitzkrieg window is less than 0.0004.

The earlier humans were present in North America, the more difficult it becomes to explain why an overkill event should have been so delayed in its initiation and so tightly constrained in its conclusion. If overkill resulted from the spread of a new technology and not a wave of novel colonists, then it should have accompanied the appearance of that same technology in Eurasia. As Stuart (1991, this volume) has pointed out, sophisticated hunting technology and behavior appeared in northwestern Europe at least as early as 35,000 yrbp—more than 20,000 years before the Eurasian megafaunal extinction crisis.

Indeed, Stuart's work has been extremely important in emphasizing that the Eurasian crisis has been greatly neglected in the extinction debate. Contrary to one oft-expressed view (e.g., Alroy, this volume), the Eurasian extinction event is not qualitatively different from the North American event. Within the limits of the available dates, the Eurasian last-occurrence data cluster within the same 13,000–10,000 yrbp time interval as do the North American data. Indeed, it seems that the only really unique aspect of the North American event is its severity (~ 73% of the genera of megafaunal herbivores, versus 28% in Eurasia; Stuart, 1991). Given that Alroy (this volume) has shown that nonanthropogenic extinction events increasingly select for large body size as the severity of the event increases, and noting that the North American herbivore megafauna was 22% more diverse than was the Eurasian, in a 30% smaller land area, the greater impact of the terminal Pleistocene extinction in North America seems increasingly unremarkable.

Given the fundamental differences in the human ecology of late Pleistocene North America and Eurasia, the only factors of which I am aware that are common to both regions are global climate change and, conceivably, the novel appearance of catastrophic disease (cf. MacPhee and Marx, 1997). Alroy (this volume) has criticized climate-based extinction hypotheses on the grounds that the Pleistocene (and indeed, the Cenozoic) has seen many prior dramatic climatic changes that failed to precipitate megafaunal extinctions. However, this argument neglects the fact that there is currently no evidence that any of the earlier Pleistocene deglaciations (much less broad Cenozoic climatic changes) exhibited the rapidity and violent reversals of the terminal Wisconsinan–Younger Dryas event. Recent work (e.g., Islebe *et al.*, 1995) measures the time scale of the Younger Dryas reversal in decades, not centuries—a rapidity that probably negates even behavioral, let alone evolutionary, adjustments in the afflicted fauna.

Work on the timing of the peopling of North America and an improving resolution of the time span of the North American and Eurasian megafaunal extinction events both point to an extended coexistence of humans and megafaunal mammals prior to a rapid and catastrophic extinction pulse. In my view, the current evidence is most parsimonious with a climate-driven extinction event linked to the unusually violent deglaciation/Younger Dryas reversal/deglaciation episode that characterized the latest Pleistocene in both North America and Eurasia. However, the issue cannot be resolved until a sufficient density of dates is available to (1) constrain the extinction event to either the Clovis interval, or a specific climatic event such as the Younger Dryas, or (2) demonstrate that the extinctions were smeared over a much more extended interval.

In conclusion, the application of statistical inference to the probabilistic determination of late Quaternary megafaunal extinction chronologies has been neglected. Results currently available from the published radiocarbon record imply that these extinctions may have been too broadly scattered through latest Pleistocene and earliest Holocene time to satisfy the strict criteria set by Martin for the blitzkrieg overkill hypothesis, but also point to the critical need to increase the density of high-precision dates for these species. Further insights await the development of a much tighter sequence of high-precision, last-occurrence dates for individual megafaunal species, and the interpretation of these dates with proper attention to statistical confidence limits. Interestingly, work reported by Graham *et al.* (1997) suggests that the North American extinction window may already have been narrowed to as little as 400 years. With the publication of the supporting data set, it may now be possible to evaluate the duration of the megafaunal extinction crisis with statistical confidence, and in doing so begin to constrain the possible causes.

References

Badgley, C. 1990. A statistical assessment of last appearances in the Eocene record of mammals. *Geol. Soc. Am. Spec. Pap.* **243**:153–167.

Beck, M. W. 1996. On discerning the cause of late Pleistocene megafaunal extinctions. *Paleobiology* **22**:91–103.

Dillehay, T. (ed.). 1997. *Monte Verde: A late Pleistocene settlement in Chile: Volume 2. The Archaeological Context.* Smithsonian Institution Press, Washington, D.C.

Gilinsky, N. L., and Good, I. J. 1991. Probabilities of origination, persistence and extinction of families of marine invertebrate life. *Paleobiology* **17**:145–166.

Graham, R. W., Stafford, T. W., and Semken, H. A. 1997. Pleistocene extinctions: chronology, non-analog communities, and environmental change. Abstract with program, "Humans and other catastrophes: explaining past extinctions and the extinction process," Spring Symposium, American Museum of Natural History, New York.

Hay, W. W. 1972. Probabilistic stratigraphy. *Eclogae Geol. Helv.* **65**:255–266.

Hoffecker, J. F., Powers, W. R., and Goebel, T. 1993. The colonization of Beringia and the peopling of the New World. *Science* **259**:46–53.

Islebe, G. A., Hooghiemstra, H., and van der Borg, K. 1995. A cooling event during the Younger Dryas Chron in Costa Rica. *Palaeogeogr. Palaeoclimatol. Palaeoecol.* **117**:73–80.

Long, A., and Martin, P. S. 1974. Death of American ground sloths. *Science* **186**:638–640.

MacPhee, R. D. E., and Marx, P. A. 1997. The 40,000-year plague: humans, hyperdisease, and first-contact extinctions, in: S. Goodman and B. Patterson (eds.), *Natural Change and Human Impact in Madagascar,* pp. 169–217. Smithsonian Institution Press, Washington, D.C.

Martin, P. S. 1986. Refuting late Pleistocene extinction models, in: D. K. Elliot (ed.), *Dynamics of Extinction,* pp. 106–130. Wiley, New York.

Meltzer, D. J. 1997. Monte Verde and the Pleistocene peopling of the Americas. *Science* **276**:754–755.

Meltzer, D. J., and Mead, J. I. 1985. Dating late Pleistocene extinctions: Theoretical issues, analytical bias, and substantive results, in: J. I. Mead and D. J. Meltzer (eds.), *Environment and Extinction: Man in Late Glacial North America,* pp. 145–173. Center for the Study of Early Man, University of Maine, Orono.

Mosimann, J. E., and Martin, P. S. 1975. Simulating overkill by paleoindians. *Am. Sci.* **63**:304–313.

Springer, M., and Lilje, A. 1988. Biostratigraphy and gap analysis: the expected sequence of biostratigraphic events. *J. Geol.* **96**:228–236.

Strauss, D., and Sadler, P. M. 1989. Classical confidence intervals and Bayesian probability estimates for ends of local taxon ranges. *Math. Geol.* **21**:411–427.

Stuart, A. J. 1991. Mammalian extinctions in the late Pleistocene of northern Eurasia and North America. *Biol. Rev.* **66**:453–562.

Vartanyan, S. L., Garutt, V. E., and Sher, A. V. 1993. Holocene dwarf mammoths from Wrangel Island in the Siberian Arctic. *Nature* **362**:337–340.

CHAPTER 6

Putting North America's End-Pleistocene Megafaunal Extinction in Context

Large-Scale Analyses of Spatial Patterns, Extinction Rates, and Size Distributions

JOHN ALROY

1. Introduction

After many decades of debate, the North American end-Pleistocene megafaunal mass extinction remains a lightning rod of controversy. The extraordinarily divergent opinions expressed in this volume show that no resolution is in sight. My own position is quite heterodox: I believe that the overkill hypothesis, at least in general terms, already has been "proven" as thoroughly as any historical hypothesis can be. All of the key evidence was available years ago, and all of it firmly refutes competing, ecologically oriented hypotheses (Martin, 1967, 1984). The event's timing, rapidity, selectivity, and geographic pattern all make good sense according to the anthropogenic model, and no sense at all otherwise. To my eyes, this assessment is so clear-cut that further "tests" (e.g., Beck, 1996) are not really necessary.

Despite this, continuing debate does have the useful side effect of encouraging researchers to ask new questions and rethink old ones, which leads to interesting discoveries about the nature of this anthropogenic extinction, and of extinction in general. Therefore, instead of rehashing earlier work this chapter will focus on three tests of the competing hypotheses that involve new data sets and analytical methods. These have to do with the geographic distribution of the extinction's survivors; the relative intensity of the extinction, as compared with earlier, natural extinctions; and the differential extinction of large-sized mammals, again as compared with the earlier Cenozoic fossil record. The first issue will be addressed using a combination of FAUNMAP data and my own data for earlier intervals of the Pleistocene; the others will be addressed using faunal and body mass data for the entire Cenozoic.

The competing predictions can be summarized very simply. The anthropogenic model predicts that (1) surviving mammal species will show no particular pattern of spatial re-

JOHN ALROY • Department of Paleobiology, Smithsonian Institution, Washington, DC 20560. *Present address:* National Center for Ecological Analysis and Synthesis, University of California, Santa Barbara, California 93101.
Extinctions in Near Time, edited by MacPhee. Kluwer Academic/Plenum Publishers, New York, 1999.

assembly because ecological interactions were not responsible for the event; (2) the extinction event will appear to have been relatively extreme compared with earlier, natural events; and (3) body mass selectivity will appear to have been far weaker in these earlier events. On the other hand, ecological models, such as those invoking disease (MacPhee and Marx, 1997), climate, and/or vegetational change (e.g., Graham and Lundelius, 1984; Guthrie, 1984), predict that (1) surviving mammal species will show radically different spatial arrangements in the Holocene because mammals are very sensitive to major environmental changes; (2) because massive climate variability is a natural feature of the entire Quaternary (e.g., Dansgaard *et al.,* 1993; Shackleton, 1995; Stager and Mayewski, 1997; Taylor *et al.,* 1997), and therefore presumably of many intervals within the Cenozoic, the extinction event will seem to have been no more or less extreme than many earlier, also naturally caused events; and (3) body mass selectivity will appear to have been similar in these earlier events. Note that the disease model (MacPhee and Marx, 1997) falls under the "ecological" category because it invokes a class of killing agents that surely were introduced to the continent throughout the Cenozoic—for mammalian diseases, there was no "first contact," but instead scores, or even hundreds of "contacts" created by repeated mammalian immigration events (e.g., see Woodburne and Swisher, 1995; Alroy, 1988a). Therefore, the disease model should be held to predict that the extinction will have the same "natural" stamp that is predicted by models invoking climate and vegetation change.

As will be shown in the following sections, the anthropogenic model is again confirmed in each case. Although some of the details were not predicted by either model, only a willfully superficial interpretation of the results could be used to argue against overkill. Nonetheless, these fine points are biologically interesting. They are that (1) some surviving mammal species did experience ecologically interesting changes in their geographical distributions—but this had no consistent relationship to known climate and vegetation changes, and involved only a tiny minority of the continental species pool; (2) there were a few extinction episodes earlier in the Cenozoic that rivaled the end-Pleistocene event—but none of them were demonstrably as rapid, they do not match up simply with known global climate change events, and none of them occurred during the greatly intensified climate cycles of the Plio-Pleistocene; and (3) there is a weak tendency for severe extinctions to differentially victimize large mammals—but it is not nearly strong enough to account for the observed selectivity, and there is no tendency for severe extinctions to focus on an extremely narrow spectrum of body sizes, as is seen with the end-Pleistocene event.

These conclusions confirm the intuition that although testing and retesting a strong hypothesis rarely weakens it, doing so often does generate new results that are of interest for other reasons. In the remaining sections I will separately discuss the data, methods, and statistical results that concern the three main tests, and then conclude by adding them to the list of widely accepted facts that bear on the end-Pleistocene mass extinction debates.

2. Changes in Surviving Assemblages

2.1. Motivation

The first test concerns the spatial distribution of North American mammal species before and after the extinction crisis. Much has been made of this subject in the past. Mammalian paleontologists realized decades ago that some late Pleistocene small mammal assemblages contained sets of species that no longer are found together in any one place (Hibbard, 1952). For example, some species that today are found in the far north, such as the heather vole, are seen together in Pleistocene faunas with other rodents that now are found in the southeastern United States, such as the eastern woodrat (Graham *et al.,* 1996). Such overlaps are called "disharmonious" (Semken, 1974) or "intermingled" (Graham, 1986).

By now these disharmonious patterns have been documented repeatedly in many fossil assemblages, and despite lingering questions about taphonomic biases that might be responsible for some apparent disharmony (Hibbard, 1952; Guilday, 1962; Semken, 1966), most workers find the evidence convincing. Disharmonious patterns usually are attributed to fundamental differences between glacial and interglacial climate regimes: If seasonal variation in climate is more equable during glacial intervals, then species otherwise kept apart by their opposed intolerances of winter and summer climate extremes might be able to coexist during glacials (Hibbard, 1955; Graham, 1976, 1986; Semken, 1984).

Nothing about disharmony, or the equability hypothesis used to explain it, necessarily bears on the issue of extinction. However, over the last decade Quaternary workers have gone beyond asking why disharmony occurred, to suggesting that this community reassembly resulted from the same, presumably climate-driven changes that caused the mass extinction (Graham and Lundelius, 1984; Guthrie, 1984; Semken, 1984; Graham *et al.,* 1996). Given the timing of the extinction around 11,000 radiocarbon years ago, the relevant climate event is presumably not a warming associated with deglaciation, but the very rapid cooling at the onset of the Younger Dryas at just about this time (Taylor *et al.,* 1997).

Essentially, the claim that disharmony and extinctions are connected rests on two types of assumptions. The first is empirical: Disharmony existed and was common, but then rapidly disappeared as reassembly ensued in tandem with the mass extinction and the climate event. The second is methodological: Given that three important ecological changes happened at the same time, one should prefer on the grounds of parsimony to explain two of them as resulting from the third. Obviously a fourth event—possibly human migration to North America, and in any event the first appearance of the Clovis hunters—also could have been caused by the climate event, but it seems ludicrous to argue the converse. Therefore, if the ecological events were causally linked to a climate event, then they were caused by it.

The logical flaw in this argument can be shown graphically (Fig. 1). There are four presumed events, all thought to have occurred at about the same time. One event (a climate change, perhaps deglaciation) is presumed to be directly causally prior to the other three. But why presume also that this allows only one causal path (Fig. 1A)? Why not allow a separate path in which spatial reassembly and human immigration both do follow directly from a climate event, but extinction follows directly from immigration (Fig. 1B)? Or a path in which extinction causes vegetation changes (Owen-Smith, 1987), which cause changes in the geographic distributions of the survivors (Fig. 1C)? The temporal coincidence and causal interconnection of all four factors is equally clear in all four scenarios.

I should note that adherents of the ecological models do not necessarily believe either that humans did appear in North America at the same time as the three ecological events, or that humans even in principle could have caused the mass extinction. In fact, this latter claim of implausibility may be what largely motivates opposition to the anthropogenic hypothesis. But what is really interesting here is the lack of any direct logical connection between evidence for disharmony and evidence for or against the anthropogenic model. By no account does spatial reassembly fall within a causal chain leading to extinction (although it may derive from it).

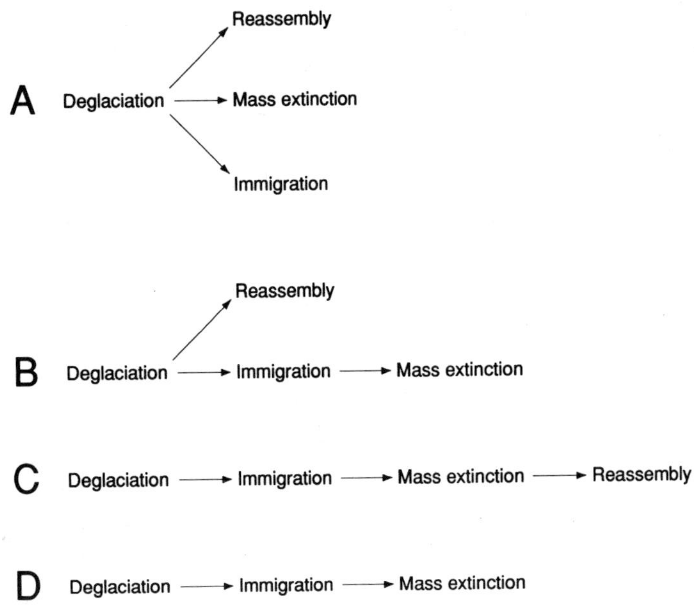

FIGURE 1. Possible causal relationships among late Quaternary events. (A) Ecological model: Deglaciation (possibly meaning either rapid cooling or warming events, or both) causes all other events, including mass extinction. "Reassembly" means geographic movements that eliminate disharmony. (B–D) Anthropogenic model: Deglaciation causes human immigration, which in turn causes mass extinction. In B and C, reassembly is assumed to occur at once during deglaciation. In C, reassembly is caused by extinction. In D (the hypothesis favored here), reassembly is treated as an independent phenomenon because it occurs in stages.

Therefore, disharmony is not really evidence for (or against) the ecological models, but instead a rhetorical issue that has to do with the plausibility of climate change as a cause of mass extinction. No one is saying that reassembly caused extinction; instead, the very fact that it existed makes it easier to believe that a climate change also could cause such a dramatic event as a mass extinction.

It is disappointing to conclude that spatial patterns are so tenuously linked to the extinction issue. To make matters worse, my purpose here will be to show that even the empirical claims supporting this rhetorical campaign are quite shaky. Analyzing all of the data that have been used to illustrate disharmony, as well as supplementary data, I will show that (1) disharmony always was present during the Pleistocene, but always was so extraordinarily rare that its disappearance can hardly be labeled a major ecological phenomenon; and (2) disharmony persisted long after the Pleistocene deglaciation, and its disappearance therefore need not be linked tightly to climate changes or mass extinction in the first place. This suggests a fourth causal model that has nothing to do with disharmony (Fig. 1D).

2.2. Data

The data used in this analysis are faunal lists. This is because disharmony always has been discussed in terms of the presence or absence of mammalian species within fossil faunas; although important in the palynological literature (e.g., Overpeck *et al.,* 1992), relative abundance has not been the primary focus of this research area. Instead, the main question for most mammalian paleoecologists is whether sets of species that were present in the same assemblages in the past no longer are (e.g., Graham, 1976; Semken, 1984).

For obvious reasons, this notion of disharmony has not been, and probably never could be, defined for extinct animals. We simply do not know enough about the past distributions of extinct species to say with certainty that a distribution at one particular time in the past truly differed from that at an immediately preceding or succeeding time: Any observed differences might merely reflect undersampling in one or both intervals. Thus, if we find species A and B together in faunas from time interval T, but not in faunas from interval $T + 1$, that always could be explained away as the result of $T + 1$ having an inadequate fossil record. On the contrary, living North American mammals are so well studied that we can have confidence, at least at the biome scale, of not just their observed presences but also their observed absences. Therefore, the present is unlike any ancient time interval in that it can serve as a template, or perhaps a "golden standard," for evaluating changes in distributions.

Unfortunately, using the Recent as a standard restricts any analysis almost entirely to small mammals—rodents, insectivorans, and a few lagomorphs. This is because the extinction event was so severe that surviving large mammals form a tiny minority of the living biota. Furthermore, bats have to be excluded because they do not preserve consistently in fossil assemblages; although they are common in cave deposits, they are very rare otherwise, and their fossil diversity is not proportional to their known Recent diversity. Marine mammals have to be excluded because their Pleistocene distributions are not well studied, and because the distributions of marine organisms do not bear on spatial patterns and ecological events in terrestrial communities.

Faunal lists for the late Quaternary are very abundant and have been compiled thoroughly by the FAUNMAP Working Group (Graham *et al.,* 1996). This compilation uses a working classification of late Quaternary time into six time intervals: early, middle, and late Wisconsinan and early, middle, and late Holocene (Table 1). Unfortunately, early and middle Wisconsinan faunas are too poorly represented to allow a thorough analysis.

My own data on late Cretaceous and Cenozoic mammal faunas (Alroy, 1996, 1998a–c, in press) include all of the remaining, pre-Wisconsinan Pleistocene records. There is controversy over the pre-Wisconsinan Pleistocene time scale (Kurten and Anderson, 1980; Lundelius *et al.,* 1987; Repenning, 1987; Alroy, 1998a), and I am not confident that the traditional subdivisions are well supported, even at the level of "Irvingtonian" (early Pleistocene) versus "Rancholabrean" (roughly middle and late Pleistocene). However, it seems useful to tentatively accept "Sangamonian" (last interglacial) age determinations from the literature, not because these are likely to be correct, but because it would be very illuminating to discover that the preceding and current interglacials exhibit different patterns. I therefore have separated the pre-Wisconsinan data into Sangamonian and pre-Sangamonian, or "early/middle" Pleistocene, categories (Table 1).

Defining Recent distributions requires choosing a scale of analysis. Some scales are too small: Hectare-scale inventories (e.g., Brown and Nicoletto, 1991) are not comparable to fossil assemblages because the latter are spatially averaged over a much larger area, especially if they are preserved in a fluvial context. County-level inventories might be more appropriate, and such data are available in printed form (e.g., Hall, 1981). However, the purpose of this analysis is to test the idea that Quaternary assemblages included species that today are not merely allopatric, but ecologically well separated. After all, this is exactly the

Table 1. Disharmony Statistics for Six Successive Intervals of the Quaternary[a]

Interval	Lists	Spp	C_D	C_H	I_D	C_E	I_R
Early/middle Pleistocene	153	114	36	1471	0.024	1157.2	0.103
Sangamonian	47	92	17	877	0.019	705.2	0.090
Late Wisconsinan	368	174	178	5273	0.033	3936.4	0.118
Early Holocene	66	99	0	1445	0.000	1276.1	0.000
Middle Holocene	253	124	18	2161	0.008	1889.3	0.062
Late Holocene	1773	165	45	4770	0.009	3820.7	0.045

[a]Lists = number of lists falling in interval; Spp = number of species falling in interval; C_D = disharmonious conjunctions; C_H = harmonious conjunctions; I_D = index of disharmony; C_E = expected number of conjunctions shared by two intervals at random; I_R = rescaled index of disharmony. See text for formulas.

kind of claim made by such workers as Graham (1976, 1986) and Graham *et al.* (1996). Therefore, it is fairer (not to mention simpler) to use biome-scale species lists as a standard: If species today do not overlap at the biome scale but in the Quaternary are found in the same fossil assemblages, this is clearly significant regardless of one's position in the mass extinction debate. Brown and Nicoletto (1991) have provided species-level lists for 21 North American biomes. With appropriate taxonomic changes to bring all three data sets into accord, these lists form the "golden standard" for this study.

2.3. Methods

Previous workers have taken two approaches to quantifying disharmony. The first involves similarity, tested in two ways (Graham *et al.*, 1996). Greater similarity among Quaternary faunas should suggest that because they included many disharmonious species, they were more homogenized than Recent faunas. Meanwhile, unrecognizable patterns of similarity among Quaternary faunas should indicate that biogeographic patterns were fundamentally dissimilar, again presumably because of disharmony.

Unfortunately, Graham *et al.* (1996) misinterpreted the results of their analyses. Their fauna-by-fauna comparison of faunal similarity and geographic distance seemed to show lesser, not greater, homogeneity of Quaternary faunas. Nonetheless, the authors persisted in explaining this pattern as the result of fundamental changes in community composition—even though greater disharmony would have equated with ecologically broader species distributions and therefore greater homogeneity.

Ironically, even their finding of lesser homogeneity was the result of a methodological error: Because climate gradients were far, far steeper during the last glacial, when half of the continent was under an ice sheet, comparing similarity and geographic distance between faunal samples should be expected to suggest steeper faunal gradients no matter what the exact composition of samples along those gradients. In other words, the result is predictable given any model of community dynamics that relates taxonomic distributions to climate in any way—"individualistic" or not. None of these models exactly specifies the steepness of faunistic gradients. The question, therefore, is not really how spatial distance compares with faunal similarity, but rather whether local Pleistocene assemblages generally have recognizable Recent analogues—the focus of the current analysis.

As for biogeographic patterns, what would seem to be most striking about the cluster analysis results of Graham *et al.* (1996) is that most major North American biomes are indeed recognizable in the Quaternary. This is surprising because the raw presence/absence data are severely affected by variation in sampling intensity, taphonomic regime, and taxonomic usage. Major groupings in a cluster analysis therefore should reflect not natural biological entities, but instead sets of faunas that share such compositional biases. To the extent that the Quaternary patterns of Graham *et al.* (1996) differ from the Recent, we cannot trust them to reflect real-world patterns instead of biases. For reasons like these, I (Alroy, 1992) have advocated abandoning similarity-based analyses in paleoecology unless uniform sampling conditions can be established (certainly not the case for any large literature-based data set).

Apart from using similarity measures, a second general approach to illustrating patterns of disharmony is to compute "areas of sympatry" (Stephens, 1960; Rhodes, 1984; Semken, 1984). By referring to Recent distribution maps, one breaks down a fossil fauna into subsets of species that today are found all together in the same place. If the entire fossil fauna can be matched to a Recent locality, it is perfectly harmonious. If most of it can, but a few species cannot and must be assigned to one or more separate, "resolved" areas of sympatry, the fauna is disharmonious. By taking such a relative approach, one avoids the very serious disadvantages of trying to separate migration patterns that result from unpredictable, individualistic species responses from those that are entirely predictable on the basis of environmental tolerances. One should note that this conflation of predictable and unpredictable migration patterns makes it impossible to evaluate the species-by-species geographic mapping favored by Graham (1986) and Graham *et al.* (1996).

The area of sympatry approach is basically valid, and my intent here is merely to extend it. The problem is that area of sympatry analyses do not express all of the patterns of geographic overlap (or "conjunction": Routledge, 1977; Alroy, 1992) among a fauna's species. These patterns can be represented with a graph in which species are points and observed overlaps in the Recent are lines (Fig. 2A), or alternately with a triangular species-by-species matrix in which cells are marked to show which species also overlap in the Recent ("H" cells in Fig. 2B) and which species do not ("D" cells). Areas of sympatry correspond to subgraphs or submatrices that are entirely filled with lines or "H" values (Fig. 2C,D). However, because each species is conventionally listed only in one area of sympatry, some of the harmonious conjunctions are not identified in this approach—so by omission, areas of sympatry fail to clearly distinguish harmonious and disharmonious Quaternary conjunctions.

It should be obvious at this point that a simpler, more comprehensive approach is to examine the conjunction matrices directly. Instead of counting conjunctions that fall into three categories—harmonious and represented by an area of sympatry, harmonious but unrepresented by an area of sympatry, and disharmonious—one should count species in two categories: harmonious (C_H) and disharmonious (C_D). This leads immediately to a simple, intuitive index of disharmony: $I_D = C_D / (C_H + C_D)$, i.e., the proportion of all conjunctions that are disharmonious. If disharmony is an important factor in Quaternary community composition, then this index should take on a high value, perhaps 0.2 or even 0.5. If not, it should approach zero.

One problem with I_D is that it cannot really be expected to take on a high value if many species in the Recent overlap geographically to start with. In the extreme case—that every pair of species on the continent overlapped at some particular point—disharmony would

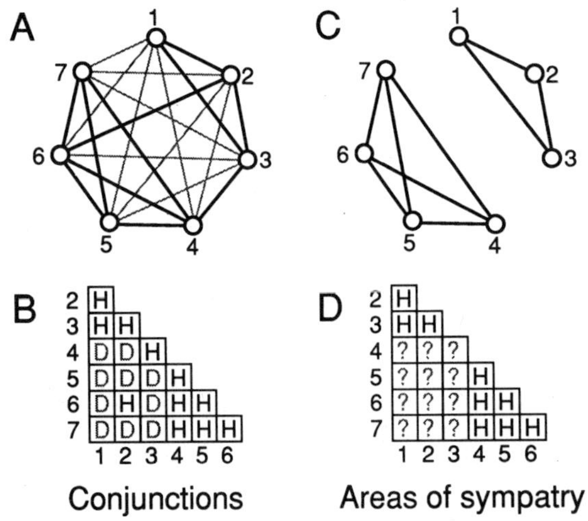

FIGURE 2. Patterns of geographic overlap (conjunctions) in the Recent among seven species found in a hypothetical Quaternary fossil assemblage. Black lines and H symbols represent conjunctions that are harmonious, i.e., still known in the Recent; gray lines and D symbols represent disharmonious conjunctions. (A) Graph of all conjunctional relationships in the assemblage. (B) Matrix showing the same pattern. (C) Conjunctional relationships illustrated by an "area of sympatry" analysis. Species 4–7 are found in a "maximal area of sympatry"; species 1–3 form a "resolved area of sympatry" (see Rhodes, 1984). (D) Matrix showing the same pattern. Because the pattern of harmonious conjunctions is complex, the areas of sympatry do not clearly distinguish some of the harmonious conjunctions from the disharmonious conjunctions (question marks).

be impossible and the index would always be zero. If instead all but one pair overlapped, finding that one pair in a local assemblage of 20 species would yield an index of $1 / ([20^2 - 20])/2) = 0.005$—even though the assemblage would be maximally unusual in this hypothetical context.

Therefore, it is helpful to rescale the index as follows. Assume that the lower value should be bounded by zero when all conjunctions are harmonious (as with I_D). The upper value should be bounded by one when the number of harmonious conjunctions is no greater than that expected if a fauna is fully randomized relative to the Recent. In such a case, the chance that a Quaternary conjunction is harmonious equals the chance of randomly "drawing" a conjunction to represent the species pair once from each subset of the data. This chance is simply the proportion of all Quaternary species pairs that are conjunct (first draw) times the proportion of all Recent species pairs that are conjunct (second draw). If we set M to $(N^2 - N)/2$, the maximum number of conjunctions among N species, then the expected number of randomly harmonious conjunctions $C_E = ([C_D + C_H]/M) \times (C_H/M)$. The rescaled index I_R is then equal to $C_D / (C_D + C_H - C_E)$.

A major advantage of the conjunction approach is that the indices need not be defined just for individual localities. Instead, one can examine at once all of the faunal samples in a particular time slice. One simply computes all of the conjunctions in each faunal list separately, recording newly encountered conjunctions as one goes along. Eventually, no further conjunctions in that time interval will remain to be discovered, and the conjunction matrix will stabilize or "converge" (Alroy, 1992). Although it is unlikely that any particular

Quaternary time interval is close to convergence in this sense, it stands to reason that similar proportions of the known and unknown conjunctions are harmonious, which means that the indices should be basically fair even if the counts are all too low.

Computing disharmony indices for entire intervals gives us a quick and unambiguous way to assess the change through time in the continent's overall faunal composition. Furthermore, it biases the statistics as strongly as possible in favor of the disharmony hypothesis: if there really are spurious conjunctions that result from time-averaging, reworking, or misidentifications, then considering all of the faunas as a set will amplify those biases, making it seem that there was much more disharmony than there really was.

2.4. Results

Conjunction indices for each Quaternary time slice are shown in Table 1 and Fig. 3. These results immediately present a striking pattern. Disharmony is at a maximum during the late Wisconsinan, which perhaps matches the predictions of the ecological change mass extinction model. However, the values are tiny: 0.033 for the simple index I_D and 0.118 for the rescaled index I_R. To put this in context, if I_D equaled 0.033 for a fossil sample with 30 species, there would be 14 disharmonious conjunctions but 421 harmonious conjunctions.

Even more strikingly, similar values are seen in each of the other intervals. This is true not only of the broad "early/middle Pleistocene" category, which includes a wide variety of both glacial and interglacial assemblages, but also of both the Sangamonian and Holocene interglacials. If the Sangamonian faunas are correctly categorized, then disharmony was nearly as common during the Wisconsinan as during an earlier interglacial that was every bit as warm as the Holocene. The climate equability hypothesis would be hard-pressed to explain such a pattern.

As for the Holocene, the poorly sampled early Holocene exhibits no disharmony, but this is based on a relatively small sample size of 1445 conjunctions (versus 5273 for the late Wisconsinan). Furthermore, the better-sampled middle and late Holocene disharmony values "rebound" to a nonzero level. These facts strongly suggest that the failure to find early Holocene disharmonies is a sampling artifact, and disharmony probably persisted without a break all the way up to the historical era.

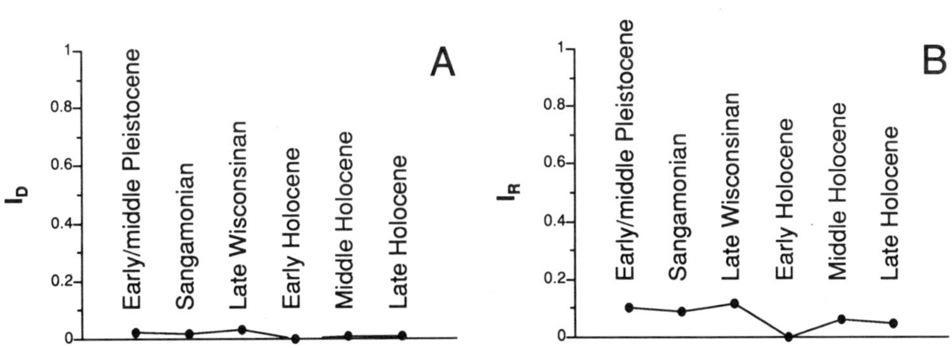

FIGURE 3. Disharmony indices for six Quaternary time intervals. (A) Raw disharmony index, I_D. (B) Rescaled disharmony index I_R. Disharmony never rises above low levels and persists into the late Holocene.

Another, slightly more complex analysis also suggests that the disharmony indices for both glacial and postglacial intervals bear the stamp of exactly the same ecological phenomenon. To show this, I will take advantage of the fact that disharmony indices can be computed not just for individual localities and sets of localities, but also for individual species. This is done by examining any one species's row in a conjunction matrix, categorizing the conjunctions in order to compute species-specific values of C_D and C_H, and by extension of I_D. What is found by going through this exercise is that the vast majority of species have miniscule I_D values that represent their disharmonious conjunctions with one, two, or most commonly no other species. However, a "core set" of species in each time interval exhibits much higher values.

The best way to investigate this core set is to look at two intervals at a time, looking to see which species are involved in at least one disharmonious conjunction in each interval. The very fact that so few such species are found in the first place immediately bolsters the impression that disharmony is very rare and not very consequential. Most pairs of intervals share six or fewer disharmonious species; just six pairs share more, ranging from 12 to 28 (Table 2). The figure of 28 disharmonious species is for the late Wisconsinan and late Holocene, the two best-sampled intervals but also a pair offset by a full 8 kyr and the most drastic climate changes of this glacial cycle.

Nonetheless, these small groups of shared species are much larger than one would expect if each interval had a completely independent, randomly selected set of disharmonious species. This can be shown with a G test of proportions in a two-by-two count matrix (Fig. 4). In each test, the rows represent the disharmonious and harmonious species in the first interval, and the columns the analogous species in the second interval. The upper left-hand cell gives the count of shared, disharmonious species; the lower right-hand cell, that of shared, harmonious species; and the off-diagonal cells, those of species that are harmonious in one interval but not in the other. The null hypothesis is that the counts in each cell can be predicted just by looking at the row and column totals, i.e., that the count of shared disharmonious species is no higher than one would expect if the species were randomly declared to be disharmonious. The test has one degree of freedom.

In four of six cases, this null hypothesis can be rejected strongly (Table 2). One of the other two comparisons (early/middle Pleistocene versus late Wisconsinan) is just barely insignificant, with $0.05 < p < 0.10$. That leaves the late Wisconsinan–middle Holocene comparison as the only real exception; but this one can be discounted because it involves only 11 species to start with. Thus, the core set of disharmonious species is largely repeated in one interval after another–despite the dramatic climate and sampling differences among the intervals.

Not only the set of disharmonious species but also the degree of disharmony shown by individual species seems to be duplicated across intervals. This can be shown by correlating the I_D scores for the shared disharmonious species (Fig. 5, Table 2). The rank-order correlations are positive in all six cases, showing that the most disharmonious species in one interval tend also to be the most disharmonious in the others. Although two of these correlations are not significant, in one case this is probably because of the very small sample size (11 species). The other case (early/middle Pleistocene–late Wisconsinan) is interesting, especially because the G tests suggested that there was only a weak overlap in the first place between their sets of disharmonious species. Thus, there is a suggestion that af-

Table 2. Comparisons of Disharmonious Species in Different Quaternary Intervals[a]

Intervals	Species	G_1	r	G_2
EMP/Sangamonian	14	10.540**	+0.741**	2.721
EMP/Late Wisconsinan	21	3.658	+0.151	16.511
Sangamonian/Late Wisconsinan	14	7.469**	+0.626*	16.028
Late Wisconsinan/Middle Holocene	11	1.591	+0.527	9.071
Late Wisconsinan/Late Holocene	28	14.673***	+0.689***	15.074
Middle Holocene/Late Holocene	13	17.613***	+0.670*	6.974

[a]Species = number of species showing at least one disharmonious conjunction in each interval; G_1 = G test of hypothesis that number of shared, disharmonious species is a random function of the number of species shared and the number of disharmonious species ($df = 1$); r = Spearman's rank-order correlation between disharmony indices of disharmonious species in each interval; G_2 = G test of hypothesis that number of disharmonious conjunctions per species is a random function of the total number of disharmonious conjunctions per species and per interval (df = number of species − 1); * = $p < 0.05$; ** = $p < 0.01$; *** = $p < 0.001$.

ter hundreds of thousands of years the set of disharmonious species does change apprecia-bly. But after a single glacial cycle, it does not.

We can go further than this, asking not just if the relative disharmony of these core species is similar through time, but if there is statistical evidence that their degree of dishar-mony changed through time at all—even if just slightly. This is easy to do with a second G test in which rows are species, columns are time intervals, and cells are counts of dishar-monious conjunctions (Fig. 6). If the row and column totals can predict the cell values, then apparent changes through time in disharmony can be explained as side effects of changes in sampling intensity and/or the overall level of disharmony across the continent. If not, then we might be able to conclude that individual species independently became more or less disharmonious, perhaps by migrating erratically. The test has $N - 1$ degrees of free-dom for N species. As it happens, for all six comparisons the observed variation is in-significant (Table 2).

To summarize these results, three things can be said with certainty. First, disharmony always was very rare. This is true not just of the Holocene, but also of the Wisconsinan glacial and of the grab-bag, climatically variable early/middle Pleistocene interval. Second, disharmony did not disappear as soon as the Wisconsinan was over. Instead, it persisted all the way through the late Holocene, if admittedly at slightly lower levels. This makes it very hard to attribute disharmony solely to climate change. Instead, it seems much more reasonable to argue that anthropogenic habitat changes during Recent times are responsi-

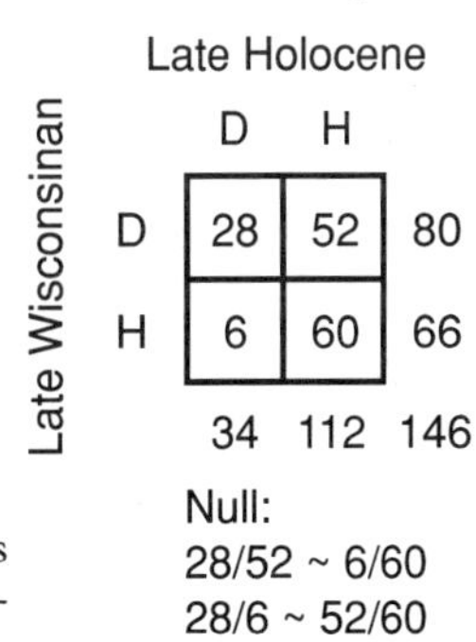

FIGURE 4. G test of the hypothesis that the number of disharmonious species shared by two intervals is a predictable function of the total number of disharmonious species in each of these intervals.

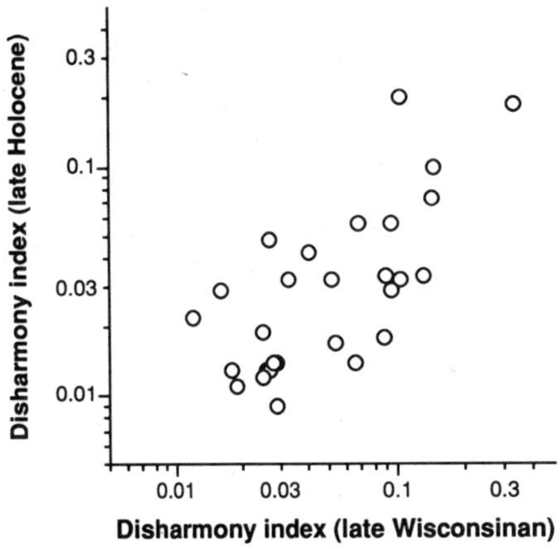

FIGURE 5. Correlation of disharmony indices for species that are disharmonious in both the late Wisconsinan and the late Holocene.

ble for restricting the ranges of many "disharmonious" species; the yellow-bellied marmot is a possible example. Finally, the core set of disharmonious species remained virtually unchanged during more than a full glacial cycle. Far from wandering "individualistically" across the landscape, these species maintained ecologically stable and predictable distributions that only gradually contracted in the face of Holocene environmental change.

All of this makes it very hard to argue that the hand of massive environmental disruption is stamped on Quaternary mammalian distributions. Instead, these distributions seem to have moved across the continent in an orderly and consistent manner as the glaciers waxed and waned. If anything, the result provides powerful support for the idea that Quaternary mammals already were well adjusted to glacial cycles long before the rapid climate changes associated with the terminal Pleistocene; any species that could not cope with such changes already were long extinct. Similar patterns of "weeding out" have been noted in such groups as marine molluscs (Valentine and Jablonski, 1991). The existence of this effect should come as no surprise because we now know that rapid and severe climate changes

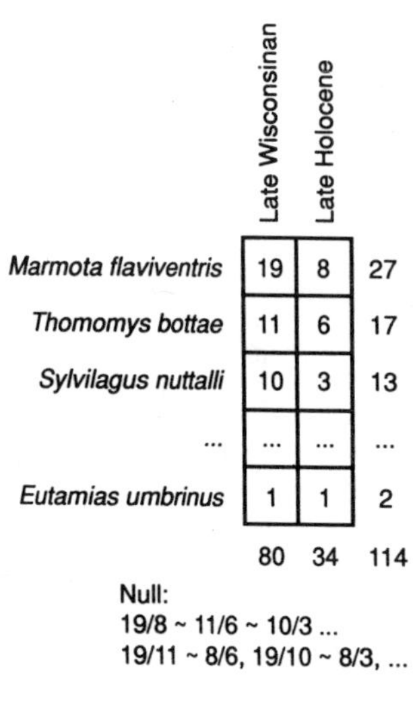

FIGURE 6. G test of the hypothesis that the number of disharmonious co junctions per species per interval is a predictable function of the totals for eac species and interval.

occurred during much of the late Quaternary, including all of the last glacial and the interglacial that preceded it (Dansgaard *et al.*, 1993; Shackleton, 1995; Stager and Mayewski, 1997; Taylor *et al.*, 1997).

3. End-Pleistocene and Cenozoic Extinction Rates

3.1. Motivation

The second test of the competing extinction models concerns this very issue of "weeding": Were mammals buffeted mercilessly by the repeated glaciation cycles of the Pleistocene, as well as by earlier dramatic climate changes in the Tertiary? Or are clear-cut extinction pulses such as the terminal Pleistocene event just not a normal feature of mammalian evolutionary history? As recognized by earlier authors (Gingerich, 1984; Webb, 1984; Barnosky, 1989), the Cenozoic North American fossil record is continuous and rich enough to provide some firm answers to these questions. If we can nail down particular large extinction pulses to particular climate changes, that might support the theory that natural climatic variability can have catastrophic consequences for mammalian communities. Furthermore, several major climate changes already are well known, making it possible to actually predict when extinction pulses should have occurred. Most importantly, it is clear that the greatest climatic variability over the last 65 my (millions of years) really was during the Pleistocene, and specifically during the last 1.0 my of the Pleistocene (Shackleton, 1995). Therefore, if any interval should exhibit intensified extinction rates, it is this one.

One crucial problem with this line of inquiry concerns the computation of extinction rates. It is well known that observed extinction rates are a function of the duration of sampling "bins" (Foote, 1994). If one waits long enough, all species eventually will become extinct. If one spaces temporal sampling bins too close together, most of these bins will document no extinctions at all. Worse still, if some intervals are short but others are long, the longer intervals will exhibit spuriously higher extinction rates, creating the impression that extinction pulses are common. Furthermore, the total number of extinctions is itself determined by the intensity of sampling: As more and more fossils are collected, more and more extinct species are described.

These problems essentially invalidate the results of earlier studies, which used nonuniform sampling intervals and made no attempt to correct for variation in sampling. Therefore, they allowed long, well-sampled intervals to be followed by short and/or poorly sampled ones, and ended up presenting artificially elevated extinction rates (Kurten and Anderson, 1980; Gingerich, 1984; Webb, 1984; Barnosky, 1989). Unfortunately, these results have been accepted uncritically by other workers (e.g., Valentine and Jablonski, 1993). In this study, I take advantage of earlier research to impose evenly spaced, 1.0-my-long sampling intervals, and to guarantee that each of these intervals is sampled in a similar way. By doing this, I will maximize the chances that the end-Pleistocene extinction rates will be unremarkably high because I will compute earlier extinction rates over a full 1.0 my, allowing some extinction episodes that may have been stepwise or gradual to appear to be single events. Thus, the test will be biased in favor of the hypothesis that the end-Pleistocene event was natural, not anthropogenic.

Apart from biased sampling and the unevenness of traditional time intervals, another, related problem is the fact that the Cenozoic time scale is not even vaguely as well resolved as the late Quaternary time scale, which relies on easily obtained carbon isotope dates with precision on the order of 1.0 kyr or less. Carbon dates cannot be computed for pre-Wisconsinan localities, leaving us with a semiquantitative time scale determined mostly by faunal turnover within the mammalian biota itself (Wood *et al.*, 1941; Woodburne, 1987; Alroy, 1992, 1994, 1996, 1998c). Because the median mammalian species lasts for about 1.6 my (Alroy, 1998c) and because even the best localities rarely include more than 20 or 30 species, such a time scale can never have better practical precision than about 0.25 my, the point at which one would expect, say, 10% of the species in a fauna to have gone extinct and been replaced. In practice, correlational error is actually on the scale of 1.0 my (Alroy, 1996, 1998c).

Being forced to use such lengthy temporal bins is not a problem in itself, because waiting for a substantial amount of background extinction to occur allows us to compute extinction rate estimates that have low binomial error. However, it does mean that the Cenozoic extinction data are not strictly comparable to the end-Pleistocene extinction rate, which is based on events that occurred in less than 2 kyr (Meltzer and Mead, 1985; Beck, 1996). The three-order-of-magnitude difference in these scales means that the Cenozoic rates are extreme overestimates: unless all of the extinction in a 1.0-my bin happens to have occurred during just 2/1000th of that interval, the total extinction rate must represent a summation of many shorter-term events that together were not of the same duration as the end-Pleistocene event. If anything, finding that Cenozoic extinction rates were comparable to or less than the end-Pleistocene rate would firmly refute the idea that the megafaunal mass extinction was "natural."

It would be helpful not just to compute unbiased extinction rates, but also to directly compare these rates and quantitative climate data. Unfortunately, appropriate data are not available at this time. The best climate proxies are oxygen isotope ratios measured in the shells of foraminiferans. By definition, these data are only available for marine drill cores. Correlating the paleomagnetic and biostratigraphic time scales used in such studies with the much coarser North American mammalian time scale is difficult, and this cross-correlational error is certainly of a similar magnitude to the entire 1.0-my duration of the sampling intervals. For isotope data after the onset of significant glaciation in the mid-Tertiary, there is a mixed signal of not just mean annual temperature, but also of ice volume; this mixed signal is hard to interpret in climatic terms (Miller *et al.*, 1987). Finally, the marine data are of questionable relevance because they concern ocean temperatures thousands of miles from the midcontinental regions that have the best fossil record of North American mammals, and do not record the variables that directly affected these terrestrial communities: terrestrial temperatures, precipitation, and continental seasonality.

A reasonable alternative would be to use leaf-margin analysis of fossil macrofloras to estimate continental temperatures and precipitation (Wolfe, 1978; Wilf, 1997; Wilf *et al.*, 1998). Unfortunately, the currently available floral data are concentrated during a few intervals, such as the early Eocene (e.g., Wilf *et al.*, 1998) and middle Miocene (e.g., Wolfe *et al.*, 1997), and it simply is not possible to compute a detailed time series at the 1.0-my scale for the entire Cenozoic. This shortcoming may eventually be rectified, but at the moment the paleobotanical record serves merely to illustrate broad climate trends.

Because of these difficulties, I will specifically focus on the five Cenozoic climate changes that were clearly the most severe and most likely to have affected communities across the globe. These were at the Paleocene–Eocene boundary; at the onset of significant Antarctic glaciation in the earliest Oligocene; during significant cooling episodes in the middle Miocene and latest Miocene; and at the point in the Pleistocene that witnessed the intensification of glacial cycles to their current extremes. If one chooses to dismiss these key events as irrelevant by assuming that they must have transpired over much longer intervals than the short-term climate shifts of the late Quaternary—which of course is not demonstrable—then one must forfeit the entire argument that climate-driven extinction events are a natural feature of mammalian evolution.

The results will show that most of these critical intervals did not exhibit elevated extinction rates, and in particular, that the entire Pleistocene before the terminal Wisconsinan event was a period of relative tranquility. A major extinction simply cannot be shoehorned into this period of 1.8 my. Furthermore, few of the 1.0-my-long sampling bins present a per-taxon extinction rate as large as that for the end-Pleistocene. Of the few bins that do exhibit relatively high rates, most fall in the Paleocene, a time of major ecological change that witnessed the replacement of an archaic suite of mammal groups by the five more evolutionarily stable orders that still dominate the North American fauna. Most of the variation in extinction rates can be accounted for just by splitting the time series at the Paleocene–Eocene boundary. Therefore, extinction in the past seems to be more a function of intrinsic evolutionary dynamics than of extrinsic, environmental forcing factors.

3.2. Data

The data discussed here have been analyzed previously (Wing *et al.,* 1995; Alroy, 1996, 1998c). They consist of extinction rate time series that are computed from a multivariate ordination of 4015 faunal lists spanning the late Cretaceous (Campanian) through Sangamonian (last interglacial). The lists can be viewed on the World Wide Web at the North American Mammalian Paleofaunal Database http://www.nceas.ucsb.edu/~alroy/nampfd.html). Together with the FAUNMAP data, these lists record the entire published fossil record of North American mammals, with the exception of nineteenth-century collections that have not been tied to specific localities. The lists do not include marine mammals (Pinnipedia, Cetacea, Sirenia) or bats, which are very infrequently preserved. They have been standardized to the latest published taxonomy using a complementary database that flags 451 invalid genera, 2692 invalid species, and 1197 invalid genus–species combinations; these can be viewed at the North American Fossil Mammal Taxonomy Database (http://www.nceas.ucsb.edu/~alroy/nafmtd.html). The two databases are tied to a set of 2415 publications.

Earlier workers, including Kurten and Anderson (1980), Gingerich (1984), Webb (1984), Barnosky (1989), Stucky (1990), and Van Valkenburgh and Janis (1993), have compiled data by reference to the traditional North American land-mammal age system, which was defined by Wood *et al.* (1941) and is generally robust. Unfortunately, this system comprises time intervals of uneven duration that sometimes are separated by known pulses of turnover and/or by hiatuses in the rock record, and sometimes are separated by arbitrary cut points. To avoid the complexities introduced by such factors, and to guarantee that the assignment of localities to time intervals is completely objective and replicable, I instead have sidestepped the entire issue of land-mammal ages by subjecting the lists to multivariate ordination. This involves new methods specifically developed to combine the information presented by faunal lists with that implied by the 186 stratigraphic sections that include 2499 (62%) of these lists (Alroy, 1992, 1994, 1996).

The basic idea is that a paleontological time scale is at its heart a relative sequence of first and last appearance events (FAEs and LAEs). At one extreme, we might place all of the FAEs together at the beginning of a sequence, and follow them with all of the LAEs, which would imply that all taxa were at some time coeval. At the other, we might alternate pairs of FAE and LAE statements so that no taxa ever co-occurred. Obviously, the truth lies in between, and it is constrained by the fossil record: We know that whenever two taxa *i* and *j* co-occur in a fossil sample, the FAE of taxon *i* (FAE*i*) must have occurred (<) before the LAE of taxon *j* (LAE*j*). Likewise, this proves that FAE*i* < LAEj. Together, these two F/L statements equate to the fact that i and j are spatially and temporally conjunct, using the earlier term.

The power of F/L statements derives from the fact that they can be inferred not just from faunal lists, but also from the superposition of fossils in stratigraphic sections: if any fossil of *i* is lower than one of *j* in any section, then FAE*i* < LAE*j*. Therefore, F/L statements give us a way to translate both stratigraphically superposed and stratigraphically isolated faunal lists into a useful, generalized biochronological data type. F/L statements have other advantages, such as their "convergence" property, their logical independence of sampling regimes, and their direct correspondence with age-range charts, that are discussed elsewhere (Alroy, 1992, 1994).

The type of constraint that a set of F/L statements places on constructing an event sequence is clear-cut: The sequence always should imply known F/L statements, and should avoid implying as-yet-unproven F/L statements whenever possible—in such cases, we should try to reverse the ordering of the FAE and LAE statements. In other words, the best relative time scale is the one that untangles as many age ranges as possible. The problem of finding an event sequence that implies a minimal number of F/L statements, and is therefore maximally parsimonious, is apparently best solved using a multivariate ordination method based loosely on correspondence analysis (Alroy, 1994). This method works mostly on a taxon-by-taxon matrix of F/L statements, generating a preliminary event sequence that is then optimized by local swapping (Alroy, 1996).

The optimal sequence is numbered from oldest to youngest, and this relative time scale or age-range chart is calibrated to numerical time using 152 geochronological age estimates based on radioisotopic dating and paleomagnetic stratigraphy. The best method for doing this involves linear interpolation, with the number of "hinge points" being determined by an F-ratio test (Alroy, 1996). In the current analysis, I use six internal hinge points and two at the extremes of 78.5 Ma (millions of years ago) and 0 Ma (Alroy, 1998c).

3.3. Methods

One of the advantages of working with a calibrated appearance event sequence is that it can be cut into uniformly spaced, 1.0-my-long sampling bins, this figure appearing to represent the best trade-off between precision and accuracy (Alroy, 1996, 1998c) and closely matching the observed error in the geochronological calibration. Standing diversity is then defined by the number of species that cross each boundary between sampling bins, e.g., the time planes at 65.0, 64.0, . . . , 1.0 Ma. Extinctions are computed by counting the number of species in each cohort that fail to survive until the next time plane. Note that this excludes single-interval species (those that appear and disappear in the same bin), which makes the analysis less vulnerable to sampling artifacts (Alroy, 1996, 1998c, in press).

It is important to note that the extinction data will be expressed as per-taxon rates. This is because extinction as a process only makes sense as a probabilistic function that independently affects each species, not as a function that arbitrarily cuts down a fixed number of actual victims regardless of the number of potential victims. There have been concerns that per-taxon figures introduce statistical dependency between standing origination and extinction (Sepkoski, 1978); however, these concerns appear to be misplaced for the current data set (Alroy, 1996, 1998c). This probably has to do with the fact that standing diversity hovers about a dynamic equilibrium throughout most of the Cenozoic, never approaching the very low diversity levels that would introduce large errors into per-taxon rates.

Two other complications concern the age ranges of genera (Alroy, 1996). If a genus crosses a time plane but no species of that genus does, it surely is present unless it happens to be polyphyletic (mammalian paleontologists have systematically avoided defining polyphyletic taxa for many decades). Therefore, standing diversity must be augmented by one whenever such a case is observed. Likewise, if two species in a genus have nonoverlapping age ranges, it is possible that the older gave rise to the younger via anagenesis, so the extinction of the older species must be discounted so as to obtain an absolutely minimal estimate. Doing so reduces the number of apparent extinctions by about 22%, a figure in accord with independent estimates of anagenetic "extinction" frequencies obtained by explicit phylogenetic analyses (Alroy, 1998c).

The final issue of data preparation concerns sampling, an underrecognized factor that often is dismissed by students of macroevolution (e.g., Sepkoski *et al.*, 1981), despite good evidence that it can control major patterns (e.g., Raup, 1976). There is a close correspondence between the number of genera that range into a bin and the number of faunal lists, or of taxonomic records (sum of lengths of faunal lists) that can be assigned to that bin. Although a variety of other methods for removing such effects are available, including gap analysis, confidence intervals, and ghost lineages, the most unbiased and least "noisy" appears to be randomized subsampling of faunal lists (Alroy, 1996, 1998c, in press). This method, which is related to rarefaction (Raup, 1975), involves drawing lists from within each bin without replacement until a standardized total number of taxonomic records is reached. The cutoff is set at the highest level that most intervals exceed, which in this case is 100 records (or about 14 lists) per interval. Only one interval falls below this cutoff (40–39 Ma, with only 79 records). The randomized subsampling is repeated 100 times, and in each case the age ranges, counts of species crossing time planes, and counts of extinctions are recomputed. Average figures across these trials are used in further analyses.

3.4. Results

The results concern three issues. First, were there any other extinction pulses during the Cenozoic that rank with the end-Pleistocene event? Second, can a large amount of the variation in extinction rates be explained by long-term temporal trends, as opposed to short-term disturbances? And finally, are the five climatic "crisis" intervals comparable to the others in terms of extinction rates, or more extreme?

All of these questions are quickly and decisively answered with the data generated by the ordination and randomized subsampling procedure (Fig. 7). It is immediately apparent that the end-Pleistocene extinction rate of 32.4% is as extreme as any other during the preceding 55 my (Fig. 7B). Admittedly, it does not rival the rate of 76.1% for the 66–65 Ma interval, which crosses the K–T boundary (not shown; the fact that the K–T extinction was more severe for North American mammals than any other in their history is disputed by some workers like Archibald [1996], but only because they have relied on the traditional time scale to compute rates). Although the 2-kyr Pleistocene rate is equaled or exceeded by some of the Paleocene 1-my rates, only a handful of post-Paleocene intervals even approach the end-Pleistocene. Examples are seen in the latest Miocene (6–5 Ma: 35.2%) and early Eocene (51–50 Ma: 32.5%).

It is important to note at this point that the end-Pleistocene extinction rate reported here is slightly higher than those given by earlier authors (Martin, 1967, 1984; Kurten and Anderson, 1980; Gingerich, 1984; Barnosky, 1989). For example, the raw FAUNMAP data for the late Wisconsinan suggest a rate of about 25%. This difference reflects the fact that the 32.4% figure depends on a sampling-standardized data set, for which only a small minority of late Pleistocene localities was drawn in each randomization trial. Because almost all of the victims were very large and therefore both easily fossilized and geographically widespread, they typically are among the first to be sampled during a subsampling trial. Hence, the rarer, surviving small mammals are underrepresented in the analysis. But this bias is a strength, not a weakness, because it shows that given a level playing field, the end-Pleistocene event stands out against most of the remaining Ceno-

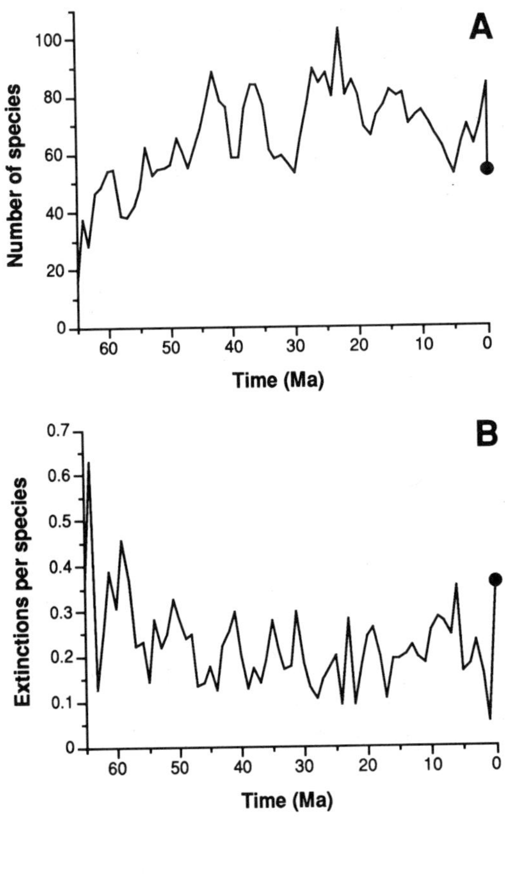

FIGURE 7. Diversity and extinction in the Cenozoic. Data are prepared by multivariation ordination and randomized subsampling (see text). (A) Number of species crossing each of 66 time planes that are 1.0 my apart. Solid circle is value after end-Pleistocene extinction. This relatively low value might have been slightly higher if sampling could have been subsampled at a higher level, but still would have been visibly different from the preceding values. (B) Per-species rate of extinction within each 1.0-my bin. Note that end-Pleistocene value (solid circle) is among the highest during the last 55 Ma.

zoic record. If any previous event had preferentially targeted widespread and well-preserved large mammals in this way, it too would have registered a similarly large extinction percentage.

The fact that so few intervals exhibit high extinction rates, and that these intervals are clustered early in the Cenozoic, suggests that much of the variation is the result of long-term trends. This can be demonstrated by regressing the rates against time itself. Excluding the end-Pleistocene data point, there is a quite significant correlation (Fig. 7B: $n = 65$; Spearman's $r = 0.292$; $t = 2.426$; $p < 0.02$). Furthermore, extinction rates in neighboring intervals are correlated (Fig. 8: $n = 64$; Spearman's $r = 0.322$; $t = 2.679$; $p < 0.01$)—despite the fact that being spaced 1.0 my apart from each other, these sampling bins are unlikely to have recorded the same short-term environmental disturbances. Note that the end-Pleistocene event escapes the correlation because it is preceded by an interval with a relatively low extinction rate.

Removing the ten Paleocene data points from the time series reduces the coefficient of variation of the extinction rates from 41.8% to 31.8%. The remaining points are still too variable to be shoehorned into a model of constant extinction probability. However, they are still far less variable than origination rates prepared in the same way (Alroy, 1996, 1998c). Thus, all of the evidence suggests that the most important control on extinction is a secular trend toward lower rates. Such a pattern could have resulted from unusual environmental disturbances during the Paleocene in the aftermath of the K–T event, but this seems unlikely in light of the available paleoclimatic data, based on either oxygen isotopes (Zachos *et al.,* 1994) or leaf-margin analysis (Wolfe, 1978; Wing *et al.,* 1995). Instead, the data are better interpreted in light of a hypothesis having nothing to do with abiotic controls: higher intrinsic rates among the dominant, archaic Paleocene groups (Multituberculata, Plesiadapiformes, assorted "condylarths") than among the five major modern orders, all of which rose to prominence during the early Eocene (Carnivora, Insectivora, Rodentia, Artiodactyla, Perissodactyla).

Finally, the presence of so many minor spikes in the time series (Fig. 7B) might suggest that the major climate changes correspond with some of the extinction pulses. This is

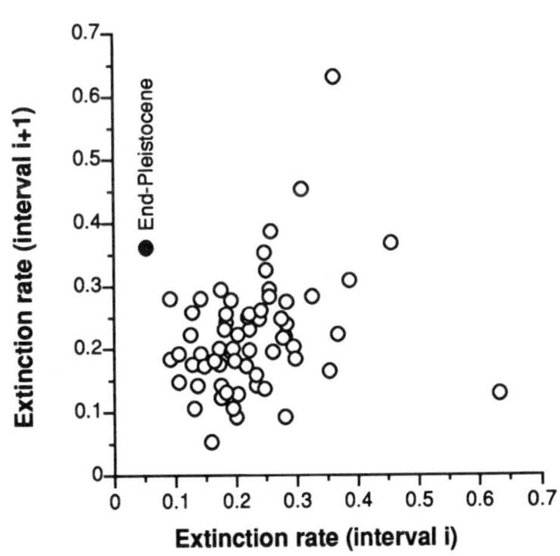

FIGURE 8. Correlation between extinction rates for neighboring 1.0-my-long sampling bins, based on the data shown in Fig. 7B. The end-Pleistocene event is unusual because it so little resembles the preceding interval.

not true. The five climate shifts discussed here are at the Paleocene–Eocene boundary (Kennett and Stott, 1991; Koch *et al.*, 1992; Bralower *et al.*, 1997: this correlates with 55.6 Ma in the current analysis), in the earliest Oligocene (Miller *et al.*, 1987; Berggren and Prothero, 1992), in the middle Miocene (Miller *et al.*, 1987; Flower and Kennett, 1994), toward the end of the late Miocene Messinian Salinity Crisis (Muller and Hsu, 1987; Hodell *et al.*, 1994), and in the late Pliocene (Dwyer *et al.*, 1995; Shackleton, 1995). Although the largest oceanic changes in the Messinian occurred at the beginning of this interval, not at the end, Cerling *et al.* (1997) believe that a shift from C3- to C4-dominated vegetation was more important than global climate changes, and that this shift occurred between the ages of the Coffee Range and Yepomera mammal localities (6.4 and 5.2 Ma in the current analysis). This falls at the end of the Messinian Salinity Crisis, and is in fact the most severe pre-Pleistocene extinction episode of the Neogene.

The results are very straightforward (Table 3). Four of five climate "crisis" intervals can be matched to extinction rates of less than 24%, with even their one-sigma binomial errors easily encompassing both the Cenozoic average of 22.2% and the post-Paleocene average of 20.2%. The one exception is the latest Miocene rate of 35.2 ± 6.3%, which is just over two standard deviations higher than the Cenozoic mean and, being the most extreme value in the post-Paleocene, surely a true extinction pulse. What is really remarkable about this interval is that it followed a long, stepwise cooling trend going all of the way back to the middle Eocene, and was followed by further cooling throughout the Plio-Pleistocene. Therefore, it would seem arbitrary to attribute this extinction pulse to global cooling per se.

What may have caused it instead is not clear. Cerling *et al.* (1997) argue that it was a vegetation change from C3 to C4 grasses, not climate change, that led directly to a taxonomic transition. However, MacFadden and Cerling (1996) have shown that by the end of the Miocene a characteristically diverse assemblage of perissodactyls in Florida (in this case horses and rhinos) already had adjusted to a mixed C3/C4 diet, so this assemblage must have survived the C3–C4 transition before perishing in the end-Miocene extinction event. This is quite important, not just because the latest Miocene extinction episode affected this previously flourishing ungulate group more than any other, but also because it conclusively refutes the idea that the C3–C4 transition happened at exactly the same time as the worst phase of this extinction crisis.

There is some disagreement over the relative importance of different climate changes during the Cenozoic. Some authors (e.g., Berggren and Prothero, 1992; Prothero and

Table 3. Major Cenozoic Climate Changes and Accompanying Extinctions Rates[a]

Interval (Ma)	Change	Species	Percent extinct
56–55	warming	42.3	23.3 ± 6.5
34–33	cooling	61.5	21.6 ± 5.3
16–15	cooling	76.9	19.2 ± 4.5
6–5	cooling	57.2	35.2 ± 6.3
3–2	cooling	69.4	23.2 ± 5.1

[a]Species counts for the beginning of each interval are averages across 100 randomized subsampling trials. One-sigma binomial errors on extinction percentages are based on the standard formula for counts, as given by Wilf (1997).

Heaton, 1996) believe that a cooling episode near the middle–late Eocene boundary (chron C17n.1n, or between 37.5 and 36.6 Ma: Cande and Kent, 1995) was much more important than the later event in the earliest Oligocene. However, paleobotanical evidence suggests that North American terrestrial communities changed relatively little until the earliest Oligocene shift (Wolfe, 1978, 1992). Furthermore, the two intervals between 38 and 37 Ma exhibit rather low extinction rates of 17 and 14%, so adding this event to the analysis would only weaken the climate–extinction link.

A final question concerns the Pleistocene record before the extinction pulse. Many authors have suggested that extinction episodes were a regular feature of the Pleistocene, but these inferences were based on lumping faunas into arbitrary categories, such as "Rancholabrean," and then computing overall extinction percentages. None of these figures were placed in perspective by standardizing sampling and imposing temporally uniform sampling intervals. The current analysis shows that extinction rates were low not just during the late Pliocene interval of 3–2 Ma, but during the two mostly Pleistocene intervals between 2 and 0.01 Ma. The 2–1 Ma interval yields a rate of 15.8 ± 4.6%, which is substantially, but not significantly, lower than the post-Paleocene mean. The 1–0.01 Ma interval, on the other hand, has a fantastically low extinction rate of 5.3 ± 2.7%, very significantly lower than the mean and, in fact, the lowest of the entire Cenozoic.

This figure should not be taken too literally, however, because the intense sampling of the late Pleistocene resulted in numerous first appearances that seem to have distorted the time scale in this interval; pre-Wisconsinan geochronologic age estimates are too sparse to detect such a bias in the calibration. Thus, the last "million" years may represent a much shorter period of a few hundred thousand years, which would explain the anomalous extinction percentage. Despite this minor effect, there is certainly no positive evidence for any kind of an extinction pulse in the late, pre-Wisconsinan Pleistocene. This is particularly surprising given that at least three deglaciation events about as severe, and almost certainly as rapid, as the latest one occurred during this interval (Shackleton, 1995).

To sum up, the end-Pleistocene megafaunal mass extinction was unlike virtually any other important crisis during the Cenozoic. It affected so many commonly preserved species that when sampled to resemble earlier intervals, it registers as a near catastrophe. It came millions of years after extinction rates stabilized and fell to a moderate average, and hundreds of thousands of years after the establishment of the extreme middle and late Pleistocene glacial cycles. Finally, although it did correspond closely with a major climate transition, only one previous event also did, and all other major climate changes went unanswered. Any ecological account of the megafaunal extinction must grapple with the fact that there is almost no credible evidence for climate-driven extinction at any other point in the fossil record of North American mammals.

4. Differential Extinction across the Size Spectrum

4.1. Motivation

The third and final test of the anthropogenic and ecological models concerns the selectivity of extinction. Perhaps the most obvious feature of the end-Pleistocene event is that it is a megafaunal extinction, not a broad-spectrum catastrophe. Small mammals passed through the North American crisis almost untouched, and there are few or no clear-cut extinctions at this time in other small vertebrates (Holman, 1995), insects (Coope, 1995), marine invertebrates (Valentine and Jablonski, 1991; Jackson, 1994), or plants, despite apparent changes in community composition (Overpeck *et al.*, 1992).

Counts of survivors and victims in different body-size classes show this pattern dramatically (Martin, 1967, 1984; Kurten and Anderson, 1980; Owen-Smith, 1987). Some of the earlier literature has obscured this pattern by setting a high bar for "small mammals" (e.g., body mass >44 kg: Martin, 1984), which results in classifying many carnivorans and even a few ungulates as "small mammals." Because of a gap in the among-species body mass distribution in both local and regional North American faunas, there is in fact a natural breakpoint between "large" and "small." However, it falls not above 10 kg, but somewhere between 1 and 10 kg (Gingerich, 1984; Alroy, 1998b), and it therefore cleanly separates the three diverse small-mammal orders (Insectivora, Lagomorpha, and Rodentia) from the ungulates and Carnivora.

In fact, the FAUNMAP data suggest extinctions for only 11 of the 165 late Wisconsinan species (7%) belonging to the three important small-mammal orders. Of these 11, 3 were quite large (*Castoroides ohioensis, Hydrochoerus holmesi, Neochoerus pinckneyi*), 2 are primarily known from the Irvingtonian and have just one questionable Wisconsinan record each (*Neofiber leonardi, Sorex scottensis*), 3 have been described from only a single locality and are therefore taxonomically suspect (*Neotoma findleyi, Neotoma pygmaea, Sylvilagus leonensis*), and 1 apparently survived into the middle Holocene (*Peromyscus nesodytes*). Graham and Lundelius (1984) list an additional 5 species, but none of them are recorded anywhere in the FAUNMAP database (Graham *et al.*, 1996), and in any event 3 of them are Pleistocene species that did not demonstrably survive into the late Wisconsinan (*Peromyscus cochrani, Tamias aristus, Thomomys orientalis*).

To sum up, just two true small-mammal species (*Peromyscus imperfectus* and *Synaptomys australis*) are unquestionable victims. Meanwhile, 50 of the 86 late Wisconsinan species (58%) belonging to the predominantly large-sized orders Artiodactyla, Carnivora, Perissodactyla, Proboscidea, and Xenarthra were victims. Apparently no more than 10 of 60 late Wisconsinan extinctions (17%) involved the three small-mammal orders, even though 165 of 252 late Wisconsinan species (65%) belonged to these orders. If anything, these figures would be made more extreme by using an arbitrary large/small cutoff of, say, 5 kg (Webb, 1984) instead of a taxonomic criterion, because the three large, extinct rodent species would be moved into the "large" category, and many surviving small carnivores would be moved into the "small" category.

The very fact that the microfauna and megafauna were so differentially affected is perhaps the best evidence in favor of the anthropogenic model. Other than size and correlates of size, there are no taxonomic, geographic, physiological, behavioral, or morphological factors that clearly differentiate the victims from the survivors. And other than size, there are no consistent factors that differentiate the mammalian targets of human predation. When faced with a choice, humans avoid hunting small animals because the net caloric payoff is too low on a kilogram-by-kilogram basis. Very large prey animals, on the other hand, represent a bonanza of easily utilized nutrition. And unlike most carnivores, humans across the world are capable of hunting any and all large terrestrial vertebrates.

Climate and vegetation change, on the other hand, are virtually the most implausible factors that one could invoke to explain the extinction of large mammals. Geographic range and physiological tolerance of climate variation are so strongly and positively correlated with size that it is hard to see how the very largest mammals could be affected by any but the most extraordinary environmental changes, whereas small mammals have such narrow dietary and climatic requirements and such small geographic ranges that it is hard to see how they survive climate cycles in the first place.

Advocates of ecological extinction models could parry this argument by pointing out that one set of ecological factors does scale in a way that penalizes large mammals during times of crisis: life history variables such as gestation interval, litter size, and first time of reproduction, all of which grow less favorable with increasing size (Eisenberg, 1981). Thus, it is more difficult for large-sized mammals to recover quickly when their population sizes are brought low by catastrophes. Based on this, one would expect that a large-sized species would be more vulnerable to stochastic extinction after falling to a very low population density in the wake of a major environmental perturbation.

Of the two predictions—that large mammals should survive because of their broad geographic distributions and climatic tolerances, or that they should perish because of their delicate demographic characteristics—I am by far more persuaded by the first. But in any case, the latter argument does present an excellent opportunity for hypothesis testing. If large mammals fare poorly during natural extinction crises, then there should be a consistent correlation among time intervals between the magnitude of extinction rates and the average size of the victims. The data and analyses presented here do weakly support this idea. However, they also show that the end-Pleistocene event followed a pattern not seen in any earlier interval: The victims were not merely large, but much, much larger than one would predict, and in fact so consistently large that they densely occupied a narrow range of the size spectrum. Earlier time intervals, on the other hand, consistently show a tendency for relatively severe extinction episodes to affect all parts of a broad size spectrum.

4.2. Data

Analyzing the size selectivity of extinction requires two types of data: the extinction rates discussed in the preceding section, and body mass estimates for these same species. It is well established in the mammalian paleontological literature that the occlusal area (length times width) of lower first molars (M_1s) can be used to generate robust estimates of body mass (Damuth and MacFadden, 1990). Lower M_1s also have the advantage of being easily identified, commonly preserved and described, and present in the dentition of all terrestrial mammals that do have teeth (there are no upper cheek teeth that meet this requirement).

Standard regression equations are available for each major, extant mammal order (Table 4). Investigations of other data have shown how to match some of the more important extinct orders to these equations, particularly the Plesiadapiformes, which scale like the Primates (Gunnell, 1989), and the Mesonychia, which scale like the Carnivora (Gunnell and Gingerich, 1996). For the others, an all-mammal regression equation seems to give reasonable results. Problem cases include the hypsodont ungulates (Perissodactyla and Artiodactyla), for which length but not area is a good predictor of size (Damuth, 1990); the Multituberculata and Xenarthra, for which the all-mammal regression is the only one available, even though M_1 size seems to be highly idiosyncratic in these groups; and the Proboscidea,

Table 4. Regression Equations Relating M_1 Size Measurements to Body Mass[a]

Group	Variable	Slope	Intercept	References
Carnivora	Area	1.653	1.897	Legendre and Roth (1988)
Insectivora	Area	1.654	1.746	Legendre (1986)
Plesiadapiformes	Area	1.490	3.577	Legendre (1986), Gunnell (1989)
Primates	Area	1.490	3.577	Legendre (1986)
Rodentia	Area	1.767	2.172	Legendre (1986)
Mesonychia	Area	1.653	1.897	Legendre and Roth (1988), Gunnell and Gingerich (1996)
Artiodactyla	Length	3.20	2.694	Damuth (1990)
Perissodactyla	Length	3.17	2.395	Damuth (1990)
Proboscidea	Area	1.47	1.658	Damuth (1990)
Other mammals	Area	1.827	1.810	Legendre (1986)

[a]Variables are log transformed. Area is based on length times width.

for which M_1s are almost never described, and which usually have only one or two lower cheek teeth in occlusion at any one time. Each of these teeth seems to function as does a consecutive series of three or more teeth in a normal mammalian dentition. Therefore, for the Proboscidea I matched M_2 measurements to the all-mammal P_4–M_2 equation of Damuth (1990), which seems to yield reasonable results.

Although M_1 measurements are widely reported, not all fossil mammal species are known from lower dentitions, and even for those that are, measurements have not always been published. Additionally, at this point in time my literature search is not complete and I believe that several hundred more measured species remain to be added. Despite these caveats, I was able to obtain size estimates for 1523 species that were based on 14,172 measurements of 2830 separate faunal samples. A single overall estimate was computed for each species because samples differed widely in temporal and spatial resolution. The data were standardized taxonomically and checked against the faunal lists to guarantee that identifications were up to date. Although the data set does include some late Cretaceous species, I restricted the analysis to the Cenozoic on the grounds that turnover rates (Alroy, 1988c) and body mass distributions (Alroy, 1998b) both shifted dramatically at the K–T boundary.

The number of measured species going extinct in each 1.0-my time interval is shown in Fig. 9. As one might expect, the data are not evenly distributed through time, with very poor sampling in the early Paleocene (65–62 Ma), Duchesnean (about 42–38 Ma), and Geringian (about 30–24 Ma). This does affect the analyses, which yield statistically significant results only after excluding the fourteen 1.0-my intervals that each include less than 10 species going extinct. Only the results excluding these intervals are reported below. The end-Pleistocene is well represented in the data set, with 21 extinct and 43 surviving taxa.

4.3. Methods

In analyzing the body mass estimate data, I will take an extremely simplistic approach in which assorted statistics will be averaged across all species going extinct during a 1.0-my

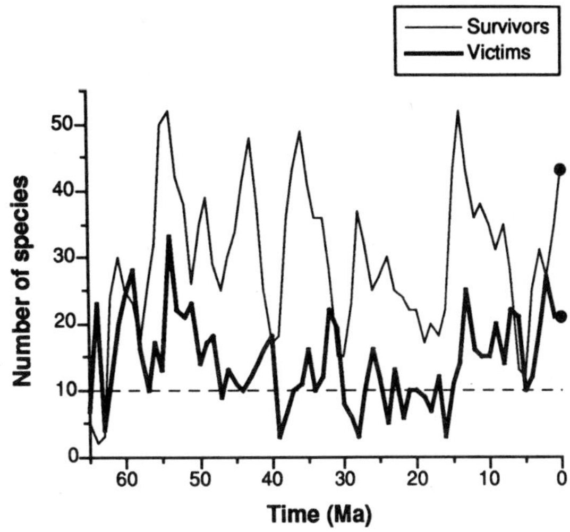

FIGURE 9. Number of species surviving and going extinct in each 1.0-my time interval for which body masses are available. Note that the data are not corrected for sampling intensity. The 14 intervals that include fewer than 10 species (dashed line) in one or both categories are omitted from the following figures.

interval (victims), or ranging beyond it into the next interval (survivors). This raises two questions. First, should all species be considered, or just those appearing in the randomized faunal subsamples that were used in the earlier analyses? Second, does it matter that the sample sizes in different intervals vary substantially? Because the goal of the computation is to compute means with the lowest possible standard errors, and because it is fair to presume that the measured species in each category are drawn randomly from the total pool, I believe that it is fair to consider all measured species regardless of how many are present in the extinction rate data set and regardless of how sampling intensity varies across intervals. Excluding data would only increase errors without moving the expected values of the averages.

A second problem concerns the statistics to be computed. Mean masses are obviously relevant, and if computed across logged values these means should not be problematic (the raw masses are extraordinarily skewed, but the logged masses begin to approach normal, or at least unskewed distributions). The variation in size within each class also is obviously relevant. This is easily quantified by computing standard deviations. Additional moment statistics such as skewness (g_1) and kurtosis (g_2) also could be computed, but going into this level of detail seems unnecessary given the clear-cut results.

One final statistic that does have the potential to tell a different story is the nearest-neighbor distance (NND). This is widely used in the paleobiological literature on morphological disparity (e.g., Foote, 1990), and should be sensitive to fine-scale patterns of spacing across the size distribution that might not have an impact on standard moment statistics. NND is computed for each species by finding the one other species with the most similar mass, and then taking the absolute value of the difference of their masses on a log scale. A mean value for all of the species in a category is then calculated across each NND estimate. Low values indicate that species are closely clumped, whereas high values indicate that they are evenly spaced across the size spectrum.

One problem with the NND measure is that it is highly dependent on the sample size (Foote, 1990). If we imagine that the true number of species in the sampling pool is very large relative to the measured subsample, we can picture the NND falling as a regular function as we measure more and more species. When just one species is measured, the statistic cannot be defined. After adding one more species, the first species's true nearest neighbor almost certainly will not have been found, so the difference should behave like a random draw from the distribution. Therefore, the mean NND should approximate twice the standard deviation at this point. As each new species is added, the NND should fall at a rate of two divided by the number of species. Eventually, the true nearest neighbor will be encountered. But until this point is reached, the statistic will continue to have a strong relationship to the sample size. Instead of worrying about the value's limit as sampling improves, I will simply scale NND to the number of species that have been found on the assumption that this will be sufficient to draw out a meaningful signal. This can be done simply by multiplying the observed NND by one less than the number of species in the sample, which results in a "dimensionless" NND that approximates the value that would be expected if each sampling pool consisted only of two species.

4.4. Results

The key pattern to be tested in each analysis is whether the victim and survivor pools differ with respect to a particular statistic, and whether these differences correlate with extinction intensity. In each of these analyses, I exclude the end-Pleistocene data point so that the resulting regression equation can be used to predict it independently.

Mean size does show an interesting pattern. There is a significant, moderately strong, and positive correlation between extinction intensity and the victim/survivor size difference (Fig. 10: $n = 51$; $r = -0.370$; $t = 2.792$; $p < 0.01$). The relationship predicts that for the end-Pleistocene event's extinction intensity of 32.4%, the differential should have been 0.66 ± 0.51 natural log unit (95% confidence limits). This is not even vaguely close to the observed value of 4.66 log units, the difference equaling a multiplicative factor of 55—and a factor of 33 even if we assume the predicted value's upper 95% confidence limit of 1.17 log units. At best, this shows only that the relationship is broadly compatible with the general idea that severe extinctions result in the differential extinction of large species, but that

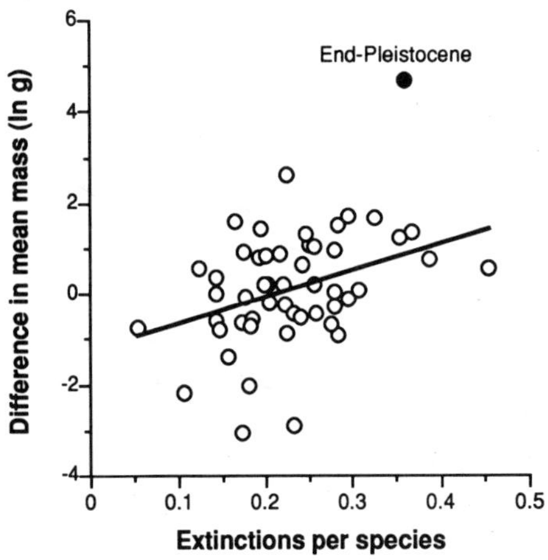

FIGURE 10. Correlation among 1.0-my intervals between extinction intensity and the difference in the mean body masses of the victims and survivors.

it fails almost totally to predict the end-Pleistocene pattern. Note that by the end-Pleistocene, both victims and survivors had evolved to much larger sizes than were typical of the early Cenozoic (Fig. 11); but this has no bearing on the fact that in earlier times, victims and survivors almost always were of similar sizes.

The results regarding variation and clumping across the distribution are just as interesting. There is no significant correlation between extinction intensity and the difference in standard deviation of size between victim/survivors (Fig. 12; $r = 0.090$; $t = 0.635$; NS). In other words, victims tend to be just as scattered across the size spectrum as survivors—and this is true also of the end-Pleistocene, where the difference of -0.61 log unit is well within the observed range of variation.

But this ignores a far more important pattern: Both the end-Pleistocene victims and survivors show dramatically narrow ranges of variation. Because mean size increased steadily during the Cenozoic, but the lower limit to size did not (Alroy, 1998b), the standard deviation of size also increased steadily. After fitting simple linear regressions to the two time-versus-standard deviation curves (Fig. 13: victims: $r = -0.739$; $t = 7.673$; $p < 0.001$; survivors: $r = -0.793$; $t = 9.116$; $p < 0.001$), one expects that during an extinction event at 0 Ma victims would have a standard deviation of 4.10 log units, and the survivors a standard deviation of 4.07 log units. These figures are close to those observed in the last 1.0 my of the Pleistocene before the event: 4.07 and 3.24 log units. But for the event itself, both numbers are far lower—2.07 and 2.67 log units. In other words, the end-Pleistocene event discriminated strongly, clustering victims into one large size class and survivors into another, smaller size class, each class showing very little variation.

The final piece of evidence has to do with the clumping of species across the size spectrum. The NND measure might be expected to be a strong function of the overall standard deviation if very small-scale clumping patterns are weak. However, the NND and standard deviation curves are uncorrelated, both for the victims ($r = 0.212$; $t = 1.518$; NS) and for the survivors ($r = 0.004$; $t = 0.027$; NS). Once again, extinction intensity shows no correlation with the difference in mean NND between victims and survivors (Fig. 14: $r = 0.182$; $t = 1.293$; NS).

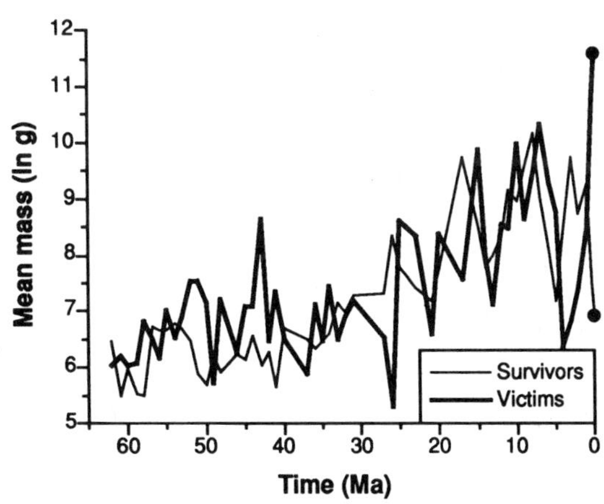

FIGURE 11. Trend through the Cenozoic in the mean body mass of victims and survivors in each 1.0-my interval.

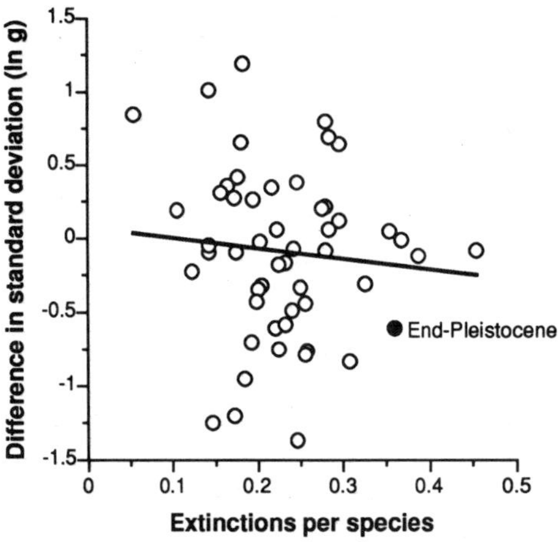

FIGURE 12. Correlation among 1.0-my time intervals between extinction intensity and the difference in the standard deviation of body mass of the victims and survivors.

But surprisingly, the NND curves fail to show a strong temporal trend, being only weakly correlated or uncorrelated with time (Fig. 15: victims: $r = -0.366$; $t = 2.755$; $p < 0.01$; survivors: $r = -0.056$; $t = 0.394$; NS). Thus, even as the overall dispersion of species across the size spectrum steadily increased throughout the Cenozoic, the spacing of similar-sized species held to the pattern established during the Paleocene. Interestingly, the pattern for surviving species seems to have shifted abruptly in the late Oligocene, a time of little climate change and just as little faunal turnover (Fig. 7).

Based on these patterns, we can predict that mean end-Pleistocene NND should have been 7.31 log units for victims and 6.38 log units (the mean across the time series) for survivors. In fact, it was 8.66 for victims and 8.40 for survivors. These predicted and observed values are not significantly different, which contrasts with the results for standard deviation. Based on this, it seems that although the end-Pleistocene event was extraordinarily selective in terms of the mean size of the victims and the narrow band of the size spectrum these victims occupied, there was little discrimination at a fine scale: Sets of similar-sized species with presumably competitive interactions were no more likely to become extinct

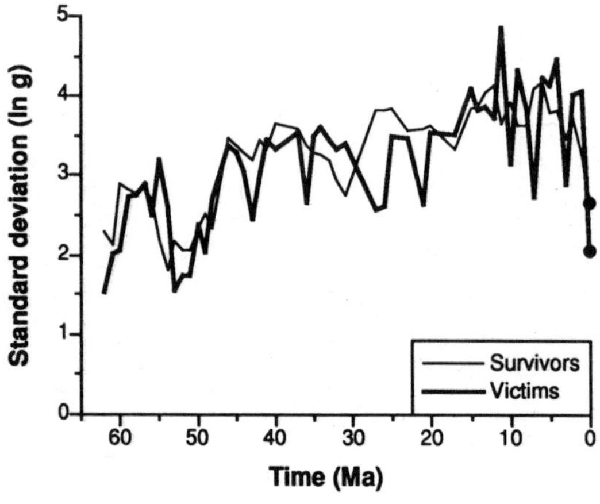

FIGURE 13. Trend through the Cenozoic in the standard deviation of body mass among victims and survivors in each 1.0-my interval.

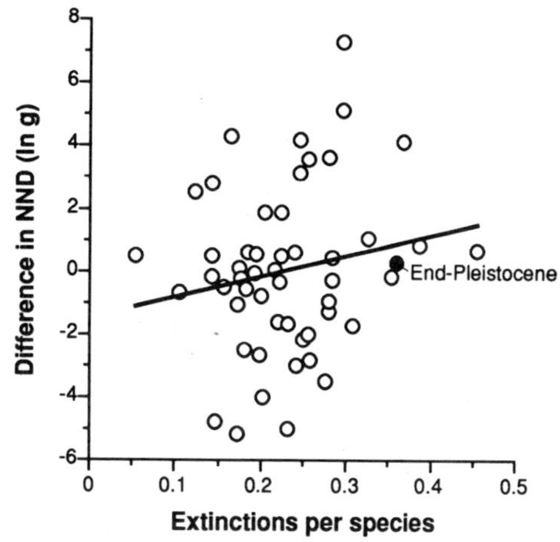

FIGURE 14. Correlation among 1.0-my time intervals between extinction intensity and the average nearest-neighbor distance between victims and between survivors.

than were isolated species having unusual sizes. This pattern has all of the marks of an unnatural, broad-brush catastrophe, with none of the telltale hints of an ecologically driven crisis similar to those that occasionally transpired during the Tertiary.

5. Discussion

5.1. In Favor of Overkill

It would be presumptuous to claim that the overkill model, even in its most flexible form, explains all of the evidence regarding the end-Pleistocene megafaunal extinction. But at this point, it should be clear that advocates of competing, ecological explanations have misinterpreted broader patterns in the fossil record (e.g., Barnosky, 1989; Graham *et al.*, 1996). In general, these oversights resulted from the cruder analytical methods that

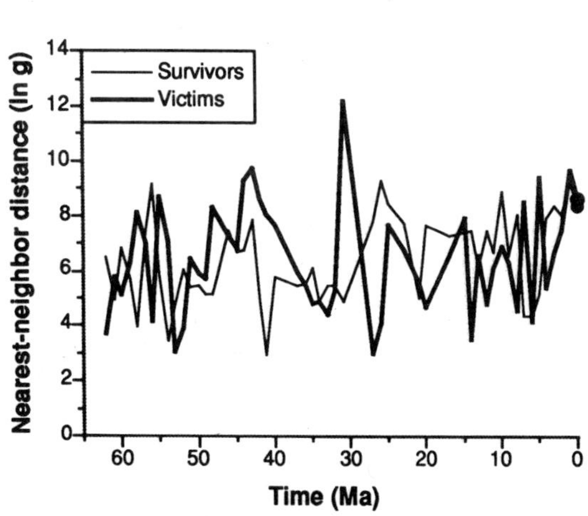

FIGURE 15. Trend through the Cenozoic in the average nearest-neighbor distance between victims and between survivors in each 1.0-my interval.

were previously available, which generated such spurious patterns as the appearance of repeated mass extinctions in the Tertiary. Together with the new results given here, I believe that there is now a very solid body of evidence to support the anthropogenic model. Because much of this has been discussed by earlier workers (e.g., Martin, 1967, 1984; Owen-Smith, 1987; MacPhee and Marx, 1997), I will simply list some of these points and add minimal commentary. Before doing so, I would like to emphasize that I have failed to identify any particular evidence that uniquely favors ecological or disease-driven models. Thus, these models are unparsimonious: They not only fail to explain some key evidence that is uniquely accounted for by overkill, but require additional ad hoc arguments to bring themselves into consistency with the facts. Their main motivation appears to be based on a few inflexible, a priori objections to overkill that I will discuss below.

1. The extinction occurred within a few hundred years, or at most 1 or 2 kyr, of the appearance of the Clovis hunting technology. There is no solid evidence for technologically sophisticated human hunters south of Alaska before this point. There is solid evidence that Clovis hunters preyed on extinct megafauna such as mammoths.
2. Almost all of the victims were either plausible human prey species or carnivores that depended on these prey species.
3. The surviving prey species mostly either (a) inhabit mountains or tundra regions, which makes them difficult to hunt (mountain goat, Dall's sheep, mountain sheep, muskoxen, caribou); (b) have pan-Holarctic distributions, had sister species with Eurasian distributions, or immigrated from Eurasia only in the very late Pleistocene, which suggests that they had long since evolved the ability to avoid human hunting (bison, wapiti, moose, Dall's sheep, mountain sheep, muskoxen, caribou); or (c) had numerous close relatives that did become extinct, which suggests that their survival was a matter of chance (peccaries, pronghorn). The only obvious exceptions to these rules are the white-tailed and mule deer. None of these lists are long compared with the standard list of more than three dozen large herbivore victims (Graham and Lundelius, 1984; Martin, 1984).
4. The surviving predator species are mostly independent of large game, because they are either (a) small enough to survive on a diet of small vertebrates (e.g., coyote, red wolf, lynx) or (b) omnivorous (black bear, brown bear). The only possible exceptions are the gray wolf, cougar, and jaguar, all of which routinely take small prey items and none of which are hypercarnivores, as were such victims as the saber-toothed cats and the American "cheetah."
5. Quaternary extinctions of large vertebrates in South America, the Caribbean, several Mediterranean islands, Madagascar, Australasia, and New Zealand occurred at several different times, but always at the same time as the first appearance of human hunters, regardless of local environmental conditions.

5.2. Against Ecological Models

The second set of observations are those that render ecological models implausible. It is important to note that points 1 through 7 cannot be held against the disease model, although point 8 is highly incompatible with it, insofar as diseases brought by immigrants are a natural phenomenon that existed throughout the Cenozoic.

1. The collapse of glaciers at this time vastly increased, not decreased, the habitable area of the continent. This is true regardless of the fact that climate vacillated during the terminal Pleistocene and early Holocene (Stager and Mayewski, 1997; Taylor *et al.,* 1997), because the glacial retreat itself was not reversed during short-term cold periods. It also is irrelevant that a few cold-tolerant species may have been trapped on mountaintops by the deglaciation, because (a) almost all of the end-Pleistocene victims were large-mammal species with large geographic ranges and significant dispersal abilities that easily could have followed the glacial retreat instead of becoming trapped; (b) the best candidate for a climate event that is temporally coincident with the extinction is the onset of the Younger Dryas, which was a cooling event, not a warming event; and (c) mammalian species that are trapped by climate change on mountaintops are known to have very low rates of extinction that lead to their persistence for thousands, not hundreds, of years (Brown, 1971).
2. Extinctions occurred at all latitudes, longitudes, and altitudes, irrespective of local climates or the local extent of climate change. If survival probabilities were enhanced anywhere, it was in already severe habitats such as the tundra.
3. Very few extinctions occurred elsewhere at this particular time, despite worldwide climate changes.
4. In stark contrast to the overkill model, simple deductions from the ecological models would predict the opposite of the observed body-size selectivity pattern: Large mammals have large geographic ranges, can migrate quickly, and can tolerate a wide variety of habitats and climates, so they not only should have been largely invulnerable to climate changes, but should have benefited even more than small mammals from the rapid expansion of habitats. Advocates of the ecological models have drawn the opposite conclusion by focusing on reproductive rates and other life history parameters, but these features also would make large mammals vulnerable to disease or overkill, and the argument begs the question of how ecological generalists could have suffered high extinction rates if those rates were elevated by strictly ecological factors.
5. As argued in Section 2, there is no evidence for serious disruption of the surviving mammal communities. Previous studies have focused on a small number of exceptional species (e.g., Graham *et al.,* 1996) instead of considering the whole pattern. Claiming that these exceptional species prove something important about Quaternary extinctions misses the point that extinction was the rule, not the exception, for Pleistocene megafauna: If reorganization had had anything to do with the extinction, it would have had to have been every bit as extraordinary.
6. Even if there had been a major ecological reorganization among small mammals, this pattern would be irrelevant because it would not fall directly in a causal path leading from ecological factors to a mass extinction.
7. As argued in Section 3, there were no extinction pulses during the preceding, equally severe glacial cycles of the Pleistocene, nor were extinction pulses clearly linked to climate changes during the Tertiary. Attempts to explain this have focused on ex-

plaining the lack of extinction during the last, Sangamonian interglacial (e.g., Graham, 1986). This ignores the fact that severe deglaciation events occurred not just twice, but repeatedly during the Pleistocene (see Shackleton, 1995). Because abrupt and massive climate shifts are found throughout the last two glacial cycles (e.g., Dansgaard *et al.,* 1993; Stager and Mayewski, 1997; Taylor *et al.,* 1997), it would be a desperate ad hoc argument to predict that if we had a detailed record going back that far, we would not find that the preceding glacial cycles also encompassed many rapid, short-term, and severe climate excursions.

8. As argued in Section 4, the weak tendency for large extinction pulses to target larger species hardly even begins to account for the end-Pleistocene event's extraordinary size selectivity. Furthermore, none of the previous, natural extinction pulses during the Cenozoic targeted a narrow range of the body-size spectrum.

5.3. Against Overkill

The third set of observations are those that still could be held against the anthropogenic model. It is important to note that none of these objections pertain directly to what happened in North America during the terminal Pleistocene, apart from one generalized, a priori assertion that itself has a weak factual basis (i.e., point 4).

1. Quaternary large-mammal extinctions in Africa and Eurasia were relatively minor, even though human hunters appeared there first.
2. There were no further North American large-mammal extinctions during the Holocene, even though human hunters were present.
3. In historical times, few large mammal species have gone extinct even when heavily hunted.
4. It is hard to believe that a few humans with primitive tools could devastate a hemisphere's large mammals in such a brief period of time.

Point 1 is easily dismissed. As explained by Diamond (1984) and Martin (1967, 1984), one would expect species evolving in the presence of primitive hominid hunters to have acquired behaviors making them hard to hunt. It is the very fact that so many megafaunal victims were Western Hemisphere endemics, and thus that so many survivors were pan-Holarctic, that suggests the importance of such instincts.

Point 2 is more subtle. I have suggested that the survivors show a pattern of being protected from hunting by behavior and/or favored habitat. Simply enough, most of the species that were vulnerable to hunting in the first place did go extinct; the survivors were buffered from such pressures. But there are a handful of exceptions to the rule. I believe there is another simple ecological explanation for this fact, namely, that if a large number of species are all depressed to low population levels at the same time, then density-dependent growth rates will make it nearly impossible to drive all of them into extinction. This should be true even if the prey species have nearly identical ecological properties; just by chance, some of these species should hold on long enough to benefit from the sudden, near-total absence of competition. In other words, as more and more species go extinct and overall prey population densities fall lower and lower, the intrinsic rates of increase of the surviving species should become extremely large.

A possible counterargument is that during such a crisis, prey species might actually have reduced, not enhanced, intrinsic rates of increase. This would be related to depensation, or Allee effects, which include inbreeding depression and the difficulty of finding a mate when there are few surviving conspecifics. But there is little evidence that real populations work this way; instead, a recent, comprehensive analysis of marine fish populations showed that virtually all of them behave according to standard logistic growth functions that do not allow for depensation (Myers *et al.,* 1995). Thus, most populations should be extraordinarily resistant to extinction when disturbances cut down their sizes.

One might naively conclude from this argument that overkill itself is implausible, because all species should be protected by density dependence. This is spurious. First, density dependence may never come into play for the victims if the few surviving prey species increase their population sizes more quickly to compensate for the victims' losses. Second, even if there is no compensation, it still follows inevitably from the standard Lotka–Volterra equations that whenever a predator can persist on at least one abundant species, all other species that it victimizes may be driven to extinction. All that is needed is for the predator's negative impact on survivorship to exceed the boost to the victims' intrinsic rates of increase afforded by density dependence. Finally, even if the argument did have merits, it would apply just as forcefully to the ecological models, which cannot ignore the fact that any cause of severe mortality will result in decreased competition for resources; as noted previously, food resources for wide-ranging, large-sized herbivores could only have increased as the ice sheets were collapsing.

Point 3 is a manipulation of the facts. The important point is not that the number of historical large-mammal extinctions is small, but that it is proportionately much larger than the number for mammals in general. Six species belonging to the ungulate orders Artiodactyla and Perissodactyla have gone extinct over the past 500 years; the Vietnamese warty pig, Arabian gazelle, red gazelle, and bluebuck (Baillie and Groombridge, 1996), and two species of hippos from Madagascar that are known from the relatively young archaeological record (MacPhee and Flemming, this volume). Baillie and Groombridge (1996) list two other species that clearly shouldn't be (Schomburgk's deer, which has been rediscovered in the wild, and the quaqqa, which is a subspecies: MacPhee and Flemming, this volume), and some of the others could be ranked as subspecies instead of full species. However, at least five more true species went extinct in the wild: Przewalski's wild horse, Pere David's deer, the Saudi gazelle, the black wildebeest, and the Arabian oryx. Several other species arguably could be added to this second list, including both species of bison. Many more species are restricted to nature preserves and surely would have gone extinct by now without heroic efforts, including all five surviving species of rhinos, the mountain zebra, several species of asses, and many ruminants.

Now if we accept a rough figure of, say, 11 ungulate extinctions in the wild, that constitutes 12% of 91 known mammalian species extinctions in historical times—but these two orders include just 5% (238/4629) of all mammal species to begin with (Wilson and Reeder, 1993). The difference is significant ($G = 6.586$; $df = 1$; $p < 0.025$), and it suggests that ungulates have a roughly doubled extinction risk compared with other mammals. Indeed, the same pattern is seen if we add endangered and critically endangered species to the counts: 10% (55/573) of extinct or endangered mammal species are artiodactyls or perissodactyls (Baillie and Groombridge, 1996). This figure is particularly important not just because it is highly significant ($G = 25.234$; $df = 1$; $p < 0.001$), but also because it is far less open to the kind of quibbling that might make a count of 11 turn into a count of 4.

The pattern of historic extinctions is particularly striking because ungulates already had suffered a major extinction in the late Quaternary, so their pool of potentially vulnerable species was diminished to start with. Furthermore, all or almost all extinct ungulates, and most endangered ungulates, dwell on continents, whereas many endangered or extinct small mammals dwell on islands. It is well known that vertebrates living on islands have suffered a greatly disproportionate rate of anthropogenic extinctions, so if we focused just on continental species the discrepancy would grow even larger. All of this shows that ungulates have indeed been extraordinarily sensitive to human impacts in historical times. I would venture that humans in fact are capable of driving virtually any large mammal species extinct, even though they also are capable of going to great lengths to guarantee that they do not.

Point 4 is perhaps the most frequently repeated and stressed, and seems to be the major motivation for the disease model (MacPhee and Marx, 1997), whose authors are willing to take on extraordinary burdens just to avoid this problem (see below). It already has been argued that a fauna that never before encountered bipedal hunting primates would not have evolved any of the natural instincts that make it possible to avoid predation (Martin, 1967, 1984; Diamond, 1984). Indeed, the lack of such instincts seems to have been universal among vertebrates dwelling on oceanic islands, which partially explains the phenomenal extinction rates of birds in the wake of human colonization (e.g., Steadman, 1989). I believe that this behavioral hypothesis is sufficient to explain why the North American extinction was possible in the first place, and why both it and several others in the late Quaternary (e.g., in Oceania) were so much more severe than extinctions in Africa and Eurasia.

5.4. Against the Disease Model

One final topic is the disease model, which has been presented as a neat solution to the drawbacks of the others (MacPhee and Marx, 1997). I already have indicated that I find this model implausible because the end-Pleistocene event is so unnatural compared with earlier, Cenozoic extinctions, particularly in terms of intensity and body size selectivity. There are at least four other reasons to discount this extraordinary hypothesis.

1. North America was not a sterile, isolated laboratory during the Cenozoic. Quite to the contrary, at least 160 mammal species immigrated to the continent during this time, and most of these survived well enough to spread across the continent and leave descendant species (Woodburne and Swisher, 1995; Alroy, 1998a). Potential carriers of pandemic disease organisms therefore appeared frequently during the Cenozoic, but no extinction resembling the end-Pleistocene event ever ensued. So far from being a "first contact" as far as disease organisms are concerned, the arrival of humans in North America probably qualified as a completely unremarkable 161st, or perhaps 201st, contact.
2. There is no such thing in the Recent as a pandemic disease that can drive virtually any large mammal species into extinction—or even cause high levels of mortality

throughout populations belonging to different mammal orders (MacPhee and Marx, 1997). Most of the well-known epidemic mammalian diseases are restricted to a single order, including rinderpest (Artiodactyla) and myxomatosis (Lagomorpha). The few diseases that can strike multiple orders (e.g., rabies, influenza) typically either have low transmission rates or produce low mortality rates.

3. Even if we do grant that a pandemic disease could strike any and all groups of mammals, this makes it difficult to explain the fact that the extinction was overwhelmingly confined to large mammals. One could raise a taxonomic argument, namely, that the victims (mostly Artiodactyla, Carnivora, Perissodactyla, Proboscidea, Xenarthra) and survivors (mostly Insectivora, Lagomorpha, and Rodentia) are easily distinguished on phylogenetic grounds; so perhaps the disease organism was simply unable to invade members of certain phylogenetic lineages. Unfortunately, this argument is weakened by the fact that both the sets of victims and survivors form polyphyletic assemblages that span all of the major branches of the Eutheria. MacPhee and Marx (1997) have tackled this problem head-on by arguing that the victimized large mammals were united by demographic and physiological characteristics that made it difficult for them to recover from disease-induced mortality. But this ignores the fact that many of North America's hundreds of small mammal species do have small geographic ranges, low population densities, low intrinsic rates of increase, and/or low basal metabolic rates (e.g., Eisenberg, 1981).
4. The disease model also fails to explain the pattern of survivorship among large mammals, in which a preference for difficult terrain (tundra, high mountains) seems to be a common factor.

5.5. Conclusion

The real lesson of late Quaternary mass extinctions is not that strange things happened in the past, a superficial impression that is the only one this ongoing debate will leave with the public. Instead, Quaternary paleoecologists should unite to publicize one of the greatest discoveries in the history of paleontology: Long before the dawn of written history, human impacts were responsible for a fantastically destructive wave of extinctions around the globe. This message should be seen as a wake-up call instead of a mere omen of disaster. Although the fossil record proves that biodiversity is far more vulnerable than we would like to think, the conservation movement's past successes show us that much can be done to save our biological heritage.

It may seem dismaying that so many Quaternary paleoecologists have overlooked such a loud and clear message. However, I believe that in previous years Quaternary paleoecologists were reasonably skeptical about the anthropogenic model because the available information on big-picture patterns was then incomplete. Now that some crucial details have finally been set straight, I hope we can agree to move on to new and more challenging problems. Regardless of the inevitable conclusion that the North American megafaunal mass extinction resulted from human impacts, Quaternary research will continue to be important because understanding the recent past is so very crucial in a world of rapid, anthropogenic climate change.

Acknowledgments

This chapter is dedicated to Paul Martin in honor of his groundbreaking studies of mass extinction. I thank Ross MacPhee and Paul Martin for their encouragement, and for their kind invitation to attend the 1996 Quaternary mass extinctions symposium. I also thank Ross MacPhee for his feedback and editorial efforts. My work benefited as well from discussions with Russ Graham, Holmes Semken, and the other symposium speakers. This work was supported by the Smithsonian Institution's postdoctoral fellowship program and by the Evolution of Terrestrial Ecosystems program.

References

Alroy, J. 1992. Conjunction among taxonomic distributions and the Miocene mammalian biochronology of the Great Plains. *Paleobiology* **18**:326–343.

Alroy, J. 1994. Appearance event ordination: a new biochronologic method. *Paleobiology* **20**:191–207.

Alroy, J. 1996. Constant extinction, constrained diversification, and uncoordinated stasis in North American mammals. *Palaeogeogr. Palaeoclimatol. Palaeoecol.* **127**:285–311.

Alroy, J. 1998a. Diachrony of mammalian appearance events: implications for biochronology. *Geology* **26**: 23–26.

Alroy, J. 1998b. Cope's rule and the dynamics of body mass evolution in North American mammals. *Science* **280**:731–734.

Alroy, J. 1998c. Equilibrial diversity dynamics in North American mammals, M. L. McKinney and J. Drake (eds.), *Biodiversity Dynamics: Turnover of Populations, Taxa and Communities.* pp. 232–287. Columbia University Press, New York.

Alroy, J. In press. Methods for removing sampling biases from diversity curves. *Paleobiology.*

Archibald, J. D. 1996. *Dinosaur Extinction and the End of an Era: What the Fossils Say.* Columbia University Press, New York.

Baillie, J., and Groombridge, B. (eds.). 1996. *The 1996 IUCN Red List of Threatened Animals.* Conservation International, Washington, D.C.

Barnosky, A. D. 1989. The late Pleistocene event as a paradigm for widespread mammal extinction, in: S. K. Donovan (ed.), *Mass Extinctions,* pp. 235–254. Columbia University Press, New York.

Beck, M. W. 1996. On discerning the cause of late Pleistocene megafaunal extinctions. *Paleobiology* **22**:91–103.

Berggren, W. A., and Prothero, D. R. 1992. Eocene–Oligocene climatic and biotic evolution: an overview, in: D. R. Prothero and W. A. Berggren (eds.), *Eocene–Oligocene Climatic and Biotic Evolution,* pp. 1–28. Princeton University Press, Princeton, N.J.

Bralower, T. J., Thomas, D. J., Zachos, J. C., Hirschmann, M. M., Rohl, U. Sigurdsson, H., Thomas, E., and Whitney, D. L. 1997. High-resolution records of the late Paleocene thermal maximum and circum-Caribbean volcanism: is there a causal link? *Geology* **25**:963–966.

Brown, J. H. 1971. Mammals on mountaintops: nonequilibrium insular biogeography. *Am. Nat.* **105**:467–478.

Brown, J. H., and Nicoletto, P. F. 1991. Spatial scaling of species composition: body masses of North American land mammals. *Am. Nat.* **138**:1478–1512.

Cande, S. C., and Kent, D. V. 1995. Revised calibration of the geomagnetic polarity timescale for the Late Cretaceous and Cenozoic. *J. Geophys. Res.* **100**:6093–6095.

Cerling, T. E., Harris, J. M., MacFadden, B. J., Leakey, M. G., Quade, J., Eisenmann, V., and Ehleringer, J. R. 1997. Global vegetation change through the Miocene/Pliocene boundary. *Nature* **389**:153–158.

Coope, G. R. 1995. Insect faunas in Ice Age environments: why so little extinction?, in: J. H. Lawton and R. M. May (eds.), *Extinction Rates,* pp. 55–74. Oxford University Press, London.

Damuth, J. 1990. Problems in estimating body masses of archaic ungulates using dental measurements, in: J. Damuth and B. J. MacFadden (eds.), *Body Size in Mammalian Paleobiology: Estimation and Biological Implications,* pp. 229–253. Cambridge University Press, London.

Damuth, J., and MacFadden, B. J. (eds.). 1990. *Body Size in Mammalian Paleobiology: Estimation and Biological Implications.* Cambridge University Press, London.

Dansgaard, W., Johnsen, S., Clausen, H. B., Dahl-Jensen, D., Gundestrup, N. S., Hammer, C. U., Hvidberg, C. S., Steffensen, J. P., Sveinbjörnsdottir, A. E., Jouzel, J., and Bond, G. 1993. Evidence for general instability of past climate from a 250-kyr ice-core record. *Nature* **364**:218–220.

Diamond, J. M. 1984. Historic extinction: a Rosetta stone for understanding prehistoric extinction equilibria, in: P. S. Martin and R. G. Klein (eds.), *Quaternary Extinctions: A Prehistoric Revolution,* pp. 824–866. University of Arizona Press, Tucson.

Dwyer, G. S., Cronin, T. M., Baker, P. A., Raymo, M. E., Buzas, J. S., and Correge, T. 1995. North Atlantic deep-water temperature change during late Pliocene and late Quaternary climatic cycles. *Science* **270**:1347–1351.

Eisenberg, J. F. 1981. *The Mammalian Radiations: An Analysis of Trends in Evolution, Adaptation, and Behavior.* University of Chicago Press, Chicago.

Flower, B. P., and Kennett, J. P. 1994. The middle Miocene climatic transition: East Antarctic ice sheet development, deep ocean circulation and global carbon cycling. *Palaeogeogr. Palaeoclimatol. Palaeoecol.* **108**:537–555.

Foote, M. 1990. Nearest-neighbor analysis of trilobite morphospace. *Syst. Zool* **39**:371–382.

Foote, M. 1994. Temporal variation in extinction risk and temporal scaling of extinction metrics. *Paleobiology* **20:**424–444.

Gingerich, P. D. 1984. Pleistocene extinctions in the context of origination–extinction equilibria, in: P. S. Martin and R. G. Klein (eds.), *Quaternary Extinctions: A Prehistoric Revolution,* pp. 211–222. University of Arizona Press, Tucson.

Graham, R. W. 1976. Late Wisconsin mammalian faunas and environmental gradients of the eastern United States. *Paleobiology* **2**:343–350.

Graham, R. W. 1986. Response of mammalian communities to environmental changes during the late Quaternary, in: J. Diamond and T. J. Case (eds.), *Community Ecology,* pp. 300–313. Harper & Row, New York.

Graham, R. W., and Lundelius, E. L., Jr. 1984. Coevolutionary disequilibrium and Pleistocene extinctions, in: P. S. Martin and R. G. Klein (eds.), *Quaternary Extinctions: A Prehistoric Revolution,* pp. 223–249. University of Arizona Press, Tucson.

Graham, R. W., Lundelius, E. L., Jr., Graham, M. A., Schroeder, E. K., Toomey, R. S., III, Anderson, E., Barnosky, A. D., Burns, J. A., Churcher, C. S., Grayson, D. K., Guthrie, R. D., Harington, C. R., Jefferson, G. T., Martin, L. D., McDonald, H. G., Morlan, R. E., Semken, H. A., Jr., Webb, S. D., Werdelin, L., and Wilson, M. C. 1996. Spatial response of mammals to Late Quaternary environmental fluctuations. *Science* **272**:1601–1606.

Guilday, J. E. 1962. The Pleistocene local fauna of the Natural Chimneys, Augusta County, Virginia. *Ann. Carnegie Mus.* **36**:87–122.

Gunnell, G. F. 1989. Evolutionary history of Microsyopoidea (Mammalia, ?Primates) and the relationship between Plesiadapiformes and Primates. *Univ. Mich. Pap. Paleontol.* **27**:1–157.

Gunnell, G. F., and Gingerich, P. D. 1996. New hapalodectid *Hapalorestes lovei* (Mammalia, Mesonychia) from the early middle Eocene of northwestern Wyoming. *Contrib. Mus. Paleontol. Univ. Mich.* **29**:413–418.

Guthrie, R. D. 1984. Mosaics, allelochemics and nutrients, an ecological theory of late Pleistocene megafaunal extinctions, in: P. S. Martin and R. G. Klein (eds.), *Quaternary Extinctions: A Prehistoric Revolution,* pp. 259–298. University of Arizona Press, Tucson.

Hall, E. R. 1981. *The Mammals of North America.* Wiley, New York.

Hibbard, C. W. 1952. Vertebrate fossils from Late Cenozoic deposits of central Kansas. *Univ. Kans. Paleontol. Contrib. Vertebr.* **2**:1–14.

Hibbard, C. W. 1955. The Jinglebob interglacial (Sangamon?) fauna from Kansas and its climatic significance. *Contrib. Mus. Paleontol. Univ. Mich.* **12**:179–228.

Hodell, D. A., Benson, R. H., Kent, D. V., Boersma, A., and Rakic-El Bied, K. 1994. Magnetostratigraphic, biostratigraphic, and stable isotope stratigraphy of an Upper Miocene drill core from the Sale Briqueterie (northwestern Morocco): a high-resolution chronology for the Messinian stage. *Paleoceanography* **9**:835–855.

Holman, J. A. 1995. *Pleistocene Amphibians and Reptiles in North America.* Oxford University Press, London.

Jackson, J. B. C. 1994. Constancy and change of life in the sea. *Philos. Trans. R. Soc. London Ser. B* **344**:55–60.

Kennett, J. P., and Stott, L. D. 1991. Abrupt deep-sea warming, palaeoceanographic changes and benthic extinctions at the end of the Palaeocene. *Nature* **353**:225–229.

Koch, P. L., Zachos, J. C., and Gingerich, P. D. 1992. Correlation between isotope records in marine and continental carbon reservoirs near the Palaeocene/Eocene boundary. *Nature* **358**:319–322.

Kurten, B., and Anderson, E. 1980. *Pleistocene Mammals of North America.* Columbia University Press, New York.

Legendre, S. 1986. Analysis of mammalian communities from the late Eocene and Oligocene of southern France. *Palaeovertebrata, Montpellier* **16**:191–212.

Legendre, S., and Roth, C. 1988. Correlation of carnassial tooth size and body weight in Recent carnivores (Mammalia). *Hist. Biol.* **1**:85–98.

Lundelius, E. L., Churcher, C. S., Downs, T., Harington, C. R., Lindsay, E. H., Schultz, G. E., Semken, H. A., Webb, S. D., and Zakrezewski, R. J. 1987. The North American Quaternary sequence, in: M. O. Woodburne (ed.), *Cenozoic Mammals of North America: Geochronology and Biostratigraphy,* pp. 211–235. University of California Press, Berkeley.

MacFadden, B. J., and Cerling, T. E. 1996. Mammalian herbivore communities, ancient feeding ecology, and carbon isotopes: a 10 million-year sequence for the Neogene of Florida. *J. Vertebr. Paleontol.* **16**:103–115.

MacPhee, R. D. E., and Marx, P. A. 1997. The 40,000-year plague: humans, hyperdiseases, and first-contact extinctions, in: S. M. Goodman and B. D. Patterson (eds.), *Natural Change and Human Impact in Madagascar,* pp. 169–217. Smithsonian Institution Press, Washington, D.C.

Martin, P. S. 1967. Prehistoric overkill, in: P. S. Martin and H. E. Wright (eds.), *Pleistocene Extinctions: The Search for a Cause,* pp. 75–120. Yale University Press, New Haven.

Martin, P. S. 1984. Prehistoric overkill: the global model, in: P. S. Martin and R. G. Klein (eds.), *Quaternary Extinctions: A Prehistoric Revolution,* pp. 354–403. University of Arizona Press, Tucson.

Meltzer, D. J., and Mead, J. I. 1985. Dating late Pleistocene extinctions: theoretical issues, analytical bias, and substantive results, in: J. I. Mead and D. J. Meltzer (eds.), *Environments and Extinctions: Man in Late Glacial North America,* pp. 145–172. Center for the Study of Early Man, University of Maine, Orono.

Miller, K. G., Fairbanks, R. G., and Mountain, G. S. 1987. Tertiary oxygen isotope synthesis, sea level history, and continental margin erosion. *Paleoceanography* **2**:1–19.

Muller, D. W., and Hsu, K. J. 1987. Event stratigraphy and paleoceanography in the Fortuna Basin (southeast Spain): a scenario for the Messinian Salinity Crisis. *Paleoceanography* **2**:679–696.

Myers, R. A., Barrowman, N. J., Hutchings, J. A., and Rosenberg, A. A. 1995. Population dynamics of exploited fish stocks at low population levels. *Science* **269**:1106–1108.

Overpeck, J. T., Webb, R. S., and Webb, T., III. 1992. Mapping eastern North American vegetation change of the past 18 ka: no-analogs and the future. *Geology* **20**:1071–1074.

Owen-Smith, N. 1987. Pleistocene extinctions: the pivotal role of megaherbivores. *Paleobiology* **13**:351–362.

Prothero, D. R., and Heaton, T. H. 1996. Faunal stability during the Early Oligocene climatic crash. *Palaeogeogr. Palaeoclimatol. Palaeoecol.* **127**:257–284.

Raup, D. M. 1975. Taxonomic diversity estimation using rarefaction. *Paleobiology* **1**:333–342.

Raup, D. M. 1976. Species diversity in the Phanerozoic: an interpretation. *Paleobiology* **2**:289–297.

Repenning, C. A. 1987. Biochronology of the microtine rodents of the United States, in: M. O. Woodburne (ed.), *Cenozoic Mammals of North America: Geochronology and Biostratigraphy,* pp. 236–268. University of California Press, Berkeley.

Rhodes, R. S., II. 1984. Paleoecology and regional paleoclimatic implications of the Farmdalian Craigmile and Woodfordian Waubonsie mammalian local faunas, southwestern Iowa. *Ill. St. Mus. Rep. Invest.* **40**:1–51.

Routledge, R. D. 1977. On Whittaker's components of diversity. *Ecology* **58**:1120–1127.

Semken, H. A., Jr. 1996. Stratigraphy and paleontology of the McPherson *Equus* beds (Sandahl local fauna), McPherson Co., Kansas. *Contrib. Mus. Paleontol. Univ. Mich.* **20**:121–178.

Semken, H. A., Jr. 1974. Micromammal distribution and migration during the Holocene. *Am. Mus. Quat. Assoc. Abstr.* **3**:25.

Semken, H. A., Jr. 1984. Paleoecology of a late Wisconsinan/Holocene micromammal sequence in Peccary Cave, northwestern Arkansas, in: H. H. Genoways and M. R. Dawson (eds.), *Contributions in Quaternary Vertebrate Paleontology: A Volume in Memorial to John E. Guilday,* pp. 405–431. Carnegie Museum of Natural History Special Publication No. 8, Pittsburgh.

Sepkoski, J. J., Jr. 1978. A kinetic model of Phanerozoic taxonomic diversity. I. Analysis of marine orders. *Paleobiology* **4**:223–251.

Sepkoski, J. J., Jr., Bambach, R. K., Raup, D. M., and Valentine, J. W. 1981. Phanerozoic marine diversity and the fossil record. *Nature* **293**:435–437.

Shackleton, N. J. 1995. New data on the evolution of Pliocene climatic variability, in: E. S. Vrba, G. H. Denton, T. C. Partridge, and L. H. Burckle (eds.), *Paleoclimate and Evolution, with Emphasis on Human Origins,* pp. 242–248. Yale University Press, New Haven.

Stager, J. C., and Mayewski, P. A. 1997. Abrupt early to mid-Holocene climatic transition registered at the equator and the poles. *Science* **276**:1834–1836.

Steadman, D. W. 1989. Extinctions of birds in eastern Polynesia: a review of the records and comparisons with other Pacific Island Groups. *J. Archaeol. Sci.* **16**:177–205.

Stephens, J. J. 1960. Stratigraphy and paleontology of a Late Pleistocene basin, Harper County, Oklahoma. *Bull. Geol. Soc. Am.* **71**:1675–1702.

Stucky, R. K. 1990. Evolution of land mammal diversity in North America during the Cenozoic, in: H. H. Genoways (ed.), *Current Mammalogy,* pp. 375–432. Plenum Press, New York.

Taylor, K. C., Mayewski, P. A., Alley, R. B., Brook, E. J., Gow, A. J., Grootes, P. M., Meese, D. A., Saltzman, E. S., Severinghaus, J. P., Twickler, M. S., White, J. W. C., Whitlow, S., and Zielinski, G. A. 1997. The Holocene–Younger Dryas transition recorded at Summit, Greenland. *Science* **278**:825–827.

Valentine, J. W., and Jablonski, D. 1991. Biotic effects of sea level change: the Pleistocene test. *J. Geophys. Res.* **96**:6873–6878.

Valentine, J. W., and Jablonski, D. 1993. Fossil communities: compositional variation at many time scales, in: R. E. Ricklefs and D. Schluter (eds.), *Species Diversity in Ecological Communities: Historical and Geographical Perspectives* pp. 341–349. University of Chicago Press, Chicago.

Van Valkenburgh, B., and Janis, C. M. 1993. Historical diversity patterns in North American large herbivores and carnivores, in: R. E. Ricklefs and D. Schluter (eds.), *Species Diversity in Ecological Communities: Historical and Geographical Perspectives,* pp. 330–340. University of Chicago Press, Chicago.

Webb, S. D. 1984. Ten million years of mammal extinction in North America, in: P. S. Martin and R. G. Klein (eds.), *Quaternary Extinctions: A Prehistoric Revolution,* pp. 189–210. University of Arizona Press, Tucson.

Wilf, P. 1997. When are leaves good thermometers? A new case for leaf margin analysis. *Paleobiology* **23**:373–390.

Wilf, P., Wing, S. L., Greenwood, D. R., and Greenwood, C. L. 1998. Using fossil leaves as paleoprecipitation indicators: an Eocene example. *Geology* **26**:203–206.

Wilson, D. E., and Reeder, D. M. 1993. *Mammal Species of the World.* Smithsonian Institution Press, Washington, D.C.

Wing, S. L., Alroy, J., and Hickey, L. J. 1995. Plant and mammal diversity in the Paleocene to early Eocene of the Bighorn Basin. *Palaeogeogr. Palaeoclimatol. Palaeoecol.* **115**:117–155.

Wolfe, J. A. 1978. A paleobotanical interpretation of Tertiary climates in the Northern Hemisphere. *Am. Sci.* **66**:694–703.

Wolfe, J. A. 1992. Climatic, floristic, and vegetational changes near the Eocene/Oligocene boundary in North America, in: D. R. Prothero and W. A. Berggren (eds.), *Eocene–Oligocene Climatic and Biotic Evolution,* pp. 421–436. Princeton University Press, Princeton N.J.

Wolfe, J. A., Schorn, H. E., Forest, C. E., and Molnar, P. 1997. Paleobotanical evidence for high altitudes in Nevada during the Miocene. *Science* **276**:1672–1675.

Wood, H. E., 2nd, Chaney, R. W., Clark, J., Colbert, E. H., Jepsen, G. L., Reeside, J. B., Jr., and Stock, C. 1941. Nomenclature and correlation of the North American continental Tertiary. *Bull. Geol. Soc. Am.* **52**:1–48.

Woodburne, M. O. 1987. *Cenozoic Mammals of North America: Geochronology and Biostratigraphy.* University of California Press, Berkeley.

Woodburne, M. O., and Swisher, C. C., III. 1995. Land mammal high resolution geochronology, intercontinental overland dispersals, sea-level, climate, and vicariance, in: W. A. Berggren, D. V. Kent, and J. Hardenbol (eds.), *Geochronology, Time Scales and Global Stratigraphic Correlations: A Unified Temporal Framework for an Historical Geology,* pp. 335–365. SEPM (Society for Sedimentary Geology), Special Publication 54.

Zachos, J. C., Stott, L. D., and Lohmann, K. C. 1994. Evolution of early Cenozoic marine temperatures. *Paleoceanography* **9**:353–387.

CHAPTER 7

Rates, Patterns, and Processes of Landscape Transformation and Extinction in Madagascar

DAVID A. BURNEY

1. Introduction

Inferring cause and effect from the fossil record is not a wholly satisfying enterprise. The evidence is stale. Many useful details are missing, perhaps never to be found. Sequential events may be collapsed together, inverted, or mixed with evidence from other times. Few relevant parameters can be measured directly. Paleoecologists must forge ahead despite these obstacles, however, as extinction and environmental change are subjects too important to ignore.

A great obstacle to wider application of information derived from paleoecological studies to the present biodiversity crisis is the obvious difficulty of inferring complex ecological patterns from proxy data. Yet the temporal perspectives offered have informed many issues in conservation biology in ways no conceivable nonhistorical approach could. Bridging the gap between methodological limits and theoretical expectations is most likely to be possible in the most isolated natural systems, such as those found on remote islands. With their severe biogeographical constraints and generally shallow human time-depth, it is sometimes possible to address globally interesting questions regarding levels of background versus human-era disturbance, biotic invasion, and extinction. As the last place on earth to experience the disappearance of a diverse mammalian megafauna, and one of the last large habitable landmasses to attract human colonizers, Madagascar can be thought of as the ultimate natural experiment in human ecology.

But how should we proceed in evaluating the results of this irreplicable experiment? Caution is certainly in order. Two conceptual approaches that may help are (1) the principles of experimental design that offer a standard frame of reference for evaluating cause and effect and (2) the emerging discipline of landscape ecology, which emphasizes the importance of appropriate scale and resolution in the collection and analysis of spatiotempo-

---

DAVID A. BURNEY • Department of Biological Sciences, Fordham University, Bronx, New York 10458.
Extinctions in Near Time, edited by MacPhee. Kluwer Academic/Plenum Publishers, New York, 1999.

ral data. In the present context it is convenient to refer to this hybrid methodology as *landscape paleoecology.*

An isolated landmass may be thought of as an "experiment in human ecology," in the sense that the prehuman condition represents background-level variation, and everything that happens after human arrival constitutes experimental treatments. In this sense, especially if there are otherwise comparable landmasses nearby with a fundamentally different human history, the experiment can be thought of as at least partially controlled (see Burney, 1996a).

Parameterization is a serious limitation: Not much progress can be made in this otherwise plausible direction unless measurements can be taken. Of particularly critical importance are methods that will allow one to establish the point in time for a particular locus that human impacts commence (i.e., "human arrival"). Often in studies of late Quaternary extinctions, this is a key point of contention (see chapters in Martin and Klein, 1984). Much of the present chapter will be devoted to establishing this time horizon at specific locations in Madagascar. In a few opportune cases I will attempt the next two critical steps: inferring how humans and the now-extinct fauna may have interacted, and for how long.

Four palynological techniques, when combined with more conventional archaeological and paleontological methods for deducing human activity, may provide proxy indications of landscape changes likely to reflect human arrival and specific types of human impacts with sufficient resolution to make useful inferences regarding the pace and distribution of the transformation process (Burney, 1993a, 1997b; Kirch and Ellison, 1994). One may infer a human presence from sedimentary evidence for (1) a large increase in ruderal pollen influx, (2) appearance of exotic pollen types, (3) a sudden and pronounced increase in microscopic charcoal particles, and (4) paleolimnological evidence for cultural eutrophication. Although none of these four alone is entirely sufficient to be certain of a human presence (ruderal and adventive taxa, wildfires, and eutrophic waters all occur through natural processes as well), various combinations of these have been compared usefully to the archaeological record of earliest detectable human presence and paleontological evidence for first appearance of human-transported species ("portmanteau biota," sensu McNeill, 1994). This approach has been used in various island areas around the world, including Hawaii (e.g., Burney *et al.,* 1995; Athens, 1997), the West Indies (e.g., Burney *et al.,* 1994), the South Pacific (e.g., Flenley *et al.,* 1991; Ellison, 1994), and Madagascar (e.g., Burney, 1993b; Burney *et al.,* 1997).

Making full use of these techniques for studying landscape paleoecology requires integration of spatiotemporal scales and disparate data types. Linkages between sites may be accomplished by (1) setting up arrays of sites as transects or grids in which the number of independent variables and confounding factors is minimized to the extent possible in "natural experiments" and (2) seeking and developing especially fortuitous "integrated" sites in which many types of relevant data may be collected from a single location and interrelated chronologically with absolute dating methods. This chapter will illustrate these two methods by first analyzing a transect of sites that span the length of the west coast of Madagascar, from the arid semidesert of the south to humid Nosy Be and the Sambirano rain forest of the far north. Second, integrated sites along the transect and elsewhere will be considered in more detail. The ultimate goal will be to compare the information gleaned from the landscape-level paleoenvironmental reconstructions thus derived with the expectations of five hypotheses (Table 1) that have been proposed for the faunal extinctions recorded for late Holocene Madagascar.

Table 1. Extinction Hypotheses Proposed for Madagascar

No.	Type	Source	Rate	Pattern	Process
H1	"Great Fire"	Humbert (1927)	Rapid	Simultaneous throughout	Landscape transformed by introduction of fire by humans; extinctions followed as a result of forest loss
H2	"Great Drought"	Mahé and Sourdat (1972)	Slow	Southern region only	Climatic desiccation, extinctions related to spread of semiarid conditions
H3	"Blitzkrieg"	Martin (1984)	Rapid	Wave across island	First-contact overkill of naive fauna, beginning at coastal settlements and moving inland
H4	"Hypervirulent Disease"	MacPhee and Marx (1997)	Very rapid	Panzootic disease pattern	Unknown pathogen(s) introduced by humans, spreading quickly; lethal to wide array of mammals, and possibly some birds and reptiles
H5	"Synergy"	This chapter	Very slow	Mosaic	Full array of human impacts played a role, but to differing extents in various regions; factors interacting to multiply effects; some amplification by background climate change

2. The West Coast Transect

The entire western coastal zone of Madagascar can be treated as a transect for present purposes (Fig. 1), a continuous belt of low-lying sedimentary terrane extending a straight-line distance of ca. 1300 km from the driest parts of the island in the southwest near Toliara (Tulear) to the Sambirano rain forest and Nosy Be in the north. The West Coast Transect (WCT) is a series of sites and site clusters along this coast that have yielded sedimentary

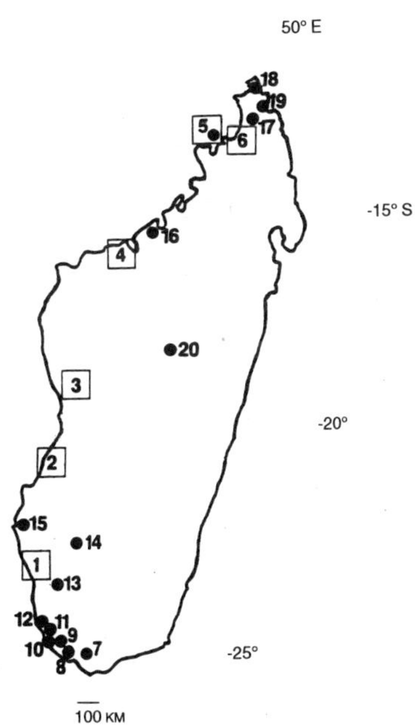

FIGURE 1. Location of the West Coast Transect and other sites mentioned. Squares denote primary sites used in transect. All of these have ^{14}C-dated chronologies for inferred human arrival, based on microscopic charcoal particle stratigraphy, and consistent with other evidence, including pollen and bone artifacts. These are (1) Ambolisatra/Andolonomby, (2) Belo-sur-mer, (3) Komango, (4) Mitsinjo, (5) Amparihibe, and (6) Benavony. Other fossil sites discussed, denoted by solid circles, are (7) Beloha, (8) Anavoha, (9) Behavoha, (10) Bemafandry, (11) Tsiandroina, (12) Itampolo, (13) Ankilitelo, (14) Ampoza, (15) Lamboharana, (16) Anjohibe/Anjohikely, (17) Ankarana, (18) Lakaton'i Anja, (19) Irodo Tafiampatsa, and (20) Ampasambazimba.

records of particular usefulness for gauging human-induced changes on the landscape, including microscopic charcoal particles, pollen records, and bones of extinct animals. Although other relevant sites near the primary study areas will provide useful interpretive insights, the emphasis here will be on those sites that have provided ^{14}C-dated stratigraphies that span the inferred period of human activity on the island (roughly the last two millennia; see MacPhee and Burney, 1991; Burney, 1993b). Six sites spread along the full length of the transect are especially useful for addressing the question of when human transformation began in a particular locality because they contain charcoal particle evidence (Fig. 2) and sometimes other datable indicators of human activity. Site characteristics and references are summarized in Table 2.

2.1. Site Characteristics and Results

2.1.1. Ambolisatra/Andolonomby

The large number of paleontological expeditions to Ambolisatra on the southwest coast (summarized in Chanudet, 1975) has resulted in many important specimens of most of the endemic "subfossil" megafauna, many bones of smaller animals, and important human artifacts as well. One of the human modified hippo femora from the site collected by Alfred Grandidier at the turn of the century yielded an AMS collagen date (acid-insoluble fraction) of 1970 ± 70 yrbp (MacPhee and Burney, 1991). This bone was cut in a distinctive way—perhaps for culinary reasons or in the process of tool-making—almost certainly in a fresh state while the bone could still be worked effectively. Hippo femora with nearly identical modifications were also collected by Grandidier and his colleagues from two other sites along the southwest coast, Lamboharana and Belo-sur-mer (MacPhee and Burney, 1991). A similarly carved elephant bird (*Aepyornis* sp.) bone was collected at Itampolo on the southwest coast by Paul Martin in 1968, and it also dates to this early period (H. James, unpublished data).

Burney (1993b) made a study of pollen and charcoal from excavation and coring in one of the dry lake beds of the Ambolisatra vicinity (at a location now called Andolonomby, near Ambolimalika village), producing a 5000-year record of fossil pollen, charcoal particles, plant macrofossils, and bones of extinct animals. Between 3500 and 2500 yrbp, precipitation apparently declined, and the site shifted from prevalently dry-forest pollen indicators to a palm savanna, and the lake level declined. At a level dated 1890 ± 90 yrbp, charcoal levels increase markedly over low background levels and palm pollen decreases, culminating within a few centuries in pollen spectra of the type presently generated by the site, which is a mosaic of grassland, succulent bushland, and dry woodland.

Thus, paleontological and palynological evidence converge to indicate a human presence along this coast by ca. two millennia ago. A later section will consider in more detail the dating of megafauna from this and other sites in the arid southwest, showing a pattern of evidence for human interaction with the megafauna at three coastal sites in the vicinity beginning ca. 2000–1700 yrbp, and survival of several now-extinct taxa for 500–1500 years after this time at some sites in the region. Two recent dates from the southwest on extinct giant lemurs (Simons, 1997) indicate survival of *Megaladapis* and *Palaeopropithecus* until ca. 500 yrbp.

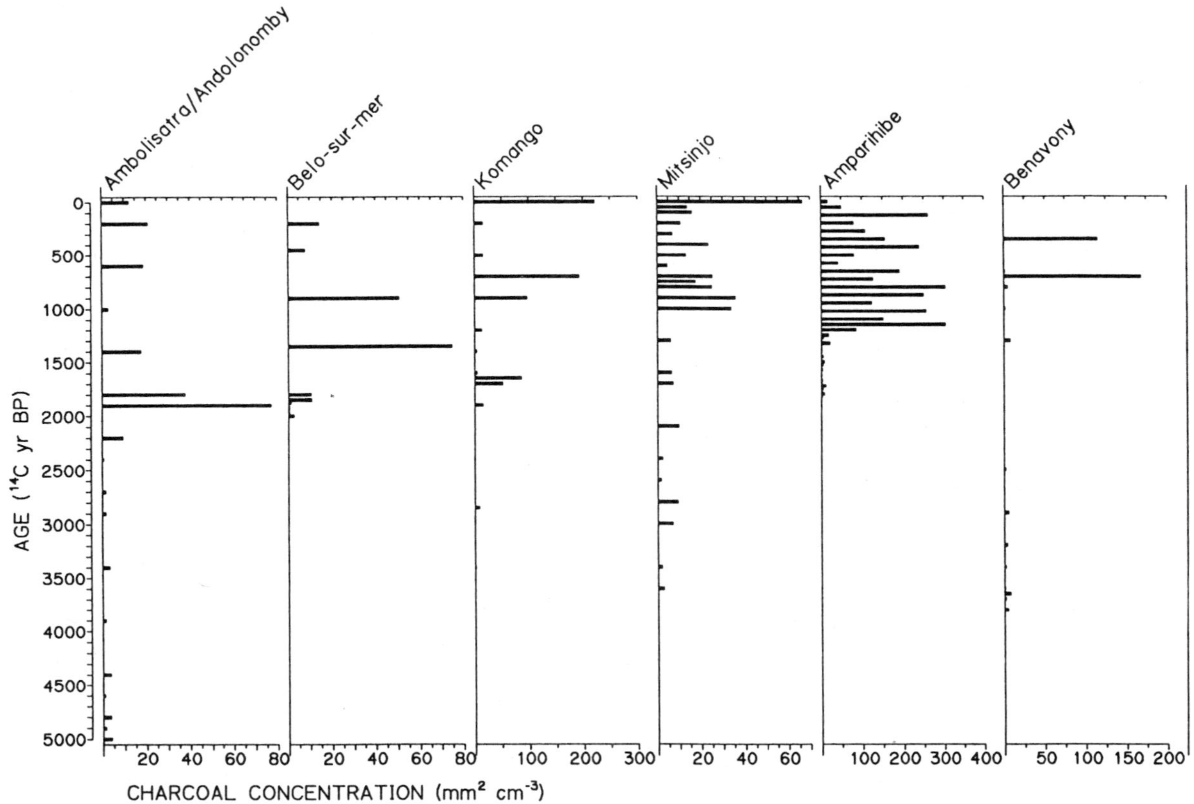

FIGURE 2. Microscopic charcoal particle stratigraphy of six ^{14}C-dated sites on the West Coast Transect arranged left to right according to increasing annual precipitation. See Fig. 1 for locations (1–6 on map), and Table 2 for site characteristics and sources. Ages of samples not dated directly were estimated from linear interpolation of ^{14}C dates for each site and extrapolation to the sediment surface. Local human activity involving firing of the landscape is inferred for each site from the first occurrence of an increase of charcoal concentration in the sediments of at least an order of magnitude. Dating of the onset of this increase suggests human impacts on the landscape a full millennium earlier in the arid southwest of the island than in wetter northwestern regions.

Table 2. West Coast Transect Site Characteristics

Site	Latitude/ longitude	Precipitation (mm yr^{-1})	Type of feature	Evidence for human arrival (yrbp)	References
Ambolisatra/ Andolonomby	S 23° 04′ E 43° 45′	400	Dry pond	Bone artifact (1970 ± 70) Charcoal and pollen (1890 ± 90)	MacPhee and Burney (1991), Burney (1993b)
Belo-sur-mer	S 20° 44′ E 44° 01′	500	Dry pond	Charcoal (1830 ± 60)	This paper, Burney *et al.* (1996), Burney and Ramilisonina (in press)
Komango	S 19° 10′ E 44° 49′	800	Shallow pond	Charcoal (1750 ± 60)	This paper
Mitsinjo	S 16° 02′ E 45° 51′	1000	Shallow lake	Charcoal, pollen (890 ± 80) Pottery (ca. 11th century AD)	Matsumoto and Burney (1994), Wright *et al.* (1996)
Amparihibe	S 13° 18′ E 48° 13′	2000	Deep crater lake	Charcoal, paleolimnology (1130 ± 50) Pottery (ca. 11th century AD)	This paper, Radimilahy (1997)
Benavony	S 13° 43′ E 48° 29′	2500	Marsh	Charcoal (710 ± 110) Pottery (ca. 11th century AD)	This paper, Radimilahy (1997)

2.1.2. Belo-sur-mer

The Cenozoic Research Group (CRG), a Malagasy–American team working through the Université d'Antananarivo, has reopened this site that yielded so many important specimens through a century of paleontological expeditions (Chanudet, 1975). It is in many ways very similar to Ambolisatra, being another interdunal swale integrated with a shallow water table (Burney and Ramilisonina, in press). Like Ambolisatra, as well as the two other sites yielding human-modified bones along this coast, the primary bone deposits near the present village of Belo-sur-mer were adjacent to the sea, near the mouth of small estuaries suitable for mooring shallow-draft sailing vessels. Unlike the others, the human-modified bone that has been collected from Belo has not been dated. However, it is cut in the same style (MacPhee and Burney, 1991) as the bones from the three other sites, and they have all yielded dates between 1970 and 1740 yrbp. Charcoal values increase in the CRG's excavations at a level dated from an *Aepyornis* eggshell fragment at 1830 ± 60 yrbp. Pollen work on the site is still in progress (T. Rakotondrazafy, unpublished data), but cursory inspection suggests that vegetation at this site has not changed markedly over the past 2000 years.

It appears that the stratigraphy at this location, 2 km northeast of the modern village of Belo-sur-mer at a place known as Ankilibehandry, is ideally suited for studying the extinctions. At a depth of 110 cm below the surface, a grayish-brown clayey sand with humic matter that dates to 1990 ± 50 yrbp overlies a yellow overwash sand with marine shells. Bones of the extinct megafauna, some showing crocodile damage, occur in the humic layer and the mottled sandy clay layer a half meter above. This mottled zone, 15 cm thick, shows increasing charcoal as noted above. Bones of the megafauna, possibly lacking some taxa but still with a good representation of tortoises, hippos, large lemurs, and crocodiles, continue up to the surface. The top 35 cm, a dark gray silty sand, is especially rich in *Archaeolemur* bones including many mandibles. The dating of these specimens will be discussed in Section 3.1.

2.1.3. Komango

About 200 km north along the coast there is another paleoecological site adjacent to a coastal river (but ca. 50 km inland), Lake Komango on the Manambolo River in the Bemaraha region. Rainfall here is ca. 800 mm, nearly twice that of Ambolisatra (ca. 400 mm) and Belo (ca. 500 mm), but still dry enough that the surrounding landscape supports only sparsely wooded savanna on the thin upland soils, and xeric trees and succulents in the limestone pinnacle karst (the Tsingy of Bemaraha) inland from the site. (Precipitation values given in this chapter are estimated from isohyets given in Humbert and Cours Darne [1965:34], as no rainfall data are available for the sites.) Megafaunal fossils and early archaeological sites are not known from this vicinity, although the remote and rugged character of this landscape has hindered investigation and determined that even today the human population density of the region is quite low.

The lake is apparently a cutoff meander of the large, braided channel of the Manambolo River. An 8-m core collected by the CRG in 1993 has yielded a basal age of 3390 ± 50 (AMS on acid-pretreated sediment). An AMS-dated wood fragment 75 cm above, from a level showing an increase of particulate charcoal above lower levels, was 1750 ± 60 yrbp.

Pollen has not been studied in this core, but this seems like a plausible date for the first human presence in the region, as these charcoal values are an order of magnitude higher than previous levels and fully half the concentration of charcoal in the highest subsequent levels and the modern sediment. Presently this savanna area is burned frequently by pastoralists.

Surveys of several caves in the Bemaraha by the CRG, including rock shelters along the Manambolo Gorge and the large caves of Anjohimbazimba near Antsalova and Kinanjao near Bekopaka, yielded no megafaunal bones except those of crocodiles. Archaeological evidence uncovered so far only goes back a few centuries, although one site just north of the region near Maintirano contained pottery of a type found elsewhere on the west coast in the early part of the present millennium (Wright and Ramilisonina, unpublished data).

It is possible that the high charcoal values around 1750 yrbp are a reflection of natural events (e.g., lightning), but they may also indicate an archaeologically undetected presence of human settlers in the vicinity at this time, or, perhaps more likely, either the spread of fires from transhuman hunter-gatherers and fishers following the large Manambolo River up from the coast, or bush fires spreading from settlements farther south in the vicinity of the Tsiribihina River.

2.1.4. Mitsinjo

A 3500-year record from pollen, charcoal particles, and other microfossils has been published from cores taken in Lake Mitsinjo (Matsumoto and Burney, 1994; Wright *et al.*, 1996), a site in the mesic but highly seasonal northwestern part of the island. The site receives ca. 1300 mm of precipitation annually, considerably more than Lake Komango, almost 400 km south. Until ca. 3000 yrbp, however, the lake level was higher and forest pollen was more prevalent. By 2000 yrbp, the site had become drier, with a decline in woody vegetation and gradual drying of the lake. Higher water levels returned by ca. 1000 yrbp. Although background levels of charcoal particles in the sediments suggest a prehuman fire regime, as in other intermediate-rainfall sites (Burney, 1996b), charcoal shows a large increase and shifts to predominantly graminoid pieces just prior to a level dated 890 ± 80 yrbp (Wright *et al.*, 1996). The record between 1000 and 2000 yrbp has been obscured by a sedimentary hiatus. However, owing to the refractory nature of charcoal, one would have expected charcoal to accumulate around the hiatus if a high influx had occurred earlier. Earliest direct evidence for humans in the area, the poorly dated Antetikala Phase ceramics, appears to match a pottery tradition dated to the eleventh to fourteenth centuries farther north (Wright *et al.*, 1996). Cultivators were well established in the area by the fourteenth century AD., and this is reflected in increasing charcoal and ruderal pollen at Lake Mitsinjo about 500 yrbp (Matsumoto and Burney, 1994; Wright *et al.*, 1996). Although there are no megafaunal sites in the immediate vicinity, Anjohibe (ca. 100 km north) and other northern cave sites that provide *Archaeolemur* dates from the beginning of the present millennium will be discussed in a later section.

2.1.5. Amparihibe

The offshore island of Nosy Be, on the extreme northwest coast of Madagascar, receives ca. 2000 mm of rainfall. It represents the westernmost extension of the Sambirano-type rain forest, and contains several deep crater lakes. The CRG has collected two cores from the largest of these, Lake Amparihibe, from water over 40 m deep, using a scaled-down Kullenberg-type sampler. In a 4-m core with a basal age of 1820 ± 60 yrbp, the fine gyttja sediments are practically devoid of microscopic charcoal particles until a level dated 1130 ± 50 yrbp (AMS ^{14}C dating on plant macrofossil), where charcoal makes an extremely abrupt increase to values more than two orders of magnitude higher. Charcoal values remain high but erratic up to the present. At the time charcoal first increased, the nutrient status of the lake apparently changed. Culturally induced eutrophication, probably from increased runoff from a burned landscape, is suggested by the appearance of fine laminae in the sediments at this time, reflecting anaerobic conditions at the sediment–water interface. Although anaerobic conditions are most likely to have been caused by increased biological oxygen demand in the water column, increased rainfall could have produced a deeper, more stratified lake at this time. Prior to the level dated 1130 yrbp, sediments were bioturbated, indicating higher oxygen availability in the bottom waters.

Like most other high-rainfall areas of Madagascar, no megafaunal sites are known from the vicinity. Archaeological evidence suggests heavy human settlement over roughly the last 500 years, but the earliest known sites for Nosy Be and the adjacent Madagascar mainland date to the eleventh century AD (Radimilahy, 1997).

2.1.6. Benavony

This marshy meadow ca. 10 km up the Sambirano River from the northwest coast is partly surrounded by highly disturbed rain forest. With local precipitation of 2500 mm, the site represents the wettest point on the WCT. Although the area is partly forested, it supports a moderately dense agrarian population and lies a few kilometers inland from the large population center of Ambanja. Benavony is probably a cutoff meander of the Sambirano. An 8-m core was collected from the marsh in 1994 by the CRG, yielding a basal peat age of 3790 ± 50 yrbp, and a peat age at a depth of ca. 4.5 m of 2980 ± 50 yrbp. Throughout the lower part of the core, charcoal particles show low to moderate background levels. An increase of nearly two orders of magnitude occurs in a peat level that dates to 710 ± 110 yrbp. This corresponds closely to the time of inferred maximum population at Mahilaka, a nearby eleventh- to fourteenth-century coastal town that was probably Madagascar's first urban center and an important stop in the Indian Ocean trade network. The walled city apparently grew rapidly in the twelfth and thirteenth centuries, but collapsed in the following century at about the time the Plague was sweeping Europe and Africa (Radimilahy, 1997).

2.2. The General Pattern

It is apparent from these results that humans were in evidence in southwestern Madagascar several centuries to almost a millennium before first evidence for their appearance along the rest of the transect. Whether these first people were permanent settlers or merely short-term visitors is not known (see Dewar and Wright, 1993), but the charcoal and pollen records show that they did effect changes on the land that continued and escalated over subsequent centuries and hint at continuous human presence during this time, at least at low densities. Landscape changes along the transect since human arrival include changes in regional fire ecology: The driest and wettest sites (i.e., opposite ends of the transect) showed little evidence for prehuman fires and a marked increase in burning after inferred human arrival. The mesic but highly seasonal middle of the transect shows a prehuman fire regime that changes only by degree after local human arrival. Humans must have effected landscape changes, among other ways, by extending fire to areas too wet or too dry to have burned frequently on their own, and perhaps also by increasing the frequency and changing the timing of fire occurrence in intermediate-rainfall areas with prehuman pyrogenic communities (Burney, 1996b). Because many of the extinct taxa apparently thrived in pyrogenic communities, however, fire alone hardly seems like a viable candidate for precipitating extinctions (Burney, 1987b). It seems more likely that fire increase may have been a consequence of declining megaherbivore populations and subsequent litter accumulation, as has been suggested previously for southwestern Madagascar (Burney, 1993b) and Australia (Flannery, 1994).

Throughout the transect, and confirmed by multiproxy indicators at several sites, there is abundant evidence for a drying of climate beginning ca. 3000 yrbp. Maximum dryness may have occurred in the period roughly 2000–1200 yrbp. The southwest has remained semiarid and probably never supported a dense human population over a large area. In the far northwest, by contrast, rainfall has been plentiful throughout the present millennium, with pastoralism and agriculture thriving. The case of the medieval walled town of Mahilaka, however, suggests that disease (and perhaps excessive rainfall) may have hindered the colonization of the region. It is possible that a lower incidence of malaria and other diseases of the humid tropics in the arid southwest permitted colonization of this less productive land well before the more humid areas were successfully penetrated. This is consistent with another settlement pattern that has been observed in Madagascar: that the cool, remote (and probably less malarious) central highlands were densely populated well before many lower and more accessible regions. People probably clustered along the coast until fairly recent centuries in western Madagascar, but palynological indicators and archaeology show that, by the beginning of the present millennium, people were well established in the cool central highlands and on the eastern escarpment (Wright *et al.*, 1992; Dewar and Wright, 1993; Burney, 1997a).

The transect approach supports the conclusion that, along Madagascar's west coast, human impacts are detectable beginning almost two millennia ago, but the proto-Malagasy required a full millennium to populate the wetter and more forested northern parts of this coast and have a detectable effect in sedimentary sequences from dense humid forest. This proxy paleodemographic approach provides useful but highly indirect results that are nevertheless in good agreement with archaeological and paleontological findings. If the peopling of even this fairly accessible coastline proceeded as such a slow wave—slow enough to take many centuries if not a full millennium or more—how did this affect the pattern of megafaunal extinction?

Based on the transect evidence presented, one might predict that (1) as this slow wave passed, the majority of large animal species were extinguished rapidly or (2) the extirpations at a given point along the transect were in fact gradual. In the latter case, one would be saying in effect that the wave was not only slow but of low amplitude, i.e., the extinction "event" took a long time to run its course even at single locations. The rapid chronology expected for a "blitzkrieg" sensu Martin (1984) is not met by the former case because the wave would have simply spread too slowly to extinguish an initially naive fauna over a large area in a short time. The latter case would represent a pattern and imply causation independent of any simple blitzkrieg mechanism. Some species might have experienced many centuries of (probably low-density) interaction with humans before disappearing. In the absence of compelling examples from elsewhere in which a technologically unsophisticated human population has caused widespread extinctions over a very long time scale, one must wonder how a low-density interaction could precipitate such major faunal losses among animals that would have long since learned to fear humans.

3. Achieving Site Integration

To get an essentially instantaneous measurement of the extinction rate at a single location along the WCT would require a well-dated site of the type sometimes known as an "integrated site." Such a site features close association (in the stratigraphic, taphonomic, and geochronological senses) of palynological, paleontological, and archaeological data (Fig. 3). With such sites one might attempt a more comprehensive reconstruction of the landscape and relevant events on the applicable spatiotemporal scales. Few sites are likely to meet all of the requirements, but Belo-sur-mer is perhaps the best Madagascar example studied so far by the author and his colleagues.

3.1. Belo-sur-mer and the "Extinction Window"

If multiple proxy indicators of human activity (Burney, 1997b) and human-modified hippo bones (MacPhee and Burney, 1991) are taken as establishing human arrival on the southwest coast two millennia ago, dates for "latest occurrences" of extinct animals and human arrival would be expected to coincide within methodological limitations if the extinction event were very rapid and directly or indirectly induced by human activities. Humans and the subsequently extinct taxa would be expected to show only a temporally brief overlap, perhaps on a scale of decades (or even less) or at most a few centuries. Radiocarbon dates from the Ankilibehandry Site at Belo-sur-mer (Table 3) show the opposite pattern,

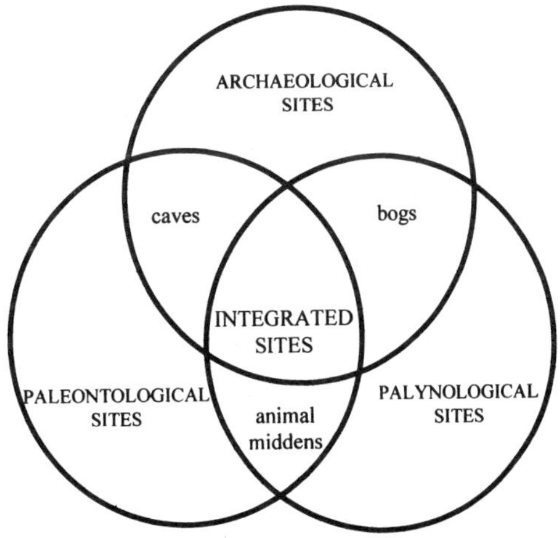

FIGURE 3. Venn diagram illustrating the integrated-site concept as applied by the author to late Quaternary sites. An integrated site could be a cave, bog, or midden, provided the site contains dated stratigraphy with paleontological, palynological, and archaeological evidence.

Table 3. ^{14}C Dates for Extinct Fauna within the Period of Inferred Human Presence in Madagascar

Site/Source	^{14}C yrbp ± 1 σ	Lab #	Material	Method/pretreatment
Ambolisatra (MacPhee and Burney, 1991)	1970 ± 50	AA-2895	*Hippopotamus lemerlei,* human-modified	AMS; collagen extraction
Ampasambazimba (Tattersall, 1973)	1035 ± 50	Pta-739	*Megaladapis* sp.	Standard; treatment unknown
Ampoza (Mahé and Sourdat, 1972)	1910 ± 120	GaK-2309	Bones, presumably. megafaunal	Standard; treatment unknown
Anjohibe (Burney *et al.* 1997)	830 ± 60	β-55060	Dung, cf. *Archaeolemur*	AMS; acid washes
	2060 ± 70	β-55095	*Archaeolemur edwardsi*	AMS; collagen extraction with alkali
Ankarana (Simons *et al.* 1995)	1020 ± 50	β-60797	*Archaeolemur* sp.	AMS; collagen extraction with alkali
Ankilitelo (Simons 1997)	510 ± 80	Not given	*Paleopropithecus* sp.	AMS; collagen extraction with alkali
	630 ± 50	Not given	*Megaladapis* sp.	AMS; collagen extraction with alkali
Behavoha (Mahé and Sourdat, 1972)	2160 ± 110	GaK-1658	*Geochelone* carapace	Standard; treatment unknown
Belo-sur-mer (this chapter)	1220 ± 50	β-90098	*Pachylemur insignis*	AMS; collagen extraction with alkali
	1280 ± 60	β-103349	cf. *Mullerornis*	AMS; collagen extraction with alkali
	1370 ± 40	β-90095	*Archaeolemur* cf. *majori*	AMS; collagen extraction with alkali
	1550 ± 50	β-90096	*Archaeolemur* sp.	AMS; collagen extraction with alkali
	1650 ± 50	β-90094	*Archaeolemur majori*	AMS; collagen extraction with alkali
	1830 ± 60	β-90099	*Aepyornis* eggshell	Standard; $CaCO_3$ with acid etching
	1990 ± 50	β-103348	Sediment at base of bone layer	AMS; acid washes
	2000 ± 50	β-90097	*Pachylemur insignis*	AMS; collagen extraction with alkali
	2050 ± 60	β-76612	*Archaeolemur majori*	AMS; collagen extraction with alkali
	2140 ± 50	β-102567	*Megaladapis madagascariensis*	AMS; collagen extraction with alkali
Beloha (Goodman and Rakotozafy, 1997)	1380 ± 90	β-63675	*Alopochen sirabensis*	AMS; collagen extraction with alkali
Bemafandry (Mahé and Sourdat, 1972)	1980 ± 90	GaK-1656	Wood adjacent to dated *Geochelone* carapace	Standard; treatment unknown
	2060 ± 150	GaK-1655	*Geochelone* carapace	Standard; treatment unknown
Irodo Tafiampatsa (Battistini and Vérin, 1966)	1150 ± 90	GaK-?	*Aepyornis* eggshell	Standard; treatment unknown
Itampolo (Mahé and Sourdat, 1972)	980 ± 200	GaK-1506	*Hippopotamus* sp.	Standard; treatment unknown

Table 3. (*Continued*)

Site/Source	^{14}C yrbp ± 1 σ	Lab #	Material	Method/pretreatment
(James, unpublished)	1880 ± 70	β-67659	*Aepyornis* bone, human-modified	AMS; collagen extraction with alkali
Lamboharana (MacPhee and Burney, 1991)	1740 ± 50	TO-1437	*Hippopotamus lemerlei*, human-modified	AMS; collagen extraction
(Mahé and Sourdat, 1972)	1220 ± 80	GaK-2310	Bones, presumably. megafaunal	Standard; treatment unknown
Southern Madagascar (Marden, 1967)	1970 ± 90	Not given	*Aepyornis* eggshell	Standard; treatment unknown
Tsiandroina (Goodman and Rakotozafy, 1997)	1980 ± 90	β-63676	*Coua primavea*	AMS; collagen extraction with alkali
50 km SW of Ft. Dauphin (Berger *et al.*, 1967)	1000 ± 150	UCLA-1893	*Aepyornis* eggshell	Standard; treatment unknown

however. The density of dates for any taxon is insufficient to state this with certainty, but the data are consistent with the notion that *Archaeolemur, Pachylemur, Mullerornis,* and perhaps other extinct forms persisted for five centuries or more after first contact with humans along this coast.

This chronology is consistent with the charcoal data from the same stratigraphy (Table 2, Fig. 2), which shows human arrival at the site, in terms of a large increase in microscopic charcoal particles, at a level that dates to 1830 ± 60 yrbp. Early archaeological sites in the Belo area are not well studied or radiometrically dated, but sherd scatters in the dunes adjacent to the site are mixed with eggshell fragments of *Aepyornis* and *Mullerornis.* These could, of course, be the remains of old sterile eggs collected for use as water containers rather than evidence of direct predation (Dewar, 1984). Archaeological survey suggests that human population density along this part of the southwest coast has probably always been low (Burney *et al.,* 1996). Pollen spectra from the site show that the vegetation has not changed much there in the last two millennia (T. Rakotondrazafy, unpublished data), and the pollen data from Ambolisatra/Andolonomby, the next site to the south in the WCT, show increasingly dry climate between 3500 and 2000 yrbp, with the major vegetation change on that site (from dry woodland to palm savanna) probably complete before human arrival. Humans along this coast probably have contributed to the expansion of bushland and grassland by woodcutting, burning, and livestock overgrazing, but large tracts of giant *Didierea* and *Euphorbia* succulents persist even today, and the region has one of the sparsest population densities of any comparably sized area of Madagascar. Legends of fantastic animals figure more prominently in the local folklore than in other areas investigated, perhaps based on traditions of past sightings of *Hippopotamus* and *Archaeolemur* (Burney and Ramilisonina, in press).

All of the available evidence points to the long persistence of at least some of the extinct megafauna at this remote site. Thus, in terms of scenarios for a human role in the extinctions, one would have to consider the possibility that humans were not necessarily effective agents of local extirpation. After centuries of humans on the landscape, it would appear that whatever they did to cause local losses was far less effective in some cases than in others. The window of time for the entire extinction event to run its course would have to be regarded as a greater than 500-year event at Belo. The duration and timing of this drawn-out event do not accord well with the expectations of any of the unicausal hypotheses for anthropogenic megafaunal extinctions in Madagascar (H1, H3, and H4 in Table 1).

3.2. Other Sites and Complex Causation

Three other sites in Madagascar (Ambolisatra/Andolonomby, Anjohibe, and Ampasambazimba) have provided opportunities for a limited degree of site integration, but none of them sample the presumed time of the extinction window with sufficient temporal precision to measure the width of the window *in situ* in the same way as Belo. However, each may shed some light on the causation issue, and taken with late occurrences from other sites around the island (Table 3) a pattern of late survival for many taxa and a drawn-out, complex extinction "event" is plausible.

3.2.1. Southwestern Sites

Megafaunal dates within the last two millennia from Ambolisatra/Andolonomby are restricted to the date that has been obtained on a human-altered *Hippopotamus* femur (MacPhee and Burney, 1991). Other dates from the site are from mid-Holocene megafaunal bone fragments in the pollen column for the site (Burney, 1993b). The author proposed an extinction scenario for the site that invoked a synergistic role for direct human hunting, landscape modification, introduced species, and climatic desiccation, perhaps beginning with a decline of 50% or more in precipitation by 2000 yrbp from maximum mid-Holocene values. With the arrival of people perhaps ca. 2000 yrbp and the modification of the landscape by fire and land clearing over the next few centuries, one can envision that human hunting might have played an important role for large terrestrial lemurs, tortoises, elephant birds, and hippos, but less so perhaps for some smaller animals such as the peculiar ungulate *Plesiorycteropus* (MacPhee, 1994). The loss of forage and water to domestic livestock was also likely to have been a factor in the decline of the endemic herbivores (through direct competition, especially in the dry season), as the density of livestock may have been relatively high in late prehistoric times (see Dewar, 1985:589).

Although the width of the extinction window for this site has not been directly determined, it is interesting to note (Table 3) that the southwest has two other sites with early dates on human-modified bones, a hippo from Lamboharana at 1740 ± 50 (MacPhee and Burney, 1991), and an *Aepyornis* from Itampolo at 1880 ± 70 (H. James, unpublished data). Older-generation dates obtained in previous decades, presumably on whole bone, show late occurrences (e.g., as recent as 980 ± 200 yrbp for a hippo bone from Itampolo; Mahé and Sourdat, 1972), but these sites need redating with modern biochemical safeguards for bone dating (Stafford *et al.*, 1991). Goodman and Rakotozafy (1997) report a late date from Beloha of 1380 ± 90 yrbp for AMS-dated collagen from *Alopochen sirabensis,* an extinct goose.

Two intriguing AMS dates (on bone collagen treated with alkali to remove humates) have been obtained recently from Ankilitelo, a site ca. 50 km inland from Toliara (Simons, 1997): 630 ± 50 and 510 ± 80 yrbp for *Megaladapis* and *Palaeopropithecus,* respectively. These dates, if confirmed, would serve to open the extinction window to 1.5 millennia for the southwestern region as a whole, and hint at the possibility of very late survival of some megafauna in the remote and inaccessible karst regions of the western interior. Calibrated at 2σ, these dates overlap the early historical period in Madagascar and serve to spark renewed interest in the oral traditions of strange beasts living in the interior recorded by Etienne de Flacourt in the mid-1600s and others more recently (summarized in Burney and Ramilisonina, in press).

3.2.2. Anjohibe and Other Northern Cave Sites

The vicinity of Anjohibe, a large cavern system northeast of Mahajanga (Majunga), has yielded information on the extinct fauna and pollen records back to ca. 40,000 yrbp (Burney *et al.,* 1997). In combination with archaeological evidence from the Mahavavy River valley (Wright *et al.,* 1996) and pollen data (Matsumoto and Burney, 1994) from Lake Mitsinjo, sites roughly 100 km to the south along the coast, a synthesis emerges for this region that suggests moderate vegetation change, an increase in burning after human arrival (but considerable natural fire before humans), and the late survival of *Archaeolemur* (830 ± 60 yrbp; Burney *et al.,* 1997). Earliest evidence for humans in this area comes from the eleventh- to fourteenth-century Antetikala Phase pottery on the Bay of Boeny (Wright *et al.,* 1996), consistent with the relatively late dates for stratigraphic charcoal increase along this coast. Archaeological sites in Anjohibe have so far yielded no evidence for human utilization of any of the megafauna except giant tortoises (Burney *et al.,* 1997). Ceramic types from sites in the cave system suggest a local human presence only in recent centuries, probably not before the fifteenth century at the earliest (H. T. Wright, personal communication).

Another karst region farther north, the Ankarana Massif, has also yielded a late date for *Archaeolemur* (1020 ± 50 yrbp; Simons *et al.,* 1995). It has been suggested that climate change to drier conditions may have caused the shrinkage of the range of the extant lemurs *Indri indri* and *Hapalemur simus* from this region some time after 4000 yrbp (Simons *et al.,* 1995; Jungers *et al.,* 1995). *Babakotia radofilai* is one of two extinct taxa that have not been dated to the late Holocene—the single date that has been obtained on this recently discovered genus is 4040 ± 100 yrbp (Simons *et al.,* 1995). Of course, one date cannot fix a species' time of disappearance, and its general absence from other fossil sites may be an indication of its rarity or localized distribution. It is only known from one site outside the Ankarana, the caverns of Anjohibe, which contain bones that date from the early Holocene on other taxa (Burney *et al.,* 1997).

Farther north, at the arid northern tip of the island, Lakaton'i Anja has yielded possible early evidence for humans (Dewar and Wright, 1993). Charcoal from the lowest occupation level of this rockshelter yielded ages of 1300 ± 80 and 1680 ± 65 yrbp. Inherent age problems with these dates, however, cannot be ruled out. A few bones of extinct fauna have been excavated here, but these have not been dated. No diagnostic pottery was found at this level, but ceramics in a level above are typical of the twelfth to thirteenth century. Another important area for early archaeological sites, Irodo on the northeast coast, has yielded a late date for *Aepyornis* eggshell, 1150 ± 90 yrbp (Battistini and Vérin, 1966). Although all megafaunal dates from prior decades should be regarded with great suspicion, owing to the now well-known problems with humate and carbonate contamination of bone samples, controlled experiments by Long *et al.* (1983) suggested that bird eggshell dating may be less problematic.

3.2.3. Ampasambazimba

This site from the central highlands is well known for its large yield of extinct forms, beginning with excavations in the late nineteenth century (reviewed in MacPhee *et al.*, 1985). Pollen and charcoal data from nearby Lake Kavitaha (Burney, 1987c) suggest that humans began to transform this landscape relatively late, probably not before the seventh or eighth century AD. This agrees well with the archaeological evidence (Wright *et al.*, 1992), and implies that any human predation on the fauna may have begun here a half millennium or more later than it did on the southwest coast. Unfortunately, Ampasambazimba appears to have sampled the fauna too early, probably between the beginning of the Holocene and ca. 4000 yrbp (MacPhee *et al.*, 1985), although Tattersall (1973) obtained a (presumably) whole-bone date of 1035 ± 50 yrbp from a *Megaladapis* bone collected on the surface of the site.

One extinct taxon, the gorilla-sized *Archaeoindris fontoynonti,* is known only from the basal sediments of this site (and no other sites). Wood from this layer dates to 8245 ± 215 yrbp (MacPhee *et al.*, 1985), raising the possibility that this species went extinct in the early Holocene, or else, perhaps like *Babakotia,* was an animal rare enough to be prone to the Signor–Lipps effect (i.e., the tendency for extremely rare taxa to "disappear" from the fossil record well before their actual extinction). The sheer size of *Archaeoindris* and the distinctive appearance of several *Babakotia* elements (Simons *et al.*, 1992) make it unlikely that they would have been overlooked in the many other sites around the island that have yielded an abundance of (mostly late Holocene) bones of other taxa.

4. Discussion

Clearly, more dating of the purified organic fractions of bones of the extinct animals is in order. A program of dating XAD-2 purified amino acids and other bone residues (Stafford *et al.*, 1991) would be appropriate. It appears that latest occurrences of several species of large extinct lemurs, as well as endemic ratites, tortoises, and hippos are well within the human period, and several sites show late survival of *Archaeolemur* in particular. Results from the integrated sites of Andolonomby/Ambolisatra and Belo-sur-mer suggest that humans and some of the taxa may have coexisted (in the taphonomic sense, at least) in the southwest for a millennium or more. On the northwest coast, a contrast is apparent in that first evidence for humans is later, but a high human population density seems to have been achieved rapidly by ca. AD 1200 in some areas, with evidence for deforestation, erosion, and perhaps cultural eutrophication at sites near good ports (Matsumoto and Burney, 1994; Wright *et al.*, 1996). The latest dates for extinct megafauna come from inland sites (Simons *et al.*, 1995; Simons, 1997; Burney *et al.*, 1997), but much more dating is needed to be certain this is not an artifact of the incompleteness of the spatial grid and the shortage of detailed ^{14}C chronologies.

Despite these inadequacies, it is tempting to propose a preliminary comparison between the empirical evidence and the predictions of the five hypotheses in Table 1. Each of these has a different combination of expectations with regard to the rates, patterns, and processes of extinction and landscape transformation. H1, H3, and H4 all require that the extinctions should occur rapidly at any particular location, and that the island wide extinction events and any causative or resultant landscape transformations should likewise be rapid, and either spread as a "wave" (H3) or more or less break out everywhere in a narrow time band (i.e., on a scale of years or decades rather than centuries or millennia). H2 was by conception postulated to explain the extinctions *in the arid south only,* but could be taken as a general climatic theory for the extinctions. The climates of the northern half and the entire east and center of the island do not appear to have deteriorated (e.g., desiccated) for any time in the period of the extinctions with sufficient severity to have generated wholesale extinction of animals that were in several cases clearly adapted to savanna and arid bushland (MacPhee *et al.,* 1985; Burney, 1987a,b, 1996b). Climate seems to be a possible contributor to ecological stress in the drier parts of the island, but even there it is hardly likely to have been more severe in the late Holocene than in other dry periods for which we have no definite extinctions (but the fossil record prior to 4000 yrbp is admittedly thin). Long-term paleoclimatic records from the WCT and elsewhere on the island show major climate changes in the late Pleistocene, and a dry early Holocene (Burney, 1997a). One site on the transect (Anjohibe), another near the transect (Ankarana) in the far northern interior, and a third (Ampasambazimba) in the central highlands show that some forest-adapted taxa underwent drastic range shrinkage since ca. 4000 yrbp in what appears to be a response to prehuman climate change (Simons *et al.,* 1995; Jungers *et al.,* 1995; Burney *et al.,* 1997). None of the many late Quaternary extinctions that have been recognized so far in Madagascar can be confidently attributed to climate change alone, although this cannot be ruled out for at least two large extinct lemurs, *Archaeoindris* and *Babakotia.*

All available evidence is consistent with the notions that (1) each major ecological region of Madagascar shows a different composite history in terms of landscape transformation and perhaps extinction chronologies (much work is still to be done here) and (2) each region studied contains some variation on the same mix of human ecological depredations (e.g., overkill, change of fire regime, exotic species introductions, deforestation, accelerated erosion) and these factors, as well as the "natural" background-level climatic variations of the late Holocene, have good potential for synergistic interactions. Diverse causes can add up to the same final result: extinction of the entire endemic megafauna as well as some smaller animals, and the biotic depauperation and exotic invasion of dry woodlands, savanna-type mosaics, and semidesert bushlands. One might expect such a multiplicity of factors to operate in a distinctly piecemeal fashion, creating a mosaic of extirpations that could only look like a single catastrophic extinction event after it had run to virtual completion throughout the island.

H5 is not a streamlined hypothesis that can be conveniently subjected to falsification tests in the sense intended by Popper (1969), but it does seem to best describe all of the evidence available. To refute this hypothesis, which is still possible, one would need a fine-scale grid of extremely well-dated sites (to minimize the Signor–Lipps effect) that show either (1) that the extinctions were rapid and approximately coeval throughout the island or at least over a broad ecological gradient (such as the WCT described above) or (2) that some single factor was sufficient to account for all of the extinctions (such as incontrovertible evidence for a "hypervirulent disease" of full transtaxonomic virulence (sensu MacPhee and Marx, 1997) or massive kill sites. Neither of these are apparent so far in Madagascar. Admittedly this lack of observations provides negative-evidence support only, but a large amount of effort by several investigators has been invested in the search for kill sites for more than a decade now, and the "smoking pistol" remains elusive. One could hazard the generalization that the extinctions and landscape transformations of Madagascar have, by and large, occurred relatively slowly (over the last two millennia, with the process still in full swing in the present decade). Furthermore, the patterns of transformation appear distinctly piecemeal or mosaiclike, with late survivals of some taxa at several sites, and pockets of late survival of intact or nearly intact ecosystems at remote locations. H5 will hold up best if this pattern is fully demonstrated. No other among these five hypotheses, or any other yet apparent, will look as compelling.

So what happened to Madagascar's late Holocene biota? Perhaps a better question to ask is whether what is happening today is very different from what happened after first contact with humans. Four hypotheses have been proposed that suggest what happened in the last two millennia in Madagascar is different than what we see happening there now. The fifth allows that the processes invoked in the other hypotheses—fire, climate change, overhunting of a naive fauna, biological invasions (possibly including disease)—were probably ingredients of the mix then as now. Humans, in manifold ways, were the key factor.

Although such combinatorial solutions are unpopular with many investigators because they are too "murky" in their expectations to test with hypothetico-deductive rigor (Popper, 1969), care should be taken that complex truths are not overlooked in the search for a simpler answer. "Science should be explained as simply as possible, but no more so," Einstein once observed.

A resounding refutation of any of the proposed hypotheses will be difficult, but it may eventually be possible with enough radiometric dates from the right places and times, and a more complete paleoecological reconstruction at the landscape level. It is conceivable that there are no distinct temporal or causal breaks between earlier and later extinctions of the human period, and current species endangerment, for that matter. If this were the case, it would be hard to dispute the importance of environmental synergy, because we see it everywhere we look in Madagascar today.

References

Athens, J. S. 1997. Hawaiian native lowland vegetation in prehistory, in: P. V. Kirch and T. L. Hunt (eds.), *Historical Ecology in the Pacific Islands,* pp. 248–270. Yale University Press, New Haven.

Battistini, R., and Vérin, P. 1966. Les datations à Madagascar par la méthode du RC-14. *C.R. Semaine Géologique 1966,* Antananarivo.

Berger, R., Ducote, K., Robinson, K., and Walter, H. 1975. Radiocarbon date for the largest extinct bird, *Nature,* **258**:709.

Burney, D. A. 1987a. Pre-settlement vegetation changes at Lake Tritrivakely, Madagascar. *Palaeoecol. Afr.* **18**: 357–381.

Burney, D. A. 1987b. Late Quaternary stratigraphic charcoal records from Madagascar. *Quat. Res.* **28**:274–280.

Burney, D. A. 1987c. Late Holocene vegetational changes in central Madagascar. *Quat. Res.* **28**:130–143.

Burney, D. A. 1993a. Recent animal extinctions: recipes for disaster. *Am. Sci.* **81**:530–541.

Burney, D. A. 1993b. Late Holocene environmental changes in arid southwestern Madagascar. *Quat. Res.* **40**:98–106.

Burney, D. A. 1996a. Paleoecology of humans and their ancestors, in: T. R. McClanahan and T. P. Young (eds.), *East African Ecosystems and Their Conservation,* pp. 17–34. Oxford University Press, London.

Burney, D. A. 1996b. Climate change and fire ecology as factors in the Quaternary biogeography of Madagascar, in: W. Lourenco (ed.), *Biogeographie de Madagascar,* pp. 49–58. ORSTOM, Paris.

Burney, D. A. 1997a. Theories and facts regarding Holocene environmental change and extinction in Madagascar, in: S. Goodman and B. Patterson (eds.), *Natural Change and Human Impact in Madagascar,* pp. 75–89. Smithsonian Institution Press, Washington, D.C.

Burney, D. A. 1997b. Tropical islands as paleoecological laboratories: gauging the consequences of human arrival. *Hum. Ecol.* **25**:437–457.

Burney, D. A., and Ramilisonina. In press. The kilopilopitsofy, kidoky, and bokyboky: accounts of strange animals from Belo-sur-mer, Madagascar and the megafaunal "extinction window." *Am. Anthropol.* 27 ms. pp.

Burney, D. A., Burney, L. P., and MacPhee, R. D. E. 1994. Holocene charcoal stratigraphy from Laguna Tortuguero, Puerto Rico, and the timing of human arrival on the island. *J. Archaeol. Sci.* **21**:273–281.

Burney, D. A., DeCandido, R. V., Burney, L. P., Kostel-Hughes, F. N., Stafford, T. W., Jr., and James, H. F. 1995. A Holocene record of climate change, fire ecology, and human activity from montane Flat Top Bog, Maui. *J. Paleolimnol.* **13**:209–217.

Burney, D. A,. Burney, L. P., Griffin, W., Wright, H. T., James, H. F., Jungers, W. L., MacPhee, R. D. E., Rafamantanantsoa, J.-G., Rakotondrazafy, T., Ramilisonina, Warren, D., and Yoder, A. D. 1996. Belo-sur-mer and other integrated sites from Holocene Madagascar, in: *Six North American Paleontological Convention Abstracts of Papers,* p. 56. 9–12 June 1996, Washington, D.C.

Burney, D. A., James, H. F., Grady, F. V., Rafamantanantsoa, J.-G., Ramilisonina, Wright, H. T., and Cowart, J. B. 1997. Environmental change, extinction, and human activity: evidence from caves in NW Madagascar. *J. Biogeogr.* **24**:755–767.

Chanudet, C. 1975. *Conditions Géographiques et Archéologiques des Subfossiles Malagaches.* Mém. Maîtrise, Université de Bretagne Occidentale, Brest.

Dewar, R. E. 1984. Recent extinctions in Madagascar: the loss of the subfossil fauna, in: P. S. Martin and R. G. Klein (eds.), *Quaternary Extinctions: A Prehistoric Revolution,* pp. 574–593. University of Arizona Press, Tucson.

Dewar, R. E., and Wright, H. T. 1993. The culture history of Madagascar. *J. World Prehist.* **7**:417–466.

Ellison, J. C. 1994. Palaeo-lake and swamp stratigraphic records of Holocene vegetation and sea-level changes, Mangaia, Cook Islands. *Pac. Sci.* **48**(1):1–15.

Flannery, T. 1994. *The Future Eaters.* George Braziller, New York.

Flenley, J. R., King, A. S. M., Teller, J. T., Prentice, M. E., Jackson, J., and Chew, C. 1991. The Late Quaternary vegetational and climatic history of Easter Island. *J. Quat. Sci.* **6**:85–115.

Goodman, S. M., and Rakotozafy, L. M. A. 1997. Subfossil birds from coastal sites in western and southwestern Madagascar: a paleoenvironmental reconstruction, in: S. Goodman and B. Patterson (eds.), *Natural Change and Human Impact in Madagascar,* pp. 257–279. Smithsonian Institution Press, Washington, DC.

Humbert, H. 1927. Destruction d'une flore insulaire par le feu. Principaux aspects de la végétation à Madagascar. *Mem. Acad. Malgache* **5**:1–80.

Humbert, H., and Cours Darne, G. 1965. Carte international du tapis végétal et des conditions écologiques à 1/1,000,000. Notice de la carte de Madagascar. *Trav. Sect. Sci. Tech. Inst. Fr. Pondichèry* hors sèrie **6**:1–162.

Jungers, W. L., Godfrey, L. R., Simons, E. L., and Chatrath, P. S. 1995. Subfossil *Indri indri* from the Ankarana Massif of northern Madagascar. *Am. J. Phys. Anthropol.* **97**:357–366.

Kirch, P. V., and Ellison, J. 1994. Palaeoenvironmental evidence for human colonization on remote Oceanic islands. *Antiquity* **68**:310–321.

Kirch, P. V., Flenley, J. R., and Steadman, D. W. 1991. A radiocarbon chronology for human-induced environmental change on Mangaia, southern Cook Islands, Polynesia. *Radiocarbon* **33**(3):317–328.

Long, A., Hendershott, R. B., and Martin, P. S. 1983. Radiocarbon dating of fossil eggshell. *Radiocarbon* **25**:533–539.

MacPhee, R. D. E. 1994. Morphology, adaptations, and relationships of *Plesiorycteropus,* and a diagnosis of a new order of eutherian mammals. *Bull. Am. Mus. Nat. Hist.* **220.**

MacPhee, R. D. E., and Burney, D. A. 1991. Dating of modified femora of extinct dwarf *Hippopotamus* from southern Madagascar: implications for constraining human colonization and vertebrate extinction events. *J. Archaeol. Sci.* **18**:695–706.

MacPhee, R. D. E., and Marx, P. A. 1997. The 40,000-year plague: humans, hypervirulent diseases, and first-contact extinctions, in: S. Goodman and B. Patterson (eds.), *Natural Change and Human Impact in Madagascar,* pp. 169–217. Smithsonian Institution Press, Washington, D.C.

MacPhee, R. D. E., Burney, D. A., and Wells, N. A. 1985. Early Holocene chronology and environment of Ampasambazimba, a Malagasy subfossil lemur site. *Int. J. Primatol.* **6**(5):463–489.

Mahé, J., and Sourdat, M. 1972. Sur l'extinction des vertébrés subfossiles et l'aridification du climat dans le Sud-ouest de Madagascar. *Bull. Soc. Geol. Fr.* **14**:295–309.

Marden, L. 1967. Madagascar, island at the end of the Earth. *Nat. Geogr.* **132**:443–487.

Martin, P. S. 1984. Prehistoric overkill: the global model, in: P. S. Martin and R. G. Klein (eds.), *Quaternary Extinctions: A Prehistoric Revolution,* pp. 354–403. University of Arizona Press, Tucson.

Martin, P. S., and Klein, R. G. (eds.). 1984. *Quaternary Extinctions: A Prehistoric Revolution.* University of Arizona Press, Tucson.

Matsumoto, K., and Burney, D. A. 1994. Late Holocene environments at Lake Mitsinjo, northwestern Madagascar. *Holocene* **4**:16–24.

McNeill, J. R. 1994. Of rats and men: a synoptic environmental history of the island Pacific. *J. World Hist.* **5**:299–349.

Popper, K. 1969. *Conjectures and Refutations.* Routledge & Kegan Paul, London.

Radimilahy, C. 1997. Mahilaka, an eleventh- to fourteenth-century Islamic port: the first impact of urbanism in Madagascar, in: S. Goodman and B. Patterson (eds.), *Natural Change and Human Impact in Madagascar,* pp. 342–363. Smithsonian Institution Press, Washington, D.C.

Simons, E. L. 1997. Lemurs: old and new, in: S. Goodman and B. Patterson (eds.), *Natural and Human-Induced Change in Madagascar,* pp. 142–168. Smithsonian Institution Press, Washington, D.C.

Simons, E. L., Godfrey, L. R., Jungers, W. L., Chatrath, P. S., and Rakotosamimanana, B. 1992. A new giant subfossil lemur, *Babakotia,* and the evolution of the sloth lemurs. *Folia Primatol.* **58**:197–203.

Simons, E. L., Burney, D. A., Chatrath, P. S., Godfrey, L. R., Jungers, W. L., and Rakotosamimanana, B. 1995. AMS ^{14}C dates for extinct lemurs from caves in the Ankarana Massif, northern Madagascar. *Quat. Res.* **43**:249–254.

Stafford, T. W., Jr., Hare, P. E., Currie, L., Jull, A. J. T., and Donahue, D. J. 1991. Accelerator radiocarbon dating at the molecular level. *J. Archaeol. Sci.* **18**:35–72.

Tattersall, I. 1973. A note on the age of the subfossil site of Ampasambazimba, Miarinarivo Province, Malagasy Republic. *Am. Mus. Novit.* **2520**:1–6.

Wright, H. T., Andrianalvoarivony, R., Bailiff, I., Burney, D., Haas, H., Raharijaona, V., Rakotovololona, Rasamuel, D., and Dewar, R. 1992. Datation absolue de sites archéologiques de Madagascar—présentation des déterminations. *Taloha* **11**:121–146.

Wright, H. T., Vérin, P., Ramilisonina, Burney, D., Burney, L. P., and Matsumoto, K. 1996. The evolution of settlement systems in the Bay of Boeny and the Mahavavy River valley, north-western Madagascar. *Azania* **31**:37–73.

CHAPTER 8

Extinctions and Local Disappearances of Vertebrates in the Western Mediterranean Islands

JOSEP ANTONI ALCOVER, BARTOMEU SEGUÍ, and PERE BOVER

1. Introduction

A landlocked sea between the Eurasian and African plates, the Mediterranean Sea may be thought of as an ocean in miniature (Margalef, 1985). Like other seas on the borderlands between contiguous plates (e.g., Antillean and Sundaland regions), the Mediterranean is rich in islands. The islands of the Mediterranean have a complex paleogeographical history. Some are properly described as "continental," but others just as clearly resemble oceanic islands, and are named oceanic-like islands (Alcover *et al.,* 1998). This is reflected in their biotic composition. On continental islands, species tend to be identical or nearly identical to those on the adjacent mainland, although much less diverse. In addition to their well-known tendency to exhibit great endemism at the species level, oceanic and oceanic-like islands also display very low diversity at higher taxonomic levels and disharmonious floral and faunal integration as compared with mainland areas of similar size (Alcover *et al.,* 1998).

In the Mediterranean Sea there are also islands that have undergone wholesale changes in their biotic composition over time, from oceanic to continental and vice versa. There are also "fossil" islands, land areas that were previously insular but are now part of a mainland. In some cases (e.g., Gargano, Baccinello) these islands were clearly oceanic in type, and possessed a commensurately endemic fauna. Finally, there are also instances of previously separate islands fusing with each other to form a new entity (e.g., Crete, Mallorca). In Table 1 we provide some basic geographical data for those Mediterranean islands that have (or once had) an oceanic character.

JOSEP ANTONI ALCOVER and PERE BOVER • Institut Mediterrani d'Estudis Avançats, 07071 Ciutat de Mallorca, Balears, Spain. BARTOMEU SEGUÍ • Departament de Ciències de la Terra, Universitat de les Illes Balears, 07071 Ciutat de Mallorca, Balears, Spain.

Extinctions in Near Time, edited by MacPhee. Kluwer Academic/Plenum Publishers, New York, 1999.

Table 1. Some Geographical Parameters of the Oceanic-like Mediterranean Islands

Island	Area (km2)	Maximum altitude (m)	Nearest distance to mainland (km)	Population (year)
Corsica	8,681	2711	89	250,000 (1990)
Sardinia	23,813	1834	186	1,650,000 (1990)
Tavolara	—	—	192	—
Mallorca	3,645	1445	165	568,065 (1991)
Menorca	701	358	195	64,860 (1991)
Eivissa	541	475	90	72,309 (1991)
Formentera	82	202	105	4,333 (1991)
Cabrera	15	172	240	10 (1997)
Dragonera	4	311	174	0
Sicily	25,461	3261	16	5,200,000 (1990)
Malta	316	253	93	355,910 (1991)
Cyprus	9,251	1951	60	764,000 (1993)
Crete	8,261	2456	114	537,183 (1991)
Karpathos	301	1215	96	4,645 (1981)
Kasos	69	601	141	—
Tylos	63	651	18	—
Rhodes	1,398	1215	18	87,831 (1981)
Amorgos	121	822	108	1,718 (1981)
Delos	3.4	—	93	19,688 (1981)
Serifos	84	585	63	16 (1981)
Naxos	428	1003	132	14,037 (1981)
Milos	151	751	105	4,554 (1981)

The existing Mediterranean islands are best thought of as comprising an intercontinental mega-archipelago (Alcover *et al.,* 1998), each segment of which has had a different geological and biotic history. For example, during the Neogene there were islands in the western and central parts of the Mediterranean that supported terrestrial faunas, but none in the eastern Mediterranean (cf. Kotsakis, 1990). It is thus unsurprising that the oldest Mediterranean insular endemics occur in the western zone, where the most ancient islands are found.

For these reasons, a simple geographical classification of Mediterranean islands into eastern, central, and western groups may be helpful. In this chapter we will focus on western Mediterranean islands that have an oceanic character, with special reference to local extinctions and disappearances among vertebrates (and other groups, such as gastropods and plants, where appropriate information exists).

For the time period under consideration here (latter part of the Quaternary, i.e., late Pleistocene and Holocene), the oceanic islands of the western Mediterranean can be grouped geographically into three archipelagos: the Pityusic (Eivissa, Formentera, and adjacent islands) and Gymnesic islands (Mallorca, Menorca, Cabrera, and adjacent islands), and the islands of the Cirno-Sardinian massif (Sardinia, Corsica, and adjacent islands). Each of these "fringing" archipelagos originated on a different Tertiary microplate that spalled off from the Iberian peninsula as the latter rotated about the axis of the Pyrenees (Alvarez, 1972; Bellon *et al.,* 1977). The Cirno-Sardinian microplate was the first to detach itself from the Iberian Peninsula. The existence of the Oschiri fauna (De Bruijn and Rümcke, 1974) implies that, in the Early Miocene, Sardinia was already a subaerial landmass composed of one or more islands. In Mallorca, the oldest vertebrate-bearing sites (Santa Margalida and Sant Llorenç; Mein and Adrover, 1982; Adrover *et al.,* 1983–1984) are somewhat younger (Middle Miocene). *Gymnesicolagus,* a member of this fauna, has been found recently in Menorca (Quintana and Moyà-Solà, personal communication). In the case of the Pityusics the most ancient insular fauna recovered so far is even younger (Late Miocene), and comes from the site of Ses Fontanelles on Eivissa (Agustí and Moyà-Solà, 1990). This chronological sequence is presumably not accidental, but instead reflects the order of microplate detachment from the central Iberian landmass.

A second feature shared by these islands is that all were affected by the so-called Messinian crisis, when the Mediterranean lost its connection with the Atlantic following collisional impact of the African and Eurasian plates (Hsü, 1972; Hsü *et al.,* 1973, 1977; Maldonado, 1985). Thereafter, the western Mediterranean basin underwent severe desiccation, as indicated by the vast deposits of evaporites laid down between 5.7 and 5.35 Ma that drape much of the seafloor in this area (Hsü *et al.,* 1973, 1977; Gautier *et al.,* 1994). One result was that the islands of the western Mediterranean became continuous with surrounding mainlands. The overland connections with the former islands would, of course, have largely consisted of salt deserts; these would have acted as strong filters, permitting relatively few immigrant species to pass (Alcover, 1987). However, some certainly got across, because the former islands experienced approximately coeval faunal turnover at the time of the salinity crisis (Agustí and Moyà-Solà, 1990). New species appear in the records of the three archipelagos at this time; several taxa are shared, indicating extensive interarchipelagic contacts. With the reopening of the Strait of Gibraltar at 5.35 Ma, the current geographical configuration of the western Mediterranean was roughly instituted and the three archipelagos went their own ways thereafter, paleogeographically and faunistically speaking.

Pleistocene glaciations also had pronounced biogeographical effects on the western Mediterranean islands. Drops in ambient sea level during pleniglacial periods facilitated faunal exchange and homogenization as islands in the same archipelago were briefly connected by landbridges (e.g., during the Last or Würmian pleniglacial around 18,000 yrbp, when sea level dropped 120 m below the present level; Fairbanks, 1989).

During the Biharian regression (about 800,000 years ago), Corsica and Sardinia suffered a faunistic turnover, although this seems to have happened without an actual land connection with the mainland (although distances were no doubt reduced). The Cirno-Sardinian turnover was marked by the extirpation of several large terrestrial mammals (*Nesogoral, Macaca, Sus*); some small mammals died out at this time as well, while others underwent profoundly rapid evolutionary change. Several immigration events also occurred. The Early–Middle Pliocene faunas of Mallorca and Menorca may well have differed, but during the glaciations they became homogenized. No further faunal turnovers occurred in this archipelago until the arrival of *Homo sapiens.* By contrast, on Eivissa and Formentera at an as-yet-undetermined moment during the Plio-Quaternary there was a massive extinction without accompanying faunal turnover (most probably during the early or middle Pleistocene).

Thus, the three archipelagos have had different faunal histories, and by the late Pleistocene–Holocene the fauna and flora of each group of islands had been notably differentiated, as arriving humans would have discovered as they began to settle in each of these unique ecosystems.

Homo sapiens colonized all of the western Mediterranean islands perhaps by the late

Pleistocene, and certainly by the Holocene (for a discussion on this topic see Cherry, 1984, 1990, 1992). The arrival of humans, together with commensals and synathropics, caused a new biotic crisis, featuring a whole host of immigrations, extinctions, expansions, and rarefactions of species as the previously existing ecological regimes were replaced by the current ones. To analyze local extinctions and disappearances on these islands, we must now turn to what we know of the flora and fauna at the time of first human settlement.

2. Upper Pleistocene Faunas and Floras

Knowledge of the late Pleistocene/Holocene biota of the western Mediterranean islands is incomplete. With the exception of gastropods (whose shells fossilize extremely well), the nonvertebrate record is very poor. There is some information on the Late Pleistocene and prehuman floras of Mallorca, Menorca, Corsica, and Sardinia (this last unpublished), but there is no comparable body of data for the Pityusic Islands (see below).

By contrast, the Quaternary vertebrate record is notably excellent. All of the large islands (surface area > 80 km^2) in the western Mediterranean contain fossiliferous deposits, as do several of the smaller islands (e.g., Cabrera, Sa Dragonera, Tavolara). These deposits, some of which are incredibly rich (e.g., Es Pouàs on Eivissa; Cova de Moleta on Mallorca), permit assessment of faunal diversity before and after anthropogenic effects. At present, more than 150 fossiliferous Quaternary sites are known, and thanks to dedicated work by a variety of scholars, we can now say that all or nearly all of the terrestrial mammals that lived on these islands prior to the arrival of people have already been identified. The record is less complete for other groups. Bats are not well known, and it is estimated that fewer than 30% of birds breeding on these islands have been identified in fossil material. Knowledge of small passeriforms is especially incomplete. The fossil history of the amphibian and reptile faunas of the Pityusic and Gymnesic islands has been fairly well investigated, but there is little fossil evidence of the past herpetofaunas of the Cirno-Sardinian islands.

Fossils are the usual route to understanding of previous vertebrate faunas, but helpful information can also come from analytical studies of endemicity (e.g., Legakis and Kypriotakis, 1994; Palmer *et al.*, in press). Such studies often permit reasonable inferences about taxa that "should" have been present in an area but have not yet been found as fossils (Myers and Giller, 1988). We are interested in applying this methodology to restructuring the herpetofauna of Corsica and Sardinia before the arrival of humans (e.g., Lanza, 1983). Time of origin of some endemic lineages can be indirectly estimated via phylogenetic studies or degree of species differentiation (e.g., Lanza and Vanni, 1987; Sbordoni *et al.*, 1990).

Tables 2 to 4 are preliminary lists of vertebrates that would have been encountered by the first human settlers on the islands of the western Mediterranean. For nonvolant taxa, the lists are probably essentially complete. For birds, which are not well known paleontologically, we have listed all species that can reasonably be inferred to have bred in the western Mediterranean in the late Quaternary, or roughly the past 30,000 years (Taylor and Moore, 1953; Rick, 1975; Driver, 1982; Florit and Alcover, 1987). There are several criteria that can be used to make such inferences (not all of equal validity): living endemics; extinct

Table 2. Upper Pleistocene Amphibians and Reptiles of the Western Mediterranean Islands

Island	Amphibia	Reptilia
Formentera	—	*Podarcis pityusensis*
Eivissa	—	*Podarcis pityusensis*
Menorca	*Alytes talaioticus*	*Podarcis lilfordi*
Mallorca	*Alytes muletensis*	*Podarcis lilfordi*
Corsica	*Euproctus montanus*	*Phyllodactylus europaeus*
	Salamandra corsica	*Algyroides fitzingeri*
	Discoglossus sardus	*Archaeolacerta bedriagae*
	Discoglossus montalentii	*Podarcis tiliguerta*
	Hyla sarda	*Coluber viridiflavus* (?)
Sardinia	*Euproctus platycephalus*	*Phyllodactylus europaeus*
	Hydromantes genei	*Algyroides fitzingeri*
	Hydromantes supramontis	*Archaeolacerta bedriagae*
	Hydromantes imperialis	*Podarcis tiliguerta*
	Hydromantes flavus	*Coluber viridiflavus* (?)
	Discoglossus sardus	
	Hyla sarda	

Table 3. Breeding or Possibly Breeding Avian Species in Western Mediterranean Fossil Communities[a]

Species	Mallorca	Menorca	Eivissa	Corsica	Sardinia and Tavolara
Calonectris diomedea	f, r	r	f, r	f, r	f, r
Puffinus mauretanicus	r	r	f, r		
Puffinus yelkouan				r	f, r
Hydrobates pelagicus	r	r	f, r	r	r
Puffinus sp.		f			
Phalacrocorax aristotelis	f, r	r	f, r	f, r	f, r
Ciconia ciconia					f, r
Anser erythropus				f	
Anser sp. indet	f	f	f		
Tadorna sp.	f, r (T.tadorna)		r (T.tadorna)	f	r (T.tadorna)
Anas platyrhynchos	r	r	r	f, r	f, r
Anas crecca	f			f	
Anas querquedula					f, r
Aythya fuligula				f	
Pandion haliaetus	r	f, r	f, r	r	
Milvus milvus	r	r	r	r	f, r
Accipiter gentillis				f, r	f, r
Accipiter nisus	f			f, r	r
Buteo buteo				f, r	f, r
Aquila heliaca				cf.	cf.
Aquila chrysaetos	f		f	f, r	f, r
Haliaeetus albicilla			f	f	f

[a]Key: f = fossil; r = reproducing currently; ra = possibly reproducing currently; bold = taxon suffered total or local extinction; italic = taxon that may have bred in the Late Pleistocene/Holocene, but not breeding currently.

(*continued*)

Table 3. (*Continued*)

Species	Mallorca	Menorca	Eivissa	Corsica	Sardinia and Tavolara
Circus aeroginosus	r	r		f, r	r
Gypaetus barbatus				r	f
Aegypius monachus	r				f
Gyps fulvus				f	f, r
Falco peregrinus	r	r	r	f, r	f, r
Falco subbuteo	ra			f, r	r
Falco eleonorae	f, r		f, r		r
Falco naumanni	cf.		f	f	r
Falco tinnunculus	f, r		f, r	f, r	f, r
Coturnix coturnix	f, r	r	f, r	f, r	f, r
Grus grus			f		
Rallus aquaticus	r			f, r	r
Rallus n. sp.			f		
Porzana porzana	f, r	r		f	
Porzana parva				f	
Porzana pusilla				f	
Crex crex			f	f	
Gallinula chloropus	r	r	r	f, r	r
Otis tarda			f		
Otis tetrax				f	r
Burhinus oedicnemus	f, r	r	f	r	r
Vanellus vanellus			f	f	
Tringa erythropus	r			f	r
Tringa totanus				f	
Limosa limosa				f	f
Numenius arquata			f		
Scolopax rusticola	f		f	f	f
Columba palumbus	r	r	f, r	f, r	f, r
Columba livia/oenas (livia)	f, r	f, r	f, r	f, r	f, r
Streptopelia turtur	r	r	r	r	f, r
Cuculus canorus	r	r	f, r	f, r	r
Tyto alba	f, r	r	r	r	f, r
Bubo insularis				f	f
Bubo bubo			f		f
Asio otus	r		r	f, r	
Otus scops	f, r	r	f, r	f, r	f, r
Athene noctua	f, ra	ra		r	aff., r
Caprimulgus europaeus	r	f, r	r	f, r	r
Caprimulgus sp.			f		
Apus apus/pallidus (both)	f, r	r	f, r	r	r
Apus melba	f, r	r	ra	f, r	f, r
Merops apiaster	r	r	r	f, r	r
Upupa epops	f, r	r	f, r	r	r
Jynx torquilla	f, r		r	r	r
Picus viridis					cf.
Dendrocopos major				f, r	r
Calandrella sp. (brachydactyla)	f, r	r	f, r	r	r
Galerida cristata (/thecklae)	r	r	f, r	f	
Lullula arborea				f, r	cf., r
Alauda arvenis	f		f	r	r
Anthus spinoletta				f, r	r

Table 3. (*Continued*)

Species	Mallorca	Menorca	Eivissa	Corsica	Sardinia and Tavolara
Hirundo rupestris	f, r		f, r	r	r
Hirundo rustica/daurica	f, r(H.r)	r(H.r), rp(H.d)	f, r(H.r)	f, r(H.r; H.d)	cf.
Lanius collurio				f, r	r
Lanius minor/excubitor	f			f	
Lanius minor	f		f	r	
Lanius excubitor	f		f		
Lanius senator	r	r	f, r	r	r
Prunella modularis	f				
Prunella collaris	f			f, r	
Sylvia atricapilla	r	r	r	f, r	r
Sylvia sarda	cf., r		f, r	r	r
Sylvia melanocephala/curruca	f, r(S.m)	r(S.m)	r(S.m)	r(S.m)	r(S.m)
Phylloscopus sp.	f			r (P.collybita)	
Ficedula sp. (F. hypoleuca)	f, ra		f		
Parus ater	f			r	r
Saxicola rubetra	f, ra			f, r	
Saxicola torquata	f, r	r	r	r	r
Phoenicurus phoenicurus				f	
Phoenicurus ochruros	f			r	
Phoenicurus phoenicurus/ochruros	f				
Phoenicurus sp.	f		f		
Oenanthe oenanthe	f, r	r	r	r	r
Erithacus rubecula	f		f	f, r	r
Luscinia megarhynchos	f, r	r	r	f, r	r
Turdus sp.	f	f	f	f, r(T.viscivorus)	f, r(T.viscivorus)
Miliaria calandra	cf., r	r	f, r	r	r
Emberiza cirlus	r		r	cf., r	r
Emberiza cirlus/cintrinella	f, r(E.cir)		r(E.cir)	r(E.cir)	r(E.cir)
Fringilla coelebs	f, r	r	r	f, r	cf., r
Fringilla coelebs/montifringilla	f, r(F.c)	r(F.c)	f, r(F.c)	r(F.c)	r(F.c)
Serinus serinus	r	r	r	r	cf., r
Serinus citrinella				f, r	r
Carduelis chloris	f, r	r	r	f, r	r
Carduelis cannabina	f, r	r	f, r	r	r
Carduelis carduelis/cannabina	f, r	r	r	r	r
Loxia curvirostra	f, r		r	r	
Pyrrhula pyrrhula	f				cf.
Coccothraustes coccothraustes	f		f	f, r	f, r
Passer domesticus/hispaniolensis	f, r(P.d)	r(P.d)	f, r(P.d)	r(P.x italiae)	r(P.h)
Petronia petronia	f, r		f, r	f, r	r
Montifringilla nivalis	f		f	r	
Sturnus vulgaris	ra			f	f
Sturnus sp.	f, ra(S.v)			r(S.u)	r(S.u)
Oriolus oriolus				f, r	
Garrulus glandarius				r	f, r
Pyrrhocorax pyrrhocorax	f	f	f		f, r
Pyrrhocorax graculus	f		f	f, r	f
Corvus monedula	f			r	f, r
Corvus corone/frugilegus	f		f	f, r(C.c)	f, r(C.c)
Corvus antecorax			f		
Corvus corax	r	r	f, r	f, r	f, r

Table 4. Late Pleistocene Mammals of the Western Mediterranean Islands[a]

Island	Mammalia
Formentera	—
Eivissa	—
Menorca	*Soriculus hidalgoi*
	Eliomys morpheus
	Myotragus balearicus
Mallorca	*Soriculus hidalgoi*
	Eliomys morpheus
	Myotragus balearicus
Corsica	*Soriculus corsicanus*
	Cynotherium sardous
	Algarolutra majori
	"Megaloceros" cazioti
	Tyrrhenicola (Microtus) henseli
	Rhagamys orthodon
	Prolagus sardus
Sardinia	*Soriculus similis*
	Homo undescribed species
	Cynotherium sardous
	Algarolutra majori
	Megalenhydris barbaricina
	Sardolutra ichnusae
	Megaloceros cazioti
	Tyrrhenicola (Microtus) henseli
	Rhagamys orthodon
	Prolagus sardus

[a]Excluding bats.

Quaternary endemics; living sedentary species not breeding on any of the islands at present, but known from Quaternary fossils; living species currently nesting on any one of the islands and known from the fossil record; species represented by remains of juveniles incapable of flight; and presence of bird medullary bone as indicator of females laying eggs. Whether all species bred on the islands throughout the late Quaternary is difficult and perhaps impossible to verify in many cases.

2.1. Eivissa and Formentera

The late Quaternary biota of the Pityusic islands should be considered somewhat anomalous for the Mediterranean during that time (Alcover *et al.*, 1994). The fossil record documents that during the Messinian this archipelago was colonized by the same fauna that colonized both Mallorca and Sardinia. The Pliocene fauna of Eivissa included three terrestrial mammals (one bovid and two glirids), a lizard, a medium-sized tortoise, birds, and at least 18 species of gastropods (Alcover *et al.*, 1981; Bour, 1985; Agustí and Moyà-Solà, 1990; Paul and Altaba, 1992). During the Early to Middle Pleistocene, all of the terrestrial mammals, the tortoise, and the majority of the gastropod species became extinct. The causes of these losses remain unknown. Florit *et al.* (1989) and Alcover *et al.* (1994) have speculated that they may have been prompted by a catastrophic event, but to date there is no evidence to support this view. What is clear is that the Pityusics were not colonized again by terrestrial mammals after this extinction episode—a remarkable fact (Florit *et al.*, 1989; Alcover *et al.*, 1994; Sondaar *et al.*, 1995).

The Late Pleistocene/Holocene fauna of the Pityusic islands is known mainly from work at Es Pouàs on Eivissa (Florit *et al.*, 1989; Sondaar *et al.*, 1995; McMinn, unpublished). Available data suggest that Pityusic ecosystems before the arrival of humans were structured in a rather peculiar way, unparalleled on any other Mediterranean island (Seguí and Alcover, in press; Palmer *et al.*, in press). Thus, the structure of the avian community shows a remarkable resemblance to that of the Hawaiian Islands, and much less agreement with that of the Gymnesic and Cirno-Sardinian islands.

The late Quaternary vertebrate fauna of the Pityusics is dominated by birds, which monopolized all ecological guilds (Seguí and Alcover, in press). The herbivore guild was dominated by a small goose (*Anser* n. sp.), while the top carnivore was the sea eagle *Haliaeetus albicilla.* Rails (*Rallus* n. sp., *Crex crex*), corvids (*Pyrrhocorax pyrrhocorax, Corvus corax*), and Balearic shearwaters (*Puffinus mauretanicus*) were common. Bird communities are generally correlated with vegetation type (Cody, 1985; Welsh and Lougheed, 1996). As the bird communities of the Pityusics and Gymnesics differed substantially during the late Quaternary, it is reasonable to infer that the vegetation of these archipelagos differed in detail (Palmer *et al.*, in press). This is helpful information, because there are virtually no data on the prehistoric flora of the archipelago. Some palynological work has been conducted on samples from Es Pouàs, but the results (as yet unpublished) have not been satisfactory (very few palynomorphs, recognizable taxa mostly anemochorous and therefore not clearly autochthonous).

In addition to birds, a few bats are known (*Myotis myotis, Rhinolophus ferrumequinum,* and *R. hipposideros*); the only other land vertebrate was the Pityusic lizard (*Podarcis pityusensis*), very abundant in sites.

Gastropoda and Diplopoda are the only invertebrates with a Quaternary record on the islands, although this record is incompletely studied. The gastropod fauna of the prehistoric Pityusic islands consists of only eight terrestrial species (*Oxychilus pityusanus, Vitrea gasulli, Trochoidea caroli, T. ebusitana, Iberellus pythiusensis, Allognathus* n. sp., *Limax majoricensis, Melanopsis* n. sp.). Palmer *et al.* (in press) have inferred the presence of several endemic taxa (so far unrecorded as fossils). It is interesting to remark here that the analysis of the endemic assemblage made by these authors demonstrates that the endemics of the Pityusics differ almost completely from their Gymnesic counterparts. Shared endemic species are few.

2.2. Mallorca and Menorca

The late Quaternary terrestrial mammal fauna of the Gymnesic islands consists of three species: *Myotragus balearicus* (Artiodactyla: Caprinae), *Eliomys morpheus* (Rodentia: Myoxidae), and *Soriculus hidalgoi* (Insectivora: Soricidae). *Myotragus* was a highly derived genus with many skeletal and dental apomorphies (e.g., single, ever-growing mandibular incisors). *Myotragus balearicus* was small (40–50 cm shoulder height) but rather heavy-bodied (40–70 kg; Alcover *et al.*, in press). Dietary inferences indicate that it was able to browse on highly abrasive food items such as *Buxus balearica* (Alcover *et al.*, 1981; Marcus, in press; Alcover *et al.,* in press); the impact of this species on the vegetation of the islands must have been substantial (Duvigneaud, 1967; Delvosalle and Duvigneaud, 1967; Alcover *et al.,* 1981; Juniper, 1984; Schüle, 1993).

The associated fauna includes an amphibian (*Alytes muletensis;* Sanchíz and Adrover, 1977; Hemmer and Alcover, 1983), a reptile (*Podarcis lilfordi;* Kotsakis, 1981), and various species of birds (Sondaar *et al.,* 1995; Seguí, 1996). The bird community seems to have differed from its Pityusic counterpart. Grazing taxa like the Pityusic goose were absent, presumably because of the landscape remodeling effects of *Myotragus.* On the other hand, true eagles (*Aquila chrysaetos*), tytonids (*Tyto alba*), and small-sized corvids (*Pyrrhocorax pyrrhocorax, P. graculus, Corvus monedula*) were all present. The prehuman gastropod assemblage consisted of a minimum of 12 species, all endemic.

Knowledge of the late Quaternary flora of the Gymnesic islands is based on only six sites (two on Mallorca and four on Menorca). Nevertheless, palynological information indicates that the vegetation of these islands was distinctly different at that time (Burjachs *et al.,* 1994; Yll *et al.,* 1994, 1997; Pérez-Obiol *et al.,* 1996), with records of the woody plants *Buxus* and *Corylus* predominating. Fewer records are found of *Juniperus, Ephedra, Erica, Acer granatensis,* and deciduous *Quercus.* The rarity of evergreen *Quercus* before 6000 yrbp is remarkable (Pérez-Obiol *et al.,* 1996), as *Quercus ilex* forest is considered to be climax for the Gymnesics at present (Bolós and Molinier, 1958). Rarity of pollen of *Pistacia* and *Olea* previous to 6000 yrbp is also very remarkable, as both taxa are major components of the current vegetation of the islands.

2.3. Corsica and Sardinia

The late Quaternary terrestrial mammal fauna of the Cirno-Sardinian islands was more diverse than their counterparts on other western Mediterranean islands. This fauna included a relatively small-sized artiodactyl "*Megaloceros*" *cazioti* (Artiodactyla: Cervidae) that was similar in size to (and probably very closely related to) the fallow deer *Dama dama* (Klein Hofmeijer, 1996). This was the biggest herbivore on the island. The other terrestrial mammals included an endemic fox (*Cynotherium sardous*), several otters (*Algarolutra majori, Megalenhydris barbaricina, Sardolutra ichnusae*), a pika (*Prolagus sardus*), a red-toothed shrew (*Soriculus similis*), and two rodents (*Rhagamys orthodon* and *Tyrrhenicola [Microtus] henseli*). According to different authors (Spoor and Sondaar, 1986; Sondaar, 1987), Sardinia may have been colonized by another human species before the arrival of *Homo sapiens.*

Table 2 lists the endemic herpetofauna, as derived from fossil evidence and biogeographical analysis. The prehuman avifauna of Corsica and Sardinia is poorly known. Nevertheless, the presence of endemic strigiforms is noteworthy (Late Pleistocene *Bubo insularis* and Middle Pleistocene *Athene angelis;* Mourer-Chauviré and Weesie, 1986). No fewer than 24 species of gastropods are known from fossil remains, or can be inferred to have been present on the basis of biogeographical principles.

The palynological record for the Late Pleistocene of Corsica has been examined by Reille (1984, 1992) and Reille *et al.* (1997). This island exhibits a high degree of environmental diversity, which complicates analysis of its past vegetation (Gamisans, 1991). In the lower and middle west of the island, forest formations were dominated by *Arbutus* and *Erica.* At higher altitudes, these were replaced by forests of deciduous *Quercus* and mixed forests with *Taxus*. As in the Gymnesics, evergreen *Quercus* made a small contribution to Corsica's vegetation until after 6000 yrbp.

3. Arrival of Humans in the Western Mediterranean Islands

The oldest possible evidence for human presence on the Gymnesics consists of *Myotragus* bones presumably displaying butchery marks that date to 7220 ± 570 rcyrbp from Cova de Canet in Esporles, Mallorca (Kopper, 1984). Evidence of fire occurred at the same level as the bones. Other Mallorcan sites have also produced evidence of early human presence (e.g., Cova de Moleta and Abric de Son Matge; Waldren, 1982).

Humans were present on Eivissa at least as early as 6400–6900 rcyrbp (Alcover *et al.*, 1994, and unpublished data). Humans probably arrived even earlier than this, since the Pityusics are closer to the mainland than Mallorca is.

On Corsica and Sardinia the earliest confirmed presence of *Homo sapiens* is somewhat more ancient, at 8500–8600 rcyrbp for Corsica (Vigne, 1989; Lanfranchi, 1996) and 8750 ± 140 rcyrbp for Sardinia (Sondaar *et al.*, 1994). Sondaar *et al.* (1984, 1986) claimed that humans may have been present in Sardinia as early as the Middle Pleistocene. They suggested that humans and other immigrants could have come in during the Biharian, when the Cirno-Sardinian massif was separated from the Tuscanan mainland by only 13–30 km (see also Spoor and Sondaar, 1986; Sondaar, 1987; Klein Hofmeijer, 1996). The oldest dated remains of *Homo* sp. (a single phalanx from Grotta Su Corbeddu, Sardinia) are 20,000 years old (Sondaar *et al.*, 1994).

4. Extinctions and Local Disappearances

Although this work is focused on extinctions that took place after human arrival on the islands, to provide a context we briefly review evidence of Plio-Pleistocene losses in the Mediterranean area.

4.1. Pliocene and Pleistocene Extinctions

During the 350,000-year Messinian crisis (Gautier *et al.*, 1994), several vertebrate faunal assemblages disappeared or were replaced on western Mediterranean islands. Thus, during this interval the *Gymnesicolagus* fauna of the Gymnesics was replaced (at least in Mallorca) by *Myotragus* and associated species. It is not improbable that on the neighboring island of Menorca the *Gymnesicolagus* fauna was replaced by a peculiar insular assemblage, which included a giant tortoise (*Cheirogaster gymnesica*), a leporid (cf. *Alilepus* n. sp.), and a dormouse (*Muscardinus cyclopeus*). This fauna is still not well dated, and its assignment to the Pliocene (and thus post-Messinian time) is only tentative. Agustí and Moyà-Solà (1990) argued that the Menorcan fauna is actually Messinian in age, but there are objections to this possibility. First, deposits containing these faunal elements are locally abundant; it is improbable that the relatively short Messinian phase would be so well represented, whereas there is nothing of Early or Middle Pliocene age. Second, the highly derived aspect of the leporids (currently under study by J. Quintana and S. Moyà-Solà) is of a nature that would have taken much longer than 350,000 years to have evolved. As we note later, Mallorcan and Menorcan faunas could have undergone mixture during the Plio-Pleistocene glaciations.

In the Pityusics, the pre-Messinian fauna of Ses Fontanelles (including cf. *Maremmia* and other endemic forms) evidently disappeared during the Messinian. However, this fauna is known from only one site and is not well studied. It was succeeded by new faunal elements, including *Eivissia* (Rodentia: Myoxidae) and a large tortoise.

Sardinia's Late Miocene fauna, which included *Oreopithecus* and *Maremmia* (Cordy and Ginesu, 1994), also became extinct during the Messinian, to be succeeded by the *Nesogoral* assemblage.

Although all of the terrestrial mammal groups then resident on western Mediterranean islands were affected by the Messinian extinctions, losses among other taxonomic groups were less severe. For example, newts (*Euproctus*) and plethodont salamanders (*Hydromantes* [= *Speleomantes*]) survived on the Cirno-Sardinian islands, as did some gastropods (see Paul and Altaba, 1992).

The Messinian crisis appears to have had a homogenizing effect on the faunas of western Mediterranean islands. This is chiefly demonstrated by the appearance of certain elements on all western Mediterranean islands after the crisis ended 5.35 Ma. Perhaps the most representative shared species is *Tudorella ferruginea* (Gastropoda: Pomatiasidae), present in the Pliocene of Eivissa (Torres and Alcover, 1981; Gasull and Alcover, 1982; Paul, 1982, 1984), Pliocene–Recent of Mallorca, and Pliocene of Sardinia (Esu, 1978) (see Table 5).

Extinctions and local disappearances also occurred on western Mediterranean islands during the Plio-Pleistocene (Table 6). Some of these losses involved many species in geologically short periods of time, and can therefore be described as "mass" extinctions. Isolated disappearances of single species also occurred (some of which, however, may have been correlated with larger losses on neighboring islands).

We currently recognize three mass extinctions in the western Mediterranean archipelagos. The first, and least well known, affected the distinctive fauna of Menorca (*Cheirogaster gymnesica,* cf. *Alilepus,* and *Muscardinus cyclopeus*). This fauna can be tentatively assigned to the Early and Middle Pliocene. Its replacement by the *Myotragus* fauna may be correlated with periods of low sea level during one or another of the major phases of Northern Hemisphere glaciation (beginning in the Late Pliocene, 2.36 Ma; Shackleton and Opdyke, 1977; Emiliani, 1992), when Menorca and Mallorca would have been united. In addition to the loss of the two terrestrial mammals and the giant tortoise, the petrel *Ptero-*

Table 5. Common Post-Messinian (Early and Middle Pliocene) Taxa Shared by Western Mediterranean Islands[a]

Taxon	Mallorca	Menorca	Eivissa	Sardinia
Tudorella	✔	✔	✔	✔
Oestophora	✔	✔	✔	✔
Podarcis	✔	✔	✔	✔
Soriculus	✔			✔
Eliomys	✔		✔	✔
Caprinae	✔		✔	✔

[a]Data from Agustí and Moyà-Soyà (1990), Alcover and Agustí (1985), Alcover *et al.* (1981), Esu (1978), Gasull (1964), Gasull and Alcover (1982), Gliozzi and Malatesta (1980), Paul and Altaba (1992), Quintana (1995), Zammit-Maempel and De Bruijn (1982).

Table 6. Main Extiction Types in the Western Mediterranean Islands

1. Mass extinctions
With mammalian faunistic turnover (viz., Menorca, Sardinia)
Without mammalian faunistic turnover (viz., Eivissa)
2. Single extinctions

dromoides minoricensis may also have disappeared at this time (Seguí *et al.,* in press). Interestingly, this extinction event had only a small effect on the gastropods; several species, including *Tudorella ferruginea, Oxychilus lentiformis, Rumina decollata, Trochoidea (Xerocrassa) nyelli, Iberellus minoricensis, Eobania vermiculata,* apparently managed to survive until human arrival (Quintana, 1995).

A second mass-extinction event and correlated turnover occurred in Sardinia during the Biharian, about 800,000 yrbp. Immigration of new species of reptiles, amphibians, and mammals (including the ancestors of *Megaloceros cazioti, Cynotherium sardous,* and *Tyrrhenicola [Microtus] henseli*) was presumably facilitated by low sea level and the consequent reduction of the distance between islands. Thus, Corsica and Sardinia, forming a single island, would have been separated by only a short distance (13 km) from Elba, which would have in turn acted as a steppingstone between Tuscany and the Cirno-Sardinian massif (Conchon, 1979; Vigne, 1983; Sondaar *et al.,* 1991). The major portion of the mammal fauna became extinct, including *Nesogoral melonii, Macaca majori,* and *Sus nanus* (*Talpa tyrrhenica, Apodemus mannu,* and *Tyrrhenoglis majori* may have disappeared at this time also). The extinction process spared *Nesiotites, Rhagamys,* and *Prolagus,* all pre-Biharian, but may have driven evolutionary changes. Thus, at this time *P. figaroi* gave rise to *P. sardus* (Lopez-Martínez and Thaler, 1975), while *R. minor* evolved into *R. orthodon* (Brandy, 1978).

As previously noted, humans may have first appeared in Sardinia during the Biharian (Sondaar, 1987), and may have prompted the extinctions that occurred at that time. Recent discovery of human presence on Flores 0.9 ± 0.07 Ma (Morwood *et al.,* 1998) bears on the possibility of early arrival in Sardinia in that it implies that the capacity to cross water gaps must have appeared relatively early in the evolution of *Homo.*

The third mass-extinction event registered in the western Mediterranean islands took place in the Pityusics during the Plio-Pleistocene. Pliocene faunas of Eivissa are well documented (Torres and Alcover, 1981; Gasull and Alcover, 1982; Paul, 1982, 1984; Alcover and Agustí, 1985; Bour, 1985; Agustí and Moyà-Solà, 1990; Paul and Altaba, 1992). The assemblage included a caprine, two glirids, a giant tortoise, a lizard, several seabirds, and no less than 18 species of gastropods. By the Late Pleistocene, the fauna had become depauperate: no giant tortoises or terrestrial mammals were left, and only 40% of the gastropods present in the Pliocene were still represented.

The timing and cause of this extinction event are not known, although it is likely that the losses took place some time between the Late Pliocene and Middle Pleistocene, perhaps as a result of a volcanic episode (Florit *et al.,* 1989; Paul and Altaba, 1992; Alcover *et al.,* 1994). In any case, the outstanding feature of the Pityusic mass extinction is that it is not correlated with faunistic turnover. After the endemic terrestrial mammals disappeared, no new ones settled on the island, and the same happened in the case of gastropods. Absence of turnover is even more noteworthy because the distance between these islands and the mainland is just 90 km (although sea currents may have militated against easy recolonization; Pinot *et al.,* 1995).

In conclusion, at least three mass-extinction events took place on western Mediterranean islands during the Plio-Pleistocene. Two were correlated with faunistic turnover processes, but the third was not. The extinctions seem to have been proportionately more extensive on small islands (Eivissa, Menorca) than larger ones (Sardinia), and affected some groups (e.g., mammals) more than others (e.g., gastropods). Pre-Biharian reptiles and amphibians survived in Sardinia, as did some small terrestrial mammals, but in Eivissa and Menorca all terrestrial mammals disappeared.

Isolated extinctions of single species can also be documented for the islands. An example of this is the extinction of the gastropod *Mastus pupa* in Mallorca and Menorca during the Riss–Würm interglacial (Cuerda, 1975). Other examples include loss of *Oestophora* sp. (Menorca, during Late Pliocene or earliest Pleistocene) and *Discoglossus* n. sp. (Mallorca and Menorca, early Pleistocene). The last two disappearances may have been coeval with the Pityusic extinction episode, but dates and causes are not known.

4.2. Extinctions and Local Disappearances during the Holocene

As already noted, human colonization of the islands of the western Mediterranean in the early Holocene (Cherry, 1984, 1992) was associated with a profound transformation of insular ecosystems. The extent of change can be evaluated by comparing the fauna and flora of the Late Pleistocene/Holocene with the present, making due allowance for inadequacies of dating.

In the Pityusic islands, "last occurrence" records for endemic birds (*Rallus* n. sp. and *Anser* n. sp.) overlapped one of the earliest dates for human presence (6130 ± 80 rcyrbp, uncalibrated). Latest records for extant species that are locally extinct in the Pityusics, such as *Pyrrhocorax pyrrhocorax, Grus grus,* and *Haliaeetus albicilla,* are of the same age. *Corvus corax* ssp., *Otis tarda, Crex crex,* and *Athene noctua,* which bred on these islands in the Late Pleistocene, no longer do so. Still others (*Calonectris diomedea, Puffinus mauretanicus,* and *Hydrobates pelagicus*) breed only in surrounding islets.

Bats also suffered several extinctions after human arrival in the Pityusics. At least one species (*Myotis myotis*) is locally extirpated, while several others have become extremely scarce. By contrast, lizards and gastropods present in the Late Pleistocene fossil record are still common (C. R. Altaba, personal communication).

The three species of terrestrial mammals inhabiting the Gymnesic islands became extinct after human colonization. In Mallorca (Cova Estreta), the most recent date for remains of *Hypnomys morpheus* is ca. 6000 yrbp (Encinas and Alcover, 1997). Occupation layers dated to ca. 4000 yrbp lack this species (Waldren, 1982), indicating that the extinction probably took place between these two dates. *Soriculus (Nesiotites) hidalgoi* may have been lost around the same time; direct dates on this species are lacking, but it co-occurs with the last *Hypnomys morpheus* at Cova Estreta. There are no data on the timing of the extinction of these two species in Menorca (but see Alcover *et al.,* 1981). The extinction of *Myotragus balaericus* is well documented in Mallorca; at Abric de Son Matge the species persists until about 2300 BC. When it finally disappeared in Menorca is not known.

Several birds common in Late Pleistocene and Holocene contexts in the Gymnesics are absent from the current avifauna. This applies to *Pyrrhocorax pyrrhocorax, P. graculus,* and *Turdus* sp., frequently encountered in fossil assemblages in both Mallorca and Menorca (Seguí *et al.,* 1997; Seguí, 1997); the small owl *Athene noctua,* very rare nowadays in Mallorca (Seguí *et al.,* 1997; Seguí, 1997); and the passerine *Montifringilla nivalis,* similar to *P. graculus* in ecological requirements and previously present in Mallorca. *Aquila chrysaetos* was probably another once-common element of the avifauna (Mourer-Chauviré *et al.,* 1977; Florit and Alcover, 1987). An unidentified small shearwater (*Puffinus* sp.), previously widespread (if rare) in Menorca, became extinct very recently (personal observation).

Podarcis lilfordi, the only Late Pleistocene saurian on Mallorca and Menorca, has disappeared from the main islands but survives on nearby islets. The zooarcheological record of Torralba den Salort indicates that this species was abundant until 200 BC (Reumer and Sanders, 1984; Sanders, 1984; Sanders and Reumer, 1985). It is rare in layers dated to ca. AD 150 and no doubt became extinct shortly thereafter. Some authors (Mertens, 1957; Kotsakis, 1981; Pérez-Mellado, 1998) consider the disappearance of *P. lilfordi* to be correlated with the introduction of snakes, while Alcover (1989) believed that the loss of this species (and the amphibians *Alytes muletensis* and *A. talaioticus*) was related to weasels, introduced by the Romans to control rabbits (Pericot, 1972).

Oddly, no gastropod extinctions can be correlated with human colonization of the Gymnesics (although some species have become very rare). *Trochoidea cuerdai* may be an exception, as this species was present in the Pleistocene of Menorca but has not been recently identified in the field (Gasull, 1963). The validity of the species is, however, uncertain (Paul, 1982).

As noted earlier, the floral aspect of the Gymnesics was rather different earlier in the Quaternary. According to the palynological record, *Buxus* and *Corylus* were abundant in Mallorca and Menorca until the middle Holocene. At present, *Buxus* occurs only in some small, isolated patches in Serra de Tramuntana (Mallorca), while *Corylus* has disappeared from the Gymnesics. Deciduous species of *Quercus* were also abundant at this time; they have since become locally extinct in Menorca and are very rare in Mallorca (there is only one locality on the latter island, and it is thought to be a recent plantation). Decline began between 6000 and 4000 yrbp (Pérez-Obiol *et al.,* 1996), after human arrival.

A major faunal turnover took place in Corsica and Sardinia after colonization of these islands by Mesolithic and Neolithic humans. Vigne (1983) has argued that *Cynotherium sardous* and *Megaloceros cazioti,* among others, disappeared before human arrival (at least on Corsica), but there is no clear evidence for this. It may be more prudent to assume that these species were still in existence at the time of first contact, but disappeared soon thereafter (cf. MacPhee and Marx, 1997). In any case, it is somewhat clearer that the first small mammals to become extinct on both islands after human arrival were *Soriculus corsicanus* in Corsica and *Soriculus similis* in Sardinia. *Prolagus sardus* was probably the last (see MacPhee and Flemming, this volume). This species is found in Roman sites in Corsica and Sardinia. If Cetti (1774) has been read correctly, this ochotonid may have survived on the small island of Tavolara until the seventeenth century, but he may have been referring to large black rats rather than *P. sardus.* Black rats are good colonizers of coastal islets and may become extraordinarily large in such contexts (Cheylan, 1988). Some endemic birds (e.g., *Bubo insularis*) became extinct on these islands after first contact. Others suffered local extinctions (e.g., *Aquila* cf. *A. heliaca, Pyrrhocorax pyrrhocorax*). However, no post-contact prehistoric extinctions of reptiles, amphibians, or gastropods have been recorded.

Corsica is the only island of the Cirno-Sardinian ridge with a well-documented and published palynological record. No species of plants have become extinct, although there have been marked changes in composition (e.g., increase of evergreen *Quercus* forest since 4500 yrbp; reduction in *Erica arborea* forest; recent increase in black pine forests [*Pinus laricio*]; Reille, 1992; Reille *et al.,* 1997).

5. Searching for Patterns and Causes

As noted, the arrival of Mesolithic and Neolithic settlers on the islands of the western Mediterranean was coincident with a number of major floral and faunal changes:

1. Extinction of all terrestrial, nonvolant mammals in precontact communities; 13 species affected.
2. Extinction or local disappearance of all large and some medium-sized predators in precontact communities; at least four species affected, including endemic *Bubo insularis* in Corsica and Sardinia and all large eagles (*Haliaeetus albicilla* in Eivissa, Corsica, and Sardinia; *Aquila chrysaetos* in Mallorca and *Aquila,* cf. *A. heliaca* in Corsica and Sardinia).
3. Extinction of all terrestrial nonpasserine endemic birds (seabirds excluded), including *Anser* n. sp. and *Rallus* n. sp. from Eivissa.
4. Range reduction affecting pelagic seabirds of genus *Puffinus.* Three species are found in the Late Pleistocene fossil record: *Puffinus mauretanicus,* previously breeding in the Pityusics, now absent from the larger islands; *P. yelkouan,* previously breeding in Tavolara; and *Puffinus* sp. indet. (possibly new), found in several sites in Menorca.
5. Extinction or local disappearance of several corvids, including *Pyrrhocorax pyrrhocorax,* locally extirpated in Mallorca, Menorca, Eivissa, and Sardinia, and *Corvus corax* ssp., no longer present in Eivissa.
6. Extinction or near extinction of all endemic herps in the Gymnesics. *Alytes talaioticus* became extinct in Menorca after Roman settlement. *Podarcis lilfordi* disappeared from Mallorca and Menorca, also after Roman settlement, and is now restricted to nearby islets. *Alytes muletensis* has a relict distribution in Mallorca limited to eight isolated mountain streams and a few other localities.
7. Local disappearances or range reductions affecting plants requiring wetter conditions than currently prevail. This applies to *Buxus balearica,* extirpated in Menorca and relict in Mallorca; *Corylus avellanarius,* extirpated in Mallorca and Menorca; and deciduous species of *Quercus,* now absent in Mallorca and Menorca.

Lack of faunal and floral changes during the Late Pleistocene as compared with documented turnover after human colonization suggests—but does not prove—that settlement prompted extinctions in some way, perhaps via environmental degradation. Available data imply that extinction of terrestrial mammals occurred more quickly on small islands (Mallorca and Menorca) than on large ones (Corsica and Sardinia). Several sites in Corsica and Sardinia allow a tentative chronology to be developed for the extinction and immigration process (e.g., Vigne and Alcover, 1985). In Mallorca and Menorca, where more sites are known, there is no clear chronology, probably because the extinction was more sudden. The extinction of endemic mammals took place in a 9000-year period in Corsica and Sardinia, but in less than 5000 years in Mallorca and Menorca.

Extinctions and local disappearances of reptiles took place later than those of mammals. The extinction of reptiles in the Gymnesics, for example, occurred more than 2000 years later than that of endemic mammals.

In our opinion, the faunal turnover in the western Mediterranean after first contact was caused by a combination of factors, although all seem to have been basically related to human colonization as the following analysis shows.

5.1. The Extinction of Terrestrial Mammals

The extinction pattern of terrestrial mammals in the western Mediterranean islands resembles that found elsewhere in the world, although the extent of extinction is unprecedented. Globally, about 35% of terrestrial mammals inhabiting oceanic and quasioceanic islands became extinct after human colonization (Alcover *et al.*, 1998). In the West Indies, 71% of insular endemic mammals disappeared; on Mediterranean islands generally, the loss figure climbed to 90%; but in the western Mediterranean islands, extinction affected 100% of terrestrial mammals.

In insular environments, as a rule meso- and mega-faunal mammals became extinct before the smaller species did. Thus, in Corsica and Sardinia, midsized *Megaloceros cazioti* and *Cynotherium sardous* became extinct before the endemic small mammals did. Highly specialized *Myotragus balearicus* may be a noteworthy exception to this rule. Although the Gymnesic islands were colonized by humans as long ago as perhaps 9000 yrbp, *M. balearicus* only became extinct about 4300 yrbp (see below).

The first micromammal to suffer extinction in the western Mediterranean islands was *Soriculus* (*Nesiotites*). The extinction of this small insectivoran (~ 35 g body weight) is curious. In general, shrews have been little affected by extinction (cf. widely distributed *Crocidura,* surviving on many islands at present; Alcover *et al.,* 1998). Among the few examples of shrew losses in recent times are red-toothed shrews (*Nesiotites*) and West Indian island-shrews (*Nesophontes*) (MacPhee and Flemming, this volume).

Causes of extinction of these species of mammals are unknown. There is evidence of human consumption of *Megaloceros cazioti, Myotragus balearicus,* and *Prolagus sardus* (Vigne *et al.,* 1981; Waldren, 1982; Klein Hofmeijer, 1996). Humans probably also hunted and ate *Cynotherium sardous* and other Sardinian otter species, but it is clear that they did not consume small rodents and insectivores. Direct impact in the form of hunting pressure could have prompted the extinction of large species, at least to some degree, but the loss of small mammals must have some other cause. The extinction of *Myotragus* may have had some unique features of its own. Rarefaction and local disappearance of *Buxus balearica* forest, a food resource for *Myotragus* (Alcover *et al.,* in press), could have played an important role. There is strong evidence that this species was stabled in caves and rock shelters (Waldren, 1982), and that the short, pointed horns were polled to prevent harm to animals kept in close quarters. This is, or is tantamount to, domestication, and the important point here is that domestication usually results in conservation of species under husbandry. Yet *M. balaericus* became extinct nevertheless—possibly the only such loss of a domestic animal on record.

5.2. Extinction of Predators

The record shows that medium- to large-sized predators (including "superpredators") on islands are prone to extinction following human arrival. Alcover and McMinn (1994) demonstrated that, in contrast to the continental situation, on islands the superpredator ecological guild is essentially monopolized by birds, as nonvolant mammals are frequently completely absent.

On both islands and continents, predators form a more vulnerable ecological assemblage than do primary consumers (Carlquist, 1965; Schoener and Spiller, 1999). Causes of their extinction are partially a reflection of their position in the trophic chain. Their populations are smaller than those of prey species, which contributes to their vulnerability. Direct hunting by humans, introduction of competitors and other predators (especially when populations are in crisis), and widespread change of trophic relations in insular environments subsequent to human arrival are all likely contributing agencies.

An interesting case is that of the sea eagle of the Pityusic islands. This species, despite being rather abundant in Late Pleistocene communities, probably never reached very high population sizes, and may have become extinct for some of the same reasons as *Haliaeetus leucocephalus/albicilla* in the Hawaiian Islands (Olson and James, 1991). The extirpation of resident prey species on Eivissa must have had an effect on population sizes of sea eagles. Another factor that may have played a role in the extinction is nest-site selection. Direct hunting by humans might have been yet another factor.

5.3. Extinction of Terrestrial Birds

Western Mediterranean islands have a low number of endemic birds species (the only extant endemics are *Sylvia sarda* and *Sitta whiteheadi*). The extinct endemic forms include *Anser* n. sp. and *Rallus* n. sp. from Eivissa, and *Bubo insularis* from Corsica and Sardinia. *Athene angelis* was described from a Middle Pleistocene site in Corsica, but might have survived until the Late Pleistocene, as did *A. cretensis* in Crete (Weesie, 1987).

Generally, passerines are less affected than are Anseriformes, Strigiformes, and Rallidae by anthropogenic changes in insular environments (Milberg and Tyrberg, 1993). According to available evidence, this rule is also applicable to the western Mediterranean islands. For example, it is clear that *Anser* n. sp. from Eivissa was an abundant breeder. Present-day populations of anseriforms in Europe breed mainly in the northern parts of the continent, especially on some islands in the North Sea (Cramp and Simmons, 1977). Predatory mammals are lacking on these islands. Until very recently, a small *Anser* bred in Eivissa. The species was often eaten by the first settlers. There is also some evidence of consumption of *Rallus* n. sp. Hunting pressure probably played a major role in the extinction of these two species. In fact, all medium- and large-sized birds suitable for human consumption might have been hunted by first humans.

5.4. Rarefaction of Procellariiforms

A frequent consequence of human colonization of islands is rarefaction and eventual disappearance of seabird colonies, involving in extreme cases the extinction of species. This has been widely documented for Pacific and Atlantic islands (e.g., Olson, 1975; Steadman, 1995; Martin and Steadman, this volume), and the islands of the western Mediterranean constitute no exception. The fossil record indicates that *Puffinus mauretanicus* was abundant throughout Eivissa. *P. yelkouan* bred in Tavolara. A small, poorly known, and not especially abundant species of this genus also bred in Menorca.

There are two main factors that probably led to the extinction, local disappearance, or rarefaction of western Mediterranean shearwaters: direct hunting and introduced competitors. There is much evidence of shearwater consumption in Eivissa. Hunting shearwaters and collecting their eggs were traditional practices in the Pityusic islands until the 1980s (see Mayol, 1977, for the analysis of the situation two decades ago). The Balearic shearwater population has decreased to 2800 pairs (Carboneras, 1992), from a much higher estimated figure (hundreds of thousands of pairs) during the Late Pleistocene.

By contrast, disappearance of *P. yelkouan* on Tavolara and *Puffinus* sp. on Menorca, as well as declining numbers of *Puffinus mauretanicus,* could be related to the introduction of *Rattus rattus.* The species came to the larger islands in company with the Romans, and eventually reached surrounding islets. The black rat is an avid consumer of eggs, especially those of small- and middle-sized birds nesting on the ground (such as the Balearic shearwater, manx shearwater, *Puffinus* sp. from Menorca, and *Rallus* n. sp. of Eivissa).

5.5. Extinction of Corvids

The extinction of medium- to large-sized Corvidae members is a constant not only on the western Mediterranean islands, but also on islands all around the world (e.g., Hawaii, James and Olson, 1991; Crete, Weesie, 1987). In view of the anthropophilic character of many corvids (the extinct species of the western Mediterranean islands included), it is difficult to explain why they have became extinct in such numbers.

6. Extinction and Local Disappearance of Herpetological Endemics in the Balearics

Small reptiles and amphibians are less prone to extinction than are mammals or large reptiles. Thus, it is surprising that one species (*Alytes talaioticus*) became extinct, whereas two others (*Alytes muletensis* and *Podarcis lilfordi*) have vanished locally on Mallorca and Menorca. There is no other known case of the extinction of a reptile the size of *P. lilfordi* on an island as large as Mallorca. Its extinction is not related to the arrival of the first human settlers (nor to the presence of other species of reptiles and amphibians introduced by humans). These species apparently coexisted with humans until at least AD 150. The cause of their extinction must be related to the introduction of the weasel (*Mustela nivalis*) by the Romans.

7. Disappearance of Forest Plant Species

As in the foregoing case concerning amphibians and reptiles, the disappearance of some arboreal species seems not to be directly tied to the first arrival of humans in the western Mediterranean islands. Palynological records from Mallorca and Menorca suggest that a major floristic change occurred between 4000 and 6000 yrbp, as marked by the expansion of species of evergreen *Quercus*. Curiously, evergreen *Quercus* forests expanded on Corsica during the same period. A climatic change has been proposed to account for replacement of *Corylus* and *Buxus* forests by *Quercus ilex* forest and *Olea/Pistacia* macchia on Mallorca and Menorca. Nevertheless, human influence cannot be excluded as a cause (Alcover *et al.,* in press).

Recent studies suggest that the expansion of evergreen *Quercus* forests may have anthropogenic causes (e.g., Gamisans, 1991). As previously noted, this hypothesis presupposes that humans would have transformed the landscape by introducing and favoring the growth of evergreen *Quercus* so as to exploit their fruits (acorns) as a food item.

Thus, the general patterns of faunistic and floristic change on the western Mediterranean islands during the Holocene can be directly related to the arrival of humans. The changes have been so important, especially in terms of species extinction and immigration, that they have transformed completely the insular ecosystem and landscape.

There are few data concerning change of landscape. On Corsica, deciduous *Quercus* and *Taxus* forests once attained the shoreline (Reille, 1984). On Mallorca and Menorca, *Buxus* forests also descended until they too attained the shoreline (Yll *et al.,* 1994). Since the arrival of humans, both kinds of formations have receded (or even disappeared in some instances; e.g., *Buxus* forest on Menorca).

In summary, the vertebrate extinctions that occurred in the western Mediterranean islands after the arrival of humans show some peculiar characteristics. These islands constitute one of the most densely populated intercontinental archipelagos in the world, and were the site of human settlement from the earliest post-Paleolithic. In some respects the extent of extinction has been greater on the Mediterranean islands than on other islands. Many of the insular extinctions in the Mediterranean area present features similar to losses on other islands, but there are some striking differences. First, after Waldren (1982), the only species of domesticated mammal that has suffered complete extinction is *Myotragus balearicus.* This species coexisted with humans perhaps for slightly more than 4500 years. The real causes of its extinction are unknown. A second peculiar trait concerns the extinction of tiny *Podarcis lilfordi* in Mallorca and Menorca. We relate this extinction to the introduction of the weasel (*Mustela nivalis*) by the Romans. Finally, a third singularity is the possible extinction of a species of *Homo* from a Mediterranean island (Sardinia). This extinction—which, of course, must be corroborated—could have a parallel in some islands of Wallacea (viz., Timor).

If extinction on islands is directly related to anthropogenic pressures, the study of losses in the western Mediterranean should help us to better predict what the future will hold for other insular faunas and floras as human-related pressures increase.

Acknowledgments

The authors wish to acknowledge the collaboration of Dr. Damià Jaume. Dr. Ross MacPhee invited us to participate in this book, encouraged us considerably, and patiently never complained for our delay.

References

Adrover, R., Agustí, J., Moyà-Solà, S., and Pons-Moyà, J. 1983–1984. Nueva localidad de micromamíferos insulares del Mioceno medio en las proximidades de Sant Llorenç en la isla de Mallorca. *Paleontol. Evol.* **18**:121–129.

Agustí, J., and Moyà-Solà, S. 1990. Neogene–Quaternary mammalian faunas of the Balaerics, in: A. Azzaroli (ed.), International Symposium on Biogeographical Aspects of Insularity (Rome, 18–22 May 1987). *Atti Convegni Lincei Accad. Nac. Lincei* **85**:459–468.

Alcover, J. A. 1987. Mamífers i illes: Síntesi de models de colonització en geografia històrica i la seva aplicació a la Mediterrània. *Paleontol. Evol.* **21**:69–74.

Alcover, J. A. 1989. El Ferreret de Mallorca. *Quercus* **39**:14–19.

Alcover, J. A., and Agustí, J. 1985. *Eliomys (Eivissia) canareiensis* n. gen., n. sp., nou glírid dell Pleistocè de la Cova de Ca Na Reia (Pitiüses). *Endins* **10–11**:51–56.

Alcover, J. A., and McMinn, M. 1994. Predators of vertebrates on islands. *BioScience* **44**:12–18.

Alcover, J. A., Moyà-Solà, S., and Pons-Moyà, J. 1981. *Les Quimeres del Passat. Els Vertebrats Fòssils del Plio-Quaternari de les Balears i Pitiüses.* Monografies Científiques, Edit. Moll **1**:1–260.

Alcover, J. A., McMinn, M., and Altaba, C. R. 1994. Eivissa: a Pleistocene ocean like island in the Mediterranean Sea. *Res. Explor.* **10**:236–246.

Alcover, J. A., Sans, A., and Palmer, M. 1998. The extent of extinctions of mammals in islands. *J. Biogeogr.* **25**:913–918.

Alcover, J. A., Pérez-Obiol, R., Yll, E. I., and Bover, P. in press. The diet of *Myotragus balearicus* Bate 1909 (Artiodactyla: Caprinae), an extinct bovid from the Balearic Islands: evidence from coprolites. *Biol. J. Linn. Soc.*

Alvarez, W. 1972. Rotation of the Corsica–Sardinia microplate. *Nature* **235**:103–105.

Bellon, H., Coulon, C., and Edel, J. B. 1977. Le déplacement de la Sardinia. Synthèse des données géochronologiques, magmatiques et paléomagnetiques. *Bull Soc. Geol. Fr.* **7**:825–831.

Bolós, O. de, and Molinier, R. 1958. Recherches phytosociologiques dans l'île de Majorque. *Collect. Bot.* **5**:699–863.

Bour, R. 1985. Una nova tortuga terrestre del Pleistocè d'Eivissa: la tortuga de la Cova de Ca Na Reia. *Endins* **10–11**:57–62.

Brandy, L.-D. 1978. Données nouvelles sur l'évolution du rongeur endémique fossile corso-sarde *Rhagamys* (F. Major, 1905) (Mammalia, Rodentia). *Bull. Soc. Geol. Fr.* Ser. 7 **20**:831–835.

Burjachs, F., Pérez-Obiol, R., Roure, J. M., and Julià, R. 1994. Dinámica de la vegetación durante ell Holoceno en la isla de Mallorca. *Trab. Pal. Bas. Apl.* 199–210.

Carboneras, C. 1992. Procellariiformes. Procellariidae, in: J. Del Hoyo, A. Elliot, and J. Sargatal (eds.), *Handbook of the Birds of the World,* Volume 1, pp. 216–257. Lynx Publications, Barcelona.

Carlquist, S. 1965. *Island Life.* The Natural History Press, New York.

Cetti, F. 1774. *I quadrupedi di Sardegna.* Sassari.

Cherry, J. F. 1984. The initial colonisation of the west Mediterranean islands in the light of island biogeography and palaeogeography. *BAR Int. Ser.* **229**:7–28.

Cherry, J. F. 1990. The first colonization of the Mediterranean islands: a review of recent research. *J. Mediterr. Archaeol.* **3/2**:145–221.

Cherry, J. F. 1992. Palaeolithic Sardinian?, in: R. H. Tykot and T. K. Andrews (eds.), *Sardinia in the Mediterranean: A footprint in the sea. Studies in Sardinian Archeology,* pp. 28–39.

Cheylan, G. 1988. Les adaptations écologiques de *Rattus rattus* á la survie dans les îlots méditerranéens (Provence et Corse). *Bull. Ecol.* **19**:417–428.

Cody, M. C. 1985. *Habitat Selection in Birds.* Academic Press, Orlando.

Conchon, O. 1979. Comme les animaux terrestres ont-ils pu atteindre la Corse? *Courr. Parc Corse* **32**:5–8.

Cordy, J. M., and Ginesu, S. 1994. Fiume Santo (Sassari, Sardaigne, Italie): un nouveau gisement à oréopithèque (Oreopithecidae, Primates, Mammalia). *C. R. Acad. Sci. Sér. 2* **318**:697–704.

Cramp, S., and Simmons, S. 1977. *Handbook of the Birds of Europe, the Middle East and North Africa,* Volume 1. Oxford University Press, London.

Cuerda, J. 1975. Los tiempos cuaternarios en Baleares. *Edit. Inst. Est. Bal.* 304 pp.

De Bruijn, H., and Rümcke, C. G. 1974. On a peculiar mammalian association from the Miocene of Oschiri (Sardinia), 2 vols. *Proc. K. Ned. Akad. Wet. Ser. B* **77**:46–79.

Delvosalle, L., and Duvigneaud, J. 1967. Voyage a Majorque. Compte rendue botanique des excursions. *Nat. Belg.* **48**:365–388.

Driver, J. C. 1982. Medullary bone as indicator of sex in bird remains from archaeological sites. *BAR Int. Ser.* **109**:251–254.

Duvigneaud, J. 1967. Majorque. Aperçu géographique et botanique. *Nat. Belg.* **48**:347–363.

Emiliani, C. 1992. *Planet Earth: Cosmology, Geology, and the Evolution of Life and Environment.* Cambridge University Press, London.

Encinas, J. A., and Alcover, J. A. 1997. El jaciment fossilífer de la Cova Estreta (Pollença). *Endins* **21**:83–92.

Esu, D. 1978. La malacofauna continentale pliopleistocenica della formazione fluvio-lacustre di Nuraghe Su Casteddu (Sardegna Orientale) e sue implicazioni paleogeografiche. *Geol. Rom.* **17**:1–33.

Fairbanks, R. G. 1989. A 17,000-year glacio-eustatic sea level record: influence of glacial melting rates on the Younger Dryas event and deep-ocean circulation. *Nature* **342**:637–642.

Florit, X., and Alcover, J. A. 1987. Els ocells del Pleistocè superior de la Cova Nova (Capdepera, Mallorca), II. Fauna associada i discussió. *Boll. Soc. Hist. Nat. Ba.* **31**:33–44.

Florit, X., Mourer-Chauviré, C., and Alcover, J. A. 1989. Els ocells pleistocènics d'Es Pouàs, Eivissa. Nota preliminar. *Bull. Inst. Cat. Hist. Nat.* **56**:35–46.

Gamisans, J. 1991. *La Végétation de la Corse.* Conservatoire et Jardin Botaniques de la Ville de Genève, Suisse.

Gasull, L. 1963. Descripciòn de unas nuevas formas del género *Helicell* (= *Xeroplexa*) de Baleares. *Bol. Soc. Hist. Nat. Bal.* **9**:83–92.

Gasull, L. 1964. Un nuevo molusco terrestre fósil para la fauna cuaternaria de Baleares: *Oestophora (O.) barbula. Bol. Soc. Hist. Nat. Bal.* **9**:81–82.

Gasull, L., and Alcover, J. A. 1982. La Cova de Ca Na Reia: desconcertant estació malacològica del Pleistocè de les Pitiüses. *Endins* **9**:41–44.

Gautier, F., Caluzón, G., Suc, J. P., and Violanti, D. 1994. Age et durée de la crise de Salinité Messinienne. *C. R. Acad. Sci.* **318**(II):1103–1109.

Gliozzi, E., and Malatesta, A. 1980. The Quaternary goat of Capo Figari (northeastern Sardinia). *Geol. Rom.* **19**:295–347.

Hemmer, H., and Alcover, J. A. (eds.). 1983. *The natural history of the Mallorcan midwife toad.* Monografies Cientifiques, Edit. Moll 3:1–255.

Hsü, K. J. 1972. Origin of saline giants: a critical review after the discovery of the Mediterranean evaporites. *Earth Sci. Rev.* **8**:371–396.

Hsü, K. J., Cita, M. B., and Ryan, W. B. F. 1973. Late Miocene dessication of the Mediterranean. *Nature* **242**:240–244.

Hsü, K. J., Montadert, L., Bernouilli, D., Cita, M. B., Erikson, A., Garrison, R. E., Kidd, R. B., Mèlierés, F., Müller, C., and Wright, R. 1977. History of the Mediterranean salinity crisis. *Nature* **267**:399–403.

James, H. F., and Olson, S. L. 1991. Descriptions of thirty-two new species of birds from the Hawaiian Islands: Part II. Passeriformes. *Ornithol. Monogr.* **46**:1–88.

Juniper, B. E. 1984. The natural flora of Mallorca, *Myotragus* and its possible effects, and the coming of man to the Balearics. *BAR Int. Ser.* **229**:145–163.

Klein Hofmeijer, G. 1996. Late Pleistocene deer fossils from Corbeddu cave. Unpublished Ph.D. thesis, Universiteit Utrecht, 436 pp.

Kopper, J. S. 1984. Canet Cave, Majorca. *BAR Int. Ser.* **229**:61–69.

Kotsakis, T. 1981. Le Lucertole (Lacertidae, Squamata) del Pliocene, Pleistocene e Olocene delle Baleari. *Boll. Soc. Hist. Nat. Bal.* **25**:135–150.

Kotsakis, T. 1990. Insular and noninsular vertebrate fossil faunas in the eastern Mediterranean islands, in: A. Azzaroli (ed.), International Symposium on Biogeographical Aspects of Insularity (Rome, 18–22 May 1987). *Atti Convegni Lincei Accad. Nac. Lincei* **85**:289–334.

Lanfranchi, F. de. 1996. Les implantations humaines préhistoriques. *Corsica antica* **2**:1–5.

Lanza, B. 1983. Ipotesi sulle origini del popolamento erpetológico della Sardegna. *Lav. Soc. Ital. Biogeogr.* **8**:723–744.

Lanza, B., and Vanni, S. 1987. Hypotheses on the origins of the Mediterranean island batrachofauna. *Bull. Soc. Zool. Fr.* **112**:179–196.

Legakis, A., and Kypriotakis, Z. 1994. A biogeographical analysis of the island of Crete, Greece. *J. Biogeogr.* **21**:441–445.

López-Martínez, N., and Thaler, L. 1975. Etude systématique et paléogéographique du groupe d'ochotonidés *Piezodus-Prolagus* (Mammalia, Lagomorpha). *Bull. Soc. Geol. Fr. Ser.* 7 **17**:850–866.

MacPhee, R. D. E., and Marx, P. A. 1997. The 40,000-year plague: humans, hyperdisease, and first-contact extinctions, in: S. Goodman and B. Patterson (eds.), *Natural Change and Human Impact in Madagascar,* pp. 169–217. University of Chicago Press, Chicago.

Maldonado, A. 1985. Evolution of the Mediterranean basins and a detailed reconstruction of the Cenozoic paleoceanography, in: R. Margalef (ed.), *Western Mediterranean,* pp. 17–59. Pergamon Press, Elmsford, N.Y.

Marcus, L. In press. Variation in selected skeletal elements of the fossil remains of *Myotragus balearicus,* a Pleistocene bovid from Mallorca. *Acta Hung.*

Margalef, R. (ed.). 1985. *Western Mediterranean.* Pergamon Press, Elmsford, N.Y.

Mayol, J. 1977. La ornitofagia tradicional en las Islas Pityusas. *Vida Silvestre* **24**:242–247.

Mein, P., and Adrover, R. 1982. Une faunule de mammifères insulaires dans le Miocène moyen de Majorque (Iles Baléares). *Geobios Mem. Spec.* **6**:451–463.
Mertens, R. 1957. Mallorca: ein herpetogeographisches Problem. *Zool. Beitr.* **3**:1–16.
Milberg, P., and Tyrberg, T. 1993. Naïve birds and noble savages: a review of man-caused prehistoric extinctions of island birds. *Ecography* **16**:229–250.
Morwood, M. J., O'Sullivan, P. B., Aziz, F., and Raza, A. 1998. Fission-track ages of stone tools and fossils on the east Indonesian island of Flores. *Nature* **392**:173–176.
Mourer-Chauviré, C., and Weesie, P. D. M. 1986. *Bubo insularis* n. sp., forme endemique insulaire de grand-duc (Aves, Strigiformes) du Pleistocène de Sardaigne et de Corse. *Rev. Paléobiol.* **5**:197–205.
Mourer-Chauviré, C., Moyà-Solà, S., and Adrover, R. 1977. Les oiseaux des gisements quaternaires de Majorque. *Nouv. Arch. Mus. Hist. Nat. Lyon* 15:61–64.
Myers, A. A., and Giller, P. S. 1988. *Analytical Biogeography: An Integrated Approach to the Study of Animal and Plant Distributions.* Chapman & Hall, London.
Olson, S. L. 1975. Paleornithology of St. Helena Island, South Atlantic Ocean. *Smithson. Contrib. Paleobiol.* **23**:1–49.
Olson, S. L., and James, H. F. 1991. Descriptions of thirty-two new species of birds from the Hawaiian Islands: Part I. Non-Passeriformes. *Ornithol. Monogr.* **45**:1–85.
Palmer, M., Pons, G. X., Cambefort, Y., and Alcover, J. A. in press. Historical processes and environmental factors as determinants of inter-island differences in endemic faunas: the case of the Balaeric islands. *J. Biogeogr.*
Paul, C. R. C. 1982. Pleistocene non-marine molluscs from Cala Salada, Ibiza. *Geol. J.* **17**:161–184.
Paul, C. R. C. 1984. Pleistocene non-marine molluscs from Cova de Ca Na Reia, Eivissa. *Boll. Soc. Hist. Nat. Bal.* **28**:95–114.
Paul, C. R. C., and Altaba, C. R. 1992. Els molluscs terrestres fòssils de les Illes Pitiüses. *Boll. Soc. Hist. Nat. Bal.* **34**:141–170.
Pérez-Mellado, V. 1998. *Podarcis lilfordi,* in: A. Salvador (ed.), *Reptiles. Fauna Ibérica* **10**:272–282.
Pérez-Obiol, R., Yll, E. I., Pantaleón-Cano, J., and Roure, J. M. 1996. Historia de *Buxus* y *Corylus* en las islas Baleares durante el Holoceno, in: P. Ramil-Rego, C. Fernandez-Rodríguez, and M. Guitián (eds.), *Biogeografía Pleistocene–Holocena de la Península Ibérica,* pp. 87–97. Xunta de Galicia.
Pericot, L. 1972. *The Balearic Islands.* Thames & Hudson, London.
Pinot, J. M., Tintoré, J., and Gomis, D. 1995. Multivariate analysis of the surface circulation in the Balearic Sea. *Prog. Oceanogr.* **36**:343–376.
Quintana, J. 1995. Fauna malacológica asociada a *Cheirogaster gymnesicum* (Bate, 1914). Implicaciones biogeográficas. *Boll. Soc. Hist. Nat. Bal.* **38**:95–119.
Reille, M. 1984. Origine de la végetation actuelle de la Corse sudorientale, analyse pollinique de cinq marais côtiers. *Pollen Spores* **16**:43–60.
Reille, M. 1992. New pollen-analytical research in Corsica: the problem of *Quercus ilex* L. and *Erica arborea* L., the origin of *Pinus halepensis* Miller forests. *New Phytol.* **122**:359–378.
Reille, M., Gamisans, J., De Beaulieu, J. L., and Andrieu, V. 1997. The late-glacial at Lac de Creno (Corsica, France): a key site in the western Mediterranean basin. *New Phytol.* **135**:547–559.
Reumer, J. W. F. 1980. Evolutie en Biogeografie van de kleine zoogdieren van Mallorca (Spanje). *Lutra* **23**:13–32.
Reumer, J. W. F., and Sanders, E. A. C. 1984. Changes in the vertebrate fauna of Menorca in prehistoric and classical times. *Z. Saeugetierkd.* **49**:321–325.
Rick, A. M. 1975. Bird medullary bone: a seasonal dating technique for faunal analysis. *Bull. Can. Archaeol. Assoc.* **7**:183–190.
Sanchíz, F. B., and Adrover, R. 1977. Anfibios fósiles del Pleistoceno de Mallorca. *Donana Acta Vertebr.* **4**:5–25.
Sanders, E. 1984. Evidence concerning late survival and extinction of endemic Amphibia and Reptilia from the Bronze and Iron Age settlement of Torralba den Salort (Alaior, Menorca), in: H. Hemmer and J. A. Alcover (eds.), *Història Biològica del Ferreret.* Monografies Científiques, Edit. Moll **3**:123–128.
Sanders, E. A. C., and Reumer, J. W. F. 1985. The influence of prehistoric and Roman migrations on the vertebrate fauna of Menorca (Spain). *BAR Int. Ser.* **229**:119–144.
Sbordoni, V., Caccone, A., Allegrucci, G., and Cesaroni, D. 1990. Molecular island biogeography. *Atti Convegni Lincei* 85:55–83.
Schoener, T. W., and Spiller, D. A. In press. Variation in the magnitude of a predator's effect from small to large islands. *Mon. Soc. Hist. Nat. Balears* **6**:35–65.

Schüle, W. 1993. Mammals, vegetation and the initial human settlement of the Mediterranean islands: a paleoecological approach. *J. Biogeogr.* **20**:399–412.

Seguí, B. 1996. Les avifaunes fòssils dels jaciments càrstics del Pliocè, Pleistocè i Holocè de les Gimnèsies. *Boll. Soc. Hist. Nat. Bal.* **39**:25–42.

Seguí, B. 1997. Avifauna fòssil del jaciment plistoholocènic de la Cova des Moro (Manacor, Mallorca). *Boll. Soc. Hist. Nat. Ba.* **40**:71–90.

Seguí, B., and Alcover, J. A., In press. Comparison of paleoecological patterns in insular bird faunas: a case-study from the western Mediterranean and Hawaii. *Smithson. Contrib. Paleobiol.* 89.

Seguí, B., Mourer-Chauviré, C., and Alcover, J. A. 1997. Upper Pleistocene and Holocene fossil avifauna from Moleta Cave (Mallorca, Balearic Islands). *Boll. Soc. Hist. Nat. Bal.* **40**:223–252.

Seguí, B., Quintana, J., Fornós, J. J., and Alcover, J. A. in press. A new genus of fulmarine petrel (Aves: Procellariiformes) from the Upper Miocene of Menorca, western Mediterranean. *Paleontology.*

Shackleton, N. J., and Opdyke, N. D. 1977. Oxygen isotope and paleomagnetic evidence of early northern hemisphere glaciation. *Nature* 270:216–219.

Sondaar, P. Y. 1987. Pleistocene man and extinctions of island endemics. *Bull. Soc. Geol. Fr.* **150**:159–165.

Sondaar, P. Y., Sanges, M., Kotsakis, T., Esu, D., and De Boer, P. L. 1984. First report on a paleolithic culture in Sardinia. *BAR Int. Ser.* **229**:29–47.

Sondaar, P. Y., Sanges, M., Kotsakis, T., and De Boer, P. L. 1986. The Pleistocene deer hunter of Sardinia. *Geobios* **19**:17–25.

Sondaar, P. Y., Martini, F., Ulzega, A., and Klein Hofmeijer, G. 1991. L'Homme pléistocènne en Sardaigne. *Anthropologie* 95:181–200.

Sondaar, P. Y., Elburg, R., Klein Hofmeijer, G., Martini, F., Sanges, M., Spaan, A., and De Visser, H. 1994. The human colonization of Sardinia: a Late-Pleistocene human fossil from Corbeddu Cave. *C. R. Acad. Sci. Ser. IIa* 320:145–150.

Sondaar, P. Y., McMinn, M., Seguí, B., and Alcover, J. A. 1995. Paleontological interest of karstic deposits from the Gymnesic and Pityusic islands. *Mon. Soc. Hist. Nat. Bal.* **3**:155–170.

Spoor, C. F., and Sondaar, P. Y. 1986. Human fossils from the endemic island fauna of Sardinia. *J. Hum. Evol.* **15**:399–408.

Steadman, D. W. 1995. Prehistoric extinctions of Pacific island birds: Bidiversity meets zooarchaeology. *Science* **267**:1123–1131.

Taylor, T. G., and Moore, J. H. 1953. Avian medullary bone. *Nature* **172**:504–505.

Torres, N., and Alcover, J. A. 1981. Presència de *Tudorella ferruginea* (Lamarck 1822) (Gastropoda: Pomatiasidae) a l'illa d'Evissa. *Bol. Soc. Hist. Nat. Bal.* **25**:185–188.

Vigne, J. D. 1983. Les mammifères terrestres non-volants du post-glaciare de Corse et leurs rapports avec l'homme: étude paléo-ethno-zoologique fondée sur les ossements. Thése 3ème cycle, Univ. P. et M. Curie, Paris.

Vigne, J. D. 1989. *Les mammifères postglaciares de Corse et leur rapports avec l'homme.* Gallia-Préhistoire, CNRS, Paris.

Vigne, J. D., and Alcover, J. A. 1985. Incidence des relations historiques entre l'homme et l'animal dans la composition actuelle du peuplement amphibien, reptilien et mammalien des îles de Méditerranée occidentale. *110e Congr. Nat. Soc. Savan. Montpellier* **2**:79–91.

Vigne, J. D., Marinval-Vigne, M. C., Lanfranchi, F. de, and Weiss, M. C. 1981. Consommation du "Lapin-rat" *Prolagus sardus* (Wagner) au Néolithique ancien méditerranéen; abri d'Araguina-Sennola (Bonifacio, Corse). *Bull. Soc. Prehist. Fr.* **78**:222–224.

Waldren, W. H. 1982. Balearic prehistoric ecology and culture. The excavation and study of certain caves, rock shelters and settlements. *BAR Int. Ser.* **149**:1–773.

Weesie, P. D. M. 1987. The Quaternary avifauna of Crete, Greece. Ph.D. thesis, University of Utrecht.

Welsh, D. A., and Lougheed, S. C. 1996. Relationship of bird community structure and species distribution to two environmental gradients in the North Boreal forest. *Ecography* **19**:194–208.

Yll, E. I., Pérez-Obiol, R., and Julià, R. 1994. Vegetational change in the Balearic Islands (Spain) during the Holocene. *Hist. Biol.* **9**:83–89.

Yll, E. I., Pérez-Obiol, R., Pantaleón-Cano, J., and Roure, J. M. 1997. Palynological evidence for climatic change and human activity during the Holocene on Minorca (Balearic Islands). *Quat. Res.* **48**:339–347.

Zammit-Maempel, G., and De Bruijn, H. 1982. The Plio/Pleistocene Gliridae from the Mediterranean islands reconsidered. *Proc. K. Ned. Akad. Wet. Ser. B* 85:113–128.

CHAPTER 9

Introduced Predators and Avifaunal Extinction in New Zealand

RICHARD N. HOLDAWAY

1. Introduction

At 270,000 km^2, New Zealand is one-thirtieth the area of Australia, one-third that of Madagascar, twice that of Cuba, and comparable in area to the British Isles, to the Philippines, and within the United States to the State of Colorado. Isolated in the southwestern Pacific, 1900 km east of Australia, New Zealand was the last major habitable landmass to be peopled, in this instance by Polynesians. Unlike the British Isles and the Philippines, which are partly on and partly off the continental shelf, New Zealand is so remote that despite zoogeographic and geotectonic evidence of a considerable antiquity, it lacked nonvolant land mammals when humans and their commensals colonized 700 years ago.

To the extent that biogeography is driven by the accidents of planetary distribution of landmasses, New Zealand is in the unusual situation of being a unique midrange landmass at least an order of magnitude larger than a large remote oceanic island such as Fiji, and more than an order of magnitude smaller than Australia, the smallest continent. The North and South Islands of New Zealand are surrounded at various distances by a variety of satellite islands and island groups that provide opportunities for natural experiments on unplanned reserves. The satellites shelter at least temporarily their native fauna (such as tuatara *Sphenodon* spp. or saddlebacks *Philesturnus* spp.) from being eliminated by the latest new mammalian predator to arrive on the main islands.

The earth sciences including Quaternary geology are well developed in New Zealand. The fossil record is well known; moas have been studied and their extinctions pondered for over 150 years since Sir Richard Owen's famous reconstruction from a piece of moa leg bone. The fossil record in Libby time is rich.

The relative clarity of the faunal record in New Zealand is attributable to the absence until recently of scavengers to disturb deposits. Since European settlement, many small sites have been lost to erosion by rabbits, possums, and sheep (Worthy and Holdaway, 1995, 1996a; Worthy, 1997). Despite the damage, the faunas in the South Island are now well

RICHARD N. HOLDAWAY • Palaecol Research, Christchurch, New Zealand.
Extinctions in Near Time, edited by MacPhee. Kluwer Academic/Plenum Publishers, New York, 1999.

known, and those of the North Island are under investigation. There are now well-curated collections of fossils in museums, and Quaternary studies have recently been strongly supported by grant agencies, fostered by a longstanding national interest in the recent past.

Historical introductions of animals and plants are limited to just over 200 years and are documented with considerable detail. Prehistoric invasions are limited to humans and the Pacific rat, which may have established itself in advance of Polynesian settlement. The radiocarbon chronology for large animals, the moas, is quite good and getting better. It is much better than that available for the large extinct vertebrates of the West Indies, Madagascar, the Philippines, and the oceanic islands of the Mediterranean.

Beyond the effects of human colonists on extinctions within the past 1000 years, there is a unique opportunity to search for evidence of prehuman extinctions at the transition from the last glacial episode (locally known as the Otiran) and the Holocene, a time when climate change is often invoked as at least partly involved in explaining Northern Hemisphere megafaunal extinction. Failure to find such evidence in New Zealand is a challenge to those who would explain late glacial extinctions elsewhere.

Permanent settlement began about AD 1300 (Anderson, 1991) with the arrival of East Polynesians, probably from in or near the Cook Islands. The arrival of people coincided with the collapse of ecosystems that had evolved in isolation during the whole Neogene (Fleming, 1962). Although the causes of the extinction event have been debated for nearly 150 years (Anderson, 1989a,b), there is broad agreement that the extinctions of many large and small vertebrates and an unknown number of large invertebrates were effected by some combination of habitat destruction, hunting, and predation by introduced mammals (McGlone *et al.,* 1994), all with an anthropogenic envelope.

To determine the cause(s) of extinction, we must establish when the fauna was exposed to each new factor, and when each extinct species vanished. The major factors that have been invoked in models of late Quaternary extinctions are climate and the arrival of humans and commensal mammals. Evidence for climate change sufficient to cause extinction is not convincing (Burrows and Greenland, 1979). The case for climate as a factor is considerably weakened by there being no evidence for an extinction event at the time of the far more extreme climatic changes at the end of the Otiran Glaciation between 14,000 and 10,000 yrbp. Although the large-scale changes in the distribution of forest, shrubland and grassland, and in the distribution of forest types (McGlone, 1988) during that period were accompanied by changes in the distribution of vertebrate faunas, no vertebrate is known to have become extinct at that time (Worthy and Holdaway, 1993, 1994a, 1995, 1996a).

In stark contrast to the continuity of faunal composition at the Pleistocene–Holocene boundary, more than 40 species of birds, a bat, at least three species of endemic frogs, and an unknown number of lizards have become extinct in the late Holocene, all—so far as is known—during the past 2000 years (Holdaway, 1989). The many extinctions and extirpations and drastic reductions in breeding range have resulted in a depauperate fauna, rich only in coastal and oceanic species whose breeding grounds are now substantially restricted to islands still lacking introduced mammals (Turbott, 1990). At present it is known only that many large vertebrates became extinct early in the period of human settlement (Anderson, 1989b; Holdaway, 1989), although future research should establish terminal dates for a range of key taxa (Holdaway, unpublished data). There is as yet no fine-scale information on when most species went extinct.

Dates for the arrival of the major external influences on the fauna and flora are known with varying precision. Human settlement occurred at about AD 1300 (Anderson, 1991).

There is no evidence for earlier permanent occupation (McGlone *et al.*, 1994), despite recent suggestions that the palynological record shows small-scale disturbance, attributable to humans, for several hundred years before that time (Sutton, 1987).

In addition to some domesticated plants, the Polynesian dog was brought deliberately by early Polynesian settlers. Undoubtedly, the settlers' large oceangoing canoes also carried, deliberately or accidentally or both, Pacific rats *Rattus exulans,* the first rodent to reach New Zealand and the first to be transported on a large scale to oceanic islands (Atkinson, 1985). Dates for the arrival of the other introduced predators are known to within a couple of decades for accidental arrivals or those that happened very early in the European era when few records were kept (e.g., *Rattus norvegicus, R. rattus, Felis catus, Sus scrofa*). Calendar year dates are known for others that were deliberately introduced by government and private organizations, such as the mustelids and the little owl. *Athene noctua* (King, 1990).

Until recently, all external influences on the New Zealand biota were presumed to have begun simultaneously with human settlement about AD 1300. However, AMS radiocarbon dating of bones from natural sites on both North and South Islands has shown that the Pacific rat reached and successfully colonized both main islands at least 1000 years before permanent human settlement (Holdaway, 1996). Although the dates have been questioned (e.g., Anderson, 1996), further work supports the rat bone gelatin chronology (Beavan and Sparks, 1998; Holdaway, in press; Holdaway and Beavan, in press). It is probable, therefore, that the New Zealand biota was exposed to a single new mammalian predator for at least 1000 years before habitat destruction and human hunting began.

There are other indications that the fauna confronting the earliest Polynesian settlers had already undergone significant change. Analysis of faunal composition in sites rich in microvertebrate remains (Worthy and Holdaway, 1994b, 1996b), and the pattern of presence and absence of species vulnerable to human predation in archaeological sites (Worthy, 1998b) suggest that several species of small birds such as the diving petrel *Pelecanoides urinatrix* and Finsch's duck *Euryanas finschi* had declined before human settlement.

In the past 30 years, there has been some controversy as to whether rodents have affected the New Zealand biota. With few exceptions (e.g., Falla, 1974; Moors, 1983; Craig, 1986), most workers now agree that on small islands the two so-called European rats (*Rattus norvegicus* and *R. rattus*) have reduced, and in some instances eliminated, species or populations (Atkinson and Bell, 1973; Bell, 1978). However, there has been a reluctance (e.g., Moors, 1983; Craig, 1986) to project these findings to the main islands, or even (Craig, 1986) to assume that the Pacific rat had any major role at all in the extinctions of New Zealand vertebrates. Craig (1986) suggested that human-induced habitat changes caused the extinctions on the northern offshore islands, and by implication on the mainland as well.

There is strong evidence that humans were not directly responsible for all pre-European extinctions. Polynesians lived on the rat-free island of Aorangi (Poor Knights group); small birds such as the fairy prion *Pachyptila turtur* and large invertebrates such as the centipede *Cormocephalus rubriceps* survived. New Zealand provides the opportunity of comparing the possible effects of humans on small and large islands together. The large (main) islands are so large and environmentally diverse that it is highly unlikely that humans were directly responsible for the extinction of species such as small lizards and flightless wrens. In addition, the prehuman faunas on the main islands are much more fully known than are those of small islands where fossil deposits—where they exist—have only recently begun to be assessed (Tennyson, personal communication).

From the present patterns of distribution of small vertebrates on offshore islands with and without populations of Pacific rats, Crook (1973), Whitaker (1978), and Atkinson (1985, 1986) have suggested that the rat has been responsible for extinctions. The absence of direct evidence for predation has caused the conclusions to be questioned or disregarded. Although some examples of exclusion from former ranges were included in the data presented, based on fossil evidence, most arguments relied on the present distributions of all species (Crook, 1973; Whitaker, 1978; Atkinson, 1985, 1986). Recent studies (Booth *et al.*, 1996, Lovegrove, 1996) have demonstrated that Pacific rats are effective predators, capable of reducing populations of small vertebrates.

Several workers (e.g., Imber, 1975; Atkinson, 1978, 1986) have suggested that the predatory ability of a rat and its effect on small vertebrate populations depend on its body size in relation to that of potential prey. For birds, it also depends on the length of the egg. Fortunately, these parameters are relatively easy to obtain for living taxa, and to estimate for extinct taxa. Hence, the relationship between predator and prey body size can be examined in the original fauna rather than just the fauna encountered by Europeans.

A key to determining whether predators forced New Zealand extinctions may lie in the fortuitous size grading in the series of predators imported to New Zealand (see Table 1). The second smallest—the Pacific rat—was introduced first. Setting aside new data that indicate an earlier arrival time, the Pacific rat was the only small mammalian predator in New Zealand for at least several centuries before the largest rodent—the Norway rat—was introduced in the late eighteenth century. The smallest rodent (the house mouse) arrived in the early nineteenth century, and cats were introduced at about the same time. Atkinson (1973) suggests that the large and most arboreal rat—the black or ship rat—did not arrive until the second half of the nineteenth century and earlier in North Island than in South Island. Mustelids were introduced in the 1880s. A series of comparisons can, therefore, be made between the compositions of the avifauna at different times roughly defined by the time of introduction of the mammals that have been suggested to be major predators. If predators were an important factor in the extinctions in each period, only the taxa vulnerable to the predators present at that time should be missing from the fauna alive in the next interval.

At present, the pre-European period cannot be usefully subdivided for faunal analysis as there is too little information on the timing of extinctions. The first major division is that between pre- and post-European contact. Only two predators were present (humans, rats) and they were so different in size they are likely to have affected different suites of species in the avifauna. The Polynesian dog is discounted as a predator during this period, as Anderson (1981, 1989b) suggests that it seldom if ever went feral, was fed almost entirely on food scraps, and the only direct effect it may have had on the fauna was as a possible hunting companion for human moa hunters. The European era can be divided more closely, because more mammals were introduced and their times of extinction are reasonably well known.

The number of predators increased rapidly in the nineteenth century. The inevitable lag between arrival of a new predator forcing bird extinctions complicates the comparison. Fortunately, the relatively long gap between the introduction of the Norway and black rats allows some measure of discrimination between the effects of these two large rodents. Atkinson (1973) suggests that the black rat arrived in the North Island 30 years before mustelids were introduced. The relative effects of the new predators should be visible in times of extinction, if the responses of prey to the predation pressure were rapid enough. The time lag in the North Island between black rat and stoat introductions is especially important because both are excellent climbers and might be expected to affect some of the same arboreal taxa. Black rats arrived in the South Island at about the same time as mustelids were introduced there, so the effects would have been simultaneous and impossible to separate now.

There are several possible approaches to gauging the effects of predation on the New Zealand biota. Investigations at the level of single species can show whether or not predators affect individual birds and perhaps populations (Booth *et al.*, 1996; Lovegrove, 1996). Studies of long-term population dynamics of single species would be very useful but are very difficult to fund. The approach chosen here is to look at and above the ecosystem level, a retrospective of what happened, when, and in the presence of what new factors. It assumes that there was some kind of dynamic equilibrium in the fauna in the short term.

In this chapter, the different vulnerabilities of different ecological and size groups in the original avifauna to different predators are used to examine the role of people and the introduced mammalian predators in the late Holocene extinctions. In particular, the possible role of the Pacific rat in the extinctions before European contact is examined.

The analysis is confined to the faunas of the three main islands (including Stewart Island) so that survival of some taxa on smaller islands does not obscure the true level of extinctions in the presence of Pacific rats. The relationship between extinction of native fauna and the presence of introduced mammalian predators can be demonstrated by comparing basic characteristics (body mass, habitat, behavior) of the groups of species going extinct in successive periods of increasing diversity of predators, with the size, habitat, and behavior of the predators present in each period.

The discussion is limited to the avifauna because at present it is difficult or impossible to prepare distributional lists of lizards (Gekkonidae, Scincidae), tuatara (Sphenodontidae), or frogs (Leiopelmatidae). Although many cryptic living species of lizard (e.g., Patterson, 1997) and at least one frog (Bell *et al.*, 1998) have been described within the past decade and some fossil taxa are known (Worthy, 1987; Worthy and Holdaway, 1995), more remain to be identified. The fossil record of these small animals is limited and difficult to interpret. Most lizards and the two tuatara cannot be identified to species from fossil remains (Worthy and Holdaway, 1995). The details of range reductions, extirpations, and extinctions in these groups are hardly known, although they are likely to be substantial.

There are too few species of endemic mammals (three bats and three seals) for meaningful patterns to be obtained, although the rarity or absence of the remains of robust short-tailed bats *Mystacina robusta* in sites with more than a few bones of Pacific rat is striking (Worthy, 1997). The species survived in European time only on rat-free islands to the southwest of Stewart Island and went extinct there shortly after the arrival of black rats in 1962 (Bell, 1978).

2. Avifauna and Predators

2.1. Possible Causes of Avifaunal Extinction—Predictions of the Predation Hypothesis

The possible role of predation in the New Zealand avifaunal extinctions may be set against the possible consequences for the New Zealand avifauna if predation was not a factor. First, given that there was no significant climate change (Burrows and Greenland, 1979) and the composition of the avifauna had been stable for thousands of years (Worthy and Holdaway, 1993, 1994a, 1995, 1996a; Worthy, 1997, 1998a), there should have been few or no extinctions in the New Zealand avifauna in the past 1000 years if humans had not arrived. Then, if, after humans arrived, neither they nor the Pacific rat or dog were significant predators and little or no environmental damage resulted from human presence, again there should have been few or no extinctions. Sutton (1987) suggested the possibility of long-term cryptic occupation by Polynesians, which implies a minimal level of disturbance, certainly below archaeological and palynological thresholds.

Again assuming that predation was not a factor, but that people did alter the environment by burning woody vegetation and other means (McGlone, 1989), any extinctions resulting should have been evenly spread across diverse phylogenetic groups. In general, environmental changes, both natural and anthropogenic, should have affected the whole avifauna. There should not have been selective extinction of particular groups in relation to body size or habitat. For example, oceanic birds nesting in western areas are unlikely to have been greatly affected by changes in the distribution of vegetation, insects, or of other organisms elsewhere in the terrestrial environment.

In contrast, predation should appear as a factor in the avifaunal extinctions as differential losses of species that would have been directly affected by the particular predators present at any one time. Other species and groups of species should have been relatively unaffected. Extinctions during the two phases of human occupation of New Zealand should have affected different groups in avifauna, because only one small (rat) and one large (human) new mammalian predator were present in the first phase, in contrast to a suite of small and medium-sized predators that arrived after European contact. The residual fauna at the end of each phase should be enriched in those taxa most resilient to the predators present during the phase, or those out of reach of the predators. In general, the range of prey available to the predators brought by Europeans should have been lacking those groups that were most susceptible to predation by small rats and humans.

2.1.1. Extinctions before European Contact—Was the Pacific Rat Involved?

Seven predictions can be made for the effects of the two predators present before European contact (Pacific rats and humans). If the Pacific rat was an important predator that had a major role in the New Zealand avian extinctions:

1. Extinctions that occurred before European contact should have affected taxa too small to be vulnerable to human hunting, and were too widespread for habitat damage to have been important.
2. Extinctions of small taxa should have been negligible in groups that inhabited wetlands or coastal habitats, which are not favored by Pacific rats.
3. Most extinctions of small taxa before European contact should be of those taxa that lived on or near the ground, with the more arboreal species being preferentially lost during European time, because Pacific rats are not as arboreal as the black rat or stoat.
4. Small taxa with a high fecundity (large clutch sizes, multiple clutches) would be more likely to have survived into European time, and some would then be vulnerable to increased predation by black rats and mustelids.
5. Taxa larger than those available to Pacific rats that went extinct before European

contact should include the large taxa most attractive to human hunters; medium-sized taxa from all habitats should be enriched in the avifauna.

6. If the Pacific rat was not a significant factor in the initial episode of extinction, there should be no significant extinction of species in the size range vulnerable to Pacific rat predation.
7. If people were not a significant influence on avian diversity before European contact, the range of taxa going extinct before 1770 should be a random sample of all taxa present at the time of human colonization: There should be no bias toward large taxa (preferred prey) that would have provided the best economic return for hunting effort.

"Small taxa" are those whose body mass was the same as or less than that of the Pacific rat populations in New Zealand, or whose eggs were within the range known to be accessible to rat predation. "Large taxa" are those of body mass greater than 5000 g, which would have been the most economic food sources for humans.

2.1.2. Extinctions after European Contact—Is There Evidence for Sequencing in Predator Introduction and Avifaunal Extinction?

After any new predator was introduced, there would necessarily be a lag stage, before any response in prey numbers would be noticed. The lag period would depend, *inter alia,* on the rate of spread of the predator, and vary between different prey species according to their size, habitat versus that of the predator, and their initial abundance, distribution, and fecundity. If predation was a major factor in the extinctions after 1770, the sequenced arrival of predators of different body masses and predatory abilities after 1770 should be reflected in a pattern of extinctions in time and space.

2.2. The Original Avifauna

Studies of extinction must be based on a firm knowledge of the fauna of interest. A major handicap to previous studies of extinctions in New Zealand has been the lack of knowledge of the complete avifauna present just before the arrival of people. For example, Anderson (1997b) uses Turbott (1990) as a main source: several taxa listed in Turbott are now known not to have existed in New Zealand, and more recent taxonomic work has shown that many taxa relegated to subspecific rank in checklists are full species. In addition, recent work on fossil deposits has revealed the past presence of species and groups not formerly known in the fauna (Worthy and Holdaway, 1993, 1994a, 1995, 1996a; Worthy, 1997, 1998a). All such changes directly affect the observed levels and rates of extinction.

A list of the bird species breeding on the three largest islands of New Zealand (North, South, and Stewart; Fig. 1) in the late Holocene before first human contact is given in Appendix 1. The list is based on one for the whole archipelago (Holdaway *et al.,* in press), but adequate lists of the original fauna on the smaller islands are not available at present. In the following discussions, "South Island" refers to both South Island and Stewart Island, unless otherwise stated.

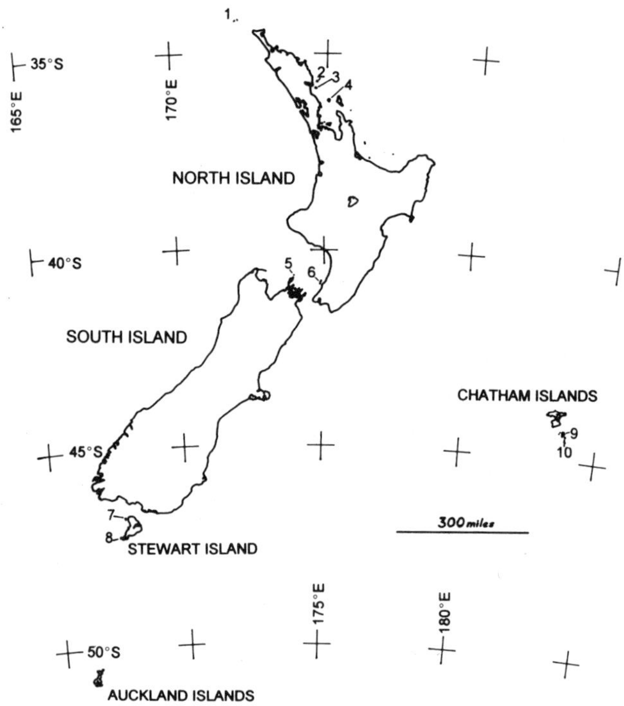

FIGURE 1. New Zealand, with major islands and island groups mentioned. 1, Three Kings; 2, Poor Knights; 3, Hen and Chickens Group; 4, Little Barrier Island; 5, Stephens Island; 6, Kapiti Island; 7, Codfish Island; 8, Big South Cape Island; 9, Pitt Island; 10, South East Island.

2.3. Parameters of Vulnerability

Few physical and ecological parameters are easily and reliably obtainable for taxa known only from fossils. The parameters used here were body mass (as an indication of availability of adults to predation), egg length (as a proxy for the vulnerability of eggs and young to predation), and "vulnerability," based on phylogenetically predictable aspects of behavior, osteological evidence for flightlessness, and preferred habitat of the major taxonomic groups.

Body Mass and Egg Length

A representative body mass (BM, Appendix 1) was selected for each species. Where available, that adopted by Heather and Robertson (1996) was used. For species with significant sexual dimorphism, the body mass for the female was used, as the sex more vulnerable to predation. For taxa known only from fossils, body masses were calculated from relationships between limb bone dimensions and body mass given in Campbell and Tonni (1983) and Anderson *et al.* (1985). Body masses for moas (Dinornithiformes) were derived from Atkinson and Greenwood (1989), modified in conjunction with unpublished information supplied by T. H. Worthy.

A representative egg length (EL, Appendix 1) was also chosen for each species, again using Heather and Robertson (1996) as the basic source. Relationships between body mass and egg length were established for seven groups of related taxa, from the data available for New Zealand taxa (Appendix 2). Values for extinct taxa were derived from these. Values for extinct taxa were checked against those for living relatives where possible. For groups such as the adzebills, which have no modern analogues, the closest living group was used for comparison (in this instance, rails).

Single representative values for body mass and egg length were used to simplify the analysis. They were considered to be satisfactory indications of the usual size of the bird and its egg, so that some impression can be obtained of the vulnerability of each bird species to predation by the several rodents introduced to New Zealand. Predators are not fixed in their ability to handle eggs, or chicks and adults for that matter, so the representative values and estimates for these basic parameters are simply useful guides.

2.4. Predators and Predatory Ability

The mammalian predators introduced to New Zealand over the past 2000 years are listed in Table 1. Comprehensive summaries of their biologies in New Zealand are available in King (1990). The size and some details of the distribution and biology for the 11 predators likely to have affected bird populations are summarized below, in approximate order of the predators' arrival.

1. Pacific rat (*Rattus exulans*). Body mass 60–80 g, up to 180 g. The small Pacific or Polynesian rat (*kiore* in New Zealand Maori) forages on or near the ground. It will climb but not as well as the black rat; it avoids water, can reach high densities in forests, and is most abundant in grasslands. Potential prey includes ground-nesters with eggs less than 57 mm long (Atkinson, 1986); small ground-nesting

Table 1. Dates (Years AD) of Colonization of Main Islands of New Zealand by Introduced Mammalian Predators

Species	Island: North	South	Stewart	Chatham
Trichosurus vulpecula	1869–1915	1858–1895	1890	1911
Erinaceus europaeus	1907–1912	1870	1930	?
Rattus exulans	c150	c150	c1400	c1350
R. norvegicus	(1772) 1792–	(1772) 1792–	1800?	1830?
R. rattus	1860	1890	Before 1911	1890?
Mus musculus	c1830	1850s	1850s?	1850s?
Felis catus	1830s	By 1850	By 1850?	Before 1840
Canis familiaris	1300	1300	1400	—[b]
Mustela furo	1882	1879	—	—
Mustela erminea	1886	1885	—	—
Mustela nivalis	1885	1885	—	—
Sus scrofa	(1773) 1792–	(1773) 1792–	1790s?[c]	early 1800s?

[a]Modified from King (1990).
[b]—, never introduced.
[c]Extinct.

and flightless birds of less than 100 g, but not species associated with swamps; mainly species with small clutch sizes and low reproductive rates.

2. Man (*Homo sapiens*). Body mass 40,000–100,000 g. The Polynesians were traditionally bird hunters of great skill, using snares, spears, and direct harvesting of burrow-nesting and colonial seabirds (Steadman, 1997). Birds were taken for food, and for feathers for decoration and ceremonial.
3. Dog (*Canis familiaris*). Body mass 2000–10,000 g. The Polynesian dog (*kuri*) seems to have been a close commensal of humans. There is some evidence that early Polynesians in New Zealand may have used dogs in hunting moas (Anderson, 1981), but wild packs reported in the nineteenth century seem to have been mainly composed of feral dogs of European strains. The Polynesian dog had probably little effect on the biota.
4. Norway rat (*Rattus norvegicus*). Body mass 200–300 g, maximum c.450 g. The Norway rat does not climb, but swims well, is often found near water, and can colonize nearshore islands unaided. It is agile enough to jump into the air and catch a flying bird (R. Hitchmough, personal communication).
5. Feral pig (*Sus scrofa*). Body mass, males 45,000–200,000 g, females 32,000–114,000 g. Was once widespread over North and South Islands in forest, shrubland, and grassland, but now patchily distributed (King, 1990).
6. Feral cat (*Felis catus*). Body mass, males 2460–3670 g, females 740–2750 g. Widespread; populations continuously augmented by releases of domestic pets. Also present, or formerly present, on several offshore islands, particularly those that had lighthouse stations.
7. House mouse (*Mus musculus*). Body mass 15–20 g, up to 30 g. The house mouse rapidly penetrated all habitats including remote forests soon after its introduction (King, 1990).
8. Black rat (*Rattus rattus*). Body mass 120–160 g, up to 225 g. The black rat is largely arboreal (King, 1990), although it will live on the ground in the absence of competitors (Atkinson, 1978). Black rats prey mostly on eggs or young of songbirds, including many forest species, but they are also likely to take adult petrels up to their own body mass and the young of larger species (Atkinson, 1978).
9. Ferret (*Mustela furo*). Body mass, mean, male 1200 g, female 600 g. Mostly in open country, including beds of braided rivers, but also in shrubland over much of the North and South Islands.
10. Stoat (*Mustela erminea*). Body mass, mean, male 324 g, female 207 g. Stoats are agile climbers and can reach most nest sites, but hole-nesting species are most vulnerable because of the restricted access and escape routes (O'Donnell, 1996; Wilson *et al.*, 1998). In addition, stoats swim well, including in the sea, and could reach most species available to the black rat.
11. Weasel (*Mustela nivalis*). Body mass, mean, male 126 g, female 57 g. The weasel is small enough to penetrate small burrows and holes among rocks, but it is rare (King, 1990). There are probably too few to affect all but the rarest species. Numbers may have been higher soon after it was introduced, but there is no evidence that it reached levels of abundance known for stoats and ferrets.

Predator Introductions and the Basis for Time Divisions

The division of time used here is based on the arrival times of the predators. Many introduction dates are known only approximately, but the general order of introductions is reasonably clear. Most importantly, the only predators present before 1770 were people and the Pacific rat, if dogs are excluded (Anderson, 1981). Before 1850, the Norway rat and the mouse were apparently the only additional rodents.

The first three rodents to arrive apparently reached the North and South Islands simultaneously. The black rat did not arrive (in the North Island first) until 1850 (Atkinson, 1973), while cats and mice were rare before 1850 (King, 1990), and pigs were widespread. By 1870, black rats were well established in the North Island and their effects should have begun to be obvious there. All of the mustelids were introduced and spread through both islands at the same time, after 1870. In the South Island, their arrival coincided with, or slightly preceded, that of the black rat (Atkinson, 1973).

By 1930, all of the introduced predators were firmly established throughout the North and South Islands and the general pattern of vegetation had stabilized. The periods 1870–1930 and 1930–1990 encompass the main periods of renewed (European) forest removal, and the removal of last remnants of original vegetation in many areas, particularly North Island.

2.5. Avian Vulnerability Groups

Based on the size, main habitat, and behavior of the main predators, each breeding species of bird was allocated to one of six vulnerability groups based on attributes such as ability to fly and principal habitat (Appendix 1). The attributes were selected so as to broadly define a priori the vulnerability of each species to the main introduced predators, including humans. Burrow- and hole-nesting species, which are particularly vulnerable to some predators (O'Donnell, 1996), are denoted by an asterisk next to the vulnerability group code in Appendix 1. The six vulnerability groups are:

1. Flightless (F): all flightless species from acanthisittid wrens (at 10–40 g) to the largest moas (at ca. 250 kg).
2. Petrels (P): oceanic birds that nest in burrows, are relatively helpless on land, require a launching platform such as a cliff edge or sloping tree on a hillside to take off, and that have a single egg clutch, long incubation and chick stages, and typically a very long adult life expectancy.
3. Coastal and freshwater (C): species including cormorants, waders, waterfowl, most rails, and egrets and bitterns that feed in or near water and that usually breed where water barriers provide some protection for the nest.
4. Arboreal (AF): all arboreal species, except predators and those that have an association with water.
5. Ground (GF): volant species that feed and nest on the ground.
6. Predators (Pr): terrestrial birds that prey largely or exclusively on vertebrates and large invertebrates, but excluding primary insectivores such as the owlet nightjar *Aegotheles novaezealandiae*.

2.5.1. Size Classes within Vulnerability Groups

Each vulnerability group was divided further, if appropriate, into subgroups based on the representative body mass and egg length. The primary division for each group was between taxa whose mass and egg length were low enough to permit predation by the Pacific rat (Atkinson, 1986), other factors such as habitat being equal. Effectively, all species in a group whose egg length was below 60 mm were included in the first size class. For flightless birds, the classes were body mass of 0–1000 g and/or egg length ≤ 60 mm (e.g., acanthisittid wrens, Finsch's duck, kakapo *Strigops habroptilus*), body mass 1001–5000 g (e.g., South Island takahe *Porphyrio hochstetteri*), and body mass over 5000 g (geese, adzebills, and moas). Petrels have comparatively large eggs for their body mass: the size divisions were 0–500 g and/or egg length ≤ 60 mm (e.g., storm petrels, diving petrels, prions, small *Puffinus* shearwaters), 501–1000 g (e.g., larger *Pterodroma* petrels), and > 1000 g (largest *Puffinus* shearwaters and *Procellaria* petrels). Freshwater and coastal species were grouped as 0–500 g and/or ≤ 60 mm (e.g., small swamp rails, shorebirds, terns, gulls), 501–1000 g (small bitterns, egrets and shags, larger shorebirds), and > 1000 g (e.g., larger gulls, egrets, bitterns, and shags). All of the arboreal species except the extinct raven *Corvus moriorum* were < 500 g, and most had a body mass of 200 g. All of the volant ground species were < 200 g. The predators included relatively small taxa such as the kingfisher *Halcyon sancta,* as well as the very large extinct Eyles's harrier *Circus eylesi* at about 2500 g, and the immense Haast's eagle *Harpagornis moorei,* in which the female reached 13,000 g (Holdaway, 1991).

In Appendix 1, Part 1 lists all taxa with an egg length ≤ 60 mm; Part 2 includes taxa (petrels) with a body mass of 201–500 g and an egg length > 60 mm; Part 3 includes taxa with a body mass of 501–1000 g and an egg length > 60 mm; Part 4 lists taxa of body mass 1001–5000 g; and Part 5 includes taxa with body mass > 5000 g.

2.5.2. Birds within the Predatory Potential of the Principal Introduced Predators

The bird species actually available to each new predator depended on when the predator arrived. For example, although black rats would have been able to catch the flightless wrens with ease, there were no flightless wrens left when stoats arrived. With this proviso, potential prey species, and groups of species, can be suggested for the main mammalian predators.

2.5.2a. Pacific Rat. Storm petrels; diving petrels; prions; *Pterodroma* petrels of less than 250 g; shearwaters of less than 250 g; subantarctic snipe (*Coenocorypha* spp.); small terrestrial ducks (*Euryanas finschi, Anas chlorotis*); the small rails *Capellirallus karamu* and *Gallinula hodgenorum,* kakapo *Strigops habroptilus* (eggs and chicks); flightless or poorly flying wrens (*Pachyplichas yaldwyni, Dendroscansor decurvirostris, Traversia lyalli*). Examples of other vulnerable species are nonvolant vertebrates under 70–120 g, including the larger leiopelmatid frogs; nocturnal ground-dwelling skinks and geckos; and the juvenile tuataras (*Sphenodon* spp.) below 150 mm snout–vent length. The surviving New Zealand leiopelmatid frogs are terrestrial, with no free-swimming tadpole stage. In addition, a range of large invertebrates (including carnivorous and herbivorous snails, giant slugs, centipedes, and large insects such as wetas and weevils) and some plants such as the megaherb *Stilbocarpa* may be expected to be vulnerable.

2.5.2b. Man. People are likely to have been most interested in the largest birds, while there was a choice. Large ground birds would return a large amount of energy per unit effort. The larger waterfowl (easily taken during the flightless molt period) would have been most sought after, but any species that could be caught was probably exploited. Some may have been favored, or avoided, for reasons of palatability, if a choice was possible (Worthy, 1998b). Species over about 2000 g in body mass are likely to have been preferred in the earliest years of occupation.

Potential prey would include large flightless (or seasonally flightless) birds including all 11 moa species; kiwis (*Apteryx* spp.); adzebills (*Aptornis* spp.); large penguins (*Eudyptes pachyrhynchus, Megadyptes antipodes*); large waterfowl (*Cnemiornis* spp., *Cygnus atratus, Mergus australis, Biziura delautouri*); pelecaniforms nesting on flat and accessible sites (*Morus serrator, Leucocarbo carunculatus*); large terrestrial rails (*Porphyrio mantelli, P. hochstetteri; Fulica prisca*); and the kakapo *Strigops habroptilus.*

2.5.2c. Dog. Feral European breeds have been shown recently to be extremely damaging to populations of larger nocturnal birds, especially kiwis (Taborsky, 1988), and probably were to kiwi, kakapo (as reported by explorers and collectors in the nineteenth century), and some ducks.

2.5.2d. Norway Rat. Ground-nesting birds, or those such as saddlebacks (*Philesturnus* spp.) that roost near the ground (Lovegrove, 1996). Eggs and young of seabirds and of terrestrial ground-nesting birds (Atkinson, 1978; Imber, 1978). Norway rats could take petrels up to twice the mass of those accessible to Pacific rats and birds up to the size of *Eudyptes* penguins. Agile songbirds, and those that nest in trees or in holes among rocks, are unlikely to be major prey items. Because Norway rats are much more tolerant of water than are Pacific rats, coastal and freshwater birds (group 3) would be at greater risk.

2.5.2e. Feral Pig. Any species of petrel breeding on accessible terrain in all but the highest rainfall areas (King, 1990).

2.5.2f. Feral Cat. Vertebrates from the size of small skinks *Oligosoma* spp. to adult rabbits *Oryctolagus cuniculus,* and can certainly kill birds of up to 2000 g, such as adult kakapo (Powlesland *et al.,* 1995).

2.5.2g. House Mouse. Very small petrels (including storm petrels), wrens, and many of the geckos and skinks are likely to be vulnerable. Indirect effects could include competition with some of the smaller vertebrates for invertebrates and seeds.

2.5.2h. Black Rat. Most taxa living on or above the ground, and up to 200 g in body mass, that had survived predation by Pacific and Norway rats. Species with larger eggs than those taken by the Pacific rat. After accidental introduction on Big South Cape Island, off the southwest coast of Stewart Island, in the early 1960s (Bell, 1978), black rats were responsible for the extinction of the snipe (*Coenocorypha iredalei*), the saddleback (*Philesturnus carunculatus*), the bush wren (*Xenicus longipes*), the robin (*Petroica australis*), and the fernbird (*Bowdleria punctata*).

2.5.2i. Ferret. Ferrets are known to prey on the eggs and young of wading birds in open country. They also take other ground-nesting birds such as the weka *Gallirallus australis,* but do not climb well and so are not likely to be serious predators of arboreal species (Beauchamp *et al.,* 1998).

2.5.2j. Stoat. Hole-nesting species (O'Donnell, 1996; Wilson *et al.,* 1998), including birds to the size of adult female kaka (*Nestor meridionalis*), with body mass at least three times that of the stoat, in the nest (Beggs and Wilson, 1991). Stoats can hunt on the ground and in burrows. Size is probably the only defense for birds, but the lower limit for vulnerability probably takes in all existing species. Pigeons (*Hemiphaga novaeseelandiae*) are taken now, so in the past prey up to at least 600 g would have been vulnerable. The stoat is strongly arboreal and would be able to reach all taxa available to black rats. In addition, stoats swim well, including in the sea, and could reach most species available to the Norway rat.

2.6. Dates of Extinction

Definitions

The definitions of extinction, extirpation, and range reduction used here are those set out by Steadman (1997). "European contact" is taken to have begun with the first visit of Lieutenant James Cook in 1769 and 1770. None of Abel Tasman's men reached shore in 1642. "Functional extinction" is the time when a species disappears from the fossil record or historic reports of its existence cease. The actual time of extinction is ever elusive. For species known only from fossils, an age estimate accurate to within a century is the most one can expect even under the most favorable dating circumstances. Even with species surviving into the European era, it is difficult to assign a true year of extinction (see MacPhee and Flemming, this volume). Instead, the year of the final report, or some extrapolated date between final report and a full review, or the present, must be adopted as the year of "functional extinction."

Few modern taxa have ever been rediscovered in viable numbers more than 20 years after the last confirmed report. Some famous cases of rediscovery have occurred, such as that of the South Island takahe (*Porphyrio hochstetteri*) in 1948 (Reid, 1974). Until recently, the species continued to decline even under intensive management.

3. Avifaunal Extinctions

At least 77 breeding bird species have become extinct on one or other of the main islands of New Zealand in the past 2000 years (Table 2). For some, extinction dates are known, albeit roughly; for most of the others, only broad ranges or limiting dates are available. Although the number of extinctions gives some idea of the faunal change accompanying human occupation of New Zealand, many other taxa underwent major contractions in range and numbers during the same period without becoming extinct (Table 3). Many species are now threatened or vulnerable. Most of the relicts that presently require intensive management would have become extinct without human intervention. Extinct species, or populations of species that have been extirpated or greatly reduced on offshore islands are not listed. The full magnitude of the faunal changes is not evident in the extinction data alone.

3.1. Extinction Rates

Adequate samples for statistical analysis of extinction rates were obtained by pooling the data for all volant terrestrial species. Extinction rates (species lost/1000 years) for pe-

Table 2. New Zealand Birds Either Globally Extinct, or with Extinct Populations on One or More of the Main Islands, since the End of the Pleistocene[a,b]

	Island			
Taxa	North	South	Stewart	Chatham
Anomalopteryx didiformis	X	X	—	—
Megalapteryx didinus	—	X	—	—
Pachyornis elephantopus	—	X	—	—
P. australis	—	X	—	—
P. mappini	X	—	—	—
Emeus crassus	—	X	—	—
Euryapteryx geranoides	X	X	X	—
E. curtus	X	—	—	—
Dinornis struthoides	X	X	—	—
D. novaezealandiae	X	X	—	—
D. giganteus	X	X	—	—
Apteryx sp. "Eastern"	—	X	—	—
A. owenii	1890	1970?	—	—
Podiceps cristatus	X	B	—	—
Poliocephalus rufopectus	B	1950	—	—
Puffinus spelaeus	—	1556–1858[c]	—	—
P. gavia	X	1634–1710[d]	—	—
P. assimilis	X	—	—	X
Pelecanoides urinatrix	X	X	X	X
P. georgicus	—	—	1291–1412[d]	X
Procellaria parkinsoni	1960s	1960s?	—	—
Pachyptila turtur	X	X	?	X
Pterodroma cooki	1930s?	1970s?	?	—
P. inexpectata	1960s?	1910s?	?	?
Storm petrel(s)[e]	X	X	?	?
Eudyptes n. sp.	—	—	—	X
Ixobrychus novaezelandiae	X	1900	—	—
Cygnus atratus[f]	X	X	X	X
Cnemiornis gracilis	X	—	—	—
C. calcitrans	—	X	—	—
Anas chlorotis	B	B?X?	1980s	X
Euryanas finschi	X	1281–1396[g]	—	—
Malacorhynchus scarletti	X	X	—	—
Pachyanas chathamica	—	—	—	X
Mergus australis	X	X	X	X
Biziura delautouri	X	X	—	—
Circus eylesi	X	X	—	—
Harpagornis moorei	—	X	—	—
Falco novaeseelandiae	B	B	B	X
Coturnix novaezealandiae	1870	1875	—	—
Gallirallus modestus	—	—	—	X(1900)
G. dieffenbachii	—	—	—	1840
Capellirallus karamu	X	—	—	—
Diaphorapteryx hawkinsi	—	—	—	X
Gallinula hodgenorum	c1769[h]	X	—	—
Porphyrio mantelli	1890?	—	—	—
Fulica chathamensis	—	—	—	X

(continued)

Table 2. (*Continued*)

Taxa	Island North	South	Stewart	Chatham
F. prisca	X	X	—	—
Aptornis otidiformis	X	—	—	—
A. defossor	—	X	—	—
Charadrius obscurus	B	1850s?	B	—
C. novaeseelandiae	1880[i]	1880[i]	?	X(B)
Coenocorypha barrierensis	(1870)	—	—	—
C. sp. A	X	—	—	—
C. sp. B	—	X	—	—
C. iredalei	—	—	X(1965)	—
C. chathamica	—	—	—	X
Sterna nereis	B	1870s	—	—
Nestor sp.	—	—	—	X
Strigops habroptilus	1906	c1980	1990s	—
Cyanoramphus novaezelandiae	1960s	1990s?	B	B
C. forbesi	—	—	—	X(B)
Sceloglaux albifacies	c1892	1914	1880s	—
Aegotheles noveaezealandiae	X	X	?	—
Xenicus longipes	1880?	1900?	1960s?(1965)	—
Traversia lyalli	X	X(1894)	—	—
Pachyplichas yaldwyni	X	X	?	—
Dendroscansor decurvirostris	—	X	—	—
Bowdleria rufescens	—	—	—	X(1900)
Notiomystis cincta	1875(B)	—	—	—
Anthornis melanocephala	—	—	—	X(1906)
Callaeas cinerea	—	1960s?	1940s?	—
Philesturnus rufusater	1940s?(B)	—	—	—
P. carunculatus	—	1940s?	1940s?(B)	—
Heteralocha acutirostris	1900s	—	—	—
Turnagra tanagra	1900?	—	—	—
T. capensis	—	1900?(1894)	—	—

[a]Time as years AD; —, no evidence for former breeding population; B, breeding population extant. Moas all assumed to be extinct by 1600. Dates for other extinct species are 68% confidence limits on the latest calibrated date for the last radiocarbon determination available; X, formerly breeding (from fossil records) but extinct before European settlement and no radiocarbon ages available; ( ), survived on small offshore island until year shown, but extinct on main island before European settlement; ?, former breeding presence uncertain, but certainly now extinct. Taxonomy based on Holdaway *et al.* (in press).
[b]Turbott, E. G. 1967. *Buller's Birds of New Zealand.* Whitcombe & Tombs, Christchurch; Turbott, E. G. (Convener). 1990. *Checklist of the Birds of New Zealand and the Ross Dependency, Antarctica,* 3rd ed. Random Century and Ornithological Society of New Zealand, Auckland.
[c]Worthy and Holdaway (1993).
[d]Author's unpublished data.
[e]"Storm petrels" includes at least *Garrodia* (*Oceanites?*) *nereis, Pelagodroma marina,* and *Oceanites* cf. *oceanicus.* The last may be the taxon names *Oceanites maorianus,* named by Oliver (1955) on the basis of three skins obtained in the nineteenth century, supposedly off the eastern coast of New Zealand. Fossil material of an *Oceanites* distinct from *O. nereis* has been collected recently. The first bones were recorded as *Fregetta tropica* in Worthy and Holdaway (1993). Comparisons have not been completed (T. H. Worthy, personal communication).
[f]Species extinct before European arrival, reintroduced naturally and by Europeans in the nineteenth century, formerly treated as an endemic species, see Worthy (1998b).
[g]Author's unpublished data.
[h]Cassels (1988).
[i]Davis (1994).

Table 3. New Zealand Bird Species Whose Populations on the North, South, Stewart, and Chatham Islands Have Undergone Major Range Contractions during the Late Holocene[a]

Species	Island			
	North	South	Stewart	Chatham
Puffinus griseus	X	X	X	X
Puffinus huttoni	—	X	—	—
Pterodroma macroptera	X	—	—	—
Pterodroma axillaris	—	—	—	X
Pterodroma magentae	—	—	—	X
Megadyptes antipodes	?	X	X	—
Eudyptes pachyrhynchus	?	X	X	—
Morus serrator	—	X	—	—
Leucocarbo carunculatus	X	X	?	—
Stictocarbo punctatus	X	X	?	—
Egretta sacra	?	X	—	—
Anas chlorotis	X	X	—	E
Falco novaeseelandiae	X	—	—	—
Gallirallus australis	X	X	—	—
Himantopus novaezelandiae	E	X	—	—
Charadrius obscurus	X	E	X	—
Coenocorypha pusilla	—	—	—	E(B)
Sterna nereis	X	E	—	—
Nestor meridionalis	X	X	—	—
Cyanoramphus auriceps	X	X	—	—
Cyanoramphus forbesi	—	—	—	X
Cyanoramphus novaezelandiae	E	X/E	—	X
Cyanoramphus malherbi	E?	X	—	—
Xenicus gilviventris	E?	X	—	—
Bowdleria punctata	X	X	—	—
Mohoua ochrocephala	—	X	E	—
Petroica traversi	—	—	—	E(B)
Prosthemadera chathamensis	—	—	—	X
Callaeas wilsoni	X	—	—	—

[a]Populations that have become extinct are not included.

trels, flightless birds, freshwater and coastal birds, and terrestrial volant birds for both islands were then compared over the periods of human occupation before (AD 1300–1770) and after (AD 1770–1990) European contact (Table 4).

During the Polynesian period, petrels and flightless birds had high rates of extinction, and both groups had the same rates on each island (petrels, $\chi^2 = 1.684$, n.s.; flightless, $\chi^2 = 0.5143$, n.s.). The extinction rate for flightless birds was about twice that for petrels before 1770. Terrestrial volant birds had the same extinction rate on both islands, but it was about half that experienced by the petrels. In contrast, freshwater and coastal birds were lost at twice the rate (19 species/1000 years) in the North Island as in the South ($\chi^2 = 3.571$, $p = 0.05$).

The total loss rate was the same for both islands (Table 4). Assuming an exponential loss of species through time, the North and South Island avifaunas both declined at 0.09% per year for the 470 years before European contact. It is unlikely that the rate remained constant throughout the period, as there is evidence (Anderson, 1989b) that the large flightless

Table 4. Extinction Rates (Species/1000 Years) for the New Zealand Avifauna, Assuming All Taxa Extinct after Polynesian Settlement in AD1300[a]

	Time span			
	AD1300–1770		AD1770–1990	
	North	South	North	South
Petrels	15	23	9	18
Flightless	32	38	14	9
Freshwater and coastal	19	9	18	32
Terrestrial and flying	9	9	32	27
Total	74.5	78.7	72.7	86.4
(early loss compensation)	78.7	81.8	83	77.3
% loss a^{-1}	0.09	0.09	0.15	0.12

[a]North, North Island; South, South and Stewart Islands. Early loss compensation, taking into account extreme reduction of *Porphyrio* spp. and *Strigops habroptilus* on North and South Islands, and scoring them as extinct before 1770, as could also be done for several other species such as the North Island population of *Apteryx owenii* that reached the nineteenth century as extreme relicts.

birds went extinct relatively early in the Polynesian period. Others, such as the large penguins and larger waterfowl, appear only in the earliest middens (Worthy, in press).

After European contact, the rate of extinction of flightless birds was again similar for the two main islands (before, $\chi^2 = 0.5143$, n.s.; after, $\chi^2 = 1.087$, n.s.). The rate for petrels, however, was somewhat higher for the South Island, but not significantly so ($\chi^2 = 3$). Both the freshwater and coastal and the volant terrestrial groups experienced different rates of extinction on the two islands. Losses among the aquatic species were faster for the South Island ($\chi^2 = 3.92, p < 0.05$). Conversely, volant terrestrial birds went extinct at a lower rate on the South Island ($\chi^2 = 4.592, p$ 0.05).

Comparison of extinction rates for the four groups within each island before and after 1770 shows that petrels suffered the same rate of extinction in both periods (North, $\chi^2 = 1.5$, n.s.; South, $\chi^2 = 0.61$, n.s.). The extinction rate for flightless birds dropped significantly for both islands, however, after 1770 (North, $\chi^2 = 6.635, p < 0.01$; South, $\chi^2 = 17.894, p < 0.005$). Fewer flightless species were left to go extinct. Some survivors, such as the two *Porphyrio,* survived in numbers that would have been next to invisible in the fossil record. For freshwater and coastal birds, the loss rate increased significantly for the South Island after 1770 ($\chi^2 = 12.902, p < 0.005$), but remained constant for the North Island ($\chi^2 = 0.027$, n.s.). The rate for the South Island increased from half to nearly twice that for the North Island (Table 4). Both islands lost terrestrial volant species at increased rates after 1770 (North, $\chi^2 = 12.902, p < 0.005$; South, $\chi^2 = 9, p = 0.005$).

The changes in extinction rate for the four groups after 1770 can be interpreted in terms of the probable effects of the predators present in each period. Taking the volant terrestrial species first, the prediction that they should have experienced relatively little predation from Pacific rats and humans seems to be borne out by the extinction rate before 1770 being half that of the petrels and flightless birds in both islands, and of the freshwater and coastal birds in the North Island. After 1770, the extinction rate for the volant terrestrial birds on both islands trebled (Table 4) to equal the rates for flightless birds before 1770. When new predators that could climb well arrived, the ecological refuge of being arboreal and able to fly away was no longer available. In the presence of the new predators, but not before, the volant terrestrial birds suffered at the same rates as the flightless birds and petrels had before.

The difference in extinction rates between flightless birds and petrels before 1770 may reflect the two-pronged attack on the flightless birds (by humans and rats) as opposed to the petrels, which seem to have largely escaped human predation on the main islands before the nineteenth century. Few remains of small petrels have been recovered from pre-European midden sites in either island, even though the bones of much smaller birds are routinely identified in the same deposits (Worthy, in press). The smaller petrels may well have been in decline from predation by Pacific rats well before Polynesian settlement (Holdaway, 1996). Freshwater and coastal birds were affected similarly through time on the North Island. When humans and Pacific rats were augmented by the suite of "European" mammals, the rate on the South Island trebled (Table 4). The relatively low total losses for the aquatic birds in comparison with the observed high rates of extinction reflect their abundance in the original fauna. The *change* in extinction rate on the South Island may indicate that lower populations of humans for most of prehistory, and of rats were less efficient at affecting bird populations. Numbers of humans and rats were restricted by the higher topographic relief and consequent higher rainfall in the parts of the island not deforested early in the Polynesian period. The predators brought by Europeans were apparently not as affected by the environmental factors and spread throughout the South Island in numbers within a few decades of their introductions.

The apparent increase in percentage loss rate after 1770 is somewhat misleading. Several taxa such as *Porphyrio mantelli, Strigops habroptilus* (North Island population), and *Ixobrychus novaezelandiae* that became extinct after 1770 were apparently in long-term decline before then. If the first two species are reallocated to the 1300–1770 suite of extinctions (Table 4, "early loss compensation"), the loss rates before and after 1770 are more nearly similar. Several other relict taxa could also be reassigned.

A second factor biasing the percentage loss rate after 1770 is the present lack of knowledge of the real time frame of the extinctions before 1770. Most of the species could well have gone extinct over a much shorter period than the 470 years used here, perhaps even over as little as 200 years, judging from the faunal representation in Polynesian midden sites (Worthy, in press). The instantaneous rate before 1770 would then be rather higher than the rate during European time.

In summary, the survey of extinction rates, even without consideration of body size of the extinct versus the surviving species, supports the view that predators were major factors in the New Zealand avifaunal extinctions. The patterns observed are not explicable in terms of models of climate change or general anthropogenic damage to the vegetation.

3.2. Patterns of Extinction with Respect to Body Mass and Egg Length

If the extinction event affected taxa regardless of their susceptibility to predation by the predators present, then all groups should have experienced similar levels of extinction over the whole period of human settlement. The species becoming extinct should not be richer in those most susceptible to predation. In fact, contrary to the prediction that absence of predation pressure should result in random extinctions throughout the vulnerability groups, patterns of extinction and survival did differ among the vulnerability groups. Only in the freshwater and coastal species was extinction not related to the body mass and egg length of the birds (Figs. 2–7).

In the petrels, all four surviving species were among the largest in the group (Fig. 2). Petrels are relatively helpless on the ground, and the eggs and chicks are left unguarded for long periods (Warham, 1990). Another factor, suggested by Imber (1975), is that smaller petrels would be more vulnerable to rats because they spend a greater proportion of their life cycle within the size range accessible to rats. All except one of the extinct species fell within the range of possible prey for the Pacific rat. *Pterodroma inexpectata,* the largest extinct species, has a body mass well above that of the rat, and its egg is marginally longer than the critical egg length as presently known (Fig. 2). The body mass of two extinct taxa was above that for which adults were likely to have been affected, except in areas such as the northern offshore islands where Pacific rats have the largest body masses of any population in New Zealand or elsewhere (King, 1990).

Only a few flightless birds in the middle of the size range, kiwis and the weka, have survived to the present (Fig. 3). Apart from the weka, all taxa whose eggs were in the size range for consumption by the Pacific rat went extinct. So did all taxa above 1000 g body mass, except the nocturnal kiwis. The kakapo (*Strigops habroptilus*), a very large nocturnal parrot, had too great a body mass for the adults to be susceptible to predation by Pacific rats, but the eggs were small enough. Its egg length was well below the expected egg length for body mass for this group of species, which was phylogenetically very diverse.

The pattern for the coastal and freshwater group was rather different (Fig. 4). Only a small proportion of the members of the group went extinct, and extinct species were of a range of body masses (Fig. 4). Several taxa went extinct on one island only; most of these were retractions of range toward the south. For example, the yellow-eyed and Fiordland crested penguins (*Megadyptes antipodes* and *Eudyptes pachyrhynchus*) and the king shag *Leucocarbo caruncalatus* breed now only on the South Island and adjacent islands. How-

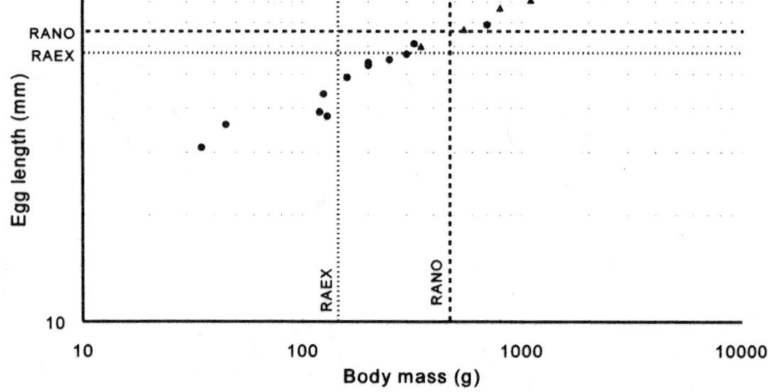

FIGURE 2. Representative egg length (mm) versus representative body mass (g) for species of petrels (Procellariiformes) breeding on the main islands of New Zealand before human contact. ▲, extant; ●, extinct; +extinction one island. RACX, RANO: values for mass, and vulnerable egg length for *Rattus exulans* and *R. norvegicus,* respectively.

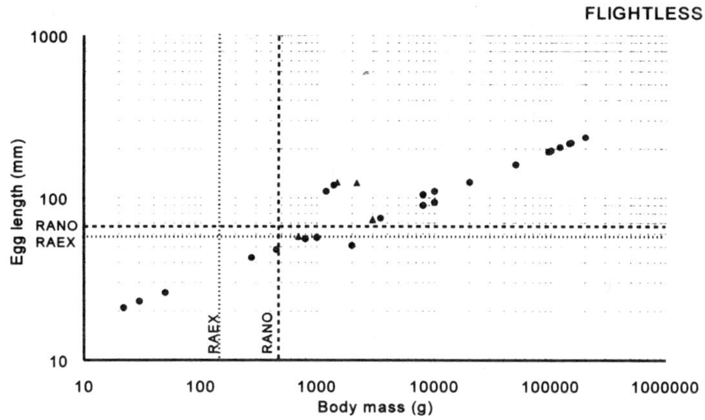

FIGURE 3. Representative egg length (mm) versus representative body mass (g) for species of flightless birds from various orders (see Appendix 1) breeding on the main islands of New Zealand before human contact with the archipelago. Symbols as in Figure 2.

ever, the dabchick *Poliocephalus rufopectus* is now extinct on the South Island. The extinctions in this group are mainly in the waterfowl, species that frequented coastal lagoons and inland lakes, and which were vulnerable to hunting by Polynesians. Most species in the group survive somewhere in the archipelago. Only two of the seven species below both the body mass and egg size limits for Pacific rat predation went extinct, one on one island only. In addition, three of the ten below the egg size limit for the Pacific rat, and below the body mass limit for the Norway rat are extinct, but only one of those is extinct on both islands. In all, only just under 12% of the taxa had become extinct on both of the main islands.

Arboreal (nonpredatory) birds suffered relatively little overall (Fig. 5). Mostly the larger taxa were affected, including the raven, most of the wattlebirds, and the piopios. Five of ten taxa above 100 g body mass are extinct. In contrast, only 4 of 19 less than 100 g went

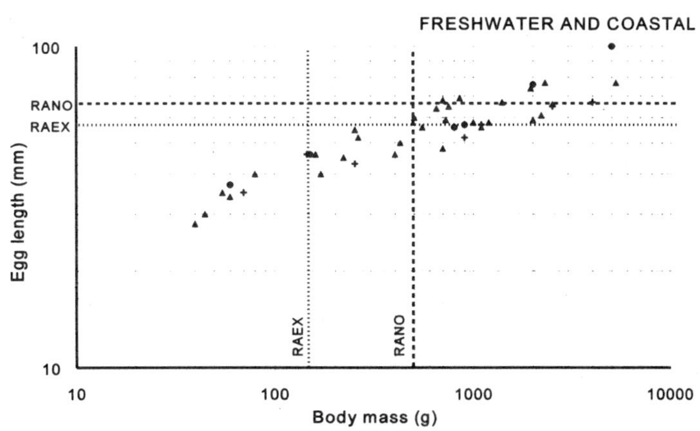

FIGURE 4. Representative egg length (mm) versus representative body mass (g) for species of coastal and freshwater birds (see Appendix 1) breeding on the main islands of New Zealand before human contact. Symbols as in Figure 2.

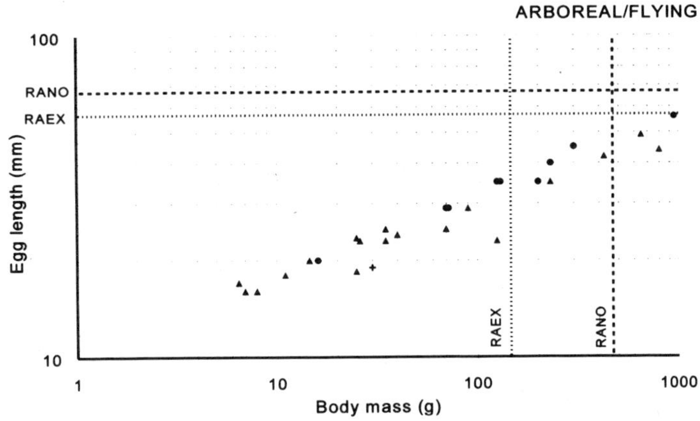

FIGURE 5. Representative egg length (mm) versus representative body mass (g) for species of flying arboreal birds (see Appendix 1) breeding on the main islands of New Zealand before human contact. Symbols as in Figure 2.

extinct, with an additional single island extinction. Of 19 taxa within the Pacific rat range, there were four extinctions plus a single island extinction (Fig. 5). Two of the five taxa within the range of the Norway rat (and therefore presumably the ship rat) went extinct. The largest extinct species, the raven *Corvus moriorum,* was, at an estimated 900 g, too large for the Pacific rat to attack as an adult, but the Pacific rat could have opened its egg.

Of the four extinct ground-living flying birds, three were *Coenocorypha* snipe, which are not water birds, as the surviving taxa all live in forest, shrubland, or tussock grasslands (Heather and Robertson, 1996). The fourth was the New Zealand quail *Coturnix novaezealandiae* whose eggs fell within the range of the Pacific rat. Survival in the presence of predation may depend ultimately on fecundity (Moors, 1983). The quail laid a clutch of 10–12 eggs (Oliver, 1955). The three surviving taxa are the New Zealand pipit (*Anthus novaeseelandiae*), the rock wren (*Xenicus gilviventris*), and the fernbird (*Bowdleria punctata*).

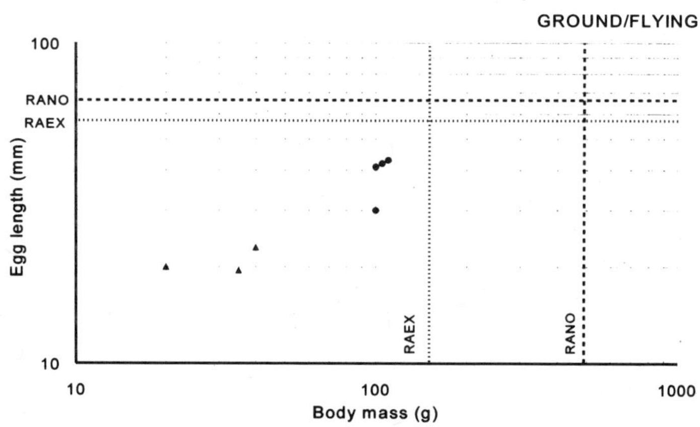

FIGURE 6. Representative egg length (mm) versus representative body mass (g) for species of ground-living, flying birds (see Appendix 1) breeding on the main islands of New Zealand before human contact. Symbols as in Figure 2.

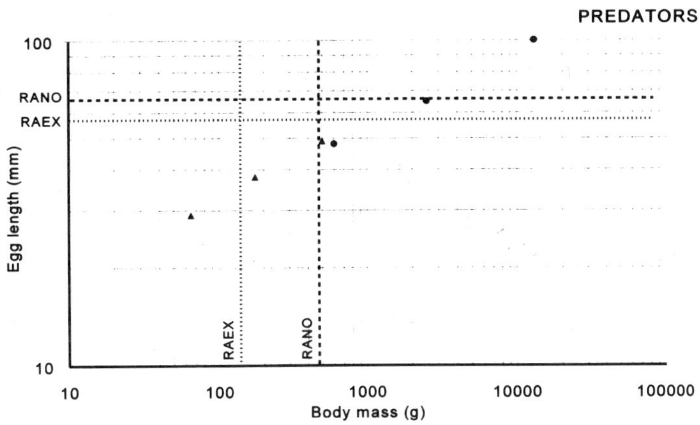

FIGURE 7. Representative egg length (mm) versus representative body mass (g) for species of avian predators (forest kingfishers, owls, raptors) (see Appendix 1) breeding on the main islands of New Zealand before human contact. Symbols as in Figure 2.

The second two live in dense closed habitats including rock piles, dense shrubland, and marsh vegetation; the first is a strong-flying songbird that has two or three clutches in a season.

Only the smallest and most generalist predatory birds survived. Haast's eagle (*Harpagornis moorei*) preyed on large birds, including the giant moa *Dinornis giganteus.* Its decline almost certainly followed closely the reduction of its main prey species. The very large

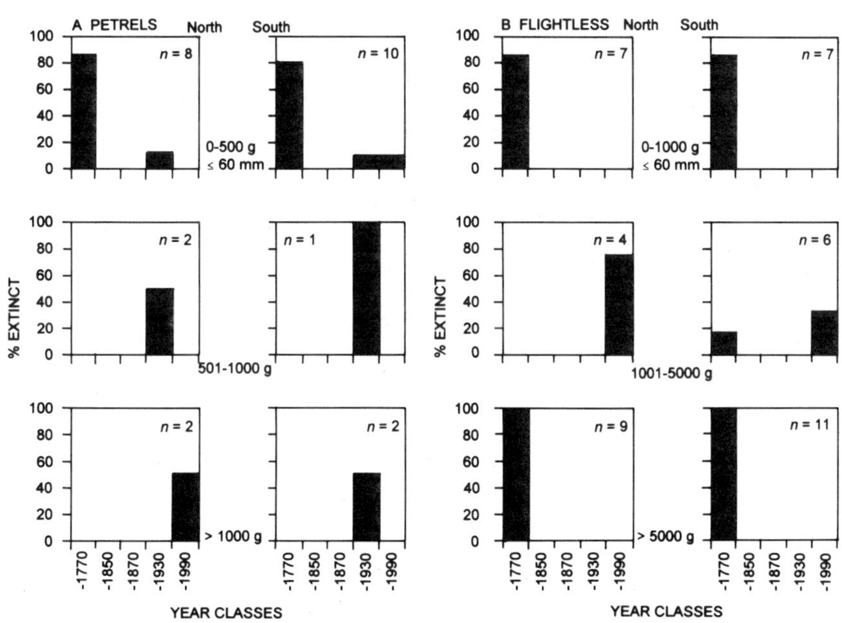

FIGURE 8. Percentage of original (prehuman contact) breeding species extinct in time periods ending at dates shown, grouped by body mass and egg length. A, petrels; B, flightless birds. North, North Island; South, South Island. -, denotes period before year.

Eyles's harrier (*Circus eylesi*) preyed on smaller birds, such as pigeons (*Hemiphaga novaeseelandiae*) and kokako (*Callaeas* spp.), so its decline is harder to explain, as the principal prey species were abundant into the nineteenth century. The laughing owl (*Sceloglaux albifacies*), a generalist feeder on forest species, made the switch to eating Pacific rats when they were available and the indigenous prey declined (Holdaway and Worthy, 1996).

3.3. Patterns of Extinction through Time

Within each vulnerability group, taxa were grouped by extinction dates within the ranges of years AD: (a) before 1770, (b) between 1770 and 1850, (c) between 1850 and 1870, (d) between 1870 and 1930, and (e) between 1930 and 1990 (Appendix 3). Broad ranges were used because the dates of extinction are seldom known with precision. Date ranges were based on the time of arrival of the main predators or groups of predators (see Section 2.4). Percentages of the original (prehuman) fauna in each group and size class becoming extinct in each of the time ranges were calculated, and the percentages (Figs. 8 and 9) and cumulative percentages (Figs. 10–12) of extinctions over time were plotted against the time ranges.

By 1770, all but two of the petrels in the small petrel class were extinct on the three main islands (Appendix 3). The last of the small class vanished before the members of the

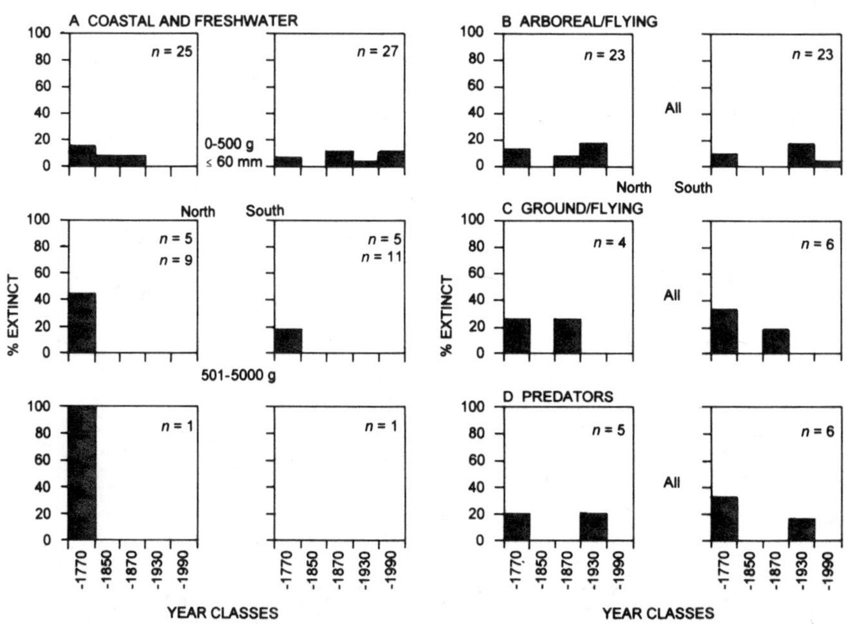

FIGURE 9. Percentage of original (prehuman contact) breeding species extinct in time periods ending at dates shown, grouped by body mass and egg length (total for predators). A, coastal and freshwater birds; B, arboreal flying birds; C, ground-living flying birds; D, predatory birds. North, North Island; South, South Island, -, denotes period before year.

two larger size classes. The small taxon that survived longest was the largest taxon of that class, the Cook's petrel (*Pterodroma cookii*) (Appendix 3). There was a clear sequence in the extinction dates for the petrels, smallest first, with the medium-sized taxa becoming extinct before the largest ones, at least on the North Island. The distinction was not seen in the South Island data, within the limits of the accuracy of the determination of extinction dates.

The pattern of extinction through time for the different size classes of flightless birds differed from that of the petrels in that there was no sequence of size through time. The smallest and largest taxa on both islands suffered most extinctions before 1770 (Fig. 8B). On the North Island, all extinctions in the intermediate size class occurred in the most recent period. On the South Island, one taxon (the undescribed eastern kiwi *Apteryx* sp.) apparently became extinct before the European era (Appendix 3), but all other extinctions in that size class were, as on the North Island, in the period from 1930 to 1990 (Fig. 8B). No extinctions of flightless birds were recorded in the intermediate time periods.

For coastal and freshwater birds, the pattern of extinction through time was different again. After European contact, aquatic birds should have been affected first when the Norway rat became established, with continuing extinction of ground-living birds. Only after 1850 would arboreal species be at risk, with the advent of black rats and stoats. Indeed, in this group a significant number of the extinctions of small taxa took place in the periods after 1770, although the timing differed between the two islands (Fig. 9A). The extinctions on the North Island occurred before those on the South. All extinctions in the two larger size classes have to date occurred only before 1770. The North Island extinction in the larger size classes have to date occurred only before 1770. The North Island extinction in the largest size class includes only the extinction on that island of the yellow-eyed penguin *Megadyptes antipodes* (Appendix 3), which is present, but endangered, on the southeastern South Island. Although its presence on the north Island as a breeding species in prehistory is not confirmed, it was present on the opposite shore of Cook Strait (Worthy, 1998b) and it is reasonable to assume that it will be discovered among the material from North Island sites when these are reanalyzed. Large penguin bones from the Washpool site in Palliser Bay attributed to *Eudyptes sclateri* (Leach, 1979) are likely to be of this species.

All of the arboreal (nonpredatory) species fit in the lowest size class, but there were size distinctions that were important within the size class (Fig. 5). Extinctions were spread through the time periods, but the more recent began earlier on the North Island, and finished later on the South Island (Fig. 9B). Extinctions in the remaining groups of flying birds, the ground-living species and the predators (Appendixes 1 and 3) occurred before European contact and again after the introduction of predators by Europeans. The single species of predator that became extinct after European arrival declined after the introduction of the larger and more capable mustelids, and at the same time on both islands, whereas the smaller ground-living species declined rapidly in the middle of the nineteenth century, when only the rodents and cats were present.

3.4. Accumulated Losses, and Prey Available to New Predators

The cumulative effects of the extinctions on the species-level diversity of the groups and the fauna as a whole are shown in Figs. 10–12. The data show that the event was well established by the time of European contact. The time base is uneven, and there is much information hidden in the pre-1770 bin that will not be fully apparent until a much more detailed chronology of the extinction event is available, but the pattern is clear.

The time lag in the extinction of petrels, smallest vanishing first, in relation to size can be seen in Fig. 10A. The effect is seen more clearly in the North Island fauna. In the South Island, the complete extinction of the middle size class is the best indication that such an effect may have been present. The almost total collapse of the small petrel fauna before 1770 is reflected in the steepness of the curve before that date. The bias in scale of the date classes greatly exaggerates the rapidity of extinctions before European contact, but the figures do emphasize the magnitude of the losses when the only known outside influences on the fauna were Polynesians and the Pacific rat. The actual track of the cumulative extinction curve for the pre-1770 period is the subject of ongoing research, to see if temporal resolution can give some indication as to which of the possible factors was more important. Results of reanalysis of faunal representation in archaeological sites on both main islands suggest that even early in the settlement period the small petrels were not available to Polynesians in the numbers that the fossil evidence suggests were present within the preceding millennium (Worthy, in press).

The overwhelming impact on the fauna before European contact can be seen in the cumulative totals for the flightless species (Fig. 10B). On both islands, the smallest taxa were decimated, presumably by rats, and the largest were eliminated, presumably by humans. Much later, the medium-sized flightless taxa declined, later and more drastically on the North Island, but total percentage losses were at least 60% on both. Flightless birds in the size class of kiwis survived while all of those in the moa size class, and almost all of the wrens, disappeared.

More limited, and mostly more recent, impacts were noted in the coastal and freshwater avifauna (Fig. 11A). The initial (pre-1770) decline was much less, in small taxa in

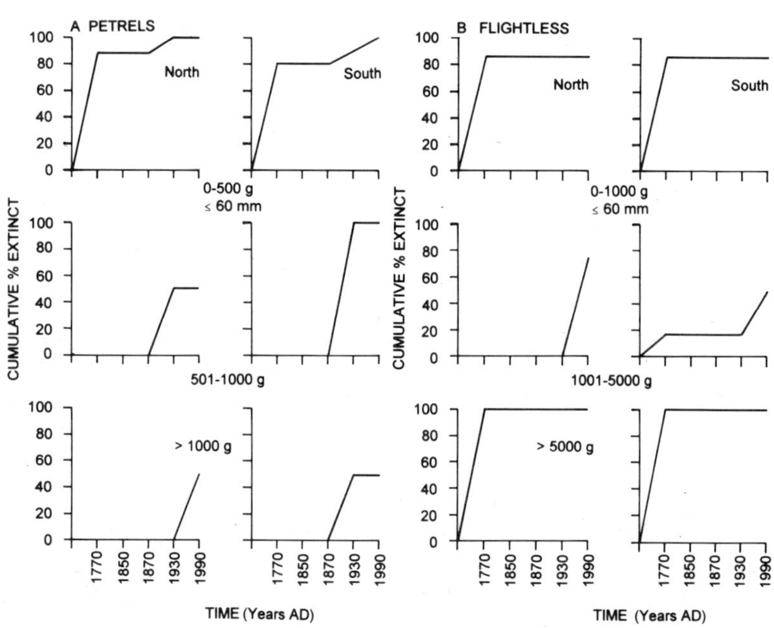

FIGURE 10. Cumulative percentage of original (prehuman contact) breeding species extinct at dates shown, grouped by body mass and egg length. A, petrels; B, flightless birds. North, North Island; South, South Island.

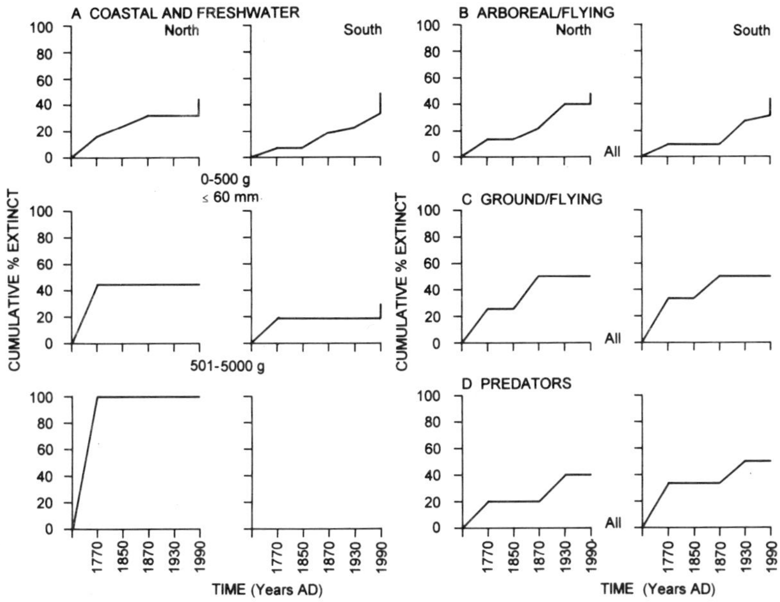

FIGURE 11. Cumulative percentage of original breeding species extinct at dates shown, grouped by body mass and egg length (total for predators). A, coastal and freshwater birds; B, arboreal flying birds; C, ground-living flying birds; D, predatory birds. North, North Island; South, South Island.

the coastal and freshwater fauna and the losses accumulated at different rates on the two islands. The total for each must be added to the taxa that are presently endangered (vertical at the right-hand end of the curves), and the total losses are similar on the two islands, at just over 40% of the original fauna. Although the numbers are small, there is some indication that losses on the North Island after 1770 started a little earlier than on the South Island, where there was a brief stasis after European contact. For the larger size classes, the initial losses before European contact were not added to later, which reflects the dominance of such species in the mainland fauna at present. The larger South Island representatives of this vulnerability group suffered less overall than the North (Fig. 11A).

The pattern for the arboreal volant taxa was similar for all three volant groups (Fig. 11B–D). In the arboreal taxa, there is a hint of a lag between the North and South Islands, with a renewed series of extinctions after 1870 on the South Island, whereas the climb in the losses began after 1850 on the North Island (Fig. 11B). Losses of arboreal species were minimal for both islands before 1770. Although the magnitudes are somewhat deceptive because of the low numbers of species involved, the ground-living flying taxa and predators experienced similar, two-stage extinctions which occurred later in the predators than in the ground birds (Fig. 11C,D).

All size classes of petrels and ground birds bore the brunt of the extinctions in the New Zealand avifauna (Figs. 11B–D, 12A–C). Both groups have been eliminated as significant components of the mainland fauna. The survival of several small taxa in the petrels and ground-living groups on offshore islands has obscured the magnitude of the losses in these groups where they have been exposed to introduced predators, in studies in which the small island faunas are pooled with those of the main islands. Even so, the magnitude of the loss-

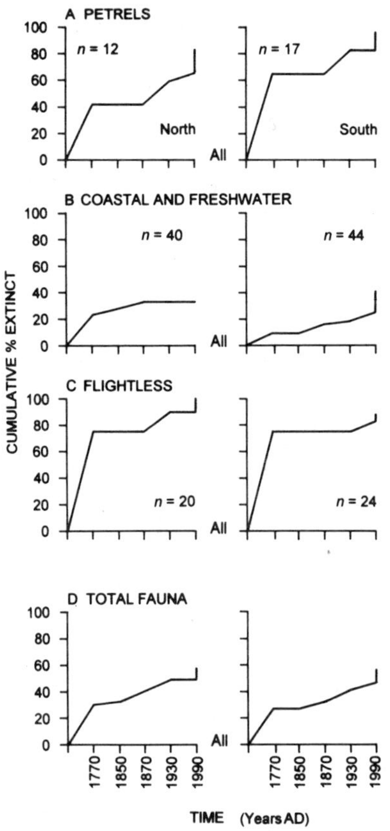

FIGURE 12. Cumulative percentage of original breeding species extinct at dates shown. A, petrels; B, coastal and freshwater birds; C, flightless birds; D, total avifauna. North, North Island; South, South Island.

es in the least affected groups (coastal and freshwater and arboreal) is still very large, approaching or exceeding 50% on both islands if endangered taxa are counted.

In total, the number of breeding species on the main islands has been reduced by about 50% in the past 2000 years, and the value approaches 60% when endangered taxa are included (Fig. 12D). The severe declines before 1770 show that most of the extinctions took place when Polynesians and Pacific rats were the only novel influences in the environment. Losses since European contact approach those of pre-1770 time in the petrels from the North Island and in coastal and freshwater and arboreal avifaunas from the North and South Islands, which contained many taxa not vulnerable to predation by people or Pacific rats.

4. Discussion

The patterns observed in the extinction data for birds on the North, South, and Stewart Islands support the hypothesis that Pacific rats played a major part in the New Zealand late Quaternary extinctions. Many of the extinctions and extirpations occurred when the Pacific rat was the only widespread novel factor in the environment. The taxa most affected were those small enough to be caught and subdued by rats that lived where rats would have been most abundant. Species associated with water, arboreal species, and all taxa outside the size range in which the rat could deal with the prey (with the exception of large flightless species) were much less affected by extinction before 1770.

4.1. Vulnerability to Mammalian Predation

New Zealand birds evolved in the presence of diurnal avian predators, which hunted by sight and sound. The mammalian predators that reached New Zealand are primarily nocturnal, and hunt largely by scent. Several New Zealand birds, including kiwis, kaka and kakapo parrots, some waders, and most petrels, are at least partially nocturnal at present. Their former habits are unknown, but the presence of only diurnal avian predators (eagle, harrier) for species larger than 1000 g would have provided strong selection pressure in favor of nocturnal behavior and cryptic plumage patterns. Nocturnality was not considered directly in the analysis because it could not be scored for extinct taxa, and behavior can vary even within genera. For example, kiwi on Stewart Island are known to be at least crepuscular and partly diurnal where undisturbed, whereas mainland populations are almost exclusively nocturnal (Heather and Robertson, 1996). Where known, it is a useful a posteriori factor to be considered for the surviving taxa for which the state is known.

Hole-nesting can be scored for most taxa on phylogenetic grounds, and this is indicated in the main list (Appendix 1). Hole-nesting species, which survived the early phases of extinction, are now among the most threatened of any group in the avifauna.

4.2. Human Impact

It seems unlikely that human activity before 1770 will account for the three genera of 20- to 50-g flightless wrens that were eliminated from the main islands (Millener and Worthy, 1991). In contrast, for the majority of large and very large species, humans were the one novel factor in the environment that could have effected extinctions. Contrary to the suggestion by Fleming (1969), Pacific rats are unlikely to have affected populations of moas by predation of adults. It is far more likely that the largest flightless birds in New Zealand were eliminated by human predation or, in some taxa, habitat destruction.

Fleming based his suggestion on the observation of rats killing adult albatrosses during incubation by opening wounds on the lower back (Kepler, 1967). The predation of albatrosses occurred during a period of severe food shortage on a small coral island, a situation that is extremely unlikely to have occurred over the 265,000 km^2 of the main New Zealand islands. Pacific rats may have affected moa reproductive rates indirectly, by eliminating populations of the large invertebrates that could have been important in chick development (Holdaway, 1989), but 10 of the 11 moa taxa survived into Polynesian settlement time, in numbers sufficient to supply the raw materials for the vast middens of the eastern South Island (Anderson, 1989b).

All 11 taxa above 5000 g in the prehuman avifauna were extinct by 1770, and all but one have been found in archaeological association (Worthy, in press). The exception, the moa *Pachyornis australis,* was recognized as a valid species relatively recently (Worthy,

1989). The extinction of the moas (Dinornithiformes) can be attributed to hunting pressure from Polynesian settlers (Anderson, 1989a,b).

Indirect effects of human intervention could not have resulted in the observed patterns of extinction. The species–area relationship might explain the loss of 10% of the species on the South Island, where 50% of the forest was removed in Polynesian times (McGlone, 1989). Even then, however, other habitats persisted, and the destruction was not complete for most of the vegetation associations. Anthropogenic deforestation could have been important for flightless species typical of the eastern South Island. It may have hastened the end of some taxa, but it is unlikely to have been critical on its own in the time frame available. Obviously, the large bird populations in the areas would have been decimated, but most of the species themselves would not have been at risk.

The virtual absence of petrels from early period middens suggests that either the human population derived most of its protein from moas and seals, eschewing the smaller birds, or that the petrels were not available, at least in significant economic numbers. Early Polynesian settlers harvested any species of bird that was available in their immediate environment, including small songbirds (Worthy, in press), and Polynesians elsewhere harvested petrels, so size or taste are unlikely to have been factors in the paucity of petrel remains in middens. For the smaller species of petrel, their absence from human contexts is likely to reflect their absence or rarity in the environment at the time of Polynesian settlement. Harvesting of petrels on offshore and outlying islands is a separate issue. In the south the practice may date from early European times (Anderson, 1997a), but the northern island populations may have been exploited from early in the Polynesian period (Moore, 1973).

Several of the small taxa that went extinct inhabited geographic areas that were outside the limits of anthropogenic habitat alteration and destruction. Fossils of the long-billed wren *Dendroscansor decurvirostris* have been found only in areas where southern beech (*Nothofagus*) forest and subalpine scrubland were the dominant vegetation types (Millener and Worthy, 1991; Worthy and Holdaway, 1994a). On the South Island, these are now the principal surviving areas of natural vegetation, albeit drastically changed in the past 100 years with the introduction of deer and other herbivores. On the West Coast, where the largest tracts of forest survived into the nineteenth century, several species of small petrel went extinct, without appearing in known Polynesian midden deposits (Worthy and Holdaway, 1993; Holdaway and Worthy, 1994). Habitat change and human predation seem unlikely to have affected these species, but they were all small enough to be taken by Pacific rats (Appendix 1).

4.3. Disease or Predation?

The specificity of the early phase of the extinction (before 1770) is sufficient to preclude disease as a major factor in the extinctions. The pattern of extinction against time in the petrels would require that the disease was size-specific through time; progressively larger taxa were more susceptible the longer the disease had been present. But such a pattern could not hold for the flightless taxa, where both the smallest and the largest size classes suffered their greatest levels of extinctions before 1770. It is the intermediate-sized species that were more resilient.

Disease can also be discounted because of the phylogenetic diversity of the taxa affected, and the phylogenetic closeness of taxa that survived. For example, the flightless Lyall's and thick-thighed wrens (*Traversia lyalli* and *Pachyplichas yaldwyni*) were both abundant on the West Coast. They both became extinct there before Europeans arrived, but the bush and rock wrens (*Xenicus longipes, X. gilviventris*) and the rifleman (*Acanthisitta chloris*) could fly and survived. In the *Nothofagus* beech forest of the central chain, was a sixth wren, the flightless long-billed *Dendroscansor decurvirostris.* It also vanished before European contact. All of the wrens belong to the same endemic family, the Acanthisittidae.

The wrens were uneconomic for humans to hunt, but small petrels have been taken for food throughout the Pacific. Any disease affecting populations on the main islands may be expected to have been passed to individuals from island populations, but the mainland colonies vanished in Polynesian times, and colonies on islands without rats have survived. The small petrels on the West Coast included the endemic Scarlett's shearwater *Puffinus spelaeus,* fairy prion *Pachyptila turtur,* and storm petrels (Worthy and Holdaway, 1993; Holdaway and Worthy, 1994). The smaller species are extinct, but the larger sooty shearwater *Puffinus griseus* and the Westland black petrel *Procellaria westlandica* survive. The pattern is not explicable by disease or human endeavor.

Disease was not a factor in the extinctions after European contact either, as the same size- and habitat-based pattern was repeated. Instead, the introduction of new predators, with different search areas and prey size ranges, was reflected in the sequential loss of taxa vulnerable to the new but not the old predators in the system. The extinctions did not affect the whole spectrum of size classes and vulnerability groups. An example of a species as vulnerable to disease as any other but whose size and vulnerability level favored a late survival is the laughing owl *Sceloglaux albifacies,* which declined rapidly in the late nineteenth century. After the owl had coexisted with, and lived on, the Pacific rat for at least 1000 and probably nearer 2000 years, and with the Norway rat for 120 years, and survived any disease carried by the many species of bird introduced in the 1860s, its population collapsed within 10 years of the introduction of stoats and ferrets (Holdaway and Worthy, 1996). Both mustelids were capable of killing adult owls as well as taking eggs and chicks, at their nests in rock outcrops and in hollow trees.

Similarly, arboreal birds that survived Pacific rat predation, only to decline and vanish after the arrival of the arboreal black rat and stoat, were those that had small clutch sizes (*Turnagra* spp.), or were hole nesters (*Heteralocha acutirostris, Mohoua ochrocephala, Nestor meridionalis*), or which had lost, or never had, predator distraction displays (*Callaeas* spp.). Taxa such as the petroicid robins retain nest defense displays and the multiple clutches that allow their relatives in Australia to produce enough young to survive predation from snakes and a range of small mammals, including rodents. The arboreal taxa include representatives of several families, and all should have been equally susceptible to disease. The presence of size and habitat patterns in the data suggests that disease was not a factor in the New Zealand extinctions.

4.4. A Parallel with Hawaii?

The role of rodents in the New Zealand extinctions begs the question of the effect of Pacific rats on the faunas of other Pacific islands. Although Atkinson (1977) argued strongly that *Rattus rattus* was responsible for many extinctions of small arboreal birds in Hawaii after 1890, the effects of predators there are still underplayed. Disease, mongooses, loss of habitat, and reduced food supply have all been presented as only or major factors in the historical declines (sources in Atkinson, 1977). The declines occurred at different times on different islands, consistent with a stepwise spread of black rats from their first established population on Oahu (Atkinson 1977).

Olson and James (1992) and James and Olson (1992) showed that the historic extinctions of birds on the Hawaiian Islands were only the latest in a series that began in Polynesian times. As in New Zealand, the extinction of most of the large herbivores can be attributed directly to human predation. The loss of many small species of songbirds is more problematic. Habitat destruction and disease could also be put forward for the prehistoric extinctions, but both are untestable. The possibility that the suite of small species eliminated before European contact included most of those in the size range and with the habitat requirements that made them susceptible to Pacific rat predation should be explored as well.

Removal of the leeward forests obviously had a profound effect on the populations of some species with restricted ranges and habitat, but there was probably sufficient area of habitat for others to have survived in the absence of predation. The taxa that survived into European time were mostly those of the mountain forests and wetter areas of the islands, which were suboptimal habitats for Pacific rats. And more arboreal species are also likely to have been less affected by Pacific rats. It may be difficult to extrapolate habitats and behavioral traits for Hawaiian birds, but a survey for each island independently might yield patterns of extinction related to differences in vulnerability similar to those observed in New Zealand.

4.5. Continental Extinctions beyond Coextinction of Scavengers or Parasitism

It is well known that the major extinction event in North America at the end of the Pleistocene did not include small taxa (Martin, 1986). Although vigorous debate surrounds the cause or causes of the event, it may be significant that *Homo sapiens* was the only new predator to arrive. Previous events, such as that in the late Hemphillian, included both large and small taxa (Hallam and Wignall, 1997). Continental biotas may be buffered against the effects of single species introductions of predators when the new predator (such as the black rat) is an addition to an array of resident predators of the same size and ability. However, at some point, a species with a completely new modus operandi may appear, as has been hypothesized for the appearance of felids in South America during the Great American Interchange (Marshall, 1988).

For North America 11,000 years ago, that novel predator was in the form of humans with Clovis technology; for New Zealand it was a small rat. In New Zealand, the Pacific rat had sufficiently different attributes (nocturnality, good sense of smell, high fecundity, omnivory) that it was able to take advantage of species that evolved in the presence of diurnal, low-fecundity predators that could not reach into nests or burrows. There is a clear parallel with the effects of the introduction of the brown tree snake to Guam (Quammen, 1996).

Australia, the closest large landmass to New Zealand, has had an entirely different history of human intervention. Earliest evidence for human contact may date from about

60,000 yrbp. (Roberts *et al.,* 1994). There is some similarity with the New Zealand record in that it appears that successive waves of immigrants brought different predators, and different technologies for themselves. For example, the earliest inhabitants apparently lacked the commensal dog variety that gave rise to the dingo that is now feral throughout the continental mainland.

Present evidence indicates that most of the large herbivores went extinct perhaps 40,000 years ago (Murray, 1984). The largest predators, the thylacine *Thylacinus cynocephalus* and Tasmanian devil *Sarcophilus harrisii,* vanished on the mainland soon after the dingo (*Canis familiaris*) arrived (Murray, 1984). Both were relict on Tasmania when Europeans arrived. Apart from the predators, which seem to have been replaced by the dingo rather than its exploiting a new niche, there were apparently few or no extinctions of small mammals before Europeans arrived.

Europeans introduced a whole suite of mammalian predators to Australia, including the red fox *Vulpes vulpes* and the domestic cat. These two species have been implicated in the reduction and extermination of numerous small mammals in the past 200 years, both marsupial and placental (Smith and Quin, 1996). The situation is complicated by the large-scale alteration of dry land habitats by rabbits, stock grazing, and removal of the Aboriginal fire regimes. Further analysis may show that the stepwise introduction of predators has produced a similar pattern to that prevailing on a much smaller geographical scale in New Zealand.

4.6. Differential Vulnerability

The patterns of extinction and survival through time of bird species in New Zealand are clearly related to the different sizes and capabilities of the predators present at any one time. In addition, a reduced suite of prey species remained for each succeeding predator to exploit. King (1984) pointed out that although earlier workers had attributed bird reductions and extinctions to mustelid predation, at that time no bird had certainly been exterminated by mustelids in New Zealand. Since then the extinction (laughing owl) and great reduction (yellowhead, kaka) of some species have been firmly attributed to predation by stoats (Holdaway and Worthy, 1996; O'Donnell, 1996; Wilson *et al.,* 1998). Ferrets are thought to be major influences on the populations of the black stilt (*Himantopus novaezelandiae*) on the South Island (Heather and Robertson, 1996), and weka on the North Island (Beauchamp *et al.,* 1998). But the fact remains that the fauna that greeted the first stoats and ferrets had already been stripped of most of its small vulnerable species. Extinction is not an experimental science.

4.7. Conservation and Predation

One of the major outcomes of the analysis may be the ability to predict whether island faunas are at risk of further damage by Pacific rats or other introduced predators. In the past, it has often been assumed that Pacific rats had little or no effect on the faunas of offshore islands (e.g., Imber, 1975). Absences of smaller, vulnerable taxa have been attributed to the presence of introduced weka rails, or to other, less obvious factors such as the supposed inability of robins *Petroica australis* and saddlebacks *Philesturnus carunculatus* to cross low mountain ranges and so colonize an island such as Codfish, off the northwestern coast of Stewart Island (Blackburn, 1968).

By examining the species list from any island with Pacific or other rats, in comparison with lists from other islands in the area that lack rats, and with the possible original fauna of the island, the time of arrival of the rat may be gauged. For example, a *Coenocorypha* snipe was present on Little Barrier Island in 1870, before cats were introduced (Miskelly, 1988). The population must have been exceedingly small as only one specimen was ever collected. The presence of the snipe implies that the rat has been on the island for perhaps less than 100 years. Interpretation of the state of the populations of other birds on the island must be made in that light.

A further example is that of Codfish Island. No snipe, robin, saddleback, or bush wren *Xenicus longipes* have ever been recorded from Codfish Island (Blackburn, 1968), where the Pacific rat was abundant before extermination operations started in 1998. All four birds were abundant on nearby Big South Cape Island before black rats colonized from fishing boats in the early 1960s, after which they all went extinct (Bell, 1978). The presence of a very small colony of South Georgian diving petrels *Pelecanoides georgicus* in the dunes on Codfish Island suggests that although the Pacific rats had been on the island long enough to have exterminated the small land birds, they had not been there long enough to have entirely removed the small petrels.

Even if Pacific rats are presently apparently coexisting with species within the rat's predatory ability as defined by its body mass and the size of egg that it can break, it is no guarantee that the situation is stable. The data presented here show that if the rat is present for long enough, all birds that are vulnerable to its predation will go extinct. Prediction of the islands and species that are most at risk may allow agencies to place priorities on expensive conservation efforts.

4.8. Importance of Simplification to Observing Pattern

Previous summaries of the New Zealand avifaunal extinctions (e.g., Cassels, 1984) have endeavored to cover the whole archipelago, and to use the global extinction of a species as the operational unit. Although such analyses rightly emphasize the extent of the catastrophe that has befallen the New Zealand fauna, the complexity of the patterns of extinction and survival allows so many exceptions that any statement of causation is seriously weakened.

Cassels (1984) used endemism and size in relation to extinct or surviving relatives as parameters, in addition to flightlessness and nocturnality, by which to interpret the reasons for faunal extinctions. Although easy to score for extinct species, these are second- or third-order attributes. Predators can take a naive bird more easily, whether that naiveté is occasioned by endemicity or by its recent evolutionary history. By limiting the present study to the islands for which there are now reliable faunal data, scoring extinctions on those islands, against the background of a known set of new influences, the complexity has been reduced, which allows the patterns of loss to be more clearly seen.

With increased knowledge of the original Holocene faunas of Norfolk Island, the Kermadecs, the northern offshore islands, the Chatham Islands, the islands off Stewart Island, and the subantarctic islands, the pattern shown here for the main three islands is likely to be repeated in each instance where introduced predators have gained access. There are already indications that the fauna of the Chatham Islands was damaged by Pacific rats as well as by humans. No rats survived the crossing to Pitt Island or the other islands south of Pitt Strait. It is on those islands that the last populations of smaller species that survived the wave of European settlement in the nineteenth century persist (Holdaway, 1994).

4.9. The Structure of the Pre-1770 Event

We know little about the timing of the extinctions that took place before AD 1770. As a first step toward a chronology of the pre-European extinctions, it would be useful to employ a cutoff date of AD 1300. It is a convenient date because it is not likely to postdate the settlement event by more than a few decades, which would be invisible in the fossil record. If any taxon is found in unequivocal association with human activity, that is evidence of survival of that species to at least AD 1300. The distinction "unequivocal" is important because some of the early sites were laid down on preexisting natural deposits.

Present evidence suggests that most of the large terrestrial species survived long enough to be incorporated in Polynesian middens. Many smaller taxa are also present. Most of the extinctions must therefore have taken place in the interval between AD 1300 and 1770. It will be important to determine which, if any, of the smaller taxa unlikely to be a target of human predation became extinct before AD 1300. For longer-lived groups, such as the petrels, extinction would have taken longer if only the eggs and chicks were affected, consequently range reductions before AD 1300 would also indicate the presence of Pacific rats.

The presence at a site of very small birds, lizards, and frogs, in addition to the usual human prey species (Worthy, in press) is evidence for a composite assemblage. It is difficult to tell from most archaeological reports which material came from which component of the deposit. Midden bone does, however, usually have a distinctive appearance and any remaining in collections can be identified (Worthy, 1998b). Work on clarifying the representation of birds in archaeological sites is proceeding, but the exact origin of many specimens will never be known (Worthy, in press).

The steep rise in the 0 to AD 1770 range of the cumulative percentage extinction plots reported here could therefore be replaced by a near-horizontal line from 0 to AD 1300, when a few small exceptionally vulnerable taxa (wrens, snipe) became extinct, followed by a very steep rise when the larger species were affected by people, and smaller taxa that reached the Polynesian era in reduced numbers, finally succumbed to long-term loss of recruitment. The 470 or so years between Polynesian settlement and European contact saw the greatest loss of vertebrate diversity in New Zealand since the beginning of the Pleistocene.

5. Conclusions

There is much circumstantial evidence that humans and other predators were responsible for avifaunal extinctions in New Zealand. When there were no people, there was no measurable level of extinction. Admittedly, massive climatic shifts 10,000 to 15,000 years ago resulted in perhaps two retractions of range from the North Island (eagle and kea), but there were no known extinctions, despite considerable shuffling of vegetation patterns. After people arrived with a small predator, there were many extinctions, even without the climate change that has been mooted as an important factor in extinctions elsewhere. However, although the circumstantial evidence is persuasive, it is not conclusive.

In the search for the cause or causes of the New Zealand extinctions, counting the bodies is not enough. It is necessary to compare the temporal and spatial distribution of extinctions with the additional factors in the environment. When this is done, the picture that emerges is of successive waves of introduced predators reducing and eliminating the bird species that they could catch, that could not breed rapidly enough to replace the losses. In the long-isolated New Zealand avifauna, that included most of the terrestrial avifauna.

In New Zealand, the synchroneity of human settlement, the arrival of new predators, and the extinction of indigenous vertebrates is clear. By using size and date of extinction, the impacts of the various predators can be separated, and those of the predators separated from human hunting. The effects of climate and disease can be discounted as not fitting the observed patterns of extinction. The sequences of predator introduction followed by extinction identified in the New Zealand chronology model what may be expected on other islands in the archipelago and elsewhere, as new predators arrive.

The Pacific rat had a profound effect on the New Zealand avifauna, and presumably on the other small vertebrates and the invertebrate fauna as well. Clarification of the role of the rat in the New Zealand extinctions now leaves the question of its role elsewhere. A survey of the literature suggests that it is perhaps the most underestimated factor in the extinctions that have befallen islands throughout the Pacific. The assumption of innocence for a new predator on an island is no longer valid. Much will be lost during the process of proving each new case; most faunas cannot wait that long.

Appendix 1: Distribution on the Main Islands and Characteristics of the Holocene Avifauna of New Zealand

Part 1

Bird species breeding on the three main islands of New Zealand before human contact, with egg length ≤60 mm (within critical range for predation by *Rattus exulans*). C, freshwater or coastal taxa; P, petrel; F, flightless; Pr, predator; GF, flying/ground; AF, flying/arboreal; *, hole-nesting.

Taxon	North	South/Stewart	Group	BM	EL	No.
Grebes						
Podiceps cristatus	Y	Y	C	1100	57	1
Poliocephalus rufopectus	Y	Y	C	250	43	2
Petrels						
Puffinus gavia	Y	Y	P*	300	57	3
Puffinus huttoni	N	Y	P*	350	60	4
Puffinus spelaeus	N	Y	P*	250	55	5
Puffinus assimilis	Y	Y	P*	200	54	6
Pelecanoides urinatrix	Y	Y	P*	130	38	7
Pelecanoides georgicus	Y	Y	P*	120	39	8
Pachyptila turtur	Y	Y	P*	125	44	9
Pterodroma pycrofti	Y	N	P*	160	49	10
Pterodroma cookii	Y	Y	P*	200	50	11

(*continued*)

Part 1 (*Continued*)

Taxon	North	South/Stewart	Group	BM	EL	No.
Garrodia nereis	Y	Y	P*	35	31	12
Pelagodroma marina	Y	Y	P*	45	36	13
Penguins						
Eudyptula minor	Y	Y	C	1100	56	14
Cormorants and shags						
Phalacrocorax melanoleucos	Y	Y	C	700	48	15
Stictocarbo punctatus	Y	Y	C	1200	58	16
Herons and bitterns						
Egretta alba	Y	Y	C	900	52	17
Egretta sacra	Y	N	C	400	46	18
Botaurus poiciloptilus	Y	Y	C	1000	51	19
Ixobrychus novaezelandiae	Y	Y	C	150	46	20
Waterfowl						
Anas superciliosa	Y	Y	C	1000	58	21
Anas gracilis	Y	Y	C	425	50	22
Anas chlorotis	Y	Y	C	500	60	23
Mergus australis	Y	Y	C	900	57	24
Euryanas finschi	Y	Y	F	800	56	25
Malacorhynchus scarletti	Y	Y	C	800	56	26
Falcons						
Falco novaeseelandiae	Y	Y	Pr	500	49	27
Gamebirds						
Coturnix novaezealandiae	Y	Y	GF	100	30	28
Rails						
Gallirallus philippensis	Y	Y	C	170	40	29
Gallirallus australis	Y	Y	F	700	58	30
Porzana tabuensis	Y	Y	C	45	30	31
Porzana pusilla	Y	Y	C	40	28	32
Fulica prisca	Y	Y	F	1000	57	33
Capellirallus karamu	Y	Y	F	275	43	34
Gallinula hodgenorum	Y	Y	F	450	48	35
Shorebirds						
Haematopus ostralegus	Y	Y	C	550	56	36
Haematopus unicolor	Y	Y	C	725	59	37
Himantopus novaezelandiae	Y	Y	C	220	45	38
Charadrius obscurus	Y	Y	C	145	46	39
Charadrius novaeseelandiae	Y	Y	C*	60	37	40
Charadrius bicinctus	Y	Y	C	60	34	41
Charadrius frontalis	N	Y	C	55	35	42
Coenocorypha barrierensis	Y	N	GF	105	42	43
Coenocorypha South Island	N	Y	GF	105	42	44
Coenocorypha iredalei	N	Y	GF	105	42	45
Gulls and terns						
Larus novaehollandiae	Y	Y	C	260	52	46
Larus bulleri	Y	Y	C	250	55	47
Sterna albostriata	Y	Y	C	80	40	48
Sterna striata	Y	Y	C	160	46	49
Sterna nereis	Y	Y	C	70	35	50
Pigeons						
Hemiphaga novaeseelandiae	Y	Y	AF	650	49	51
Parrots						
Strigops habroptilus	Y	Y	F*	2000	51	52

(*continued*)

Part 1 (*Continued*)

Taxon	North	South/Stewart	Group	BM	EL	No.
Nestor meridionalis	Y	Y	AF*	425	42	53
Nestor notabilis	Y	Y	AF*	800	44	54
Cyanoramphus novaezelandiae	Y	Y	AF*	70	25	55
Cyanoramphus auriceps	Y	Y	AF*	40	24	56
Cyanoramphus malherbi	Y	Y	AF*	35	23	57
Cuckoo						
Chrysococcyx lucidus	Y	Y	AF	25	18.5	58
Eudynamys taitensis	Y	Y	AF	125	23	59
Kingfishers						
Halcyon sancta	Y	Y	Pr*	65	29	60
Owls						
Ninox novaeseelandiae	Y	Y	Pr*	175	38	61
Sceloglaux albifacies	Y	Y	Pr*	600	48	62
Owlet-nightjars						
Aegotheles novaezealandiae	Y	Y	AF	200	35	63
Songbirds						
Acanthisitta chloris	Y	Y	AF	7	16	64
Xenicus longipes	Y	Y	AF	16	20	65
Xenicus gilviventris	Y?	Y	GF	20	20	66
Pachyplichas yaldwyni	Y	Y	F	50	26	67
Traversia lyalli	Y	Y	F	22	21	68
Dendroscansor decurvirostris	Y	Y	F	30	23	69
Anthus novaeseelandiae	Y	Y	GF	40	23	70
Bowdleria punctata	Y	Y	GF	35	19.5	71
Mohoua albicilla	Y	N	AF	14.5	20	72
Mohoua ochrocephala	N	Y	AF*	25	23.5	73
Mohoua novaeseelandiae	N	Y	AF	11	18.5	74
Gerygone igata	Y	Y	AF	6.5	17	75
Rhipidura fuliginosa	Y	Y	AF	8	16	76
Petroica macrocephala	Y	Y	AF	11	18	77
Petroica australis	Y	Y	AF	35	25	78
Notiomystis cincta	Y	N	AF*	30	19	79
Anthornis melanura	Y	Y	AF	26	23	80
Prosthemadera novaeseelandiae	Y	Y	AF	90	29	81
Callaeas cinerea	N	Y	AF	230	40	82
Callaeas wilsoni	Y	N	AF	230	35	83
Philesturnus carunculatus	N	Y	AF	70	29	84
Philesturnus rufusater	Y	N	AF	70	29	85
Heteralocha acutirostris	Y	N	AF*	300	45	86
Turnagra capensis	N	Y	AF	130	35	87
Turnagra tanagra	Y	N	AF	130	35	88
Corvus moriorum	Y	Y	AF	950	56	89

Part 2

Bird species breeding on the three main islands of New Zealand before human contact, with egg length >60 mm (outside critical range for predation by *Rattus exulans*) and body mass of 201–500 g. C, freshwater or coastal taxa; P, petrel; F, flightless; Pr, predator; GF, flying/ground; AF, flying/arboreal; *, hole-nesting.

Taxon	North	South/Stewart	Group	BM	EL	No.
Petrels						
Pterodroma inexpectata	Y	Y	P	325	61	90
Pterodroma macroptera	Y	Y	P	500	67	91

Part 3

Bird species breeding on the three main islands of New Zealand before human contact, with egg length >60 mm (outside critical range for predation by *Rattus exulans*) and body mass of 501–1000 g. C, freshwater or coastal taxa; P, petrel; F, flightless; Pr, predator; GF, flying/ground; AF, flying/arboreal; *, hole-nesting.

Taxon	North	South/Stewart	Group	BM	EL	No.
Petrels						
Puffinus griseus	Y	Y	P	800	77	92
Procellaria parkinsoni	Y	Y	P	700	69	93
Waterfowl						
Hymenolaimus malacorhynchos	Y	Y	C	750	65	94
Aythya novaeseelandiae	Y	Y	C	650	64	95
Gulls and terns						
Larus dominicanus	Y	Y	C	850	69	96
Sterna caspia	Y	Y	C	700	68	97

Part 4

Bird species breeding on the three main islands of New Zealand before human contact, with egg length >60 mm (outside critical range for predation by *Rattus exulans*) and body mass of 1001–5000 g. C, freshwater or coastal taxa; P, petrel; F, flightless; Pr, predator; GF, flying/ground; AF, flying/arboreal; *, hole-nesting.

Taxon	North	South/Stewart	Group	BM	EL	No.
Kiwis						
Apteryx Eastern South Island	N	Y	F	1400	120	98
Apteryx australis	Y	Y	F	1500	125	99
Apteryx owenii	Y	Y	F	1200	110	100
Apteryx haastii	N	Y	F	2200	124	101
Petrels						
Procellaria westlandica	N	Y	P	1100	81	102
Penguins						
Eudyptes pachyrhynchus	Y	Y	C	4000	67	103
Gannets						
Morus serrator	Y	Y	C	2300	77	104
Shags and cormorants						
Phalacrocorax carbo	Y	Y	C	2200	61	105

(continued)

Part 4 *(Continued)*

Taxon	North	South/Stewart	Group	BM	EL	No.
Phalacrocorax varius	Y	Y	C	2000	59	106
Leucocarbo carunculatus	Y	Y	C	2500	65	107
Leucocarbo chalconotus	Y	Y	C	2500	66	108
Waterfowl						
Cygnus atratus	Y	Y	C	5000	104	109
Tadorna variegata	Y	Y	C	1400	67	110
Biziura delautouri	Y	Y	C	2000	76	111
Rails						
Porphyrio hochstetteri	N	Y	F	3000	74	112
Porphyrio mantelli	Y	N	F	3500	75	113
Hawks and eagles						
Circus eylesi	Y	Y	Pr	2500		114
Gulls and terns						
Catharacta skua	Y	Y	C	1950	74	115

Part 5

Bird species breeding on the three main islands of New Zealand before human contact, with egg length >60 mm (outside critical range for predation by *Rattus exulans*) and body mass of >5000 g. C, freshwater or coastal taxa; P, petrel; F, flightless; Pr, predator; GF, flying/ground; AF, flying/arboreal; *, hole-nesting.

Taxon	North	South/Stewart	Group	BM	EL	No.
Moas						
Anomalopteryx didiformis	Y	Y	F	50,000	160	116
Megalapteryx didinus	N	Y	F	20,000	125	117
Pachyornis elephantopus	N	Y	F	145,000	216	118
Pachyornis australis	N	Y	F	120,000	205	119
Pachyornis mappini	Y	N	F	20,000	125	120
Emeus crassus	N	Y	F	50,000	160	121
Euryapteryx geranoides	Y	Y	F	95,000	192	122
Euryapteryx curtus	Y	N	F	20,000	125	123
Dinornis struthoides	Y	Y	F	100,000	195	124
Dinornis novaezealandiae	Y	Y	F	150,000	218	125
Dinornis giganteus	Y	Y	F	200,000	236	126
Penguins						
Megadyptes antipodes	Y	Y	C	5,250	77	127
Waterfowl						
Cnemiornis gracilis	Y	N	F	8,000	105	128
Cnemiornis calcitrans	N	Y	F	10,000	110	129
Gruiforms						
Aptornis defossor	N	Y	F	10,000	94	130
Aptornis otidiformis	Y	N	F	8,000	90	131
Hawks and eagles						
Harpagornis moorei	N	Y	Pr	12,000		132

Appendix 2: Calculations of Body Mass and Egg Length

Parameters

Vulnerability to predation was measured by comparing a representative body mass for the taxon, with that of the possible predator. Any species of the same or lower body mass as a predator was assumed to be potentially available to that predator, other things being equal. For vulnerability of reproduction to predation, a representative egg length for each taxon was compared with the greatest egg length known to be taken by the particular predator.

Body masses and egg lengths for the indigenous birds were obtained from standard texts where information was available. Body masses for extinct species were calculated from limb (leg) bone dimensions according to the relationships developed by Campbell and Tonni (1983) and Anderson *et al.* (1985). Relationships between egg length (mm) and body mass (g) were derived for key taxonomic groups from literature values for extant species. Egg length was calculated from the expected body masses, and checked back so far as possible using data from living species of about the same body mass, where these exist. In all instances, a single representative value for body mass and for egg length (usually the mean) are used for each taxon. As the bounds for the ability of predators to handle eggs, and indeed adults and young birds themselves, are not rigid, the estimates for both body mass and egg length are useful guides, and are conservative in most instances. No greater precision was possible or necessary for the arguments developed here.

Relationships of Representative Values of Egg Length (mm) and Body Mass (g)

Where E is egg length (mm) and B is body mass (g), the relationships were as follows:

Moas: $\ln E = 2.05862 + 0.27931 \times \ln B$; $n = 9$ individuals; $r^2 = 0.8967$
Petrels: $\ln E = 2.5769 + 0.25575 \times \ln B$; $n = 36$ species; $r^2 = 0.9760$
Penguins: $\ln E = 2.2357 + 0.25098 \times \ln B$; $n = 8$ species; $r^2 = 0.8950$
Waterfowl: $\ln E = 2.6207 + 0.2263 \times \ln B$; $n = 10$ species; $r^2 = 0.8064$
Rails/gruids: $\ln E = 2.57227 + 0.21456 \times \ln B$; $n = 8$ species; $r^2 = 0.9701$
Parrots: $\ln E = 2.3315 + 0.2176 \times \ln B$; $n = 7$ species; $r^2 = 0.9465$
Songbirds: $\ln E = 2.2771 + 0.2549 \times \ln B$; $n = 23$ species; $r^2 = 0.9181$

Appendix 3: Time of Extinction of New Zealand Birds by Size Class

Size groups and time of extinction for birds on the North, South, and Stewart Islands of New Zealand. Years given are those by which the taxon can be considered to have been extinct; *, endangered; 1998, present.

		North Island						South and Stewart Islands					
Size class	Taxon	1770	1850	1870	1930	1990	1998	1770	1850	1870	1930	1990	1998
	Petrels												
0-200	*Puffinus gavia*	X						X					
≤ 60 mm	*Puffinus huttoni*							X	X	X	X	X	X
	Puffinus spelaeus							X					
	Puffinus assimilis	X						X					
	Pelecanoides urinatrix	X						X					
	Pelecanoides georgicus							X					
	Pachyptila turtur	X						X					
	Pterodroma pycrofti	X											
	Pterodroma cookii	X	X	X	X			X	X	X	X		
	Garrodia nereis	X						X					
	Pelagodroma marina	X						X					
201–500	*Pterodroma inexpectata*	X	X	X	X			X	X	X	X		
	Pterodroma macroptera	X	X	X	X	X	X*						
501–1000	*Puffinus griseus*	X	X	X	X	X	X*	X	X	X	X	X	X*
	Procellaria parksoni	X	X	X	X	X		X	X	X	X		
1001–5000	*Procellaria westlandica*							X	X	X	X	X	X*
	Flightless												
0–1000	*Pachyplichas yaldwyni*	X						X					
≤ 60 mm	*Traversia lyalli*	X						X					
	Dendroscansor decurvirostris							X					
	Gallirallus australis	X	X	X	X	X	X*	X	X	X	X	X	X
	Fulica prisca	X						X					
	Capellirallus karamu	X											
	Gallinula hodgenorum	X						X					
	Euryanas finschi	X						X					
1001–5000	*Apteryx* Eastern South Island							X					

	Apteryx australis	X	X	X	X	X	X*	X	X	X	X	X	X
	Apteryx owenii	X	X	X	X			X	X	X	X	X	
	Apteryx haastii							X	X	X	X	X	X
	Porphyrio hochstetteri							X	X	X	X	X	X*
	Porphyrio mantelli	X	X	X	X								
	Strigops habroptilus	X	X	X	X			X	X	X	X	X	
5001–	*Anomalopteryx didiformis*	X						X					
	Megalapteryx didinus							X					
	Pachyornis elephantopus							X					
	Pachyornis australis							X					
	Pachyornis mappini	X											
	Emeus crassus							X					
	Euryapteryx geranoides	X						X					
	Euryapteryx curtus	X											
	Dinornis struthoides	X						X					
	Dinornis novaezealandiae	X						X					
	Dinornis giganteus	X						X					
	Cnemiornis gracilis	X											
	Cnemiornis calcitrans							X					
	Aptornis otidiformis	X											
	Aptornis defossor							X					
	Coastal/freshwater												
0–500	*Podiceps cristatus*	X						X	X	X	X	X	X
≤ 60 mm	*Poliocephalus rufopectus*	X	X	X	X	X	X	X	X	X	X	X	
	Eudyptula minor	X	X	X	X	X	X	X	X	X	X	X	X
	Ixobrychus novaezelandiae	X	X					X	X	X			
	Phalacrocorax melanoleucos	X	X	X	X	X	X	X	X	X	X	X	X
	Anas superciliosa	X	X	X	X	X	X	X	X	X	X	X	X
	Anas gracilis	X	X	X	X	X	X	X	X	X	X	X	X
	Anas chlorotis	X	X	X	X	X	X*	X	X	X	X	X	

Note 1: An initial distribution for which there is an extinction (no X) in the first column indicates presence on that island during the Holocene, but does not indicate presence or absence during the Pleistocene.
Note 2: Three taxa colonized New Zealand after Polynesian settlement and before European contact. These were *Circus approximans, Porphyrio melanotus*, and *Anas rhynchotis*. Neither they, nor the taxa such as *Zosterops lateralis* that colonized at various times in the European era are considered here. There is evidence that at least some of these taxa were kept from colonizing by competition from indigenous species before environmental changes resulting from human presence allowed the creation of new niches, and old niches were emptied by extinction (Holdaway and Worthy, 1997; Diamond and Veitch, 1981; McCallum, 1982).

(*continued*)

Appendix 3 (*Continued*)

Size class	Taxon	North Island 1770	North Island 1850	North Island 1870	North Island 1930	North Island 1990	North Island 1998	South and Stewart Islands 1770	South and Stewart Islands 1850	South and Stewart Islands 1870	South and Stewart Islands 1930	South and Stewart Islands 1990	South and Stewart Islands 1998
	Mergus australis	X						X					
	Malacorhynchus scarletti	X						X					
	Gallirallus philippensis	X	X	X	X	X	X	X	X	X	X	X	X*
	Porzana tabuensis	X	X	X	X	X	X	X	X	X	X	X	X
	Porzana pusilla	X	X	X	X	X	X	X	X	X	X	X	X
	Haematopus unicolor	X	X	X	X	X	X	X	X	X	X	X	X
	Haematopus ostralegus	X	X	X	X	X	X	X	X	X	X	X	X
	Himantopus novaezelandiae	X	X	X				X	X	X	X	X	X*
	Charadrius obscurus	X	X	X	X	X	X*	X	X	X	X		
	Charadrius bicinctus	X	X	X	X	X	X	X	X	X	X	X	X
	Charadrius novaeseelandiae	X	X					X	X	X			
	Charadrius frontalis							X	X	X	X	X	X
	Sterna albostriata	X	X	X				X	X	X	X	X	X
	Sterna striata	X	X	X	X	X	X	X	X	X	X	X	X
	Sterna nereis	X	X	X	X	X	X*	X	X	X			
	Larus novaehollandiae	X	X	X	X	X	X	X	X	X	X	X	X
	Larus bulleri							X	X	X	X	X	X
	Egretta sacra	X	X	X	X	X	X	X	X	X	X	X	X*
	Egretta alba	X						X	X	X	X	X	X*
501–1000	*Botaurus poicioptilus*	X	X	X	X	X	X	X	X	X	X	X	X
	Hymenolaimus malacorhynchos	X	X	X	X	X	X	X	X	X	X	X	X
	Aythya novaeseelandiae	X	X	X	X	X	X	X	X	X	X	X	X
	Larus dominicanus	X	X	X	X	X	X	X	X	X	X	X	X
	Sterna caspia	X	X	X	X	X	X	X	X	X	X	X	X
1001–5000	*Eudyptes pachyrhynchus*	X						X	X	X	X	X	X*
	Morus serrator	X	X	X	X	X	X	X	X	X	X	X	X
	Phalacrocorax carbo	X	X	X	X	X	X	X	X	X	X	X	X
	Phalacrocorax varius	X	X	X	X	X	X	X	X	X	X	X	X
	Leucocarbo carunculatus	X						X	X	X	X	X	X*
	Leucocarbo chalconotus							X	X	X	X	X	X
	Stictocarbo punctatus	X	X	X	X	X	X	X	X	X	X	X	X

	Cygnus atratus	X						X					
	Tadorna variegata	X	X	X	X	X	X	X	X	X	X	X	X
	Biziura delautouri	X						X					
	Catharacta skua							X	X	X	X	X	X
	Megadyptes antipodes	X						X	X	X	X	X	X*
	Arboreal/flying												
0–500	*Hemiphaga novaeseelandiae*	X	X	X	X	X	X	X	X	X	X	X	X
≤ 60 mm	*Nestor meridionalis*	X	X	X	X	X	X*	X	X	X	X	X	X*
	Nestor notabilis							X	X	X	X	X	X
	Cyanoramphus novaezelandiae	X	X	X	X			X	X	X	X	X	
	Cyanoramphus auriceps	X	X	X	X	X	X	X	X	X	X	X	X
	Cyanoramphus malherbi	X						X	X	X	X	X	X*
	Chrysococcyx lucidus	X	X	X	X	X	X	X	X	X	X	X	X
	Eudynamys taitensis	X	X	X	X	X	X	X	X	X	X	X	X
	Aegotheles novaezealandiae	X						X					
	Acanthisitta chloris	X	X	X	X	X	X	X	X	X	X	X	X
	Xenicus longipes	X	X	X				X	X	X	X		
	Mohoua albicilla	X	X	X	X	X	X						
	Mohoua ochrocephala							X	X	X	X	X	X*
	Mohoua novaeseelandiae							X	X	X	X	X	X
	Gerygone igata	X	X	X	X	X	X	X	X	X	X	X	X
	Rhipidura fuliginosa	X	X	X	X	X	X	X	X	X	X	X	X
	Petroica macrocephala	X	X	X	X	X	X	X	X	X	X	X	X
	Petroica australis	X	X	X	X	X	X	X	X	X	X	X	X
	Notiomystis cincta	X	X	X									
	Anthornis melanura	X	X	X	X	X	X	X	X	X	X	X	X
	Prosthemadera novaeseelandiae	X	X	X	X	X	X	X	X	X	X	X	X
	Callaeas cinerea							X	X	X	X		
	Callaeas wilsoni	X	X	X	X	X	X*						
	Philesturnus carunculatus							X	X	X	X		
	Philesturnus rufusater	X	X	X	X								
	Heteralocha acutirostris	X	X	X	X								
	Turnagra capensis							X	X	X	X		
	Turnagra tanagra	X	X	X	X								

(continued)

Appendix 3 (*Continued*)

Size class	Taxon	North Island						South and Stewart Islands					
		1770	1850	1870	1930	1990	1998	1770	1850	1870	1930	1990	1998
	Corvus moriorum	X						X					
	Ground/flying												
0–500	*Coturnix novaezealandiae*	X	X	X				X	X	X			
≤ 60 mm	*Coenocorypha barrierensis*	X											
	Coenocorypha South Island							X					
	Coenocorypha iredalei							X					
	Xenicus gilviventris							X	X	X	X	X	X
	Anthus novaeseelandiae	X	X	X	X	X	X	X	X	X	X	X	X
	Bowdleria punctata	X	X	X	X	X	X	X	X	X	X	X	X
	Predators												
0–1000	*Falco novaeseelandiae*	X	X	X	X	X	X	X	X	X	X	X	X
≤ 60 mm	*Halcyon sancta*	X	X	X	X	X	X	X	X	X	X	X	X
	Ninox novaeseelandiae	X	X	X	X	X	X	X	X	X	X	X	X
	Sceloglaux albifacies	X	X	X	X			X	X	X	X		
1001–5000	*Circus eylesi*	X						X					
5000	*Harpagornis moorei*							X					

References

Anderson, A. J. 1981. Pre-European hunting dogs in the South Island, New Zealand. *N.Z. J. Archaeol.* **3**:15–20.

Anderson, A. J. 1989a. Mechanics of overkill in the extinction of New Zealand moas. *J. Archaeol. Sci.* **16**:137–151.

Anderson, A. J. 1989b. *Prodigious Birds: Moas and Moa-Hunting in Prehistoric New Zealand.* Cambridge University Press, London.

Anderson, A. J. 1991. The chronology of colonization in New Zealand. *Antiquity* **65**:767–795.

Anderson, A. J. 1996. Research report: Was *Rattus exulans* in New Zealand 2000 years ago? AMS radiocarbon ages from Shag River mouth. *Archaeol. Oceania* **31**:178–184.

Anderson, A. J. 1997a. Historical and archaeological aspects of mutton-birding in New Zealand. *N.Z. J. Archaeol.* **17**:35–55.

Anderson, A. J. 1997b. Prehistoric Polynesian impact on the New Zealand environment: Te Whenua Hou, in: P. V. Kirch and T. L. Hunt (eds.), *Historical Ecology in the Pacific Islands: Prehistoric Environmental and Landscape Change,* pp. 271–283. Yale University Press, New Haven.

Anderson, J. F., Hall-Martin, A., and Russell, D. A. 1985. Long-bone circumference and weight in mammals, birds and dinosaurs. *J. Zool. (A)* **207**:53–61.

Atkinson, I. A. E. 1973. Spread of the ship rat (*Rattus r. rattus* L.) in New Zealand. *J. R. Soc. N.Z.* **3**:457–472.

Atkinson, I. A. E. 1977. A reassessment of factors, particularly *Rattus rattus* L., that influenced the decline of endemic forest birds in the Hawaiian Islands. *Pac. Sci.* **31**:109–133.

Atkinson, I. A. E. 1978. Evidence for the effects of rodents on the vertebrate wildlife of New Zealand islands, in: P. R. Dingwall, I. A. E. Atkinson, and C. Hay (eds.), *The Ecology and Control of Rodents in New Zealand Nature Reserves,* pp. 7–30. New Zealand Department of Lands and Survey Information Series no. 4.

Atkinson, I. A. E. 1985. The spread of commensal species of *Rattus* to oceanic islands and their effects on island avifaunas, in: P. J. Moors (ed.), *Conservation of Island Birds,* pp. 35–81. ICBP Technical Publication No. 3.

Atkinson, I. A. E. 1986. Rodents on New Zealand's northern offshore islands: Distribution, effects and precautions against further spread, in: A. E. Wright and R. E. Beever (eds.), *The Offshore Islands of Northern New Zealand,* pp. 75–83. New Zealand Department of Lands and Survey Information Series No. 16.

Atkinson, I. A. E., and Bell, B. D. 1973. Offshore and outlying islands, in: G. R. Williams (ed.), *The Natural History of New Zealand,* pp. 372–392. Reed, Wellington.

Atkinson, I. A. E., and Greenwood, R. M. 1989. Relationships between moas and plants. *J. Ecol. Soc. N.Z.* **12**(Suppl.):67–96.

Beauchamp, A. J., van Berkum, B., and Closs, M. J. 1998. The decline of North Island Weka (*Gallirallus australis greyi*) at Parekawa Bay, Bay of Islands. *Notornis* **45**:31–43.

Beavan, N., and Sparks, R. J. 1998. Factors influencing ^{14}C ages of the Pacific rat *Rattus exulans. Radiocarbon.* **40**(1):601–614.

Beggs, J. R., and Wilson, P. R. 1991. The kaka, *Nestor meridionalis,* a New Zealand parrot endangered by introduced wasps and mammals. *Biol. Conserv.* **56**:23–28.

Bell, B. D. 1978. The Big South Cape Islands rat irruption, in: P. R. Dingwall, I. A. E. Atkinson, and C. Hay (eds.), *The Ecology and Control of Rodents in New Zealand Nature Reserves,* pp. 33–40. New Zealand Department of Lands and Survey Information Series no. 4.

Bell, B. D., Daugherty, C. H., and Hay, J. M. 1998. *Leiopelma pakeka,* n. sp. (Anura: Leiopelmatidae), a cryptic species of frog from Maud Island, New Zealand, and a reassessment of the conservation status of *L. hamiltoni* from Stephens Island. *J. R. Soc. N.Z.* **28**:39–54.

Blackburn, A. 1968. The birdlife of Codfish Island. *Notornis* **15**:51–65.

Booth, A. M., Minot, E. O., Fordham, R. A., and Innes, J. G. 1996. Kiore (*Rattus exulans*) predation on the eggs of the little shearwater (*Puffinus assimilis haurakiensis*). *Notornis* **43**:147–153.

Burrows, C. J., and Greenland, D. E. 1979. An analysis of the evidence for climatic change in New Zealand in the last thousand years: evidence from diverse natural phenomena and from instrumental records. *J. R. Soc. N.Z.* **9**:321–373.

Campbell, K. E., and Tonni, E. P. 1983. Size and locomotion in teratorns (Aves: Teratornithidae). *Auk* **100**:390–403.

Cassels, R. 1984. The role of prehistoric man in the faunal extinctions of New Zealand and other Pacific islands, in: P. S. Martin and R. G. Klein (eds.), *Quaternary Extinctions: A Prehistoric Revolution,* pp. 741–767. Arizona University Press, Tucson.

Cassels, R. T. S., Jones, K. L., Walton, A., and Worthy, T. H. 1988. Late Prehistoric subsistence practices of Pavewanui, Lower Rangitikei River, New Zealand. *N.Z. J. Archaeol.* **10**:109–128.

Craig, J. L. 1986. The effects of kiore on other fauna, in: A. E. Wright and R. E. Beever (eds.), *The Offshore Islands of Northern New Zealand,* pp. 75–83. New Zealand Department of Lands and Survey Information Series no. 16.

Crook, I. G., 1973. The tuatara, *Sphenodon punctatus* Gray, on islands with and without populations of the Polynesian rat, *Rattus exulans* (Peale). *Proc. N.Z. Ecol. Soc.* **20**:115–120.

Davis, A. 1994. Status, distribution, and population trends in the New Zealand shore plover *Thinornis novaeseelandiae. Notornis* **41**(Suppl.):179–194.

Falla, R. A. 1974. Rare birds and conservation, in: C. J. R. Robertson (ed.), *Birds in New Zealand,* pp. 121–134. Reed, Wellington.

Diamond, J. M., and Veitch, C. R. 1981. Extinctions and introductions in the New Zealand avifauna: cause and effect? *Science* **211**:499–501.

Fleming, C. A. 1962. The extinction of moas and other animals during the Holocene period. *Notornis* **10**:113–117.

Fleming, C. A. 1969. Rats and moa extinction. *Notornis* **16**:210–211.

Hallam, A., and Wignall, P. B. 1997. *Mass Extinctions and Their Aftermath.* Oxford University Press, London.

Heather, B. D., and Robertson, H. A. 1996. *The Field Guide to the Birds of New Zealand.* Viking, Auckland.

Holdaway, R. N. 1989. New Zealand's pre-human avifauna and its vulnerability. *J. Ecol. Soc. N.Z.* **12**(Suppl.):11–25.

Holdaway, R. N. 1991. Systematics and palaeobiology of Haast's eagle (*Harpagornis moorei* Haast, 1872) (Aves: Accipitridae). Unpublished Ph.D. thesis, Department of Zoology, University of Canterbury, Christchurch, New Zealand. 558 pp.

Holdaway, R. N. (ed.). 1994. *Chatham Island Ornithology. Notornis* **41**(Suppl.). 208 pp.

Holdaway, R. N. 1996. The arrival of rats in New Zealand. *Nature* **384**:225–226.

Holdaway, R. N. In press. A spatio-temporal model for the invasion of the New Zealand archipelago by the Pacific rat *Rattus exulans. J. R. Soc. N.Z.*

Holdaway, R. N., and Beavan, N. R. In press. Reliable ^{14}C AMS dates on bird and Pacific rat *Rattus exulans* bone gelatin, from a $CaCO_3$-rich deposit. *J. R. Soc. N.Z.*

Holdaway, R. N., and Worthy, T. H. 1994. A new fossil species of shearwater *Puffinus* from the Late Quaternary of the South Island, New Zealand, and notes on the biogeography and evolution of the *Puffinus gavia* superspecies. *Emu* **94**:201–215.

Holdaway, R. N., and Worthy, T. H. 1996. Diet and biology of the laughing owl *Sceloglaux albifacies* (Aves: Strigidae) on Takaka Hill, Nelson, New Zealand. *J. Zool.* **239**:545–572.

Holdaway, R. N., Worthy, T. H., and Tennyson, A. J. T. In press. A list of breeding birds of the New Zealand region before human intervention. *N. Z. J. zool.*

Imber, M. J. 1975. Petrels and predators. *Bull. Int. Council Bird Pres.* **12**:260–263.

Imber, M. J. 1978. The effects of rats on the breeding success of petrels, in: P. R. Dingwall, I. A. E. Atkinson, and C. Hay (eds.), *The Ecology and Control of Rodents in New Zealand Nature Reserves,* pp. 67–71. New Zealand Department of Lands and Survey Information Series No. 4.

James, H. F., and Olson, S. L. 1992. Descriptions of thirty-two new species of birds from the Hawaiian Islands: Part II. Passeriformes. *Ornithol. Monogr.* **46**:1–88.

Kepler, C. B. 1967. Polynesian rat predation on nesting Laysan albatrosses and other Pacific seabirds. *Auk* **84**:427–430.

King, C. M. 1984. *Immigrant Killers. Introduced Predators and the Conservation of Birds in New Zealand.* Oxford University Press, London.

King, C. M. 1990. *The Handbook of New Zealand Mammals.* Oxford University Press, London.

Leach, B. F. 1979. Maximizing minimum numbers: avian remains from the Washpool Midden site, in: A. J. Anderson (ed.), *Birds of a Feather: Osteological and Archaeological Papers from the South Pacific in Honour of R. J. Scarlett,* pp. 103–114. New Zealand Archaeological Association Monograph No. II, BAR Int. Ser. 62.

Lovegrove, T. G. 1996. A comparison of the effects of predation by Norway (*Rattus norvegicus*) and Polynesian rats (*Rattus exulans*) on the saddleback (*Philesturnus carunculatus*). *Notornis* **43**:91–112.

Marshall, L. G. 1988. Land mammals and the Great American Interchange. *Am. Sci.* **76**:380–388.

Martin, P. S. 1986. Refuting late Pleistocene extinction models, in: D. K. Elliott (ed.), *Dynamics of Extinction,* pp. 107–130. Wiley, New York.

McCallum, J. 1982. The penetration of exotic passerines into modified forests on Little Barrier Island, northern New Zealand. *Tane* **28**:37–51.

McGlone, M. S. 1988. New Zealand, in: B. Huntley and T. Webb (eds.), *Vegetation History,* pp. 557–599. Kluwer, Dordrecht.

McGlone, M. S. 1989. The Polynesian settlement of New Zealand in relation to environmental and biotic changes. *N.Z. J. Ecol.* **12**(Suppl.):115–129.

McGlone, M. S., Anderson, A. J., and Holdaway, R. N. 1994. An ecological approach to the Polynesian settlement of New Zealand, in: D. G. Sutton (ed.), *The Origins of the First New Zealanders,* pp. 136–163. Auckland University Press, Auckland.

Millener, P. R., and Worthy, T. H. 1991. Contributions to New Zealand's Late Quaternary avifauna. II. *Dendroscansor decurvirostris,* a new genus and species of wren (Aves: Acanthisittidae). *J. R. Soc. N.Z.* **21**:179–200.

Miskelly, C. M. 1988. The Little Barrier Island snipe. *Notornis* **35**:273–281.

Moore, P. R. 1973. Evidence of former Maori occupation of the Aldermen Islands. *Tane* **19**:21–29.

Moors, P. J. 1983. Predation by mustelids and rodents on the eggs and chicks of native and introduced birds in Kowhai Bush, New Zealand. *Ibis* **125**:137–154.

Murray, P. 1984. Extinctions Downunder: a bestiary of extinct Australian Late Pleistocene monotremes and marsupials, in: P. S. Martin and R. G. Klein (eds.), *Quaternary Extinctions: A Prehistoric Revolution,* pp. 600–628. Arizona University Press, Tucson.

O'Donnell, C. F. J. 1996. Predators and the decline of New Zealand forest birds: an introduction to the hole-nesting bird and predator programme. *N.Z. J. zool.* **23**:213–219.

Oliver, W. R. B. 1955. *New Zealand Birds,* 2nd ed. Reed, Wellington.

Olson, S. L., and James, H. F. 1992. Descriptions of thirty-two new species of birds from the Hawaiian Islands: Part I. Non-passerines. *Ornithol. Monogr.* **45**:1–88.

Patterson, G. B. 1997. South Island skinks of the genus *Oligosoma,* description of *O. longipes* n. sp. with redescription of *O. otagense* (McCann) and *O. waimatense* (McCann). *J. R. Soc. N.Z.* **27**:439–450.

Powlesland, R. G., Roberts, A., Lloyd, B. D., and Merton, D. V. 1995. Number, fate, and distribution of kakapo (*Strigops habroptilus*) found on Stewart Island, New Zealand, 1972–1992. *N.Z. J. zool.* **22**:239–248.

Quammen, D. 1996. *The Song of the Dodo: Island Biogeography in an Age of Extinctions.* Hutchinson, London.

Reid, B. 1974. Sightings and records of the takahe (*notornis mantelli*) prior to its "official rediscovery" by Dr G. B. Orbell in 1948. *Notornis* **21**:277–295.

Roberts, R. G., Jones, R., Spooner, N. A., Head, M. J., Murray, A. S., and Smith, M. A. 1994. The human colonisation of Australia: optical dates of 53,000 and 60,000 years bracket human arrival at Deaf Adder Gorge, Northern Territory. *Quat. Sci. Rev.* (*Quat. Geochronol.*) **13**:575–583.

Smith, A. P., and Quin, D. G. 1996. Patterns and causes of extinction and decline in Australian conilurine rodents. *Biol. Conserv.* **77**:243–267.

Steadman, D. W. 1997. Extinctions of Polynesian birds: reciprocal impacts of birds and people, in: P. V. Kirch and T. L. Hunt (eds.), *Historical Ecology in the Pacific Islands: Prehistoric Environmental and Landscape Change,* pp. 51–79. Yale University Press, New Haven.

Sutton, D. G. 1987. A paradigmatic shift in Polynesian prehistory: implications for New Zealand. *N.Z. J. Archaeol.* **9**:135–155.

Taborsky, M. 1988. Kiwis and dog predation observations in Waitangi State Forest. *Notornis* **35**:197–202.

Turbott, E. G. (Covener). 1990. *Checklist of the Birds of New Zealand and the Ross Dependency, Antarctica,* 3rd ed. Random Century and Ornithological Society of New Zealand, Auckland.

Warham, J. 1990. *The Petrels, Their Ecology and Breeding Systems.* Academic Press, London.

Whitaker, A. H. 1978. The effects of rodents on reptiles and amphibians, in: P. R. Dingwall, I. A. E. Atkinson, and C. Hay (eds.), *The Ecology and Control of Rodents in New Zealand Nature Reserves,* pp. 75–86. New Zealand Department of Lands and Survey Information Series No. 4.

Wilson, P. R., Karl, B. J., Toft, R. J., Beggs, J. R., and Taylor, R. H. 1998. The role of introduced predators and competitors in the decline of kaka (*Nestor meridionalis*) populations in New Zealand. *Biol. Conserv.* **83**:175–185.

Worthy, T. H. 1987. Osteology of *Leiopelma* (Amphibia: Leiopelmatidae) and descriptions of three new subfossil *Leiopelma* species. *J. R. Soc. N.Z.* **17**:201–251.

Worthy, T. H. 1989. Validation of *Pachyornis australis* Oliver (Aves: Dinornithiformes), a medium sized moa from the South Island, New Zealand. *N.Z. J. Geol. Geophys.* **32**:255–266.

Worthy, T. H. 1997. Quaternary fossil fauna of South Canterbury, South Island, New Zealand. *J. R. Soc. N.Z.* **27**:67–162.

Worthy, T. H. 1998a. The Quaternary fossil faunas of Otago, South Island, New Zealand. *J. R. Soc. N.Z.* **28**:421–521.

Worthy, T. H. 1998b. A remarkable fossil and archaeological avifauna from Marfell's Beach, Lake Grassmere, South Island, New Zealand. *Rec. Canterbury Mus.* **12**(1):79–176.

Worthy, T. H. In press. What was on the menu—avian extinction in New Zealand. *N.Z. J. Archaeol.*

Worthy, T. H., and Holdaway, R. N. 1993. Quaternary fossil faunas from caves in the Punakaiki area, West Coast, South Island, New Zealand. *J. R. Soc. N.Z.* **23**:147–254.

Worthy, T. H., and Holdaway, R. N. 1994a. Quaternary fossil faunas from caves in Takaka Valley and on Takaka Hill, northwest Nelson, South Island, New Zealand. *J. R. Soc. N.Z.* **24**:297–391.

Worthy, T. H., and Holdaway, R. N. 1994b. Scraps from an owl's table—predator activity as a significant taphonomic process newly recognised from New Zealand Quaternary deposits. *Alcheringa* **18**:229–245.

Worthy, T. H., and Holdaway, R. N. 1995. Quaternary fossil faunas from caves on Mt Cookson, North Canterbury, New Zealand. *J. R. Soc. N.Z.* **25**:333–370.

Worthy, T. H., and Holdaway, R. N. 1996a. Quaternary fossil faunas, overlapping taphonomies, and palaeofaunal reconstruction in North Canterbury, South Island, New Zealand. *J. R. Soc. N.Z.* **26**:275–361.

Worthy, T. H., and Holdaway, R. N. 1996b. Taphonomy of two Holocene microvertebrate deposits, Takaka Hill, Nelson, New Zealand, and identification of the avian predator responsible. *Hist. Biol.* **12**:1–24.

CHAPTER 10

Late Quaternary Extinctions in Australasia

An Overview

T. F. FLANNERY and R. G. ROBERTS

1. Introduction

The purpose of this chapter is to document some of the data and thinking that have arisen over the last 20 years concerning late Quaternary Australasian extinctions and vegetation change. It also gives a brief overview of the Australasian megafaunal bestiary, and the state of understanding of the timing of human arrival in Australasia. It concludes by developing further the hypothesis first elaborated in Flannery (1994a) concerning the nature, timing, and consequences of megafaunal extinction in Australasia, and presents a new, testable hypothesis.

Data accumulation over the past two decades concerning these issues has been rapid. Since 1983, entirely unknown extinct Quaternary mammal faunas have been recovered from New Guinea, the Moluccas, New Ireland, the Solomon Islands, and Tonga. In Australia, new species and even new genera have been added to the continental Quaternary bestiary.

New data have become available concerning the nature and timing of prehistoric changes in flora and fire regimes, while new studies on the timing of megafaunal extinction have begun to throw doubt on the idea that any species survived after about 35,000 radiocarbon years before present (rcyrbp).

The body of data presented here is divided into three geographic and/or temporal units corresponding to three phase of extinction: Australian Pleistocene, New Guinea Pleistocene, and the Holocene of Australia, New Guinea, the Moluccas, and the Melanesian island arcs. Each of these phases is temporally associated with the arrival of a new wave of humans or their commensal animal companions.

2. A Previous Dominant Paradigm

Perspectives on the timing, magnitude, and effects of Quaternary extinctions in Australasia have changed dramatically over the past decade. Until the mid-1990s, one dominant

T. F. FLANNERY • Mammal Section, Australian Museum, Sydney, New South Wales 2000, Australia.
R. G. ROBERTS • School of Earth Sciences, La Trobe University, Melbourne, Victoria 3083, Australia.
Extinctions in Near Time, edited by MacPhee. Kluwer Academic/Plenum Publishers, New York, 1999.

paradigm was that humans and megafauna had coexisted for millennia, or possibly tens of millennia, in Australia and New Guinea. Megafaunal extinction was widely attributed to fire and vegetation change, as well as increased aridity associated with the last glacial maximum.

This view developed following the publication of a series of papers containing radiocarbon dates that implied a long period of overlap between humans and megafauna in Australia. Key among these was the date of 26,600 ± 650 rcyrbp for the Lancefield Swamp megafaunal assemblage and apparently associated human artifacts (Gillespie *et al.,* 1978). This work suggested that at least five species of Australian megafauna survived until about the time of the glacial maximum, and were contemporaneous with humans. There were also data suggesting that elements of New Guinea's megafauna survived until at least 14,000 rcyrbp (Flannery *et al.,* 1983).

A human presence in Australia had been firmly dated at 20,000 rcyrbp in 1965, and over 30,000 by 1970 (Mulvaney, 1986). Thus, a human presence in Australia dating to before the time of the last glacial maximum at around 22,000 calendar yrbp (18,000 rcyrbp) has been firmly established for at least 30 years.

By the 1980s, dates for an even earlier human presence were being presented. In New Guinea, stone axes had been found associated with raised coral terraces that were dated using U-series and TL (thermoluminescence) dating to 45,000–53,000 calendar yrbp (Groube *et al.,* 1986). By 1994, TL dates from Malakunanja and Nauwalabila rock-shelters in the Northern Territory suggested that humans had arrived in Australia by between 50,000 and 60,000 calendar yrbp (Roberts *et al.,* 1990, 1994). Even older dates (more than 116,000 calendar yrbp) for human occupation in Australia have recently been put forward (Fullagar *et al.,* 1996) (see below).

Evidence for a dramatic, east-coast-wide (and possibly continentwide) change in vegetation structure and fire regime in Australia began to accumulate after 1985. Analysis of pollen and charcoal from the Lake George core in southern New South Wales (Fig. 1) documented the local elimination of fire-sensitive species such as southern pines and tree ferns, and an increase in the number of fire-tolerant species such as eucalypts. This coincided with a dramatic rise in the number of charcoal particles in the sediment. This change is not attributable to climate change (Singh and Geissler, 1985) and was suggested, based on $S^{18}O$ matching, to have occurred about 120,000 calendar yrbp. Wright (1986), however, argued that the change could, on the basis of sedimentation rates, just as plausibly be dated at about 60,000 rcyrbp.

Evidence for similar changes elsewhere on the continent was not long in coming. Kershaw (1986) published analyses of pollen taken from a core from Lynch's Crater in northeastern Queensland. There, the local change was dated to about 38,000 rcyrbp. More recently, analysis of cores taken off the Queensland coast indicates a change in vegetation, from "dry" rain-forest communities to sclerophyll, possibly at 100,000 and 150,000 calendar yrbp (Kershaw *et al.,* 1993), again based on $S^{18}O$ matching. Because the pollen from the offshore sediments comes from an extensive catchment, here we have evidence not for localized, but widespread vegetation change.

Uncertainty as to the timing of this change persists. The dates would, if taken at face value, indicate a diachronous change in flora, which occurred at various locations, at least on the east coast, at varying times between 120,000 and 40,000 calendar yrbp. If at least some of the dates are in error, however, it is possible that the change was virtually instantaneous along the east coast, if not continentwide.

FIGURE 1. Map of toponyms mentioned in text.

This vegetation change, from dry rain-forest-dominated to sclerophyll-dominated communities, is associated with an increase in fire. It has been widely supposed that this change helped hasten the extinction of the megafauna.

Various proponents of this hypothesis have placed varying emphases on the relative importance of humans, fire, and climate as major factors affecting the megafauna. None, however, subscribes to a blitzkrieg-style extinction attributable solely to human hunting. Most would be comfortable with the idea that increased aridity at the height of the glacial maximum some 22,000 calendar yrbp was the major climatic factor involved (e.g., Horton, 1984).

It is now time to examine in more detail the nature of the Australasian megafauna, and recent developments in dating the arrival of humans, and of select, critical sites containing megafauna.

3. Australian Pleistocene

3.1. Introduction

Fossils of extinct megafaunal marsupials were discovered in Australia as early as 1830 (Rich and Archbold, 1991) and Sir Richard Owen had described a variety of extinct species as early as 1838 (Owen, 1838). Despite this early start, there is still no compendium of Australian Pleistocene megafaunal sites, although they must now number in the hundreds, as those recorded from the state of South Australia alone numbered 68 by 1980 (Williams, 1980). Pleistocene megafaunal sites are now known from each state, and represent each major environmental type, with the possible exception of rainforest.

3.2. Extinct Species

Currently, some 28 genera and approximately 55 species of vertebrates have been identified that are thought to have become extinct toward the end of the Pleistocene. These include approximately 16% of the continent's total mammal fauna (including all of the largest carnivores and herbivores), gigantic reptiles (matsoiid snakes, varanid lizards, crocodiles, and horned turtles), and a number of flightless or poorly flighted birds, including several weighing a kilogram or less. Many of these species were widespread, and fossils of at least some taxa have been found in all regions of Australia, including Tasmania and some smaller offshore islands.

Some of these extinct species, such as the wren-sized pilot bird *Pycnoptilus,* seem anomalous in the context discussed here, as such species hardly constitute megafauna. Their extinction may nonetheless be related to that of larger species. The sole surviving pilot bird is a symbiont of the lyre bird and forages in the leaf litter excavated by the lyre bird as it feeds. The extinct pilot bird may have had a similar association with a large, extinct species such as the giant flightless cuckoo *Centropus colossus.*

The largest of the extinct marsupials (*Diprotodon optatum*) weighed about one metric tonne (Murray, 1991), while the smallest (echidnas of the genus *Megalibgwilia* and the primitive wombat *Warenja wakefieldi*) weighed about 10 kg. The extinct taxa were diverse in morphology and ecology (see Appendix 1). Indeed, it is easier to characterize the surviving larger marsupials than their extinct relatives. The surviving large marsupials are either fast (e.g., the species of large kangaroos of the genus *Macropus*), fossorial (the wombats), or arboreal (koala and tree kangaroos). Many of the surviving larger marsupials have experienced dwarfing (Marshall and Corruccini, 1978; Flannery, 1994a).

Some ecological niches were left entirely empty in Australia following Pleistocene extinction, including that of the large browsers (the largest surviving browsers weigh less than 20 kg; Strahan, 1995) and large terrestrial carnivores.

3.3. Timing of Extinction

Baynes (1992) has analyzed radiocarbon dates from sites containing megafauna for Western Australia (which comprises the western third of the continent). He notes that if the criteria of Meltzer and Mead (1985) are applied, then there is not a single reliable date for megafauna in Western Australia (extinct megafauna there includes koalas and wombats, although they survived in eastern Australia). He suggests that all became extinct before 35,000 rcyrbp.

Sites in eastern Australia that were thought to have provided reliable dates for megafauna have recently been critically reappraised. The Lancefield site, long thought to have provided firm evidence for the co-occurrence of human and megafauna at 26,000 rcyrbp (Gillespie *et al.,* 1978), is now thought to have resulted from the redeposition of megafaunal remains (van Huet, 1995), which may be much older. The Spring Creek site (Flannery and Gott, 1985), which contains a diverse megafaunal assemblage originally dated to 19,800 rcyrbp, is now thought to be older than 35,000 rcyrbp (White and Flannery, 1995).

New excavations undertaken at the Cuddie Springs site in northern central New South Wales suggest that humans and megafauna may have overlapped there about 30,000 rcyrbp. The megafauna recovered thus far is diverse, including *Diprotodon* sp., *Genyornis* sp., and the crocodile *Palimnarchus* sp. Only preliminary data have so far been published (Dodson *et al.,* 1993)—insufficient to fully evaluate the significance of the site and the reliability of the radiocarbon dates. Of possible significance, however, is a major stratigraphic unconformity lying at the base of the layers containing evidence of human occupation.

Stratigraphically conformable, apparently reliable amino acid racemization (AAR) dates have recently been obtained on *Genyornis* eggshell from Lake Eyre in northern South Australia (Magee *et al.,* 1995), suggesting that at least this element of the megafauna survived until about 50,000 rcyrbp. A slightly later date is indicated by a 14C age of 37,700 rcyrbp on *Genyornis* eggshell from an overlying level, but this date should be considered a minimum age estimate for the survival of *Genyornis,* given that similar 14C dates were obtained from much older materials in the same sequence. In more recent levels, *Genyornis* eggshell is absent, even though emu eggshell is common. This suggests that *Genyornis* went extinct, at least locally, by 40,000–50,000 rcyrbp.

With this single exception, the Australian fossil record is yet to supply a reliable terminal date for any Pleistocene megafauna. Indeed, the Australian Pleistocene remains poorly known by international standards. Land-mammal stages have yet to be recognized and no more than a handful of early Pleistocene sites have been located. Of these, few have been reliably dated.

3.4. Timing of the Arrival of Humans

In contrast to the difficulties experienced establishing reliable terminal dates for megafauna, the presence in Australia of humans by at least 35,000 rcyrbp is well documented. A series of well-dated rock-shelter and open sites from disparate regions, e.g., Tasmania, Western Australia, and New South Wales, attest to a probable continentwide human presence by that time (Dortch, 1979; Allen, 1989; Cosgrove, 1989; Jones, 1989). The data supporting a human presence in Australia by at least 35,000 rcyrbp are now so compelling that they constitute one of the few points of consensus among Australian archaeologists.

Older dates are still contested, but evidence from luminescence dating techniques is building a strong case for human presence at least as old as 50,000 calendar yrbp in northern Australia. The TL dates from Malakunanja rock-shelter in the Northern Territory suggest a human presence by between 52,000 and 61,000 calendar yrbp (Roberts *et al.*, 1990), and this claim has been strengthened by optical dates of between 53,000 and 60,000 calendar yrbp for initial human occupation of nearby Nauwalabila rock-shelter (Roberts *et al.*, 1994).

Recently, even older claims for a human presence in Australia have been made. Evidence from TL dating, suggesting initial human occupation sometime between 116,000 and 176,000 calendar yrbp at Jinmium rock-shelter, eastern Kimberley, Northern Territory, have been published (Fullagar *et al.*, 1996). These dates are, however, highly contentious, and recent reappraisals suggest that human arrival at Jinmium may be much more recent, possibly Holocene in age (Gibbons, 1997; Roberts, 1997).

At the present time, therefore, there is a large corpus of dates (both radiocarbon and luminescence) of up to 40,000 yrbp for the human colonization of central and southern Australia. Somewhat older dates, of up to 60,000 calendar yrbp, have been proposed for human arrival in northern Australia, but no archaeological sites have thus far been reliably dated to the start of the last interglacial—a period when some researchers have postulated a major vegetation change took place. At Malakunanja, the absence of artifacts in sediments dated to between 120,000 and 60,000 calendar yrbp and their appearance in all later sedimentary layers, prompted Roberts *et al.* (1990) to suggest that people did not make landfall in northern Australia until 60,000 calendar years ago. It remains a matter of continuing research to determine if people colonized the entire continent within a few millennia of first landfall, as some have suggested, or whether the southern and central regions were not colonized for a further 15,000–20,000 years, as the present data imply (Roberts, 1997).

4. New Guinea Pleistocene

4.1. Introduction

The Pleistocene vertebrate fossil record of New Guinea remained almost entirely unknown until 1983 (Flannery, 1995b). In that year, descriptions of three species of large extinct Pleistocene marsupials, whose remains had been found in Nombe rock-shelter, Chimbu Province, were published (Flannery *et al.*, 1983). Recent research has suggested that one of these, *Dendrolagus noibano* may in fact be the gigantic ancestor of the dwarfed *Dendrolagus dorianus notatus,* and thus is not technically extinct (Flannery, 1996). Two additional taxa have since been described from montane Irian Jaya (Flannery, 1992a), another from the Irian lowlands (Hardjasasmita, 1985), and yet another from the midmontane of Papua New Guinea (Flannery and Plane, 1986). Additional materials of previously described taxa have been described from a number of new localities (Menzies, 1994; Flannery, 1995b).

With only six extinct Pleistocene taxa known (Appendix 2), the New Guinean megafaunal assemblage is much smaller than that of Australia. Proportionately, it represents only 2.7% of the total indigenous, prehuman mammal fauna of the island, a smaller percentage than that of any other landmass affected by Pleistocene/Holocene extinction. It could be argued that the New Guinean megafauna described so far is only a small percentage of the true total. Several arguments militate against this, however, including low supraspecific diversity, concordance between Pliocene and Pleistocene faunas, and the fact that the same taxa keep turning up in newly discovered sites (Flannery, 1994b). Indeed, Flannery (1994b) argued it is likely that, in contrast to Australia, the New Guinean megafauna was not particularly diverse, and that most New Guinean megafaunal species have already been described.

4.2. Extinct Species

The six Pleistocene megafaunal marsupials recorded from New Guinea belong to just two subfamilies: Macropodinae and Zygomaturinae. The Macropodinae are represented by three species of *Protemnodon.* These large, stocky wallabies ranged in size from an estimated 30 to 50 kg—well within the range of living Australian kangaroos.

The Zygomaturinae, in contrast, are represented by three genera, two of which are monotypic. They ranged in weight from an estimated 100 to 300 kg, and were rather bearlike or wombatlike in morphology. All were browsers.

Elevational zonation is a strongly marked characteristic of the extant New Guinean mammal fauna, and the Pleistocene megafauna shows evidence for equally well-marked zonation. The lowlands (below 1000 m) is presently a distinct zone with a unique flora and fauna. During the late Pleistocene, the midelevation zone may have intruded into the upper part of the lowland zone via a zonational depression associated with the glacial maxima.

A single taxon, *'Kolopsis' watutense,* is known from the lowlands. This was a large tapir-sized marsupial. The type specimen (a dentary) was recovered from sediments near Nimboran, Lake Sentani area, Irian Jaya (Hardjasasmita, 1985). Its age is uncertain, although it is thought to be Pleistocene. It is indistinguishable in morphology from dentaries of Pliocene age from the Wau area of Papua New Guinea, some 800 km to the east (Flannery, 1994b). Recently, another dentary of this species was discovered in Pleistocene sediments near Kainantu in Papua New Guinea (J. Menzies, personal communication, 1997).

The midelevation zone (1000–2000 m) has remained substantially forested throughout the Pleistocene (Hope and Hope, 1976). It has produced a series of fossil localities, including rock-shelters, swamps and lacustrine deposits, all from Papua New Guinea. These have yielded a relatively uniform megafauna, the most common element of which is the zygomaturine diprotodontid *Hulitherium thomasettii.* This 100- to 200-kg marsupial, which may have been rather pandalike in morphology (Flannery and Plane, 1986), is known from four localities (Menzies, 1994; Flannery, 1995b). Two species of *Protemnodon* are also known from this zone. Both were browsers: *P. tumbuna* weighing about 50 kg, *P. nombe* about 40 kg.

The high-elevation zone (above 2000 m) has yielded a complex of fossil localities in the Kwiyawagi area of central Irian Jaya (Hope *et al.,* 1993), and a single site on Mt. Kare, Papua New Guinea (Menzies, 1994). The Irian Jayan sites are at about 2900 m elevation, and include both cave and fluviatile deposits, and have yielded numerous remains, all of which represent just two megafaunal taxa. *Maokopia ronaldi* is a small (ca. 100 kg) zygomaturine, while *Protemnodon hopei* is a large wallaby (est. 50 kg). Palynological data suggest that both species inhabited alpine habitats above the tree line (Hope *et al.,* 1993). The few remains from Mt. Kare have been tentatively identified as belonging to the genera *Maokopia* and *Protemnodon,* and possibly another zygomaturine (Menzies, 1994; Flannery, 1995b).

4.3. Timing of Extinction in New Guinea

With two exceptions, the dates associated with New Guinean Pleistocene megafauna are older than 38,000 rcyrbp. The Pureni site complex in Papua New Guinea's Southern Highlands has yielded associated skeletons of both *Hulitherium thomasettii* and *Protemnodon tumbuna* preserved in bog deposits. Radiocarbon dates on associated plant material gave an age of 38,600 ± 2500 rcyrbp (Williams *et al.*, 1972). This is probably a minimum date, for TL dating of similar deposits in the same basin suggests a date of at least 50,000 calendar yrbp (Haberle, 1994). The associated skeleton of *H. thomasettii* found at Telefomin was recovered from sediments estimated to be at least 50,000 years old (Menzies, 1994).

The remains of *M. ronaldi* and *P. hopei* (the latter including an associated if not articulated hindlimb) have also been found in sediments associated with peats dated to 39,860 ± 770 rcyrbp in the West Baliem River, and the cellulose fraction of associated wood that has been dated at 42,740 ± 1020 rcyrbp (Hope *et al.*, 1993). These 14C determinations are best regarded as minimum ages.

The only age determinations associated with megafaunal remains that postdate these are those for bones from Nombe rock-shelter, Chimbu Province. Sediments in the shelter have yielded remains of *Protemnodon tumbuna, P. nombe,* and an unidentified diprotodontid. A few bones of *P. nombe* have been found in sediments dating to 24,000–14,000 rcyrbp (based on 14C dates on flowstones; Gillieson and Mountain, 1983). No articulated remains have been found in the rock-shelter and it is possible that bones have been redeposited higher in the stratigraphic column as a result of sediment disturbance.

By way of summary, we can note that the articulated or associated skeletons of megafaunal species have been found at three sites, all of which are associated with minimum 14C ages of at least 38,000 yrbp. Nonarticulated, probably redeposited, remains of *Maokopia* and *Protemnodon* at Kelangurr Cave have been associated with TL dates of 20–25000 yrbp (Flannery, in press). Association of megafaunal fossils with the more recent dates (24,000–14,000 rcyrbp) from Nombe rock-shelter is poorly substantiated.

The discovery of a moderately diverse marsupial megafauna, which is distributed over a broad elevational range, from lowland rain forest to periglacial tundra, poses a further difficulty for those who suggest that aridity may have led to megafaunal extinction in Australasia. Because New Guinea is mountainous, Pleistocene climatic change has seen vegetation zones migrate a kilometer or two vertically, but has not forced dramatic spatial reorganizations such as are seen in Australia. It is extremely unlikely that the drought stress hypothesis of megafaunal extinction developed for Australia could explain megafaunal extinction in New Guinea.

4.4. Human Arrival in New Guinea

A human presence dating back to before the last glacial maximum has been known since White (1970) published details of the Kosipe archaeological site, where stone tools were associated with a date of about 26,000 rcyrbp. More recently, Groube *et al.* (1986) have argued that humans have been present in New Guinea for at least the past 40,000 calendar years (see above).

Redating of the coral reef terraces by state-of-the-art uranium-series methods (Chappell *et al.,* 1996), and a reassessment of the published TL data, indicate a probably age of 47,000 to 61,000 calendar years for the oldest artifact-bearing tephras on the Huon Peninsula (Roberts, 1997). On the adjacent island of New Britain, 14C ages of between 32,600 and 35,600 yrbp have been obtained from lumps of charcoal associated with chert artifacts (Pavlides and Gosden, 1994), further indicating a minimum date of about 40,000 calendar years for human arrival in this region.

5. The Holocene of Australia, New Guinea, Melanesia, and the Moluccas

5.1. Introduction

Recognition that at least local Holocene extinctions have taken place in the Australasian region began with the discovery of the remains of the thylacine (*Thylacinus cynocephalus*) and Tasmanian devil (*Sarcophilus harrisii*) in Holocene deposits in mainland Australia. Although the late survival of these species on the mainland had long been known, confirmation of their Holocene age came soon after the use of radiocarbon dating in Australia.

By 1963 the remains of *Thylacinus* had been recognized in late Pleistocene/early Holocene deposits in New Guinea (Van Deusen, 1963). Further evidence for Holocene extinction was revealed when Hope (1981) published a description of a now-extinct wallaby (genus *Thylogale*) from 5000-year-old sediments in montane Irian Jaya.

Until 1990, the fossil record of mammals from the Melanesian island arc region was almost entirely unknown. Flannery and Wickler (1990) described two large, extinct murids from Buka Island, Solomon Islands, and speculated that they were terrestrial in habit. Since then, extinct mammals have been described from New Ireland (Flannery and White, 1991) and Halmahera (Flannery *et al.,* 1995), and locally extinct bats have been recorded from the Cook Island (Steadman, 1991; Kirch *et al.,* 1992) and 'Eua, Tonga (Rainey and Pierson, 1992).

It is worth noting here that the only Pleistocene fossil record for the Melanesian islands is that from Balof Cave, New Ireland (Flannery and White, 1991). The fauna from a basal clay appears to predate a human presence, at least locally, yet the deposit preserves only mammal species that are present in the higher (human) levels.

5.2. Extinct Species

The species that experienced Holocene extinction in the region (see Appendix 3) differ greatly from those that became extinct in the Pleistocene. In Australia, two large carnivores and one flightless bird became extinct on the mainland. In New Guinea, one carnivore and two species of wallaby of the genus *Thylogale* (from the high mountains of Irian Jaya) became extinct, while a species of *Dorcopsis* wallaby and a bandicoot (genus uncertain) became extinct on Halmahera.

Curiously, two species of *Thylogale* became extinct in the mountains of Irian Jaya at about the time the dog arrived, while no similar extinction is discernible in the mountains of Papua New Guinea, nor in Australia. Remains of thylacines have never been found in Irian Jaya, despite the discovery of extensive Pleistocene/Holocene faunal assemblages. It seems possible that the thylacine never inhabited Irian Jaya, and that the extinction of the two species of *Thylogale* there when the dog arrived was the result of the wallabies having evolved in the absence of a doglike predator.

On more distant islands marsupials were largely absent, and it was the endemic murids that suffered. Two large, apparently terrestrial, rats (genus *Melomys* and *Solomys*) became extinct in the northern Solomons, while endemic species of *Rattus* vanished from New Ireland and probably Morotai.

5.3. Timing of Extinction

Terminal dates for those taxa that became extinct during the Holocene in Australia are still uncertain. The thylacine is thought to have become extinct on the mainland by about 2000 rcyrbp (Rounsevell and Mooney, 1995), but a 14C date of 430 yrbp for *Sarcophilus* remains from the southwest of Western Australia suggests that the Tasmanian devil survived later (Jones, 1995). The only other well-documented Holocene extinction on mainland Australia is that of the Tasmanian native hen (*Gallinula mortierii*), a flightless grazing bird whose remains have been found in Holocene deposits in the Australian Alps. All of these species survived in Tasmania until historic times.

The timing of Holocene extinctions in New Guinea is less well understood. The remains of *Thylogale christenseni* and *Thylogale* sp. have been located in Billingeek rockshelter, where charcoal associated with their remains has been dated at 3250 ± 80 rcyrbp (Hope *et al.,* 1993), this being the latest date for both species. The thylacine, however, is less securely dated, the remains of late Pleistocene/early Holocene age reported on by Van Deusen (1963) seemingly being the only securely dated occurrence.

The extinct murids from the Melanesian islands show different patterns. *Melomys spechti* and *Solomys spriggsarum* seem to have survived until at least 6700 rcyrbp. A hiatus in deposition means that there is no record on Buka between this time and about 1800 rcyrbp, by which time these taxa had apparently become extinct (Flannery and Wickler, 1990). By this time also, a variety of introduced mammals had reached the island, including *Phalanger orientalis, Rattus exulans, R. praetor, Sus scrofa,* and *Canis familiaris.*

Rattus sanila seems to have survived on New Ireland until about 3000 rcyrbp (Flannery and White, 1991). *R. exulans* and *R. praetor* both reached New Ireland at this time (but not pigs or dogs), and the endemic *R. sanila* became extinct soon after.

The remains of both a *Dorcopsis* wallaby and a bandicoot (neither of which could be identified to species) have been recorded from Halmahera (Flannery *et al.,* 1995). At least the *Dorcopsis* seems to have survived until about 1800 rcyrbp, after which time dogs, pigs, and other placental mammals were introduced. Two species of large (estimated weight 200–400 g), apparently extinct, *Rattus* have recently been located in archaeological sediments on Morotai, and the remains of a *Dorcopsis* wallaby, possibly *Dorcopsis muelleri mysoliae,* on Gebe, but nothing more is known of these taxa at present.

In summary, the Holocene extinctions in the region seem to have occurred predominantly, if not entirely, within the last three or four millennia. They occurred at about the time the dog (*Canis familiaris*) reached the region, and the Australian extinctions have long been attributed to that cause. In New Guinea, the Solomon Islands, New Ireland, and Morotai, they coincide roughly with the introduction of dogs, pigs, and *Rattus* of Asian origin, and the arrival of Lapita people into the region.

5.4. Historic Extinctions

Some historic extinctions in the region seem to represent a continuation of Holocene extinction trends. The extinction of the thylacine in Tasmania by about 1936 is one clear case of this. Extinctions of murids (many terrestrial) have continued in the Solomon Islands into historic times, with at least four taxa (see Appendix 3) disappearing from the central Solomons over the past century (Flannery, 1995a; see also MacPhee and Flemming, this volume). These likewise seem to represent a continuation of Holocene extinction trends.

A major Australian extinction event occurred in the twentieth century, resulting in the loss of 23 species (6.9% of total, pre-Quaternary extinction mammal fauna, and 8.2% of the historic mammal fauna). With the exception of the thylacine and possibly the Toolache wallaby (*Macropus greyi*), all weighed 5 to 0.05 kg. These are known as the "critical weight range" mammals. Most were desert dwelling.

This extinction event has been blamed, variously, on the introduction of rabbits or stock, the fox, and/or changed fire regimes. Flannery (1994a) attributed it in part to the "aftershock" of megafaunal extinction. The fact that only critical weight range mammals from the desert regions and woodlands were affected, but not (with a single possible exception, the paradise parrot) birds, reptiles, or smaller mammals, is indicative of a highly selective mechanism.

The case for this extinction being the result of a long-delayed aftershock of megafaunal extinction revolves around the idea that the critical weight range mammals are all dependent on a heterogeneous environment that provided both food and cover. Environmental heterogeneity, this hypothesis suggests, was created by megafaunal browsing and grazing. Following megafaunal extinction, environmental heterogeneity was maintained through "firestick farming" by Aborigines. With the discontinuation of fire management by Aborigines over the past 200 years, intense fires have burned vast areas of Australian vegetation, resulting in homogeneous, uniform-age stands. The immediate consequence of such fires is increased predation, and loss of adjacent feeding and sheltering places.

6. A New, Testable Paradigm

Flannery (1994a) developed a new, testable hypothesis concerning the extinction of the Australasian megafauna. The hypothesis centered around the concept that megafaunal extinction was rapid and continentwide, and was caused by the arrival of humans. He further suggested that the changes in fire regime and vegetation documented at several sites across the continent were the result of megafaunal extinction, and occurred because of an increase in fuel loads. He posited that because these events occurred in the distant past, it would probably not be possible to separate them temporally, given the resolution of existing dating techniques. Thus, they would appear to have occurred simultaneously.

At the time this hypothesis was developed, reliable dates for human arrival, megafaunal extinction, and vegetation change were too few to speculate meaningfully as to when these events might have occurred. The recent data summarized above, however, allow for an informed guess to be made.

The luminescence (TL and OSL) dates for a human presence at Malakunanja and Nauwalabila between 50,000 and 60,000 calendar yrbp appear to provide a fairly reliable estimate for the first appearance of humans in the far north of the continent. The amino acid racemization dates on *Genyornis* eggshell at Lake Eyre likewise appear to give a reliable estimate for the disappearance of at least one species of megafauna at, or shortly after, 50,000 rcyrbp. With this new information in mind, we venture to suggest that human arrival, megafaunal extinction, and vegetation change all occurred sometime between 50,000 and 60,000 calendar yrbp. An implicit assumption is that people colonized the entire continent rapidly once landfall had been made on the northern shores.

A few crucial lines of inquiry provide important tests of this hypothesis. The Cuddie Springs site is foremost among them. If it can be demonstrated that humans and megafauna overlapped for a substantial period of time between 30,000 and 40,000 rcyrbp, then clearly the hypothesis is refuted.

The early dates for a human presence at Jinmium is clearly also a challenge to our hypothesis. If they receive further support, then clearly our hypothesis is disproved also.

Other work in progress that will test the hypothesis concerns a program of OSL dating of Pleistocene megafaunal sites from throughout Australia and New Guinea. This is currently being conducted by the authors. We hope that this will provide the skeleton of a chronology for the last quarter of the Pleistocene. This should allow us to assess whether megafaunal extinctions occurred simultaneously across the continent toward the end of the Pleistocene, or whether faunal attenuation or differential extinction prevailed. We hope it will also provide some reliable terminal dates for a variety of megafaunal taxa.

7. Summary

The Australasian region has suffered three major phases of mammalian extinction during the Quaternary. The first affected large terrestrial reptiles and mammals, as well as flightless birds in Australia and (mammals only) New Guinea. These extinctions almost certainly occurred more than 35,000 rcyrbp. We hypothesize that they occurred 50,000–60,000 calendar yrbp, simultaneously with the arrival of humans and a dramatic change in vegetation.

The second phase of extinction affected large terrestrial carnivores and some terrestrial herbivores. It began less than 4000 radiocarbon years ago and affected Australia, New Guinea, and the Melanesian islands, but not Tasmania. It coincides with the introduction of *Canis familiaris* into the region and, in the Melanesian island arcs, with the introduction of pigs and exotic murids.

Recent extinctions have been a mixture of a continuation of Holocene extinction trends and an "aftershock" event resulting ultimately from Pleistocene extinctions, possibly accentuated by introduction of the fox.

Appendix 1: Australian Taxa Thought to Have Become Extinct in the Latter Part of the Pleistocene

Taxon	Weight (kg)	Niche
Reptilia		
Meiolanidae*		
Ninjemys oweni	200	herbivore
Meiolania sp.	200	herbivore
Varanidae		
Megalania prisca	1000	carnivore
Boiidae??		
Wonambi narracourtensis	100	carnivore
Crocodylidae		
Palimnarchus pollens	1000	carnivore
Quinkana fortirostrum	200	carnivore
Aves		
Dromornithidae*		
Genyornis newtoni	100	?herbivore
Megapodiidae		
Progura gallinaceae	4–7	?herbivore
Centropodinae		
Centropus colossus	1	?insectivore
Orthonychidae		
Orthonyx hypsilophus	<1	insectivore
Acanthizidae		
Pycnoptilus sp.	10 g	insectivore
Accipitridae		
gen. et sp. indet.	???	carnivore
Mammalia		
Monotremata		
Tachyglossidae		
Megalibgwilia ramsayi	ca. 10	insectivore
?Zaglossus hacketti	ca. 30	insectivore
Marsupialia		
Diprotodontidae*		
Diprotodon minor	<1000	browser
Diprotodon optatum	>1000	browser
Euryzygoma dunense	ca. 500	browser
Euwenia grata	ca. 500	browser
Nototherium mitchelli	ca. 500	browser
Zygomaturus trilobus	ca. 500	browser
Palorchestidae*		
Palorchestes azael	ca. 500	?browser
Palorchestes parvus	ca. 100	?browser
Vombatidae		
Lasiorhinus angustidens	ca. 50	grazer
Phascolomys medius	ca. 50	grazer
Phascolonus gigas	ca. 200	grazer

(continued)

Weights are very approximate estimates, based primarily on dental size and estimated from near relatives. Asterisk denotes extinct family.

Appendix 1. *(Continued)*

Taxon	Weight (kg)	Niche
Ramsaya magna	ca. 100	grazer
Vombatus hacketti	ca. 30	grazer
Warenja wakefieldi	ca. 10	grazer
Thylacoleonidae*		
Thylacoleo carnifex	ca. 60	carnivore
Potoroidae		
Propleopus oscillans	ca. 40	carnivore
N. gen.	ca. 5	???
Macropodidae		
Procoptodon goliath	ca. 250	grazer
Procoptodon rapha	ca. 150	grazer
Procoptodon pusio	ca. 75	grazer
Procoptodon texasensis	ca. 150	grazer
Simosthenurus pales	ca. 150	browser
Simosthenurus orientalis	ca. 75	browser
Simosthenurus occidentalis	ca. 50	browser
Simosthenurus brownei	ca. 50	browser
Simosthenurus maddocki	ca. 50	browser
Simosthenurus gilli	ca. 30	browser
Sthenurus stirlingi	ca. 150	browser
Sthenurus tindalei	ca. 100	browser
Sthenurus atlas	ca. 150	browser
Sthenurus oreas	ca. 100	browser
Sthenurus andersoni	ca. 50	browser
Kangurus congrous	ca. 40	?grazer
Troposodon minor	ca. 40	browser
Macropus pearsoni	ca. 150	grazer
Macropus ferragus	ca. 150	grazer
Macropus thor	ca. 30	browser
Macropus piltonensis	ca. 30	browser
Protemnodon anak	ca. 100	browser
Protemnodon roechus	ca. 100	browser
Protemnodon brehus	ca. 100	browser

Appendix 2: New Guinean Taxa Thought to Have Become Extinct in the Latter Part of the Pleistocene (See Appendix 1)

Taxon	Weight (kg)	Elevational range (m)	Niche
Marsupialia			
Diprotodontidae*			
'Kolopsis' watutense	ca. 300	sea level	browser
Hulitherium thomasettii	150	1000–2000	browser
Maokopia ronaldi	100	3000	browser
Macropodidae			
Protemnodon hopei	50	3000	browser
Protemnodon tumbuna	50	1000–2000	browser
Protemnodon nombe	40	1000–2000	browser

Appendix 3: Melanesian Taxa Thought to Have Become Extinct during the Holocene (Including Historic Extinctions)

Taxon	Weight	Distribution	Terminal date	Niche
Marsupialia				
Thylacinus cynocephalus	30	Australia, New Guinea	10,000 yrbp	carnivore
Thylogale christenseni	2	New Guinea	3250 yrbp	browser
Thylogale sp.	8	New Guinea	3250 yrbp	browser
Dorcopsis sp.	10	north Moluccas	1800 yrbp	browser
?Echymipera sp.	1	north Moluccas	???	insectivore
Placentalia				
Muridae				
Solomys spriggsarum	1	north Solomons	6700 yrbp	herbivore
Solomys salamonis	0.5	Nggela	AD1800	herbivore
Solomys sp.	0.5	Malaita	AD1900	herbivore
Melomys spechti	0.1	north Solomons	6700 yrbp	herbivore
Rattus sanila	0.1	New Ireland	3000 yrbp	herbivore
Uromys imperator	1	Guadalcanal	AD1880	herbivore
Uromys porculus	0.1	Guadalcanal	AD1880	herbivore
Chiroptera				
Nyctimene sanctacrucis	0.1	Santa Cruz	ca.AD1900	frugivore
Pharotis imogene	0.01	New Guinea	ca.AD1890	insectivore

References

Allen, J. 1989. When did humans first colonise Australia? *Search* **20**:149–154.

Baynes, A. 1992. Australian Pleistocene megafauna: when and where. *The Beagle* **9**(1):259 (abstract).

Chappell, J., Omura, A., Esat, T., McCulloch, M., Pandolfi, J., Ota, Y., and Pillans, B. 1996. Reconciliation of late Quaternary sea levels derived from coral terraces at Huon Peninsula with deep sea oxygen isotope records. *Earth Planet. Sc. Let.* **141**:227–236.

Cosgrove, R. 1989. Thirty thousand years of human colonisation in Tasmania: new Pleistocene dates. *Science* **243**:1706–1708.

Dodson, J., Fullagar, R., Furby, J., Jones, R., and Prosser, I. 1993. Humans and megafauna in a late Pleistocene environment from Cuddie Springs, north western New South Wales. *Archaeol. Oceania* **28**:94–99.

Dortch, C. 1979. Devil's Lair, an example of prolonged cave use in southwestern Australia. *World Archaeol.* **10**:258–279.

Flannery, T. F. 1992a. New Pleistocene marsupials (Macropodidae, Diprotodontidae) from subalpine habitats in Irian Jaya. *Alcheringa* **16**:321–331.

Flannery, T. F. 1992b. A revision of the *Thylogale brunii* complex (Macropodidae: Marsupialia) in Melanesia, with description of a new species. *Aust. Mammal.* **15**:7–24.

Flannery, T. F. 1994a. *The Future Eaters: An Ecological History of the Australasian Lands and People.* Reed Books, Chatswood.

Flannery, T. F. 1994b. The fossil land mammal record of New Guinea: a review. *Sci. New Guinea* **20**:39–47.

Flannery, T. F. 1995a. *Mammals of the South West Pacific and Moluccan Islands.* Reed Books, Chatswood.

Flannery, T. F. 1995b. *Mammals of New Guinea.* 2nd ed. Reed Books, Chatswood.

Flannery, T. F. 1996. Evolution (68–74) in: T. F. Flannery, R. Martin, A. Szalay, and P. Schouten (eds.), *The Tree-Kangaroos: A Curious Natural History.* pp. 1–202. Reed Books, Chatswood.

Flannery, T. F. In press. Pleistocene mammals from Kelangurr Cave, Irian Jaya.

Flannery, T. F., and Gott, B. 1985. The Spring Creek locality: a late Pleistocene megafaunal site from southwestern Victoria. *Aust. Zool.* **21**:385–422.

Flannery, T. F., and Plane, M. 1986. A new late Pleistocene diprotodontid (Marsupialia) from Pureni, Southern Highlands Province, Papua New Guinea. *Bur. Miner. Resour. J. Aust. Geol. Geophy.* **10**:65–76.

Flannery, T. F., and White, J. P. 1991. The zoogeography of New Ireland mammals, including evidence for the earliest deliberate human translocation of fauna. *Natl. Geogr. Res. Explor.* **7**:96–113.

Flannery, T. F., and Wickler, S. 1990. Quaternary murids (Rodentia: Mammalia) from Buka Island, Papua New Guinea, with descriptions of two new species. *Aust. Mammal.* **13**:127–139.

Flannery, T. F., Mountain, M.-J., and Aplin, K. 1983. Quaternary macropodids from Nombe Rockshelter, Simbu Province, Papua New Guinea with comments on the nature of megafaunal extinction in the New Guinean Highlands. *Proc. Linn. Soc. N.S.W.* **107**:75–98.

Flannery, T. F., Bellwood, P., White, J. P., Moore, A., Boeadi, and Nitihaminoto, G. 1995. Fossil marsupials (Macropodidae, Peroryctidae) and other mammals of Holocene age from Halmahera, North Moluccas, Indonesia. *Alcheringa* **19**:17–25.

Fullagar, R. L. K., Price, D. M., and Head, L. M. 1996. Early human occupation of northern Australia: archaeology and thermoluminescence dating of Jinmium rock-shelter, Northern Territory. *Antiquity* **70**:751–773.

Gibbons, A. 1997. Doubts over spectacular dates. *Science* **278**:220–223.

Gillespie, R., Horton, D. R., Ladd, P., Macumber, P. G., Rich, T. H., Thorne, A. G., and Wright, R. S. V. 1978. Lancefield Swamp and the extinction of the Australian megafauna. *Science* **200**:1044–1048.

Gillieson, D., and Mountain, M.-J. 1983. The environmental history of Nombe rockshelter, Papua New Guinea highlands. *Archaeol. Oceania* **18**:53–62.

Groube, L., Chappell, J., Muke, J., and Price, D. 1986. A 40,000 year-old human occupation site at Huon Peninsula, Papua New Guinea. *Nature* **324**:453–455.

Haberle, S. G. 1994. Late Quaternary environmental history of the Tari Basin, Papua New Guinea. Unpublished Ph.D. thesis, Australian National University, Canberra.

Hardjasasmita, H. S. 1985. Fossil diprotodontid: Zygomaturus Owen, 1859, dari Nimboran, Irian Jaya. *Pertemuan Ilmiah Arkeologi* **3**:999–1004.

Hope, G., Flannery, T. F., and Boeadi. 1993. A preliminary report of changing Quaternary mammal faunas in subalpine New Guinea. *Quat. Res.* **40**:117–126.

Hope, J. H. 1981. A new species of *Thylogale* (Marsupialia: Macropodidae) from Mapala rock shelter, Jaya (Carstensz) Mountains, Irian Jaya (Western New Guinea), Indonesia. *Rec. Aust. Mus.* **33**:369–387.

Hope, J. H., and Hope, G. S. 1976. Palaeoenvironments for man in New Guinea, in: R. L. Kirk and A. G. Thorne (eds.), *The Origins of the Australians,* pp. 29–54. Human Biology Series No. 6. Australian Institute of Aboriginal Studies, Canberra.

Horton, D. R. 1984. Red kangaroos: last of the Australian megafauna, in: P. S. Martin and R. G. Klein (eds.), *Quaternary Extinctions: A Prehistoric Revolution,* pp. 639–680. University of Arizona Press, Tucson.

Jones, M. 1995. Tasmanian devil, in: R. Strahan (ed.), *The Mammals of Australia,* pp. 82–84. Reed Books, Chatswood.

Jones, R. 1989. East of Wallace's line: issues and problems in the colonisation of the Australian continent, in: P. Mellars and C. Stringer (eds.), *The Human Revolution: Behavioural and Biological Perspectives on the Origins of Modern Humans,* pp. 743–782. Edinburgh University Press, Edinburgh.

Kershaw, A. P. 1986. Climatic change and Aboriginal burning in north-east Australia during the last two glacial–interglacial cycles. *Nature* **322**:47–49.

Kershaw, A. P., McKenzie, G. M., and McMinn, A. 1993. A Quaternary vegetation history of northeastern Queensland from pollen analysis of ODP Site 820. *Proc. Ocean Drilling Progr. Sci. Results* **133**:107–114.

Kirch, P. V., Flenley, J. R., and Steadman, D. W. 1992. Ancient environmental degradation. Prehistoric human impacts on an island system: Mangaia, central Polynesia. *Nat. Geogr. Res. Explor.* **8**:166–179.

Marshall, L. G., and Corruccini, R. S. 1978. Variability, evolutionary rates and allometry in dwarfing lineages. *Palaeobiology* **4**:101–119.

Magee, J. W., Bowler, J. M., Miller, G. H., and Williams, D. L. G. 1995. Stratigraphy, sedimentology, chronology, and palaeohydrology of Quaternary lacustrine deposits at Madigan Gulf, South Australia. *Palaeogeogr. Palaeoclimatol. Palaeoecol.* **113**:3–42.

Meltzer, D. J., and Mead, J. I. 1985. Dating Pleistocene extinctions: theoretical issues, analytical bias and substantive results, in: J. I. Mead and D. J. Meltzer (eds.), *Environments and Extinctions: Man in late Glacial Age North America,* pp. 1–209. University of Maine Center for the Study of Early Man, Orono.

Menzies, J. I. 1994. Some new records of Pleistocene megafauna from New Guinea. *Sc. New Guinea* **20**:113–139.

Mulvaney, J. 1986. Archaeological retrospect. *Antiquity* **60**:96–107.

Murray, P. 1991. The Pleistocene megafauna of Australia, in: P. V. Rich, J. M. Mongahan, R. F. Baird, and T. H. Rich (eds.), *Vertebrate Palaeontology of Australasia,* Pioneer Design Studio, Melbourne.

Owen, R. 1838. Appendix, in: T. L. Mitchell, *Three Expeditions into the Interior of eastern Australia.* T. W. and Boone, London.

Pavlides, C., and Gosden, C. 1994. 35,000 year-old sites in the rainforests of West New Britain, Papua New Guinea. *Antiquity* **68**:604–610.

Rainey, W. E., and Pierson, E. D. 1992. Distribution of Pacific island flying foxes, in: D. E. Wilson and G. L. Graham (eds.), *Pacific Island Flying Foxes: Proceedings of an International Conference,* pp. 111–123. Fish and Wildlife Service Biological Report 90(23). U.S. Department of the Interior, Washington, D.C.

Rich, P. V., and Archbold, N. W. 1991. Squatters, priests and professors: a brief history of vertebrate palaeontology in Terra Australis, in: P. V. Rich, J. M. Monaghan, R. F. Baird, and T. H. Rich (eds.), *Vertebrate Palaeontology of Australasia,* pp. 1–1437. Pioneer Design Studio, Melbourne.

Roberts, R. G. 1997. Luminescence dating in archaeology: from origins to optical. *Radiation Measurements* **27**:819–892.

Roberts, R. G., Jones, R., and Smith, M. A. 1990. Thermoluminescence dating of a 50,000 year-old human occupation site in northern Australia. *Nature* **345**:153–156.

Roberts, R. G., Jones, R., Spooner, N. A., Head, M. J., Murray, A. S., and Smith, M. A. 1994. The human colonisation of Australia: optical dates of 53,000 and 60,000 years bracket human arrival at Deaf Adder Gorge, Northern Territory. *Quat. Geochronol.* **13**:575–583.

Rounsevell, D. E., and Mooney, N. 1995. Thylacine, in: R. Strahan (ed.), *The Mammals of Australia,* pp. 164–165. Reed Books, Chatswood.

Singh, G., and Geissler, E. A. 1985. Late Cainozoic history of vegetation, fire, lake levels and climate, at Lake George, New South Wales, Australia. *Philos. Trans. R. Soc. London* **311**:379–447.

Steadman, D. W. 1991. Extinct and extirpated birds on Aitutaki and Aitu, southern Cook Islands. *Pac. Sci.* **45**:254–257.

Strahan, R. (ed.). 1995. *The Mammals of Australia.* Reed Books, Chatswood.

Van Deusen, H. M. 1963. First New Guinea record of *Thylacinus. J. Mammal.* **44**:279–280.

van Huet, A. M. S. 1995. Taphonomy of a late Quaternary site: Lancefield Swamp, Victoria. M.Sc. thesis, Earth Sciences, Monash University, Victoria.

White, J. P. 1970. Ol Tumbuna: archaeological excavations in the eastern central highlands, Papua New Guinea. Canberra, The Australian National University, Research School of Pacific Studies, *Terra Australis 2.*

White, J. P., and Flannery, T. F. 1995. Late Pleistocene fauna at Spring Creek, Victoria: a re-evaluation. *Aust. Archaeol.* **40**:13–17.

Williams, D. L. G. 1980. Catalogue of Pleistocene vertebrate fossils and sites in South Australia. *Trans. R. Soc. South Aust.* **104**:101–115.

Williams, P. W., McDougall, I., and Powell, J. M. 1972. Aspects of the Quaternary geology of the Tari-Koroba area, Papua. *J. Geol. Soc. Aust.* **18**:333–347.

Wright, R. S. V. 1986. How old is zone F at Lake George? *Archaeol. Oceania* **21**:138–139.

CHAPTER 11

Late Pleistocene Megafaunal Extinctions

A European Perspective

ANTHONY JOHN STUART

1. Introduction

This chapter focuses on the accelerated extinction of megafauna that occurred within the late Pleistocene, ca. 50,000 to 10,000 radiocarbon years before present, with some reference to earlier (Pleistocene) and later (Holocene) events. It is important to distinguish between, on the one hand, extinction of a species because of replacement by its evolutionary descendant, or by a less closely related competitor (the normal processes of faunal turnover), and on the other, extinction without ecological replacement. It is the latter that concerns us here.

In this chapter, radiocarbon dates before present (given uncalibrated) are presented as "C14yrbp"; other (older) dates in years before present are given as "yrbp."

The extinctions of the late Pleistocene were a global phenomenon, unprecedented in the geological record, in that they primarily affected terrestrial megafauna, mostly large mammals plus a few large birds and reptiles, leaving marine biota unscathed (Martin, 1984). The term *megafauna* is used here for species with estimated mean adult body weights of 40 kg or over. Generally speaking within the megafauna, the larger the animal, the more it was at risk of extinction, and very few extinctions without replacement occurred among species weighing less than 40 kg. Megafaunal extinctions were very variable in their severity from one zoogeographical region to another (Martin, 1984; Stuart, 1991; Fig. 1), with the greatest effects in North America, South America, and Australasia, and less marked but still major impact in Northern Eurasia. In contrast, sub-Saharan Africa and southern Asia were hardly affected. As discussed below, the severity of late Pleistocene extinctions in Northern Eurasia has not been generally appreciated. Strikingly, in the four first-named regions all of the animals of 1000 kg (1 metric tonne) or more became extinct. In contrast, all species of 1 tonne or more present in the late Pleistocene of sub-Saharan Africa and southern Asia survive to the present day (Table 1).

ANTHONY JOHN STUART • Norwich Castle Museum, Norwich NR1 3JU, United Kingdom.
Extinctions in Near Time, edited by MacPhee. Kluwer Academic /Plenum Publishers, New York, 1999.

FIGURE 1. A global selection of terrestrial mammals of the Late Pleistocene. Species surviving today (Africa, southern Asia) are shaded, the rest are extinct (either entirely or in their respective zoogeographical regions). Taxa of 1000 kg mean body weight or more are indicated by an asterisk. North America: Columbian mammoth (*Mammuthus columbi**); woolly mammoth (*Mammuthus primigenius**); horses (*Equus* spp.); extinct camel (*Camelops hesternus*); mastodon (*Mammut americanum**); ground sloth (*Glossotherium harlani*); extinct tapir (*Tapirus californicus*); glyptodont (*Glyptotherium floridanum*); sabertooth (*Smilodon fatalis*). South America: giant ground sloth (*Megatherium**); litoptern (*Macrauchenia*); glyptodont (*Glyptodon**); notoungulate (*Toxodon**); horses (*Equus* spp.); sabertooth (*Smilodon*); ground sloth (*Mylodon*). Northern Eurasia: cave bear (*Ursus spelaeus*); giant deer (*Megaloceros giganteus*); woolly rhinoceros (*Coelodonta antiquitatis**); woolly mammoth (*Mammuthus primigenius**); hippopotamus (*Hippopotamus amphibius**); "narrow-nosed" rhinoceros (*Stephanorhinus hemitoechus**); straight-tusked elephant (*Palaeoloxodon antiquus**); spotted hyaena (*Crocuta crocuta*). Africa: African elephant (*Loxodonta africana**); white rhinoceros (*Ceratotherium simum**); black rhinoceros (*Diceros bicornis**); spotted hyena (*Crocuta crocuta*). Southern Asia: Indian rhinoceros (*Rhinoceros unicornus**); Asian elephant (*Elephas maximus**). Australasia: diprotodon (*Diprotodon optatum**); short-faced kangaroo (*Procoptodon goliath**); marsupial "lion" (*Thylacoleo carnifex*).

2. The Search for a Cause

Almost every researcher who has seriously considered late Pleistocene extinctions has his or her own particular variant on what may have caused these dramatic losses. To the contending hypotheses of overkill by prehistoric human hunters (Martin, 1984), and climatic/environmental change (Guthrie, 1984, 1990; Graham and Lundelius, 1984; Graham, 1990), we can now add hyperdisease (MacPhee and Marx, 1997)—an option not seriously examined by most previous authors. Extraterrestrial causes, such as asteroid or comet impact or other sudden catastrophe, can be ruled out because extinctions were not synchronous globally (Martin, 1984), and there is no evidence of any such event from this intensively studied and well-known part of the geological record.

Although it is obviously useful and informative to research the extinction of a single species such as woolly mammoth *Mammuthus primigenius,* it is important not to lose sight of the wider picture. We should be looking for an underlying global cause for accelerated late Pleistocene extinctions, even though the history of extinction for each species is different in detail, and the severity and tempo of extinctions vary among zoogeographical regions. Indeed, it is these differences that provide the means of distinguishing the processes that are at work.

Table 1. Geographical Distribution of Late Pleistocene Megafauna of 1000 kg or More Mean Body Weight[a]

North America (Nearctic)
**Mammuthus columbi,* Columbian mammoth
**Mammuthus primigenius,* woolly mammoth[b]
**Mammut americanum,* American mastodon
**Eremotherium rusconii,* Rusconi's ground sloth
South America (Neotropical—provisonal list)
**Glyptodon,* glyptodont
**Eremotherium,* a giant ground sloth
**Megatherium,* a giant ground sloth
**Toxodon,* a notoungulate
**Cuvieronius,* a mastodon
**Haplomastodon,* a mastodon
**Notiomastodon,* a mastodon
**Stegomastodon,* a mastodon
Australasia
**Diprotodon optatum,* diprotodon
**Megalania prisca,* giant monitor lizard (? 1 tonne or more)
Northern Eurasia (Palearctic)
**Palaeoloxodon antiquus,* straight-tusked elephant
**Mammuthus primigenius,* woolly mammoth[b]
**Coelodonta antiquitatis,* woolly rhinoceros
**Stephanorhinus hemitoechus,* narrow-nosed rhinoceros
**Stephanorhinus kirchbergensis,* a rhinoceros
**Hippopotamus amphibius,* hippopotamus[c]
Africa (Ethiopian)
Loxodonta africana, African Elephant
Diceros bicornis, black rhinoceros
Ceratotherium simum, white rhinoceros
Hippopotamus amphibius, hippopotamus
Southern Asia (Oriental)
Elephas maximus, Asian elephant
Rhinoceros unicornis, Indian rhinoceros
Rhinoceros sondaicus, Javan rhinoceros

[a]Data from Martin (1984), Stuart (1991), and Murray (1991).
[b]*M. primigenius* occurred in both North America and northern Eurasia.
[c]*H. amphibius* survives in Africa today.
*Extinct species.

2.1. Overkill

Few doubt that Holocene extinctions of terrestrial vertebrates, large and small, on islands from New Zealand to the Mediterranean resulted from human colonization (e.g., Martin, 1984). Hunting, habitat destruction, and introduction of alien species have devastated island faunas across the globe. Moreover, most extinctions on continents during the Holocene probably can be ascribed to the same processes.

However, the role of humans in Late Pleistocene extinctions is less clear. First, it is difficult to see how hunting with upper paleolithic technologies could have led to the extinction of so many species with wide geographical distributions such as woolly mammoth *Mammuthus primigenius,* which ranged across the northern half of North America and most of Northern Eurasia (Stuart, 1991, Fig. 7). Overkill would predict marked extinctions closely following the arrival of modern humans *Homo sapiens* (with upper paleolithic technologies) in a given region, or perhaps the first appearance of humans in regions colonized by earlier *Homo* species. However, in Europe major megafaunal extinctions correlate neither with the first arrival of *Homo* nor with the arrival of *Homo sapiens* (see below).

2.2. Climatic Change

Proponents of the climatic/environmental change hypothesis point to the marked climatic warming and resulting major disruption of biota that occurred toward the end of the Pleistocene from about 15,000 to 10,000 C14yrbp (Graham and Lundelius, 1984; Guthrie, 1984, 1990; Graham, 1990). Major changes in North America and Northern Eurasia included the replacement of vast areas of open vegetation with grasses and herbs—"mammoth steppe"—by forests (mainly boreal conifer and temperate deciduous forests). If these changes were responsible for major extinctions, then clearly the transition from the Last Cold Stage to the Holocene must have been unique in the Pleistocene, because no such extinction episode occurred previously.

However, many sources including deep-sea oxygen isotope records, which can be plausibly related to Milankovitch cycles (Bassinot *et al.,* 1994) and long pollen records of vegetational change from lake sediments in deep basins (Watts, 1980; Beaulieu and Reille, 1992; Tzedakis and Bennett, 1996), show a complex pattern of climatic/environmental changes through the Pleistocene, in particular the last ca. 780,000 years (Middle and Late Pleistocene). In the deep-sea cores at least 19 major warm phases (interglacials), apparently similar to the Holocene, can be seen within this period, and at the beginning of several of these, the curves show a similar pattern with rather rapid temperature rise, as in the transition from the Late Pleistocene to the Holocene. In the fragmentary, but paleontologically rich, Middle and Late Pleistocene record from northwestern Europe, the essential similarity of the floral and faunal sequences from successive interglacials is well known (West, 1980; Stuart, 1982, 1995; Zagwijn, 1992; Jones and Keen, 1993). These show repeated development of temperate forests with accompanying molluscan, beetle, and vertebrate faunas. The transition from the penultimate cold stage to the Last Interglacial (correlated with Oxygen Isotope Stage 5e) shows a very similar pattern to the transition from the Last Cold Stage to the Holocene, except that the former was actually more rapid and the Last Interglacial was overall rather warmer than the Holocene (Maslin and Tzedakis, 1996). However, at least for Europe and North America, where such detailed evidence is available, no megafaunal extinctions occurred at this transition.

Detailed oxygen isotope analyses on the Greenland ice cores indicate that, in contrast to the stability of the Holocene, climates of the Last Cold Stage were unstable, fluctuating on a time scale of a few thousand to a few hundred years (Dansgaard *et al.,* 1993). This evidence does not support earlier ideas (Graham and Lundelius, 1984; Guthrie, 1984, 1990; Graham, 1990) that Late Pleistocene vertebrate faunas had coevolved into ecologically interdependent communities, which would imply long periods of stability. Climatic change at the end of the Pleistocene is supposed to have disrupted these communities, resulting in extinctions. Apparent evidence from Greenland ice cores that the Last Interglacial was similarly unstable (GRIP Members, 1993), perhaps indicating that the Holocene was unique in its stability, can be attributed to disturbances in the lower part of the cores. Nevertheless, some deep pollen sequences from southern Europe provide evidence of two oscillations within the Last Interglacial (Maslin and Tzedakis, 1996). However, it should be stressed here that it is the transition from cold to temperate stage that is critical for the extinctions debate, not the nature of the temperate stage itself.

All of this strongly indicates that climatic/environmental changes of the kind experienced in the Pleistocene would not in themselves produce accelerated extinctions of megafauna (although they very probably drove the much more gradual "background," extinctions of large and small mammals, and plants, which occurred through the European Pleistocene, see below). Of course, it would be necessary only to show that one earlier transition from cold stage to interglacial was essentially similar to that at the end of the Pleistocene to invalidate the climatic/environmental hypothesis for megafaunal extinction.

A final point here is that Late Pleistocene extinctions were demonstrably not synchronous on a global scale, which argues against a global climatic cause.

2.3. Hyperdisease

MacPhee and Marx (1997) postulate that extinctions could have been driven by pathogens introduced by humans as they spread around the globe. Like Martin's "blitzkrieg" hypothesis, hyperdisease should have resulted in a "dreadful syncopation," with extinctions closely following first contact with humans. A staggered pattern of extinctions as seen in Northern Eurasia (see below) would not fit with this hypothesis. The hyperdisease hypothesis is testable in that the responsible pathogens might be detected in the bones or other tissues of extinct megafauna.

3. The European Evidence

The extent of late Pleistocene extinctions in Northern Eurasia (here Europe, plus former Soviet Asia) has been generally underestimated, and the region tends to be overlooked in the extinctions debate (e.g., MacPhee and Marx, 1997, Fig. 7.1). Although losses were less marked than in North America, they were still impressive, claiming about 26% (11 out of 43) species of 40 kg or over, and significantly all 6 species of 1000 kg or more (Tables 1 and 2).

3.1. Extinctions ca. 780,000 to 50,000 yrbp

Because of problems of correlation, dating, and taxonomy, it is not yet possible to fully assess the pattern of extinctions through the Pleistocene for Northern Eurasia, or even for Europe alone. Considering just western Europe, because of its relatively well-known Pleistocene record, nine megafaunal species disappeared without replacement within the period 50,000 to 10,000 C14yrbp (including regional, but not global, extinctions of hippopotamus and spotted hyena), whereas perhaps ten disappeared without replacement in the

Table 2. Extinctions of Large Mammals (of 40 kg or More) in the European Late Pleistocene[a]

	Estimated mean weight
Carnivora	
Ursus spelaeus Rosenmüller and Heinroth, cave bear	400 kg
Crocuta crocuta (Erxleben), spotted hyena	85 kg
Proboscidea	
Palaeoloxodon antiquus (Falconer and Cautley), straight-tusked elephant	6000 kg
Mammuthus primigenius (Blumenbach), woolly mammoth	5000 kg
Perissodactyla	
Stephanorhinus hemitoechus (Falconer), narrow-nosed rhinoceros[b]	1600 kg
Stephanorhinus kirchbergensis (Jaeger), a rhinoceros[b]	2000 kg
Coelodonta antiquitatis (Blumenbach), woolly rhinoceros	2000 kg
Artiodactyla	
Hippopotamus amphibius Linnaeus, hippopotamus	2500 kg
Megaloceros giganteus (Blumenbach), giant deer	700 kg
Bison priscus (Bojanus), extinct bison[c]	900 kg

[a]Extinct magafaunal species known from northern Asia but not Europe include spiral-horned antelope, *Spiroceros kiakhtensis* Pavlova; one or more species of horse *Equus,* and *Camelus knoblochi* Nehring (Vereschagin and Baryshnikov, 1984). The latter is doubtfully distinct from the living Bactrian camel *Camelus ferus* of Central Asia.
[b]Previously called *Dicerorhinus hemitoechus* and *D. kirchbergensis,* respectively.
[c]Probably ancestral to the living European bison, *Bison bonasus*—not a case of extinction without replacement.

preceding 730,000 years. Megafaunal extinctions from ca. 780,000 to 50,000 yrbp (Middle and Late Pleistocene) include the giant deer *Megaloceros verticornis* and *Megaloceros dawkinsi;* the bovids *Soergelia elisabethae* and *Hemitragus stehlini;* a water buffalo *Bubalus murrensis;* one or more species of horse *Equus;* giant hyena, *Pachycrocuta brevirostris;* sabertooth, *Homotherium latidens;* and early muskox, *Praeovibos priscus.* The three last-named species may have eventually succumbed to competition in conjunction with environmental changes, following substantial periods of overlap with ecologically similar species (respectively: spotted hyena *Crocuta crocuta;* lion *Panthera leo;* and muskox *Ovibos moschatus*). The available figures are very approximate, but it is clear that there is marked increase in extinction rate, in the late Pleistocene, from an average of about one per 73,000 years for the period 780,000 to 50,000 yrbp, to about one per 4000 years from 50,000 to 10,000 C14yrb, an 18-fold increase. Significantly, within the period 780,000 to 50,00 yrbp, no species over 1000 kg was lost without replacement. In contrast, six such species disappeared after that date.

Pleistocene extinctions in Europe ca. 780,000 to 50,000 yrbp also included smaller mammals (less than 40 kg), e.g., shrews *Beremendia fissidens* and *Sorex savini,* a beaver-like rodent *Trogontherium cuvieri,* and macaque monkey *Macaca sylvanus.* The latter is a regional extinction only, as the species is still extant in North Africa. The presence of *M. sylvanus* in Europe today (Gibraltar) is the result of recent introduction. These "background" extinctions of both large and small mammals were very probably driven by the repeated climatic changes through the Pleistocene, involving several episodes of major glaciation. These episodes resulted in a progressive decline in faunal diversity in the European record (Stuart, 1993). Extinctions, at a low rate, in molluscs and plants can be attributed to the same cause. Plant extinctions (regional, but not global) include hemlock *Tsuga* (Early Pleistocene); water fern *Azolla filiculoides* (Middle Pleistocene); a water lily *Brasenia purpurea* (Middle/Late Pleistocene); and a fern *Osmunda claytoniana* (Middle Pleistocene) (Godwin, 1975). Significantly, all of these survive today in North America and/or east Asia. As pointed out by many authors, east–west mountain ranges in Eurasia presented serious barriers to migration in the face of major climatic fluctuations, whereas in North America north–south ranges allowed much freer migrations in response to climatic changes.

3.2. Extinctions ca. 50,000 to 10,000 C14yrbp

Data, including available radiocarbon dates, relevant to Late Pleistocene extinctions in Northern Eurasia were reviewed in some detail by the author (Stuart, 1991). The dates relevant to extinctions that have been published since have not substantially altered the previous picture, with the notable exception of evidence for Holocene survival of woolly mammoth *Mammuthus primigenius* on Wrangel Island, in the Arctic Ocean off northeast Siberia. Radiocarbon dates on surface finds of mammoth molars indicate that mammoths survived on this island until after 4000 C14yrbp, having died out about 10,000 C14yrbp on the mainland (Vartanyan *et al.,* 1993, 1995; Long *et al.,* 1994). These findings have inevitably fueled the extinctions debate. Vartanyan *et al.* (1993) argued that mammoths survived latest on Wrangel Island because of the continued presence there of relict steppe—tundra vegetation. Supporters of overkill suspect that the Wrangel mammoths had several thousand years' grace, surmising that humans did not reach this remote area until sometime after 4000 C14yrbp. It would be useful to have much more data for Wrangel Island, including fossils of mammoths and other animals in stratified contexts, and records of late Pleistocene and Holocene vegetational history and archaeology.

It is clear that in Europe, and Northern Eurasia generally, megafaunal extinctions in the Late Pleistocene, although greatly accelerated in comparison with earlier extinctions, nevertheless were staggered over tens of thousands of years. It is useful to distinguish two phases of extinction (Stuart, 1991, 1993; Martin and Stuart, 1995). The earlier phase involved species that were widespread in the warm Last Interglacial (ca. 113,000 to 127,000 yrbp, Oxygen Isotope Stage 5e), and also in previous interglacials, typically in association with regional temperate deciduous forest (Stuart, 1982, 1995). These species retreated to southern Europe in response to cooling at the onset of the Last Cold Stage (Fig. 2), and the replacement of forests by treeless "steppe–tundra" across most of Europe and Northern Asia. Paleobotanical evidence (Bennett *et al.,* 1991; Zagwijn, 1992) indicates that local areas of temperate deciduous forest persisted south of the Alps and Pyrenees throughout the Last Cold Stage, from which areas thermophilous trees were able to reexpand in the Holocene. These refugia were probably vital in sustaining the interglacial megafauna both at this period and through previous cold phases. The data are generally rather imprecise for this period (much of it beyond the range of radiocarbon dating), but straight-tusked elephant *Palaeoloxodon antiquus* and hippopotamus *Hippopotamus amphibius* probably survived until about 40,000–50,000 yrbp (Fig. 2). The rhinos *Stephanorhinus hemitoechus* and *S. kirchbergensis* persisted until at least 20,000–25,000 C14yrbp. Evidently all four had gone by ca. 20,000 C14yrbp—the time of the onset of the main glaciation of Northern Europe and the Alps (Stuart, 1991, 1993; Martin and Stuart, 1995). These extinctions follow the arrival of modern humans *Homo sapiens,* and may have resulted from overkill of animals already undergoing severe reduction in population

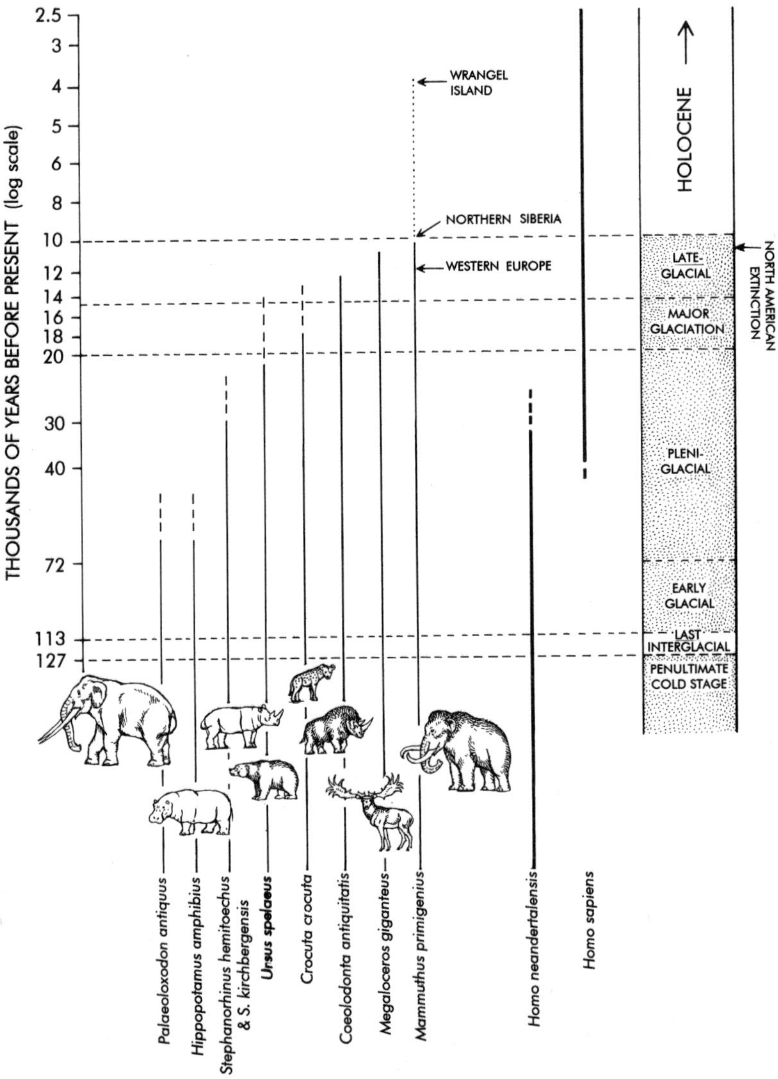

FIGURE 2. The pattern of megafaunal extinctions in northern Eurasia. Chart showing estimates of latest survival of selected large mammals in Europe and Siberia, in relation to major environmental changes and replacement of Neanderthals (*Homo neanderthalensis*) by modern humans (*Homo sapiens*). Note how the inferred extinction dates are very staggered compared with the strong pulse of extinctions in North America (indicated on the right of the diagram). Note also that species widespread in interglacials (straight-tusked elephant *Palaeoloxodon antiquus;* hippopotamus *Hippopotamus amphibius;* and the rhinos *Stephanorhinus hemitoechus* and/or *S. kirchbergensis*) became extinct (having retreated to the south of Europe) prior to the main glaciation of ca. 15,000–20,000 C14yrbp. In contrast, species widespread in periods of cold "steppe–tundra" environments (cave bear, *Ursus spelaeus;* spotted hyena, *Crocuta crocuta;* woolly rhinoceros, *Coelodonta antiquitatis;* giant deer, *Megaloceros giganteus;* and woolly mammoth, *Mammuthus primigenius*) survived into the Late Glacial, a period of general warming transitional to the Holocene. With one known exception, all had gone by 10,000 C14yrbp. Woolly mammoth died out significantly earlier in western Europe (by 12,000 bp) than in northern Siberia (ca. 10,000 bp), and survived on Wrangel Island until after 4000 C14yrbp. It is clear that extinctions correlate neither with the arrival of *Homo sapiens* nor with episodes of climatic change.

size and geographical distribution by climatic cooling and accompanying large-scale vegetational changes.

The second phase of extinctions occurred after the main glaciation, in the Late Glacial ca. 14,000–10,000 C14yrbp (Fig. 2; Stuart, 1991, 1993; Martin and Stuart, 1995). This phase affected more or less cold-adapted species widespread in Northern Eurasia in the Last Cold Stage: woolly mammoth *Mammuthus primigenius;* woolly rhino *Coelodonta antiquitatis;* giant deer *Megaloceros giganteus;* cave bear *Ursus spelaeus;* and spotted hyena *Crocuta crocuta.*

The Late Glacial was a period of major environmental change, transitional to the Holocene. In marked contrast to the apparent situation in North America, the Northern Eurasian extinctions were clearly nonsynchronous. In particular, giant deer survived in Europe some 1500 years later (to ca. 10,500 C14yrbp) than did woolly mammoth (to ca. 12,000 C14yrbp) (Fig. 2). It is also apparent that within a species, populations disappeared earlier in some areas than in others, notably woolly mammoth, which lasted some 2000 years later in mainland Siberia that it did in Europe, and at least another 6000 years on Wrangel Island. According to reconstructed vegetation maps (based on radiocarbon-calibrated pollen profiles) published by Huntley and Birks (1983), in western Europe extensive areas of steppe-tundra were present well after 12,000 C14yrbp, and were still substantial at 11,000 C14yrbp (i.e., prior to reexpansion of open vegetation with the renewed cold of the Younger Dryas). The data suggest that suitable mammoth habitat was present more than a millennium after the demise of these animals in this region (Stuart, 1991), arguing against environmental change as the sole cause of megafaunal extinction. The second phase of extinctions may also have been due to overkill by modern humans, but in this case acting on cold-adapted species undergoing major reductions in population size and geographical range in response to ameliorating climates and encroaching forests in the Late Glacial, as the Pleistocene gave way to the Holocene.

In addition to the extinctions discussed above, two species widespread in the Late Pleistocene—fallow deer *Dama dama* and lion *Panthera leo*—had much more restricted ranges in the Holocene, although both avoided outright extinction within Northern Eurasia (Stuart, 1991). Fallow deer, a temperate species, failed to expand northward from its southern European refuge at the beginning of the Holocene, while the lion, which ranged across most of Northern Eurasia (and beyond) in both cold and temperate stages, was confined to the southeast of Europe in the Holocene. Today it survives only in sub-Saharan Africa and a small area of India, but was present in North Africa and across much of the Near and Middle East until the nineteenth century. The lion probably persisted until at least 10,500 C14yrbp in Northwest Europe, significantly later than mammoth, in association with surviving open steppe—tundra vegetation (Stuart, 1991). The restricted Holocene distributions of both lion and fallow deer may be related to the spread of dense forests in the Late Glacial and early Holocene of Central and Northern Europe, in combination with hunting pressure, as both species avoid closed forests today. In previous interglacials the feeding activities of elephants, rhinos, and other large herbivores appear to have resulted in a mosaic of open and forest habitats, which supported a more diverse fauna than the more uniform forests of the Holocene, from which all herbivores of 1 tonne or more were absent. This process may be compared with Owen-Smith's "keystone" hypothesis (Owen-Smith, 1987, 1988, this volume).

The loss of giant deer, which like the lion also outlived the mammoth in Europe, may also have resulted from range shrinkage due to this process, again in combination with hunting pressure. Subsequent to the extinctions of the Late Pleistocene, no further megafaunal extinctions occurred until the last five millennia of the Holocene, when an ass *Equus hydruntinus* (totally extinct), aurochs *Bos primigenius* (surviving today only as domestic cattle), muskox *Ovibos moschatus* (surviving in Greenland and Arctic Canada), and lion *Panthera leo* (now Africa and India) disappeared from Northern Eurasia (Stuart, 1991). In addition, the isolated relict population of mammoth *Mammuthus primigenius* on Wrangel Island became extinct sometime after 4000 C14yrbp.

Throughout the Middle and Upper Pleistocene the European fauna underwent repeated reciprocating expansion and contraction in ranges, the cold phases favoring one set of species and the temperate stages the other. In each cold stage the interglacial animals retreated to southern Europe, and recolonized Central and Northern Europe in the warm stages. Conversely, species widespread in the cold phases, retreated northwards to the tundra and/or eastwards to the steppe, or to higher altitudes during the interglacials. This cycle was disrupted significantly only in the last 50,000 years or so, by accelerated megafaunal extinctions.

Of major importance to the extinctions debate is good evidence that in Northern Eurasia significant megafaunal extinctions were staggered and correlate neither with climatic change nor with human migrations (Fig. 2). There is no accelerated megafaunal extinction in the Early or Middle Pleistocene although humans *Homo* sp. first appeared ca. 500,000 yrbp in the north, and probably before 800,000 yrbp in the south (Dennell and Roebroeks, 1996). Neither did extinctions coincide with the arrival of *Homo sapiens* (ca. 40,000 yrbp; Stringer, 1995). Extinctions occurred only at times (ca. 40,000 to 20,000 C14yrbp, and 14,000 to 10,000 C14yrbp) when populations were already stressed and reduced in numbers and geographical range by climatic changes. Such climatic changes had occurred many times previously in the Pleistocene, but the one new factor is the arrival of *Homo sapiens* with upper paleolithic technologies, and hunting techniques. The pattern argues against both hyperdisease (MacPhee and Marx, 1997) and "blitzkrieg" (Martin, 1984) in that extinctions were staggered and do not correlate with "first contact" by *Homo sapiens.* The staggered pattern probably results from the interplay of climatic change and overkill by human hunters.

Overkill occurring only at times of major climatic changes explains how extinctions could have affected animals such as woolly mammoth with wide geographical ranges. As long as these wide distributions were maintained, the species were safe. Only when ranges were severely reduced by major climatic and vegetational changes might hunting pressure have become critical, resulting in extinctions.

4. North America

Although significantly smaller in area (only about 70% of the larger region), Late Pleistocene North America supported some 54 species of mammal over 40 kg, compared with 43 in Northern Eurasia. This difference probably reflects a lower overall extinction rate for the Middle and Late Pleistocene of North American than for Northern Eurasia (Stuart, 1993), but produced a situation where many more species of megafauna were at risk of extinction in the Late Pleistocene. However, losses in North America were much more severe than might have been expected from this factor alone: 70% (38 out of 54 species) compared with 26% for Northern Eurasia.

Not only the severity but also the timing and pattern of extinctions were very different in North America, with most species apparently going extinct about 11,000 C14yrbp (Martin, 1984; Stuart, 1991; Martin and Stuart, 1995), i.e., later than most extinctions in Northern Eurasia. This severe and sudden, strongly pulsed, extinction can be plausibly attributed to the arrival of Clovis hunters within a period of major climatic changes, when habitats and animal ranges were undergoing profound rearrangement. The impact was probably further enhanced by the naiveté of the fauna in the face of this novel and aggressive predator (Martin, 1984).

5. Concluding Remarks

At first sight for North America and Northern Eurasia, especially Europe, there is a wealth of data on climatic change, the archaeological record, faunal history, and extinction chronologies—all actually or potentially dated by radiocarbon. However, even for these relatively well-studied regions, far more information is required, in particular radiocarbon-dated records (preferably in good stratigraphic context) of Late Pleistocene extinct and extant megafauna from different areas. With much more data from Northern Eurasia there is the potential to follow detailed changes/shrinkages in range prior to total extinction, determining where a species made its "last stand." It should be possible, eventually, to relate this pattern of faunal changes and extinctions to climatic and vegetational changes and to the archaeological record, across this vast region.

In the case of North America, was the megafauna really wiped out everywhere in a few hundred years or less as the present data suggest? Alternatively, are we at least partly seeing the departure of species from particular regions (where the dating has been done), related perhaps to climatic/environmental changes, when populations may have survived significantly later elsewhere?

This chapter has focused on Northern Eurasia and North America because these are the regions with the best available data, especially radiocarbon dates, relevant to extinctions in the Late Pleistocene. The general paucity of such data from the rest of the world—Australasia (Flannery and Roberts, this volume), South America (for which a few dates are now available), Africa, and southern Asia—means that at present we cannot be confident of even the outline chronologies for extinctions in these regions, and therefore are in no position to test the rival hypotheses on the possible causes of these extinctions. In my view, obtaining these data is vital for eventually resolving the extinctions debate. The key to finding a fundamental global cause for Late Pleistocene extinctions is likely to be found by comparing and contrasting extinction patterns across the different zoogeographical regions.

Acknowledgments

I am grateful to Dr. Adrian Lister and to an anonymous referee for their very helpful comments on the manuscript. The figures are by Patricia J. Wynne.

References

Bassinot, F. C., Labeyrie, L. D., Vincent, E., Quidelleur, X., Shackleton, N. J., and Lancelot, Y. 1994. The astronomical theory of climate and the age of the Brunhes–Matuyama magnetic reversal. *Earth Planet. Sci. Lett.* **126**:91–108.

Beaulieu, J. L., and Reille, M. 1992. Long Pleistocene pollen sequences from the Velay Plateau (Massif Central, France), 1, Rabains Maar. *Vegetation Hist. Archaeobot.* **1**:223–242.

Bennett, K. D., Tzedakis, P. C., and Willis, K. J. 1991. Quaternary refugia of north European trees. *J. Biogeogr.* **18**:103–115.

Dansgaard, W., Johnsen, S. J., Clausen, H. B., Dahl-Jensen, D., Gunderstrup, N. S., Hammer, C. U., Hvidberg, C. S., Steffensen, J. P., Sveinbjornsdottir, H., Jouzel, J., and Bond, G. 1993. Evidence for general instability of past climate from a 250-kyr ice-core record. *Nature* **364**:218–220.

Dennell, R., and Roebroeks, W. 1996. The earliest colonization of Europe: the short chronology revisited. *Antiquity* **70**:535–542.

Godwin, H. 1975. *History of the British Flora: A Factual Basis for Phytogeography,* 2nd ed. Cambridge University Press, London.

Graham, R. W. 1990. Evolution of new ecosystems at the end of the Pleistocene, in: L. D. Agenbroad, J. I. Mead, and L. W. Nelson (eds.), *Megafauna and Man,* pp. 54–60. North Arizona Press, Flagstaff.

Graham, R. W., and Lundelius, E. L. 1984. Coevolutionary disequilibrium and Pleistocene extinctions, in: P. S. Martin and R. G. Klein (eds.), *Quaternary Extinctions: A Prehistoric Revolution,* pp. 223–249. University of Arizona Press, Tucson.

Greenland Ice Core Project Members (GRIP). 1993. Climatic instability during the Last Interglacial period recorded in the GRIP core. *Nature* **364**:203–207.

Guthrie, R. D. 1984. Mosaics, allochemicals and nutrients, in: P. S. Martin and R. G. Klein (eds.), *Quaternary Extinctions,* pp. 259–298. University of Arizona Press, Tucson.

Guthrie, R. D. 1990. Late Pleistocene faunal revolution: a new perspective on the extinction debate, in: L. D. Agenbroad, J. I. Mead, and L. W. Nelson (eds.), *Megafauna and Man,* pp. 42–53. North Arizona Press, Flagstaff.

Huntley, B., and Birks, H. J. B. 1983. *An Atlas of Past and Present Pollen Maps for Europe: 0–13,000 Years Ago.* Cambridge University Press, London.

Jones, R. L., and Keen, D. H. 1993. *Pleistocene Environments in the British Isles.* Chapman & Hall, London.

Long, A., Sher, A. V., and Vartanyan, S. L. 1994. Holocene mammoth dates. *Nature* **369**:364.

MacPhee, R. D. E., and Marx, P. A. 1997. The 40,000 year plague: humans, hyperdisease, and first contact extinctions, in: S. M. Goodman and B. D. Patterson (eds.), *Natural Change and Human Impact in Madagascar,* pp. 169–217. Smithsonian Institution Press, Washington, D.C.

Martin, P. S. 1984. Prehistoric overkill: a global model, in: P. S. Martin and R. G. Klein (eds.), *Quaternary Extinctions: A Prehistoric Revolution,* pp. 354–403. University of Arizona Press, Tucson.

Martin, P. S., and Stuart, A. J. 1995. Mammoth extinction: two continents and Wrangel Island. *Radiocarbon* **37**(1):7–10.

Maslin, M., and Tzedakis, C. 1996. Sultry Last Interglacial gets sudden chill. *Eos Trans. Am. Geophys. Union* **77**(37):353–354.

Murray, P. 1991. The Pleistocene megafauna of Australia, in: P. Vickers-Rich, J. M. Monaghan, R. F. Baird, and T. H. Rich (eds.), *Vertebrate Palaeontology of Australasia,* pp. 1071–1164. Pioneer Design Studio, Victoria.

Owen-Smith, N. 1987. Pleistocene extinctions: the pivotal role of megaherbivores. *Paleobiology* **13**(3):351–362.

Owen-Smith, N. 1988. *Megaherbivores: The Influence of Very Large Body Size on Ecology.* Cambridge University Press, London.

Stringer, C. B. 1995. The evolution and distribution of later Pleistocene human populations, in: E. S. Vrba and G. H. Denton (eds.), *Paleoclimate and Evolution: with Emphasis on Human Origins.* Yale University Press, New Haven.

Stuart, A. J. 1982. *Pleistocene Vertebrates in the British Isles.* Longman, London.

Stuart, A. J. 1991. Mammalian extinctions in the Late Pleistocene of northern Eurasia and North America. *Biol. Rev.* **66**:453–562.

Stuart, A. J. 1993. The failure of evolution: late Quaternary mammalian extinctions in the Holarctic. *Quat. Int.* **19**:101–107.

Stuart, A. J. 1995. Insularity and Quaternary vertebrate faunas in Britain and Ireland, in: R. C. Preece (ed.), *Island Britain: A Quaternary Perspective,* pp. 111–125. *Geol. Soc. Spec. Publ.* No. 96.

Tzedakis, P. C., and Bennett, K. D. 1996. Interglacial vegetation succession: a view from southern Europe. *Quat. Sc. Rev.* **14**:967–982.

Vartanyan, S. L., Garrutt, V. E., and Sher, A. V. 1993. Holocene dwarf mammoths from Wrangel Island in the Siberian Arctic. *Nature* **382**:337–340.

Vartanyan, S. L., Arslanov, K. A., Tertychnaya, T. V., and Chernov, S. 1995. Radiocarbon dating evidence for mammoths on Wrangel Island, Arctic Ocean until 2000 BC. *Radiocarbon* **37**(1):7–10.

Vereschagin, N. K., and Baryshnikov, G. F. 1984. Quaternary mammalian extinctions in northern Eurasia, in: P. S. Martin and R. G. Klein (eds.), *Quaternary Extinctions: A Prehistoric Revolution,* pp. 483–516. University of Arizona Press, Tucson.

Watts, W. A. 1980. A long pollen record from Laghi di Monticchio, southern Italy: a preliminary account. *J. Geol. Soc. London* **142**:491–499.

West, R. G. 1980. Pleistocene forest history in East Anglia. *New phytol.* **85**:571–622.

Zagwijn, W. H. 1992. Migration of vegetation during the Quaternary in Europe. *Courier Forschungsinst. Senckenberg.* **153**:9–20.

CHAPTER 12

The Quiet Crisis

A Preliminary Listing of the Freshwater Fishes of the World that Are Extinct or "Missing in Action"

IAN J. HARRISON and MELANIE L. J. STIASSNY

1. Introduction

Throughout most of human history, freshwater resources have been more than adequate to serve human needs while maintaining the integrity and biological diversity of Earth's ecosystems. However, an exponentially increasing human population is placing ever greater demands on Earth's limited supply of fresh waters. Already more than half of all accessible surface fresh water is put to use by humanity with concomitant degradation or loss of habitat such that Earth's freshwater ecosystems have been altered more profoundly than have terrestrial ecosystems (Allan and Flecker, 1993; Postel *et al.*, 1996; Stiassny, 1996, in press; Pimental *et al.*, 1997; Vitousek *et al.*, 1997). Some particular examples serve to illustrate the gravity of the situation. In the United States, 98% of an estimated 5.2 million km of streams are sufficiently degraded to be unworthy of federal designation as wild or scenic rivers (Benke, 1990); industrial agriculture around the Aral Sea in the last 30 years has resulted in an approximate halving of the lake's surface area and depth, and a tripling of its salinity (Mainguet and Létolle, 1997; Pimental *et al.*, 1997); only 2 of Japan's 30,000 rivers are neither dammed nor modified in some way (McAllister *et al.*, 1997). Undoubtedly these types of habitat degradation, often coupled with the deleterious effects of the introduction of exotic species (Courtenay and Moyle, 1992), have profound impacts on the resident biota. Although the precise degree of freshwater impoverishment remains to be fully documented, there can be little doubt that the losses are already great.

The aim of this study is to document the species of freshwater fish thought to have become extinct in "recent" history and note the probable causes for these extinctions. Kotte-

---

IAN J. HARRISON and MELANIE L. J. STIASSNY • Department of Ichthyology, American Museum of Natural History, New York, New York 10024.

Extinctions in Near Time, edited by MacPhee. Kluwer Academic/Plenum Publishers, New York, 1999.

lat (1994) and MacPhee and Flemming* (this volume) have warned against speculating over the processes that drive extinction, and they have stressed that such investigations must be founded on reliable databases. Early on in this study it became evident that the standard lists of extinct fish taxa, such as those of the World Conservation Monitoring Centre (WCMC; 1992) and the World Conservation Union (IUCN) *Red List* (1996), either did not provide enough data or were inaccurate. Magin *et al.* (1994), referring to earlier editions of the IUCN *Red List,* indicated that many scientists and conservationists find the IUCN lists subjective and unrepresentative of real extinction risks. MacPhee and Flemming (this volume) faced these problems when they made a census of mammal extinctions and this prompted them to review the data for themselves.

Similarly, we decided to review listings of fish extinctions by checking primary literature sources, in certain instances by checking museum-held material, and by contacting experts with direct field experience. Kottelat (1994) has criticized *"Desktop-Naturschutz"* (desktop conservation), which amounts to the production of "lists (which have no impact at all on the conservation of the habitats)" and stressed that proactive survey work in the field could provide a more reliable supply of ecological data. Although we certainly agree that fieldwork must form the core of any assessment of biodiversity, these field data themselves are of little utility unless there is some system ensuring that the interpretation of the data is factually accurate rather than being merely speculative; on this point we are in agreement with Kottelat's (1994) desire to see more facts and fewer approximations as the basis of conservation planning.

We recognize that our extinction lists are provisional and will undoubtedly suffer from errors of omission and some misinterpretation. However, following MacPhee and Flemming (this volume), we consider that the main contribution of this study is the creation of a framework ensuring that data have been assessed as rigorously as possible, and are presented in a form facilitating further review and testing. We hope this list will also aid in the recognition of the main causes of recent freshwater fish extinctions.

2. Definition of Terms

2.1. Fishes

Our study interprets the term *fishes* in its broadest sense, namely, aquatic craniates that possess gills that are used throughout life and limbs in the the shape of fins (Nelson, 1994). Only taxa that are known to have had representatives in "recent history" (defined below) are considered in this study; these include the "jawless fishes" (Agnatha), the "cartilaginous fishes" (Chondrichthyes), and the "bony fishes" (i.e., nontetrapod Sarcopterygii and Actinopterygii).

*MacPhee and Flemming (this volume) have devised an excellent system of categories for assessing recent mammalian species-level extinctions and they kindly provided us with an early draft of the manuscript. There are important differences in the state of knowledge of mammals and fishes and while the extinction categories we present in this chapter are based on those of MacPhee and Flemming, ours have been somewhat modified to make them more applicable to studies of freshwater fishes.

2.2. Extinction, Extirpation, Endemics, and Natives

Unfortunately, there is little consistency in the terminology used to denote the conservation status and distributional data applied to species in much of the pertinent literature. For example, species are often referred to or listed as extinct in a given area; for example, Povz (1995) listed *Chondrostoma genei* as extinct, and Keith and Allardi (1996) listed *Aphanius iberus* (= *Lebias iberus,* following Lazara [1995]) and *Valencia hispanica* as extinct. However, in each of these cases, individuals are to be found elsewhere within their natural ranges, and are thus only "regionally extinct." For clarity we suggest the universal adoption of the term *extirpated* to denote regional extinction and restriction of use of the term *extinct* for cases of global extinction. Similarly, use of the term *endemic* has not been consistently applied. Species are correctly described as endemic to a region *only* when their global geographical range is restricted to that region, such as the endemic Lake Victoria cichlids (Stiassny, 1996). Unfortunately, the term *endemic* has also been applied less exclusively. For example, Chen and Cui (1993) refer to the disappearance of the endemic cyprinid *Aspiorhynchus laticeps* from Lake Bositen, Xinjiang, China, following the introduction of European perch. This terminology implies that *A. laticeps* is now extinct because it has disappeared from its entire range. However, in this example the term *endemic* has been misapplied as the taxon is "native" (i.e., naturally occurring or indigenous) to Lake Bositen but is not restricted to it. In fact, *A. laticeps* was originally described from the Kashgar River. The opposing couplet to the term *native* is *exotic* or *introduced,* denoting species that have been introduced into a region that is not part of their natural range.

2.3. Extinctions in "Recent History"

Following MacPhee and Flemming (this volume), we take the period of "recent history" to span AD 1500 until the present.

3. Proof Extinct or Proof Extant?

Unless one knows with certainty all aspects of the habitat and range of a species, it is impossible to provide proof of an extinction. This is particularly pertinent for fishes, which are naturally inconspicuous and their distribution and ecology are often incompletely documented (Miller *et al.*, 1989; Greenwood, 1992; Stiassny, 1996). In many cases, an extinction can only be posited on the assumption that the absence of information about its continued existence actually represents the absence of the species in nature. There are two different approaches to this problem. The first recognizes the potential fallibility of an assumption that an absence of documentation equals extinction, and sets criteria for a more structured adjudication of an extinction. Thus, a species is to be presumed "extant unless proven extinct" (Diamond, 1987). This approach may be appropriate for a well-studied species where any decline in numbers or possible threat of extinction is likely to have been documented. However, Diamond (1987) noted that other species that are less well-documented, either through a lack of fieldwork or through poor knowledge of their identification and distribution, could easily become extinct without being documented as such.

A second approach that may be more appropriate for more poorly known groups assumes that the absence of documentation equals extinction, that is, the species is "extinct unless proven extant." This approach is comparable to Viederman and colleagues' (1994) "Precautionary Principle," which was adopted by the Biodiversity Convention; it states that an absence of scientific proof of a threat to biological diversity should not be allowed to delay the implementation of conservation measures if there is some indication that this threat exists (McAllister *et al.,* 1997). Bruton (1995:23) noted that the numbers of threatened or extinct fishes may be grossly underestimated and there is a need "to bring the real situation to the attention of a wider audience." Diamond's more inclusive policy of listing any species as extinct unless proven extant would rectify this by producing a much more extensive listing of extinct taxa, thereby drawing more attention to what has been referred to as the "quiet crisis . . . taking place beneath the surface of the world's rivers and lakes" (Richter *et al.,* 1997:1082).

Notwithstanding the "quiet crisis," which is a very real event in some aquatic habitats, an "extinct unless proven extant" approach can be expected to include species that are still extant but have not been collected or observed for some period of time. An example of this is provided by a West African goby, *Sicydium brevifile* Ogilvie-Grant. Harrison (1993) stated that this species was only known from the holotype, collected from the Cameroons in 1884, despite numerous collecting expeditions to this region throughout the twentieth century. According to the "extinct unless proven extant" rationale, one could legitimately cite this species as extinct. However, recent reexamination of museum-housed material indicates that additional material of *S. brevifile* exists, including specimens that were collected in 1991 (Harrison, in preparation), thereby confirming the species was still extant at that time.

It seems clear that any extinction list based on the "extinct unless proven extant" approach would quickly become viewed as poorly substantiated and unreliable, if not alarmist. The extinction list could not be taken seriously by scientists or conservation policymakers and, rather than aiding the implementation of conservation strategies, the list could prove detrimental to the support and development of such strategies (Mann, 1991; Kottelat, 1994). To produce an extinction list that can be viewed with some confidence and, hence, has some practical value, it is necessary to set rigorous criteria that more appropriately establish an extinction, therefore we favor the more conservative approach of "extant unless proven extinct."

4. Establishing Systematic Validity

The first stage in adjudicating an extinction is to ascertain the systematic validity of the species under consideration. MacPhee and Flemming (this volume) indicate that the species' name should be acceptable according to the *International Code of Zoological Nomenclature* (1985), the species should be taxonomically well-characterized and accepted as such in the current systematic literature, and the species should, preferably, have been examined as part of a systematic revision.

MacPhee and Flemming do not include subspecies in their extinction lists, on the basis of difficulties in interpreting this taxonomic rank. Although many subspecies may represent perfectly good evolutionary species, the allocation of subspecific rank is a matter of convenience, often applied to sidestep systematic issues that are difficult to resolve (Wiley, 1981). Until these issues are resolved through taxonomic revision, a subspecies can only be interpreted as a variant population of an evolutionary species that was either formerly more widespread or still remains widespread (Table 8). Kottelat (1995, 1997) discusses the application of the "phylogenetic species concept" and "pragmatic species concept" as the most objective methodologies for taxonomic reviews and he states that these concepts overcome the problem of assessing arbitrarily defined subspecies. We have adopted Kottelat's approach where it has previously been applied in taxonomic reviews. Under other circumstances, we make reference to subspecies but do not categorize them until more detailed taxonomic analyses establish the subspecies "are clearly and unambiguously distinct and occur in discrete geographical areas" (Kottelat, 1995). We recognize that this may temporarily underestimate the number of documented extinctions.

5. Establishing a Date of Extinction

Having confirmed the systematic validity of a species, it is necessary to establish a date after which the species is believed to be extinct. MacPhee and Flemming (this volume) refer to this as the "effective extinction date" (EED) and we adopt their terminology here. MacPhee and Flemming give three categories of acceptable evidence for an EED. Two of these concern taxa that are known only as fossils or were never the subject of an observation in the living state and these categories do not apply to the recent extinctions of fishes reported herein. For our purposes, the most important category is as follows.

Collection-Based Evidence

An EED should be supported by collection-based evidence for the last record of the taxon, after subsequent unsuccessful efforts to locate it. Collection-based evidence refers to specimens where the identity and date of collection are reliable, and accompanied by ancilliary field data (e.g., locality of capture and collectors). In the absence of vouchered material, reported sightings of a given species should be noted but are best treated as unsubstantiated.

An EED should be as precise as possible, citing the year when the species was last reliably collected. The aquatic habitat of fishes makes them generally inconspicuous and many species are likely to be observed only if there is a particular effort to catch them. Moreover, it is very difficult to collect from some types of aquatic habitats, such as deep or torrential waters where species may still evade even the most concerted efforts to locate them. Therefore, recognition of the absence of a species from an area may occur only several years, or decades, after its last collection. In such cases, it is reasonable to cite the EED as being sometime between the last collection and the first suggestion of extinction. When the extinction can be linked, with some certainty, to a particular event of known date such as the introduction of an exotic species or the damming or draining of a river, then this date should also be noted as the probable EED. For example, several species of endemic cyprinids, described from Lake Lanao in the Philippines between 1924 and 1932, were first noted as being absent from the lake in 1982 and their hypothesized extinction was linked to a combination of factors, culminating in the introduction of competitive and predatory species in both 1963 and 1975. Under these circumstances, the EED for the Lanao species can be said to have occurred between 1932 and 1982, and most probably from 1963 onwards. This is represented in the extinction tables as "1932–1982 [1963–1982]."

6. Extinction Criteria

An EED, by itself, is not sufficient proof of an unequivocal extinction. Additional criteria are necessary to substantiate an extinction as a real event rather than the absence of documentation pertaining to the species. Most previous authors have relied on one of two criteria for adjudicating an extinction: either the species must have been absent for a significant period of time before it can be accepted as extinct or there must have been significant yet unsuccessful attempts to relocate the species after notification of its absence.

We consider that each of these criteria represents an important test of the likelihood that a species is extinct. However, the development of these criteria as more precise, *quantitative* assessments of the probability of extinction for diverse taxa is difficult, if not arbitrary, and neither criterion can be easily applied to every extinction scenario (see further discussion in Sections 6.1 and 6.2). Consequently, we believe that these criteria should be applied together rather than in isolation to resolve an extinction. Additionally, we propose a third criterion (Section 6.3) that reduces the possibility of listing a rarely observed, but extant, species as extinct.

6.1. Extinction Implied by Lengthy Period of Absence

Prior to 1994, the IUCN stipulated that a species could only be accepted as extinct if it had not been located in the wild during the past 50 years (Groombridge, 1993). The disadvantage of this ruling is that it can underestimate the total number of documented extinctions by discounting those species for which there is good evidence for their extinction within the last 50 years (Magin *et al.,* 1994). For this reason, WCMC (1992) did not incorporate the 50-year ruling and in its later recommendations IUCN (1994) also dropped the 50-year rule and defined their "Extinct" category as "when there is no reasonable doubt that the last individual has died."

IUCN's (1994) criteria placed greater emphasis on the role of sampling to establish an extinction. However, as discussed below, in many cases it can be difficult or impossible to judge when sufficient sampling has been conducted to confirm the absence of a species. One solution is to readopt the 50-year rule and assume that a species might be seen or collected purely by chance if it is still extant. MacPhee and Flemming (this volume) state that the IUCN/CITES recommendation of 50 years is arbitrary but allows a reasonable time period during which a species might be found if it was not extinct. Hence, they incorporate the 50-year rule as a criterion for defining their category of a "resolved extinction." We concur with MacPhee and Flemming and also apply the 50-year rule to resolve extinctions. The advantage gained by adhering to a system that ensures species are not prematurely listed as extinct outweighs the disadvantage, in the short term, of underestimating the total number of extinctions.

Nevertheless, an extant species may remain undetected for significantly more than 50 years if its habitat has not been satisfactorily sampled (Diamond, 1987). Again, this is particularly pertinent for fishes that utilize habitats that are often inaccessible and frequently difficult to sample (Miller *et al.,* 1989). For these reasons, Menon (1994) and Collar (1998) have indicated that the 50-year rule should be applied with extreme caution. Accordingly, we propose the application of this rule in conjunction with other criteria.

6.2. Extinction Implied by Representative Sampling

Ponniah (1994) correctly noted that sampling effort provides a more realistic test for extinction than does a simple period of absence of data. The possible extinction of some species that inhabit very restricted biotopes can be ascertained by a few days of sampling rather than a 50-year wait. For example, the possible extinction of some cyprinodontiform fishes that live in single, small, desert pools could be established following a relatively simple sampling of that habitat.

IUCN (1994:13) recognized the importance of representative sampling and stated that the extinction of a species in the wild can be presumed "when exhaustive surveys in known and/or expected habitat, at appropriate times (diurnal, seasonal, annual) throughout its historic range have failed to record an individual. Surveys should be over a time frame appropriate to the taxon's life cycle and life form." This repeat sampling is important as it increases the probability that the species had not been overlooked by chance, or that it was not just absent because of some temporary, regional disappearance occurring as part of the normal life cycle of the species.

How does one define "exhaustive surveys"? In some cases it is relatively easy to judge when sufficient sampling has been conducted, such as the case of desert cyprinodontiform fishes mentioned above. By contrast, there are many more cases where it is difficult, if not impossible, to conduct the level of "exhaustive survey" that could definitively establish the presence or absence of a species. This is a particular problem where the habitat of the species covers a very large or inaccessible area, such as a large or deep lake, which cannot be sampled thoroughly. For example, a population of the small coregonine *Prosopium coulteri* was first detected in Lake Superior in 1952, despite the fact that this lake had been well-studied for decades prior to this (Eschmeyer and Bailey, 1955). Thus, some inevitable effects of using sampling data as an adjudication of extinction are as follows:

- Many species of fish occupy ranges that are too large or habitats that are too inaccessible to allow anything more than a subjective assessment of the available sampling data (in these cases there is an inherent value in incorporating the 50-year rule, thereby providing an extensive time period to corroborate sampling data).
- The level of sampling necessary to ascertain an extinction for one species may be significantly different from that required by a different species in a markedly different habitat; thus, it is difficult to define some standard measure of sampling for all fishes.

Rather than try to define a precise measure for representative sampling that can be uniformly applied, we propose a more qualitative approach. Sampling should be as extensive as possible and there should be an indication of one or both of the following:

- Sampling efforts were targeted at the species in question, in an attempt to relocate it in its known or expected range
- Sampling efforts, although perhaps not directly targeted at the species in question, would have been likely to collect the species from its known or expected range.

Such a policy enables a more informed judgment to be made as to the presence or absence of a species in its natural range, and thereby facilitates a level of accuracy within a process that is, by default, subjective in its nature.

To resolve an extinction, sampling data must indicate that the species is absent from its known or expected range (in the appropriate habitat) following the hypothesized EED. However, although the absence of a species after repeated collection attempts can be strongly suggestive of extinction, in most cases this cannot be considered irrefutable evidence for the extinction. South Africa, which has no confirmed fish extinctions, provides two useful examples to illustrate this. The minnow *Labeo quathlambae* was originally described from specimens collected from the Umkomazana River, Natal, in 1938 and at the time it was recorded as showing a dramatic decline in population size following the introduction of trout to its habitat (Barnard, 1938). Greenwood and Jubb (1967) assumed the species was "probably extinct" on the basis of Barnard's (1938) documentation of population decline and environmental threat, and the fact that attempts to collect the species from its type locality and similar streams in the region had been unsuccessful after 1938. However, Pike and Tedder (1973) documented the rediscovery of the species, in 1970, from the Tsoelikana River, Lesotho, some 50 km southwest of the type locality (Miller, 1971). Similarly, the river pipefish *Syngnathus watermeyeri* was collected in 1963 from only three South African estuaries and was not seen again despite intensive sampling efforts in the region between 1989 and 1992 (Whitfield and Bruton, 1996). Consequently, *S. watermeyeri* was presumed to have become extinct as a result of river impoundment (Whitfield, 1997), and the species was listed as extinct in the 1994 IUCN *Red List* (Groombridge, 1993). One year later a single specimen was captured from the type locality and an additional population was discovered at a new locality in 1996 (Whitfield, 1997). Miller *et al.* (1989) have provided numerous comparable examples of North American freshwater fishes presumed extinct and yet subsequently rediscovered. Thus, whereas sampling effort is a practical way of testing for an extinction, it is not without problems and should not be relied on as the sole means of adjudicating an extinction. These examples illustrate the value of maintaining the 50-year rule in combination with sampling criteria.

6.3. Extinction Implied by Population Decline or Environmental Threat

There are many examples of fishes that have been collected more than 50 years ago and have not been recorded since, despite subsequent collecting in their known ranges. Examples include the west African gobiid *Ebomegobius goodi,* known from only one specimen collected from Cameroon in 1940 (Miller, 1988), or the African cyprinid *Raiamas longirostris* known from a single specimen collected in Zaire in 1901 (Howes, 1991). If it were only necessary for species to meet extinction criteria 6.1 and 6.2 above, then these species would reasonably be listed as extinct; however, there is no evidence for any causation for the hypothesized extinction of these species. MacPhee and Flemming (this volume) have noted that species such as these may be extinct, or they may simply be rare, or evasive. The gobiid *Sicydium brevifile* (discussed above) is another example of a species that would have been classified as extinct if it only needed to meet criteria 6.1 and 6.2 but is now known to be extant. Accordingly, we adopt the protocol that there should be some evidence that the purportedly extinct species was undergoing a decline in population size prior to the hypothesized EED, such that it was already considered "threatened" according to IUCN (1994) categories (i.e., "critically endangered," "endangered," or "vulnerable"), or the species faced an environmental catastrophe that can reasonably be assumed to have caused its sudden extinction.

As noted above, evaluation of extinctions based on sampling data necessitates some degree of subjective assessment. There are cases where sampling data are not fully exhaustive, as when sampling does not cover the entire geographic range of the species. Nevertheless, additional, ecological information may exist that adds further "weight" to the sampling data. Lake Victoria cichlids provide a very good, contemporary example of this. Many endemic cichlids are becoming locally extirpated from parts of the lake (Witte *et al.*, 1992a; Seehausen and Witte, 1995). The quantitative evidence for the possible extinctions of cichlids is based on surveys that are very recent and/or do not adequately cover the potential geographic ranges of the species (see Section 11.3.1). Nevertheless, there is considerable evidence to show that Lake Victoria is in a state of ecological transformation and disruption (Witte *et al.*, 1992a; Kaufman, 1992; Kaufman and Ochumba, 1993) that clearly poses a serious environmental threat to the endemic cichlids. Although the catastrophic regional declines in populations as evidenced by sampling are not, per se, conclusive evidence of extinctions, the weight of the additional ecological evidence is enough to aid a judicious assessment of the possibility of extinctions based on the corresponding sampling data. However, it is important to note that direct proof of an extinction must still be established by a confirmed absence of the fish themselves; the ancillary evidence of environmental threat can corroborate the hypothesized extinction but not prove it. In summary, evidence of the following is required to resolve a recent extinction:

1. The species' name is accepted as taxonomically valid according to the rules of zoological nomenclature (ICZN, 1985) and, preferably, the species' taxonomic status has been considered in a recent systematic revision.
2. There is evidence for an EED within recent history (i.e., after AD 1500) but not more recently than 50 years ago; and the EED can be ascribed according to collection-based evidence.
3. There have been unsuccessful attempts, with appropriate sampling (discussed in Section 6.2), to relocate the species subsequent to the EED.
4. There is some indication that, prior to the EED, the species was threatened by a decline in population size or it faced an environmental threat that can reasonably be assumed to have caused its sudden extinction.

7. Extinction Categories

The criteria discussed above establish a system for adjudicating a recent extinction as accurately as is practically possible. Ideally, all proposed extinctions would either fit all of these criteria and be accepted, or not fit the criteria and be rejected. However, our analyses of putative extinctions of freshwater fishes indicate that there are very many cases where some element of the supporting evidence may be lacking or open to question but the extinction is still strongly suggested by other evidence. Rather than disregarding species that have not met all extinction criteria, we would rather prioritize them for further attention. Therefore, a survey of extinctions should account for these less well-documented extinctions, delimit them from more definitive extinctions that meet all of the criteria set out above, and thereby draw attention to them as needing further investigation and, where possible, instigate remedial conservation action.

Mace (1994), in a discussion of the role of "threatened species lists," emphasized the importance of establishing an objective and standardized system for categorizing these species according to the perceived threat. Similar argumentation can be applied to putatively extinct species, and MacPhee and Flemming (this volume) have devised a system that categorizes extinctions according to the extent of supporting evidence. Their system incorporates a category of "Resolved Extinctions" for those species that meet all of their criteria for defining an extinction, and several categories of "Unresolved Extinctions" that describe the strengths and weaknesses of the evidence supporting other putative extinctions. We follow MacPhee and Flemming's system and have adopted their terminology, with some modification for our own extinction lists for fishes.

7.1. Category A: Resolved Extinctions, AD 1500–1948

Species are considered eligible for this category if they meet the four criteria listed above that define a "recent" extinction.

7.2. Category B: Unresolved Extinctions, AD 1500–1948—Missing Data

Species are placed in this category when *both* of the following criteria apply:

1. There is evidence that the extinction occurred during the period AD 1500–1948.
2. The extinction cannot be fully resolved *either* because the taxonomy of the species is ambiguous *or* because some element of the evidence supporting the EED is ambiguous.

Species that show ambiguity in taxonomy and do not have some element of evidence indicating an EED cannot be reliably placed in any extinction category and are allocated to category U (see Section 8.1). Because several species fit into category B, it has been subdivided to identify where the supporting data are ambiguous, as follows.

7.2.1. Category B1: EED Substantiated/Taxonomy Unsubstantiated

Species are placed in this subcategory when their EED is substantiated in some way (either by sampling after the EED or by population decline/environmental threat prior to the EED) but the taxonomic validity of the species is ambiguous. This includes:

- Species that have not been adequately reviewed, or the professional literature includes some question as to the taxonomic validity of the species
- Species that have not been formally named or diagnosed according to the rules of zoological nomenclature (ICZN, 1985)

It seems reasonable that until a formal description of a species appears, one cannot begin to talk about whether it is extinct. Clearly, if the taxonomy of the species can be resolved, its status as extinct or extant is open to reevaluation.

7.2.2. Category B2: Taxonomy Substantiated/EED Unsubstantiated by Sampling

Species are placed in this subcategory when there is no problem with their taxonomy, but an EED between 1500 and 1948 has been cited which is not fully substantiated because attempts to relocate the species in nature after the hypothesized EED do not adequately indicate that the species is absent from its known or expected range and habitat. However, this subcategory requires that when there is no evidence for representative sampling, then the species must show some evidence of a population decline or environmental threat prior to the hypothesized EED.

7.2.3. Category B3: Poorly Defined Extinctions with Taxonomy Substantiated/EED Unsubstantiated by Population Decline or Environmental Threat

Species are placed in this subcategory when there is no problem with their taxonomy but an EED between 1500 and 1948 has been cited that is not fully substantiated because there is no indication of population decline or environmental threat that might have resulted in the extinction of the species. However, this subcategory requires that when there is no evidence of a population decline or environmental threat prior to the hypothesized EED, then the EED must be supported by rigorous evidence of thorough sampling throughout the species' range and habitat after the EED. Some particular problems associated with this category are discussed in Section 10.2.

Species that have their EED supported neither by representative sampling nor by population decline/environmental threat cannot be reliably placed in any extinction category and are allocated to category U (see Section 8.1).

7.3. Category C: Unresolved Extinctions—Extinction after 1948 but with No Other Ambiguity

Species are considered eligible for this category if they meet the four criteria listed above that define a "recent" extinction, with the exception that the evidence for the extinction date is within the last 50 years (i.e., after 1948, rather than prior to 1948). The proposed extinctions of these species are supported by a great deal of evidence and they would be considered as category A "resolved extinctions" were it not for the 50-year rule. Nevertheless, for reasons given in Section 6.1, we opt to include them as unresolved extinctions until a suitable waiting period has elapsed.

7.4. Category D: Unresolved Extinctions—Extinction after 1948 and with Missing Data

Species are placed in this category when the following two criteria apply:

1. There is evidence that extinction occurred after 1948.
2. The extinction cannot be fully supported *either* because the taxonomy of the species is ambiguous *or* because some element of the evidence supporting the EED is ambiguous.

As with category B, species that show ambiguity in taxonomy and do not have some element of evidence indicating an EED cannot be reliably placed in any extinction category and are allocated to category U (see Section 8.1). Several species would fit into category D, and therefore it has been subdivided to identify the areas where the supporting data are ambiguous as follows.

7.4.1. Category D1: EED Substantiated/Taxonomy Unsubstantiated

Species are placed in this subcategory when their EED is substantiated in some way (either by sampling after the EED or by population decline/environmental threat prior to the EED) but the taxonomic validity of the species is ambiguous (see category B1 for further details on taxonomic validity).

7.4.2. Category D2: Taxonomy Substantiated/EED Unsubstantiated by Sampling

Species are placed in this subcategory when there is no problem with their taxonomy but an EED after 1948 has been cited that is not fully substantiated because attempts to relocate the species in nature after the hypothesized EED do not adequately indicate that the species is absent from its known or expected range and habitat. However, this subcategory requires that when there is no evidence for representative sampling, then the species must show some evidence of a population decline or environmental threat prior to the hypothesized EED.

7.4.3. Category D3: Poorly Defined Extinctions with Taxonomy Substantiated/EED Unsubstantiated by Population Decline or Environmental Threat

Species are placed in this subcategory when there is no problem with their taxonomy but an EED after 1948 has been cited that is not fully substantiated because there is no indication of population decline or environmental threat that might have resulted in extinction of the species. However, this subcategory requires that when there is no evidence of a population decline or environmental threat prior to the hypothesized EED, then the EED must be supported by rigorous evidence of thorough sampling throughout the species' range and habitat after the EED. Some particular problems associated with this category are discussed in Section 10.2.

Species that have their EEDs supported neither by representative sampling nor by population decline/environmental threat cannot be reliably placed in any extinction category and are allocated to category U (see Section 8.1).

8. Problematic Categories

8.1. Category U—Unclassifiable Species

Species are placed in this category when current evidence for their extinction is not strong enough to assign them to categories A–D, but there is some information that gives reasonable cause to flag the species for further analysis. This category includes some species that have been described as being under severe environmental threat or being "virtually extinct," with little or no additional information. It would be inappropriate to assign these species to any of the extinction categories on the basis of the information that is currently available. However, this category presents an important opportunity to draw attention to these species so that other researchers may make more detailed investigations.

8.2. Extinct in the Wild

IUCN (1994) included the category "Extinct in the Wild" for taxa that survive only in cultivation or captivity. MacPhee and Flemming (this volume) recognized the value of this category for conservation purposes but did not incorporate it into their system because they were only interested in truly "extinct" species where all individuals, wild or captive, have died. We agree with MacPhee and Flemming's reasoning; moreover, we have some additional concerns about the application and validity of the "Extinct in the Wild" category as applied to fishes.

The IUCN (1994:13) definition of "Extinct in the Wild" included species represented by "a naturalized population (or populations) well outside the past range." However, the rising frequency, through the twentieth century, of local extirpations and translocations of freshwater taxa means that their current distributions can be significantly different from their natural distributions (see Bianco, 1995, for a discussion of "zoogeographic pollution" in Italian waters). The lake trout *Salvelinus namaycush,* which had a native distribution extending northwest from the northeastern United States to Alaska (Martin and Olver, 1980), provides an interesting example involving a series of introductions and extirpations. Native populations of the lake trout were eliminated from parts of the Great Lakes, and perhaps totally extirpated from Lake Michigan, by the introduction of the sea lamprey *Petromyzon marinus* (Smith, 1972; Behnke, 1980a). However, *S. namaycush* was itself introduced to lakes in Europe, South America, and New Zealand (Martin and Oliver, 1980). Thus, it is conceivable that some species of fish (notably cultured or sport fish) may become represented by naturalized populations outside their natural range, although these species are not under threat of extinction in the wild.

The "Extinct in the Wild" category can also be misleading by indicating a greater conservation concern for species held only in captivity than for species confined to highly restricted populations in the wild (e.g., a single locality). A species that is still extant in the wild can be at a similar or perhaps greater risk of extinction than a species that is extant only in an artificial habitat.

There are, therefore, various reasons why we hesitate to use the "Extinct in the Wild" category in any context. It seems that a more informative assessment of threat to species, whether wild or captive, can be made using the other IUCN (1994) *Red List* categories (e.g., "critically endangered," "endangered," "vulnerable," "lower risk").

9. Key to Categories

The system presented here is designed to be a practical and informative method for defining and categorizing extinctions. There are, evidently, diverse criteria for evaluating extinctions and this is reflected in the number of categories and subcategories. We recognize that this system might, initially, be criticized as overly complex. Unfortunately, the simpler the system, the less informative it is. To help clarify the assignment of a species to a category, we present the following key.

1a. EED at least 50 years ago →2
1b. EED less than 50 years ago →7

2a. Species taxonomically ambiguous (requiring review or not fully described/ named) →3
2b. Species taxonomically valid and fully described/named →4

3a. EED supported by evidence of (1) appropriate but unsuccessful sampling after the EED and/or (2) population decline/environmental threat before the EED →Category B1
3b. EED without supporting evidence, that is, *neither* (1) appropriate but unsuccessful sampling after the EED *nor* (2) population decline/ environmental threat before the EED →12

4a. EED *fully* supported by evidence of *both* (1) appropriate but unsuccessful sampling after the EED *and* (2) population decline/environmental threat before the EED →Category A
4b. EED *incompletely* supported (insufficient sampling after the EED and/or lack of information on population decline/environmental threat before the EED) →5

5a. EED partially supported by evidence of *either* (1) appropriate but unsuccessful sampling after the EED *or* (2) population decline/environmental threat before the EED →6
5b. EED lacking evidence of *both* (1) appropriate but unsuccessful sampling after the EED *and* (2) population decline/environmental threat before the EED →12

6a. EED lacking evidence of appropriate but unsuccessful sampling after the EED, but supported by evidence of population decline/environmental threat before the EED →Category B2
6b. EED lacking evidence of population decline/environmental threat before the EED, but supported by evidence of appropriate but unsuccessful sampling after the EED →Category B3

7a. Species taxonomically ambiguous (requiring review or not fully described/ named) →8
7b. Species taxonomically valid and fully described/named →9

8a. EED supported by evidence of (1) appropriate but unsuccessful sampling after the EED and/or (2) population decline/environmental threat before the EED →Category D1
8b. EED without supporting evidence, that is, *neither* (1) appropriate but unsuccessful sampling after the EED *nor* (2) population decline/ environmental threat before the EED →12

9a. EED *fully* supported by evidence of *both* (1) appropriate but unsuccessful sampling after the EED *and* (2) population decline/environmental threat before the EED →Category C
9b. EED *incompletely* supported by evidence (insufficient sampling after the EED and/or lack of information on population decline/environmental threat before the EED) →10

10a. EED partially supported by evidence of *either* (1) appropriate but unsuccessful sampling after the EED *or* (2) population decline/environmental threat before the EED →11
10b. EED lacking evidence of *both* (1) appropriate but unsuccessful sampling

after the EED *and* (2) population decline/environmental threat before the EED →12

11a. EED lacking evidence of appropriate but unsuccessful sampling after the EED, but supported by evidence of population decline/environmental threat before the EED →Category D2

11b. EED lacking evidence of population decline/environmental threat before the EED, but supported by evidence of appropriate but unsuccessful sampling after the EED →Category D3

12. Some information associated with species gives reasonable cause to highlight it for further analysis →Category U

10. Notes on Data Compilation

10.1. Difficulties on Vague Data and Terminology

One notable difficulty encountered while compiling this preliminary assessment and categorization of extinction data was evaluating extinctions that are discussed in imprecise terms. The supporting data for extinctions can be incomplete or vague (e.g., *Eudontomyzon* sp.; see Table 7), or the extinction is proposed in guarded terminology. Often, for example, species are suggested to be extinct but not categorically stated as such; phrases like "possibly extinct," "seems to be extinct," "not known if they still exist" are commonplace (Collar, 1998). This problem has arisen partly because of the lack of any practical system for assessing the relevant data and defining an extinction and, in the absence of such a system, many authors are understandably cautious about claiming that a species is extinct. However, the result is that our allocation of these vaguely defined extinctions to their respective categories is necessarily provisional pending the availability of more precise information.

Allocation of species to categories D1–3 or U was particularly problematical and we had to exercise a great deal of subjective analysis of the value of existing sampling data in an effort to determine which species merited inclusion and where. This was a significant problem in our assessment of the Lake Victoria cichlids (Section 11.3.1).

10.2. Special Problems with Categories B3 and D3

Both of these categories represent cases where there is no evidence for causation of an extinction and the extinction is only implied by the absence of specimens from collections after a certain date (taxon's EED). These categories provide holding stations for those problematic species that are known from the type collection but no subsequent collections, and feared to be extinct. Diamond (1987) had used such species known only from type collections to illustrate the importance of the "extinct unless proven extant" rationale. MacPhee and Flemming (this volume) warn of the dangers of making judgments on the extinction of species known from restricted "hypodigms" (sensu MacPhee and Flemming, this volume; see Section 11), and they avoid making any judgments when they are not confident that post-EED sampling has been adequate. They state that these species might be absent not because they are extinct but because they are unusually rare, cryptic, or can evade collection. Alternatively, the collected specimen or specimens could represent hybrids or morphologically aberrant forms. So as to reduce the risk of rare but extant species being incorrectly placed in categories B3 and D3, we recommend that species are only included in these categories when there is precise evidence of efforts to relocate the species in nature after the EED. There are many species that are represented only by small type collections (Stiassny, in press) and that might be assigned to these categories because there is some evidence of post-EED sampling in their type localities. However, much of the time the sampling efforts do not seem to be particularly thorough or targeted at the species in question. We need to have very specific information on the quality of post-EED sampling before a judicious decision can be made to include species in these categories. Therefore, these species represent special cases that require careful, independent scrutiny, preferably with advice from specialists who are well acquainted with the taxonomic group and the history of fieldwork in the area under consideration.

Ebomegobius goodi and *Raiamas longirostris* (discussed in Section 6.3) are two examples of species known only from their holotypes that, on first consideration, might be placed in categories B3 or D3 because there is evidence of subsequent collections from their type localities. However, more detailed scrutiny of ancillary information pertaining to these species suggests that they may not be extinct. For example, Miller (1988:271) suggested that *E. goodi* "is likely to be widely distributed" and Howes (1991) stated that the specialized habits and unsampled habitats of *R. longirostris* may have enabled it to escape collection. Because of this ancillary information, these species are not included in the B3 category.

We do not consider the current survey to be the place for these more detailed investigations and we have avoided placing candidate species into categories B3 and D3 except for one instance, the preliminary placement of *Chasmistes muriei* in category D3 (Table 6, see note *a*). Although the evidence of representative sampling for *C. muriei* is not as rigorous as we would ideally require for this category (see Section 7.4.3), we believe that the region of the type locality has probably been representatively sampled by North American fisheries surveys. We presume that more detailed analyses of other species represented only by small hybodigms will identify which other species can justifiably be assigned to categories B3 and D3. However, we view this category as representing very poorly defined extinctions and hope that further sampling efforts, more detailed analyses of the ecology of the sampled regions, and comparative taxonomic analyses of purportedly extinct B3/D3 species with other taxa might provide data allowing reallocation of these species to the better defined categories.

10.3. Importance of Published Extinction Evidence

Invariably, we have been more confident about assigning species to extinction categories on the basis of published, rather than unpublished, information. This opinion is also shared by MacPhee and Flemming (this volume). Published data provide a source of primary information that is referable and thus accessible to anyone wishing to review it (cf. the IUCN *Red Lists,* which do not cite publications, making it very difficult to cross-check the extinction data). Although the publication of data does not guarantee their technical accuracy, it is reasonable to assume that, in most cases, the authors have felt suitably confident about the accuracy of their data to publish them. Furthermore, when the data appear in scientifically edited and reviewed publications, there have been other specialists who have judged the quality of the work. Thus, by virtue of being published, the data have already undergone some assessment of reliability.

Where possible, we have attempted to use published information as the principal data for assigning species to extinction categories, and unpublished information as corroborative data. Only 4 of the 70 assignments of species that we make to our extinction categories A–D have not been supported by significant published information: *Brycon acuminatus, Henochilus wheatlandi,* and *Platytropius siamensis* were assigned to category C (see Tables 3 and 13), and *Rhizosomichthys totae* assigned to category D2 (see Table 5). The remaining 66 assignments, although aided by information provided as personal communications from colleagues, were ultimately based on published information. Our category U (Table 7) includes a number of species that have been suggested as possibly extinct in personal communications and other unpublished sources, but we cannot confidently assign them to another category until that information has been further substantiated and published.

The foregoing discussion indicates the importance of ensuring that precise extinction data are published and easily located in the scientific literature. Nevertheless, it is not common for authors to address extinctions as a primary issue. This is perhaps because there is no specific forum designed for the publication of data pertaining to species extinctions. Consequently, references to extinctions are usually only secondarily addressed within discussions of more general issues about taxonomy (*Rheocles sikorae;* see Table 3) or conservation (e.g., two Turkish species of *Phoxinellus* are noted as extinct in a discussion on the effects of introduced species; Crivelli, 1995, and see Table 5). In some cases the necessary information is dispersed throughout several different publications (e.g., *Coregonus fera,* Lake Lanao cyprinids; see Tables 3 and 4). Therefore, much of the relevant data for extinctions is very difficult to locate in the literature and this makes tracking down extinction data an arduous process. We believe that the future accuracy and consistency of extinction requires the following:

- An official forum designated for the publication of extinction data as its primary purpose and according to precise rules and standards
- A categorization system, such as the one proposed here, for accurately evaluating the data according to a precise set of criteria

The forum would govern the publication of extinction data so that they are presented in a style that is comprehensive, precise, and consistent, much in the same way that the International Commission on Zoological Nomenclature governs the naming of taxa. Species would be assigned to the different extinction categories via this forum by a proposal and panel-review. The proposal would be required to include all necessary supporting data to meet the criteria for the proposed extinction category. This process would ensure that researchers can expediently publish data on extinctions, and would guarantee that extinctions are reliably categorized and all relevant information can be easily located in the literature. Discussion in this chapter and in MacPhee and Flemming (this volume) has indicated the importance of reliable taxonomic and nomenclatural information for the assessment of proposed extinctions of species. Accordingly, the framework for the forum is currently being designed as a collaborative project between systematists and other biologists based at the American Museum of Natural History and several other, international scientific institutions and conservation organizations. A detailed discussion of this forum will be presented in a forthcoming publication.

11. Results

The lists of species that are included in our extinction categories A–D (Tables 1–6) and ?D1 and ?D2 for Lake Victoria cichlids (Tables 9 and 10; see Section 11.3.1) represent a reassessment of the available data. We have sought to amass information for all species that are thought to be extinct or are threatened with extinction, and we have attempted to include recent data on any previously undocumented extinctions. However, a survey of global extinctions is a formidable task and we recognize that we may have overlooked pertinent data that might suggest additions, removals, or changes to different categories. Although we have attempted to be as consistent and systematic as possible when allocating species to particular extinction categories, we recognize that our decisions have an element of subjectivity, and there are species that are perhaps not placed in the most appropriate category. Where possible, we have included information on the number of specimens available for study for those species that have been assigned to the aforementioned extinction categories (Tables 1–6, 9 and 10). MacPhee and Flemming (this volume) refer to this material as the "hypodigm" and state that such information indicates how well the species has been collected, and thus how reliably or extensively the biology, taxonomy, and conservation status of the species can be described.

As noted in our Introduction, we consider the main contribution of this study not to be a definitive statement of extinctions, but instead a prototype of a format from which more accurate lists can be generated. We are presenting this information as a "work in progress" and, now that we have presented a practical system for judging and categorizing extinctions according to an explicit set of criteria, we invite specialists in ichthyological taxonomy and biogeography to review these data. The great value of this system is that it allows species to be moved from one category to another as new information becomes available. Ideally, further information should allow the continual reassessment of unresolved extinctions (categories B–D) and species in our "problematic taxa" category U, such that, ultimately, there will be sufficient evidence to have either the extinctions accepted as fully resolved (category A) or the species shown to be extant and appropriate conservation actions considered and hopefully implemented.

Given that the data herein are provisional or may be incomplete, it would be premature to draw from them any detailed conclusions. However, some preliminary observations and comment may be of interest.

11.1. Comparison with IUCN Extinction Accounts

Compared with previous systems such as the IUCN *Red Lists,* the extinction categories proposed herein provide a more detailed account of possible extinctions in terms of their numbers and the reliability of the underlying data. The 1996 IUCN *Red List* noted 93 freshwater fish as "Extinct" or "Extinct in the Wild" but provided no further qualifying information. According to our criteria, only three species qualify as category A (Tables 1 and 14) and can thus be viewed as fully "resolved extinctions." Our survey lists a further 67 species assigned to the "unresolved extinction" categories B–D (Tables 2–6 and 14).

Fifty of the ninety-three species listed as "Extinct" or "Extinct in the Wild" in the 1996 IUCN *Red List* are haplochromine cichlids noted as formerly distributed in Lake Victoria (incorrectly in the case of *Haplochromis [Prognathochromis] vittatus;* see Table 12). We have disqualified nine of these proposed haplochromine extinctions from our categories (Table 12). We include a large number of Lake Victoria cichlids in our survey but consider their proposal as extinct species to be highly questionable, even compared with the other species treated in this survey. Accordingly, we list and discuss the Lake Victoria cichlids in their own categories (marked with "?") for separate consideration (see Section 11.3.1). We place 40 of the haplochromines that IUCN (1996) listed as "Extinct" or "Extinct in the Wild" in our category "?D2" (Table 10) and 1 species in category "?U" (Table 11). None of the 48 species of Lake Victoria cichlids that we place in category "?D1" were listed as extinct in the 1996 IUCN *Red List.* Fourteen of the 54 species of Lake Victoria cichlids that we place in category "?D2" were not listed as extinct in the 1996 IUCN *Red List.*

Considering the noncichlid taxa, we have placed 6 of the species listed as extinct by IUCN (1996) in our category U (Table 7) because we have found insufficient data to categorize the hypothesized extinctions of the species. We disqualify 30 noncichlid taxa (Table 8) that have recently been noted by various authors as "extinct" or "extinct in the wild," including 6 that were listed as extinct or extinct in the wild by IUCN (1996).

Our extinction categories A–D (excluding Lake Victoria cichlids) include 39 species that were not listed as extinct in the 1996 IUCN *Red List* (Table 13). None of these species has sufficient evidence to categorize it as a fully resolved extinction (category A). However, 12 are assigned to the relatively well supported category C. Twenty are assigned to categories B1 and D1, indicating some taxonomic ambiguity; 10 of these are members of the Lake Lanao cyprinid radiation.

11.2. Numbers of Species in the Extinction Categories

Our survey places 172 species of freshwater fish into the extinction categories and subcategories A to D, including the Lake Victoria cichlids in categories "?D1" and "?D2" (Tables 1–6, 9, 10, and 14). The majority of these species fall into categories D1/?D1 (39%) and D2/?D2 (37%) (Table 14), suggesting that taxonomic ambiguity and insufficient sampling data represent almost equally the greatest problems in categorizing extinctions. However, these results are heavily skewed by data for the cichlids found in Lake Victoria. As noted above, the proposal of these species as extinct is highly speculative. Discounting the Lake Victoria cichlids, our results indicate that the greatest number of hypothesized extinctions fall into category C (50%). This highlights the importance of creating a category that accounts for possible extinctions within the last 50 years. Indeed, 93% of all hypothesized extinctions of freshwater fish proposed in categories A–D are believed to have occurred within the last 50 years, and this figure rises to 97% if we include the Lake Victoria cichlids. Figure 1 shows the numbers of recent extinctions reported over the last 100 years and highlights the increasing rate of extinction over the last 50 years.

A relatively large number of hypothesized extinctions in categories A–D are found in category D1 (27% of extinctions, exclusive of Lake Victoria cichlids; Table 14). The results for the latter category are biased by another species flock, the Lake Lanao cyprinids. This flock includes five species assigned to category C (see Table 3; *Cephalakompsus, Mandibularca, Ospatulus,* and *Sprattlicypris*), and ten species of *Puntius* that have some ambiguity associated with their taxonomy (Reid, 1980) and are assigned to category D1 (Table 4). The Lanao species of *Puntius* represents 53% of the species included in the D1 category.

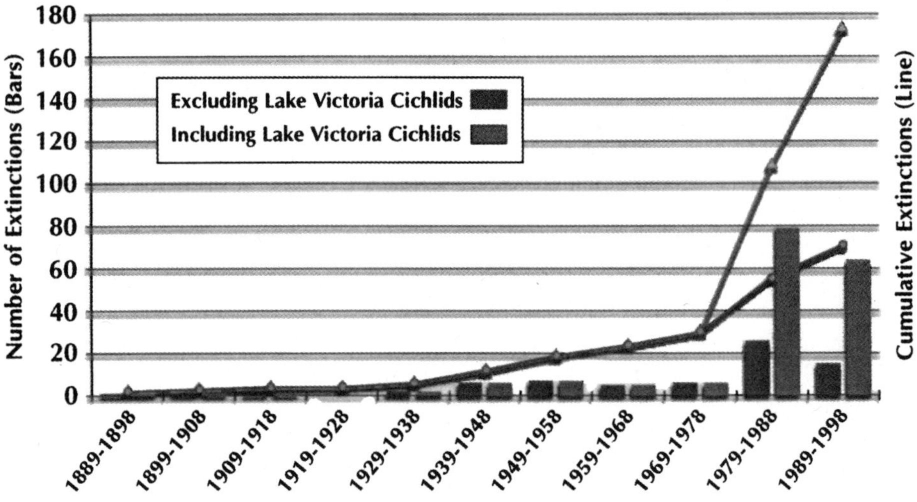

FIGURE 1. Freshwater fish extinctions 1889–1998.

Our results indicate that many species might be extinct, but very few (three) can be shown unequivocally to be extinct. There are either not enough data, or the data are too imprecise and incomplete to fully support many of the proposed extinctions. This has been a particular problem in assessing the possible extinction of Lake Victoria cichlids (see Section 11.3.1). Additionally, we list 43 species in category U (Table 7) and 30 species of Lake Victoria cichlids in category "?U" (Table 11). In these cases, the available data suggest the species are extinct, but are currently inadequate to allow assignment to any extinction category.

The total number of potentially extinct or seriously threatened species of freshwater fish included in this survey is 245 (all categories A–D and U, including categories for Lake Victoria cichlids). This represents only 2% of the approximately 10,500 species of fish known to spend part or all of their life in fresh water (Nelson, 1994). Moreover, 132 of the 245 species (54%) are cichlids from a single lake, Lake Victoria, for which the extinction data re very weakly resolved (see Section 11.3.1). Hence, the interpretation of the results is important. From a purely systematic viewpoint, the results are not indicative of a large species loss relative to the overall biodiversity of freshwater fish species, even allowing for the fact that these 245 species may be an underestimate of the real number of potentially extinct freshwater species (Bruton, 1995; and see Sections 3 and 4). However, from a conservation viewpoint the results indicate a steady increase in the number of possible extinctions over the last 50 to 100 years, apparently resulting from the geographically widening impact of human activity on freshwater ecosystems. Accordingly, systematists and ecologists alike have noted the implication that we face an incumbent "crisis of biodiversity" (Stiassny, 1994:66). In effect, the "quiet crisis" (see Section 3) that is occurring through some of our lakes and rivers is expected to spread through more and diverse aquatic habitats, unless some action is taken to alter this trend. The results from the current study make one point very clear, namely, our understanding of any potentially developing biodiversity crisis in fresh waters is extremely limited. Two important conclusions can be drawn from this.

Foremost, there is an urgent need to conduct further field analyses and improve the quality and scope of published data so as to allow for more informed policies of conservation management. Second, there is a need to exercise caution when interpreting the limited data we currently have available.

11.3. Taxonomic Distribution of Possible Extinctions

The distribution of A–D, ?D1, and ?D2 category extinctions in fish families is shown in Table 14. Fifty-nine percent of these concern Lake Victoria cichlids and, as indicated above, these require special discussion.

11.3.1. Possibly Extinct Lake Victoria Cichlids

Concern over the declining numbers of haplochromine cichlids endemic to Lake Victoria was raised in the 1980s following an explosive increase in the catches of the introduced predator *Lates niloticus* (Hughes, 1983; Barel *et al.,* 1985). Subsequently, Witte *et al.* (1992b) published the results of a detailed survey of the Mwanza Gulf, in the southern part of Lake Victoria. They concluded that, of the 123 haplochromine species recorded from the region in 1979, more than 80 (ca. 70%) had apparently disappeared by 1990. Witte *et al.* (1992b) extrapolated from this to estimate that some 200 species of haplochromines had disappeared or were threatened with extinction lakewide. There are a number of problems with this conclusion. The first is that Witte *et al.* (1992b) implied, perhaps unintentionally, that any species missing from their surveys is extinct. However, some of the species that Witte *et al.* (1992b) listed as absent have subsequently been rediscovered (see Tables 11 and 12). Similarly, Seehausen (1996) and Seehausen *et al.* (1997b) noted that several species have been absent from collections for 1 or 2 years only to reappear in subsequent collections. This may indicate that the population size of such species has significantly declined and that those species are now so rare that they are difficult to collect regularly; but in such cases, the species themselves are extant. Furthermore, some species may also be overlooked because they have undergone ecological shifts in response to new conditions associated with the changing ecology of the lake as a whole. Seehausen *et al.* (1997b) provided an example of this in *Haplochromis tanaos,* which disappeared from catches in 1987 and was noted as absent (under the name *Haplochromis* "double stripe") by Witte *et al.* (1992b). The species was rediscovered in 1993 but was found to be occupying a different area and habitat. Hence, it is probably injudicious to assume a species of Lake Victoria cichlid is extinct simply because it has not been collected for a few years in a particular region.

An additional problem is the assumption that just because a species has disappeared from the Mwanza Gulf it is not present elsewhere in another part of the lake. Witte (1984) noted that Lake Victoria cichlids have traditionally been described as ecologically less specialized (eurytopic) and more widespread than cichlids in Lakes Malawi and Tanganyika. Oijen *et al.* (1981) indicated that almost one-quarter of the 250+ *Haplochromis* species known at that time from Lake Victoria may have lakewide distributions, and Witte (1984) noted that some species include morphologically diverse populations that are spatially isolated from each other. Our comparison of species lists given for faunal surveys in the north of the lake (Kaufman and Ochumba, 1993) and the south of the lake (Witte *et al.,* 1992b;

Seehausen, 1996; Seehausen *et al.*, 1997b) confirmed the presence of apparent conspecifics in both regions. There is, however, growing indication that ecological specialization and habitat restriction (stenotopy) is more strongly developed in Lake Victoria cichlids than first thought (Witte, 1984). Seehausen (1996) and Seehausen and Bouton (1998) have subsequently indicated that there are many, previously unrecognized, stenotopic rock-dwelling species which apparently have regionally restricted distributions in Lake Victoria. Nevertheless, our concern is that habitat restriction has been taken also to imply an absolute geographic restriction (see discussion of Eldredge and Cracraft, 1980:306). For example, Ribbink (1991:43) stated "it is now apparent that there are clear geographic limits to the distribution of numerous cichlid species (Witte, pers. comm.)." However, Witte (1984:162) had stated that "a lake-wide survey of the specific habitat of a species is necessary to confirm its lake-wide distribution"; our knowledge of the current distribution and diversity of the haplochromine cichlid fauna of Lake Victoria seems to be based on surveys that, although detailed, are not representative of the entire lake. Although there have been some surveys from Kenyan and Ugandan sites in the northern part of the lake (Harrison *et al.*, 1989; Ogutu-Ohwayo, 1990, 1993; Kaufman and Ochumba, 1993), and Oijen (1991) has indicated collections from some scattered localities in northern and southern parts of the lake, most of the detailed surveys have been focused on a restricted area around the Mwanza Gulf and adjacent waters in the south (Witte *et al.*, 1992b; Seehausen and Witte, 1995; Seehausen, 1996; Seehausen *et al.*, 1997b). There are large parts of the lake that have not been sampled sufficiently, if at all, but might harbor those species that have apparently disappeared from the sampled regions. Witte (1984) refers to examples of rock-frequenting cichlids that have left their substrate and migrated to new localities. Hence, although habitat restriction may make geographic restriction more probable, the latter should not be assumed. Moreover, recent studies indicate that Lake Victoria cichlids, despite being stenotopic, can shift habitat (see notes above for *H. tanaos*), so they might be found in regions where they would otherwise be unexpected (Seehausen *et al.*, 1997b).

As a further point, it would be unwise to make generalized conclusions concerning the regional ecology of Lake Victoria. The lake, as a whole, has a complex biogeographic history and ecology. The level of complexity is evident from the fact that it has been possible for a species flock of possibly more than 500 cichlids, assembled into 16 diverse trophic groups, to evolve in a remarkably short period of time (Seehausen, 1996). Recent environmental events and ecological changes within the lake have created more complexity (Kaufman, 1992; Witte *et al.*, 1992a; Kaufman and Ochumba, 1993). In the absence of a detailed knowledge of the former ecology of the unsampled parts of the lake, we should not assume that these areas are undergoing similar changes as those that are well-studied. Hence, it is perhaps imprudent to assume that because the Mwanza Gulf might have lost a large proportion of its haplochromine fauna, all other parts of the lake are in the same potential state of faunal decline.

The sampling data for Lake Victoria cichlids indicate local extirpations of many species, and are strongly suggestive of extinctions in certain case. The data, however, are not sufficiently representative for the whole lake to allow us to confidently assign any species to our extinction categories. Our principal concern is that quantitative analyses of the data for Lake Victoria have only been possible since the 1980s, that is, the data published by Witte *et al.* (1992b); hence, it is premature to make definitive judgments on this very large and complex ecosystem. Moreover, the taxonomy of many of the Lake Victoria cichlids is also very complex and requires careful review (Oijen *et al.*, 1991; Seehausen,

1996; Kaufman *et al.,* 1997). It would perhaps be more prudent to postpone judgment on the potential extinctions until a more complete database has been assembled, including sampling from additional localities and further taxonomic investigation. However, we also recognize that there is considerable evidence indicating that Lake Victoria is undergoing a catastrophic series of ecological and limnological changes that clearly present an immediate and serious environmental threat to the endemic cichlids, and to overlook this would be a major omission. Accordingly, we have attempted to review the available data and provisionally assign species to appropriate extinction categories. Hence, we prefix the Lake Victoria cichlid extinction categories with "?" and list them separately (see Tables 9–11). In this way we hope to present a summary for Lake Victoria cichlids that may be suspected as being threatened or extinct, and so establish a basic reference and guide from which more detailed assessments may be developed.

11.3.2. Extinctions in Other Taxonomic Groups

Once the Lake Victoria cichlids have been excluded from our lists, 70 putative extinctions of freshwater species of fish remain in categories A–D. These 70 possible extinctions are distributed among 10 of the 36 orders of fish that include species that occasionally or permanently live in fresh water (Nelson, 1994). Some 51% of all category A–D extinctions are cypriniforms, a huge order of primary freshwater fishes, comprising about 25% of all freshwater species (Nelson, 1994). However, a large number of the putative cypriniform extinctions come from a single, endemic radiation, the Lake Lanao cyprinids of the genera *Cephalakompsus, Mandibularca, Ospatulus, Puntius,* and *Sprattlicypris* (Tables 3 and 4). There is some doubt over the taxonomic validity of several of these Lanao cyprinids (Reid, 1980; and see Table 4 notes); however, if all are valid, then they account for 42% of cypriniform extinctions and 21% of all possibly extinct freshwater fishes listed in categories A–D.

Siluriforms and characiforms include 22 and 13% of all freshwater species. Yet despite the large size of these groups, they each account for only two putative extinctions (one schilbeid and one trichomycterid siluriform; two characids), representing a total of only 6% of possible extinct freshwater fishes in categories A–D. Salmoniforms, on the other hand, include less than 1% of all freshwater fish but account for 14% of extinctions in categories A–D. The salmonid extinctions are of either *Coregonus* or *Salvelinus,* both of which include isolated populations in postglacial lakes of North America and northern Europe (Behnke, 1972, 1980a; Kottelat, 1997). The taxonomy of these populations is exceptionally complex and they have been referred to variously as distinct species, subspecies, or conspecific populations of a widespread species on the basis of other minor morphological differences. Many of the species listed in this survey are of questionable taxonomic and extinction status and appear in our categories B1 and D1 (Tables 2 and 4).

Cyprinodontiforms, comprising 8% of all freshwater fishes, account for 11 possible extinctions and represent 16% of the extinctions in categories A–D. Ten of the cyprinodontiform extinctions meet all of the extinction criteria but are thought to have occurred within the last 50 years and, therefore, are placed in category C. *Characodon garmani* and *Cyprinodon latifasciatus* should perhaps be placed in category A as there is evidence to suggest that their extinction occurred sometime in the 1930s, along with the cyprinid *Stypodon signifer,* as a result of pollution and modification, through industrial activity, of their spring habitats in the Parras Valley, Mexico (Miller *et al.*, 1989). However, the absence of these three species was first confirmed by a field survey in 1953 (Miller, 1964; Miller *et al.*, 1989). and the species are therefore placed in category C (Table 3) at least until the year 2003. Another cyprinodontiform, *Empetrichthys merriami,* was also considered absent in 1953 (Miller *et al.*, 1989) and so will be transferred from category C to category A in 2003. Only one cyprinodontiform is not assigned to category C and this is the Monkey Spring pupfish, *Cyprinodon* sp. This species remains unnamed but the details of its habitat and the events leading to its extinction in 1971 are well documented by Minkley *et al.* (1991), and hence, the species is assigned to category D1.

The remaining proposed extinctions are single examples in nine diverse families (Table 14). Our survey indicates that extinctions are not uniformly distributed among taxonomic groups. Including the Lake Victoria cichlids, we find that the Cichlidae (Section 11.3.1), Cyprinidae, and Cyprinodontidae account for 84% of all of the possible extinctions proposed in categories A–D and ?D1/?D2. However, as noted above, our results represent a preliminary survey of the data and our summary of the taxonomic distribution of possible extinctions is by no means definitive.

11.4. Relating Ecology and Biogeography to Extinctions

Freshwater habitats have been likened to ecological islands of water surrounded and isolated by land, and these "aquatic islands" are extremely vulnerable to ecological perturbation and may have high rates of extinction, as has been documented for other insular ecosystems (WCMC, 1992; Magin *et al.*, 1994; Stiassny, 1996, in press). Stiassny (in press) states that the most obvious reasons for the vulnerability of island species, namely, a restricted range and low population numbers, also apply to freshwater fish that may have their ranges limited to a single watershed, lake, or river. Myers (1996) noted that the inherent complexity of ecosystems may enable them to absorb stress and mask symptoms of damage for a long period of time before the ecosystem's buffering capacity is finally pushed beyond its level of resilience and an abrupt collapse of the system follows. It is reasonable to assume that, in many cases, the small size of restricted ecosystems and their resident populations also reduces their buffering capacity and it is easier for them to be pushed to the level of collapse.

If one excludes the Lake Victoria cichlids, 16% of the extinctions reported in categories A–D are of species that inhabited small pools or springs (Fig. 2). These habitats represent extreme examples of "aquatic island" faunas. Minor changes in these drainages will have a large effect on the absolute volume and surface area of a pool which, in turn, has many diverse consequences. Small pools will also be particularly susceptible to the influx of chemicals or nutrients, through agricultural, industrial, and domestic pollution. Any deleterious environmental change within such a pool may affect the entire population of species when its range is restricted to the pool. Therefore, a crash in the species' population represents a threat of extinction rather than extirpation according to terminology we advocate here (Section 2.2). The large percentage of possible cyprinodontiform extinctions can be attributed, in part, to the fact that nine of the ten possible extinctions were restricted to species inhabiting small pools and springs where the ecology was drastically changed, if not totally destroyed.

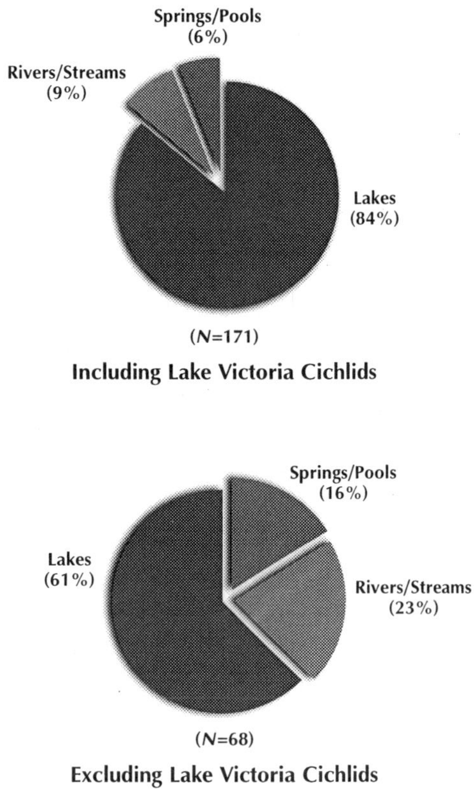

FIGURE 2. Percentage of extinctions from different habitats in categories A–D or ?D1/?D2 for Lake Victoria cichlids.

The effects of changing water levels on species are not necessarily restricted to small pools. *Cottus echinatus,* a sculpin formerly restricted to the relatively large Utah Lake, is also thought to have become possibly extinct because of a drop in water level of the lake (Table 5). Indeed, lakes represent the habitat most commonly associated with extinctions of freshwater fishes. Again, excluding the Lake Victoria cichlids, 61% of the possible extinctions reported in categories A–D are lacustrine (Fig. 2), and this number rises to 84% if the Lake Victoria cichlids are included. This latter figure includes species from several very large lakes, such as North American Great Lakes, and as noted, Africa's Lake Victoria. The heavy bias in the number of proposed extinctions toward cichlids and cyprinids (Sections 11.3.1 and 11.3.2) is, in each case, related to ecological events affecting an entire species flock within a lake: endemic haplochromine cichlids in the case of Lake Victoria (Tables 9–11) and endemic cyprinids in the case of Lake Lanao (Tables 3 and 4). The principle of an "aquatic island" can also be applied to lakes, particularly because the biodiversity of even the largest lakes tends to show characteristics that are the product of their biogeographic history as an independent water body. However, a function of the large size of the lakes is that their ecosystems may be complex and, therefore, the processes that caused any proposed extinction are also often complex.

Insular species are also vulnerable to extinction because of their biology (WCMC, 1992), and particularly because they may have evolved in the absence of certain predators and competitors and so are highly vulnerable to introductions of these kinds. The heavy bias of the total number of proposed freshwater fish extinctions toward Cichlidae and Cyprinidae is partly the result of introduction of the predatory Nile perch *Lates niloticus* into Lake Victoria (Tables 9–11; and see Section 11.5), and the introduction of the goby *Glossogobius giurus* and eleotrid *Hypseleotris agilis* into Lake Lanao (Tables 3 and 4).

Species adapted to particular ecotopes are also particularly vulnerable when these ecotopes are disturbed and the species is unable to replenish a declining population. *Moxostoma lacerum* is an example of an extinct species (category A, Table 1) that, although once widespread throughout the central United States, was restricted within this geographical range to deep pools with clear water over a rocky bedrock in moderate to large streams (Miller *et al.,* 1989). The fish had large eyes and modified lips that apparently facilitated searching for and eating molluscs. Increasing siltation and turbidity of the clearwater pools reduced both the numbers of molluscs present and the ability of the fish to visually locate them, thereby causing the decline of regional populations. Populations of *M. lacerum,* although linked by river systems, could not form ecologically or genetically interlinked "metapopulations." Instead, they were ecologically restricted and discrete, "insular" populations that were unable to replenish each other in the face of local extirpations and eventual extinction.

McDowall (1996) has commented on the prevalence of diadromous fishes, and rarity of nondiadromous fishes, in rivers of New Zealand that have undergone a previous defaunation as a result of volcanic ash flows destroying the river habitats. Apparently the diadromous species were able to recolonize the river systems by marine dispersal from unaffected neighboring localities, whereas the nondiadromous species, restricted to fresh waters, could not move from one isolated river system to another via a marine phase in their life cycle.

The current survey indicates that, if one excludes Lake Victoria cichlids, there are approximately equal numbers of category A–D extinctions in tropical and temperate regions. If the Lake Victoria cichlids are included, then tropical extinctions become approximately four times more prevalent than temperate ones. Stiassny (1996, in press) states that whereas documentation for tropical freshwater biodiversity is less than for temperate regions, species loss has the potential to be far worse there as fish diversity and future environmental threats are greatest in the tropics. A geographical comparison of extinction data, including data for species placed in category U, indicates that the majority of species come from Africa; however, this reflects the large number of haplochromine cichlids from Lake Victoria. If these haplochromines are removed from the analysis, then only 6 species, from a total of 113 placed in categories A–D or U, are from Africa or Madagascar (5% of the total). Most of the remaining species in categories A–D or U are Asian (33 species, or 30% of the total when the Lake Victoria cichlids are excluded). Central, North America, and Europe have similar numbers (18, 17, and 16% of the total, respectively) and fewer extinctions are reported for South America (10%).

The quality of the data is highly variable between continents. Some 58% of the categorized species from North America represent well-supported extinctions, that is, they are assigned to categories A or C. In contrast, only 24% of the Asian species are well-supported (category C) and 64% are either taxonomically ambiguous (category D1; Table 4) or unclassifiable (category U; Table 7). South American extinctions include 64% that lack complete sampling data or are unclassifiable, and European extinctions include 94% that are either taxonomically ambiguous or unclassifiable.

Various authors (Smith *et al.*, 1993; Magin *et al.*, 1994) have commented on the problems of interpreting extinction data when the taxonomic and geographic content is biased by a differential focus of scientific attention on some groups or regions more than others. Smith *et al.* (1993) noted that the apparent rise in the extinction rate of fishes over the past 50 years is at least partially explained by an increased investigative effort on aquatic ecosystems. Our results indicate that the geographic distribution of reported extinctions in freshwater fish is as closely linked to the history of the study of systematic ichthyology in these regions as it is to the biology of the region itself. Greenwood (1992) noted that the fish fauna of North America is well-studied, and this is confirmed by the quality and quantity of information available for threatened and extinct fishes reported here and by previous authors (e.g., Williams *et al.*, 1989; Miller *et al.*, 1989). Greenwood (1992) also classified the western European fauna as potentially well-known, although more recently Kottelat (1997:3) commented on the "astonishingly poor state of European fish systematics and nomenclature." Our results seem to agree more with Kottelat's opinion than Greenwood's, in terms of the quantity of substantiated extinction data. Greenwood (1992:133) referred to the freshwater faunas of Africa and Asia as "sorely incomplete" and we have found the available extinction data to be frequently incomplete or inprecise, as indicated by the number of species in categories D1 and U. Similarly, the South American freshwater ichthyofauna, although probably exceptionally diverse compared with other parts of the world, is relatively poorly known (Böhlke *et al.*, 1978; R. Vari, personal communication), and we have found most of the reported extinctions to be unresolved (category D2) or unclassifiable (category U).

The taxonomic and geographic distributions of reported extinctions are also linked to the effectiveness with which any particular habitat can be sampled. Although the size and ecology of small pools and streams may put their faunas at greater risk of extinction, the large proportion of extinctions reported from these habitats is probably also associated with the fact that, being spatially closed or geographically restricted ecosystems, they can be sampled more easily than larger systems and the absence of any particular species is more likely to be detected. The extensive documentation of extinctions of cyprinodontiforms from small pools and desert springs is an example of this. Sixty percent of possible freshwater fish extinctions from Central America (categories A–D and U) are assigned to category C; however, this is only because the results are heavily biased by data from restricted, and hence easily sampled sites such as small streams and pools. The greater part of the biodiversity and conservation status of the Central American freshwater ichthyofauna is not well known, with numerous species that are undescribed or whose status is uncertain.

Larger lakes and rivers form less spatially restricted or tightly bounded systems than pools and springs and, therefore, extinctions will only be reported for these habitats following more intensive sampling. The most extreme example is provided by marine ecosystems. These are very much more open and unbounded than freshwater ecosystems, potentially allowing greater opportunities for recolonization of degenerated areas by neighboring populations (Chen and Cui, 1993). Consequently, marine ichthyofaunas may be less threatened with extinction than freshwater ones (Beverton, 1992) and to date no marine species of fish are reported as recently extinct, although massive population declines and "economic extinctions" are increasingly prevalent in the inshore realm (McAllister, 1995). Interestingly, some of the largest freshwater lakes become almost comparable to marine ecosystems in terms of the difficulty with which hypothesized extinctions are substantiated by representative sampling. This has been a particular problem in the evaluation of data for the declining populations of haplochromine cichlids in Lake Victoria (Section 11.3.1). Evaluation of the conservation status of threatened salmonids (e.g., species of *Coregonus* and *Salvelinus*) in postglacial European lakes and, more particularly, the Great Lakes of the United States and Canada, would have been impossible had these species not been the object of commercial and sports fisheries operations, which meant they were regularly and extensively sampled. It is ironic that fishing was responsible for both the demise of those species as well as for the documentation of that demise.

11.5. Major Processes Responsible for Extinctions

Smith *et al.* (1993) and Magin *et al.* (1994) stated that most extinctions are associated with human activity. Our survey confirms this for freshwater fishes. Regardless of the preferred habitat or geographic distribution of the species, all of the 172 possible extinctions listed in categories A–D, ?D1, and ?D2 (i.e., including the Lake Victoria cichlids) can be attributed to human activity. As noted in Section 11.2 and shown in Fig. 1, most category A–D extinctions fall into the C and D categories for fishes that are thought to have become extinct within the last 50 years. This can be attributed, in part, to the enormous increase in available data resulting from ichthyofaunal surveys that have been conducted over the last 50 years. This is perhaps most notable for parts of Asia, which have many species assigned to categories D1 and D2. Nevertheless, the relatively large number of apparent extinctions over the last 50 years is also a consequence of the greater impact of an exponentially increasing human population on limited global freshwater resources (Stiassny, 1996, in press).

The causes of the possible extinctions of Lake Victoria cichlids are too complex and insufficiently documented for us to treat species individually (and we have therefore not included these in the summary figure 3), although we make some general comments below. For the remaining 70 species listed in extinction categories A–D, we find the following cited causes for particular extinctions (see Fig. 3): habitat modification in 71% of species; competition and/or predation from introduced species in 54% of species; overfishing in 29% of species; pollution in 26% of species; hybridization in 4% of species; disease or parasites also in 4%; and deliberate eradication in 1% (the petromyzontid, *Lampetra minima;* Table 3). Habitat degradation is a broad category, including the effects of dams, channelization, siltation, water extraction, eutrophication, and so forth (see Bruton, 1995), which perhaps explains why it is implicated in so many extinctions, including the "resolved extinction" of *Moxostoma lacerum* (Table 1 and Section 11.4). Many of these factors have risen through the twentieth century as a result of greater human appropriation of fresh waters to serve intensified agricultural, industrial, and domestic requirements (Allan and Flecker, 1993; Postel *et al.,* 1996; McAllister *et al.,* 1997; Pimental *et al.,* 1997). The severity of the effects of dams and water regulation has been addressed by various authors; Dynesius and Nilsson (1994) provided a comprehensive statistical overview for the northern third of the world, and Roberts (1993) discussed some of the common negative impacts that dams can have on local ichthyofauna. However, Kottelat (1994) notes that land use and dam construction have, in the past, been blamed as endangering species whose distributions are more extensive than the area affected by the land use or dam. As a result, Kottelat called for more precise assessments of individual cases, based on detailed fieldwork.

Introduced plants and animals prove deleterious for many reasons and have become

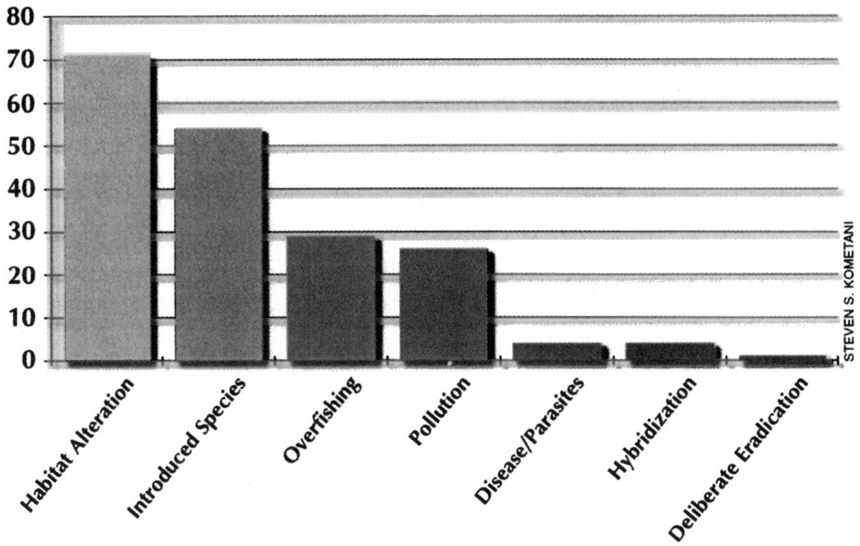

FIGURE 3. Percentages of extinctions that can be attributed to particular causes (for species in categories A–D, and excluding Lake Victoria cichlids).

internationally recognized as presenting a threat that is almost as severe as human-mediated habitat change (Beverton, 1992; Moyle, 1997). Our results indicate that competition and predation from introduced species is implicated in more than half of all extinctions listed in categories A–D. Introduced species that are closely related to the native fauna might hybridize with the native forms and eliminate them through gradual genetic dilution of the endemic genotype; this has been cited as a contributing factor to the "resolved" extinction of *Salvelinus agassizi,* as well as one other salmonid (Table 4) and two North American cypriniforms (Tables 3 and 6). When species are introduced to a new locality, they can also carry disease or parasites to which the native fauna has little or no resistance. Introduced diseases and parasites are noted as a possible cause of the extinction of the New Zealand grayling *Prototroctes oxyrhynchus* (Table 1), and the two endemic species (*Adrianichthys kruyti* and *Weberogobius amadi*) of Lake Poso Sulawesi (Table 5).

Pollution and overfishing are factors, like habitat modification, that are closely linked to the demands of a growing population. Although overfishing should not be dismissed as a cause of possible extinctions, the data should be interpreted carefully. As noted previously, the available data for hypothesized extinctions will be biased toward those species that are well-documented through appropriate sampling. Whereas commercially important "target species" such as species of *Coregonus* and *Salvelinus* in lakes of Europe and North America do face a greater threat of extinction than nontarget species, there is also a greater probability that any decline in population size or extinction of the target species is more likely to be documented than for a nontarget species.

The degradation of fresh waters as a clean resource is discussed by Wetzel (1992) and Stiassny (in press). Roberts (1993:107) noted that whereas different chemical pollutants might, individually, be relatively innocuous to fishes, the pollutants can be lethal when found together. Roberts engagingly referred to this process as "the double whammy." Myers (1996:1028) described the same process as "synergism," that is, the uniting of energies of two discrete mechanisms such that together they have an amplified effect. Many extinctions of freshwater fishes seem to involve complex synergisms. For example, our extinction tables include reference to many species of fish where habitat modification, pollution, and introductions of exotic species or overfishing appear to have worked either simultaneously or sequentially to bring about the eventual extinction of the species. Examples are provided by apparently extinct or threatened salmonid species of *Coregonus* and *Salvelinus* in the postglacial European lakes and the North American Great Lakes (Tables 3, 4, and 7); populations that might normally have withstood competition from introduced species or the effects of pollution have been unable to do so because their numbers have already been weakened by overfishing.

The declining numbers and possible extinction of many cichlid species in Lake Victoria are related to the introduction of the Nile perch *Lates niloticus* in many instances, as widely documented in the literature (Kaufman, 1992; Witte *et al.*, 1992a,b; Kaufman and Ochumba, 1993; Seehausen and Witte, 1995; Seehausen, 1996; Seehausen *et al.*, 1997b; Kaufman *et al.*, 1997). However, the introduction of *L. niloticus* has not been the only contributing factor to the haplochromine decline. The aforementioned authors have documented the combination of population explosions in the 1980s of *L. niloticus,* fishing pressure (cichlids are caught as bait for the Nile perch fishery, and are caught as bycatch), increasing lakeside agriculture, urbanization, and deforestation, resulting in the decline of cichlids and chronic eutrophication and anoxia in parts of the lake. Oijen and Witte (1996), citing Seehausen and Witte (1994), refer to a "second wave of extinction in Lake Victoria" occurring after the 1980s and caused by factors including increased turbidity, spread of the introduced water hyacinth *Eichhornia crassipes,* and the increased use of fish poison in shallow waters. Floating carpets of the water hyacinth in inshore waters reduce light intensities, oxygen concentrations, wave action, and water turbulence, causing water stagnation in these regions, and the accumulation of rotting plant material among rocks cause further oxygen depletion (Seehausen, 1996). Oijen and Witte (1996) stated that the combination of the effects of the water hyacinth and the increased use of fish poison in shallow waters may have made areas of the lake uninhabitable for haplochromines. Seehausen *et al.* (1997a) indicated that the increasing turbidity of littoral waters caused by recent eutrophication is constraining mate recognition systems, and this has resulted in decreased color diversity and a greater probability of hybridization. Seehausen *et al.* (1997a) stated that this increasing turbidity could be responsible for the declining diversity of rock-dwelling cichlids in Lake Victoria, although they did not specifically refer to any named extirpations or extinctions. The decline of Lake Victoria cichlids is, therefore, part of a complex series of changes in the ecology of the lake, involving diverse, synergized interactions (Stiassny, 1996; Seehausen *et al.*, 1997b).

Extinctions are clearly extremely complex events, caused by the interaction of often many, diverse factors. The relative importance of these causal factors should be assessed thoroughly with minimum speculation, as noted by Kottelat (1994). Perhaps most importantly, it is essential to have a reliable and workable system for assessing the evidence that supports the initial proposal of the extinction and categorizes the extinction according to the strengths and weaknesses of these data. The system of criteria and categories that we have adopted here is, in our opinion, a significant step in that direction and has enabled us to present a more comprehensive and informative inventory of recent extinctions of freshwater fishes than would have otherwise been possible.

12. Notes for Tables

Tables 1–13 include the following information.

Hypodigm descriptions
These are based on MacPhee and Flemming (this volume); see Section 11 for discussion.

H, holotype only
H+, 2 or 3 assigned specimens including olotype
‡, 4 or more specimens
?, information currently unavailable or unretrieved

Cause of extinction

1, habitat alteration
2, pollution
3, competition/predation from introduced species
4, overfishing
5, disease/parasites
6, hybridization
7, deliberate eradication

IUCN (1996) status
Abbreviations follow the 1996 IUCN *Red List*

Continually updated information on modern-era extinctions is available at www.creo.org, the website of the Committee on recently Extinct Organisms.

Acknowledgments

We are very grateful for the valuable information on particular species and regional fauna provided by G. Arratia, E. Capuli, K. E. Carpenter, M. de Carvalho, A. M. Colon, L. J. V. Compagno, N. Cox, L. De Vos, C. J. Ferraris, R. Froese, D. Golani, K. Hosoya, L. Kaufman, I. Kornfield, M. Kottelat, A. Kreuzberg, S. Kwang-Tsao, K. Lazara, P. Loiselle, G. M. Reid, J. I. Mojica, S. Molur, E. Mukhina, S. M. Norris, H.-J. Paepke, G. Pequeno, R. Petiyagodda, M. de Pinna, E. P. Pister, T. R. Roberts, P. H. Skelton, G. G. Teugels, M. Toledo-Piza, R. Vari, and S. Weitzman. O. Seehausen kindly provided us with proof copies of manuscripts in press. Use of these colleagues' data does not indicate they necessarily share our opinions. D. Catania (CAS), K. Hartel (MCZ), S. Jewett (USNM), and R. Winterbottom (ROM) kindly loaned specimens for examination. For translation of relevant articles we thank R. Britz, A. Cliff, T. Skelton, X. Wang, and S. Winick. We are grateful to R. D. E. MacPhee and C. Flemming for inviting this contribution to the symposium and for a stimulating cross-flow of ideas. Informative and enthusiastic discussion was also provided by G. J. Howes, and by M. R. DeJong who also conducted valuable on-line bibliographic searches. R. Vari and I. Kornfield kindly reviewed the manuscript and provided useful comments. I.J.H. is grateful to the American Museum of Natural History for scientific fellowship.

Table 1. Category A—Resolved Extinctions

Family	Species	Common name	Range	EED	Hypodigm	Causes	References	Notes
Catostomidae	*Moxostoma lacerum*	harelip sucker	Eastern US	1893–1940	‡	1	1,2	
Retropinnidae	*Prototroctes oxyrhynchus*	New Zealand grayling	New Zealand	1920s [1923–]	H+ (‡?)	1,3,5	3	*a*
Salmonidae	*Salvelinus agassizi*	silver trout	New Hampshire, US	1930–1939	‡	3,4,6	1,4,5	*b*

References
1, Miller *et al.* (1989); 2, Williams and Nowak (1993); 3, Faragher and Harris (1994); 4, Behnke (1980a); 5, Bailey and Oliver (1939).

Notes
[a]Day (1981) noted rumors of the species survival in remote parts of the South Island and McDowall (1989) considered the extinction is not properly understood; WCMC (1992) included *P. oxyrhynchus* in their list of extinct taxa but note there may still be some chance of its survival. However, there appears to be no confirmed evidence of the species's survival; the absence of reliable records of this, formerly common species, for over 70 years suggests its inclusion in category A is reasonable.
[b]Behnke (1980a) considered *S. agassizi* might have been overlooked in other lakes and "hopes" that it still persists. However, no specimens have been confirmed since the 1930s despite reasonable sampling activity (Bailey and Oliver, 1939; Behnke, 1980a) and, therefore, the assignment of this species to category A seems reasonable.

Table 2. Category B1—Unresolved Extinctions

Family	Species	Common name	Range	EED	Hypodigm	Causes	References	Notes
Salmonidae	*Salvelinus inframundus*	Orkney char	Hoy Island, UK	1862–1909	H+	3?	1	*a*
Atherinidae	*Chirostoma compressum*	Cuitzeo silverside	Lake Cuitzeo, Mexico	1940–1974 [1941]	H	1	2–4	*b*

References

1, Kottelat (1997); 2, Barbour (1974); 3, Mayden *et al.* (1992); 4, Lyons *et al.* (1998).

Notes

[a]Kottelat's (1997:153) checklist of European freshwater fishes recognizes *Salvelinus inframundus* as a valid species according to his "pragmatic species concept." However, the validity of the species should be regarded tentatively; the taxonomy of *Salvelinus* is exceptionally complex and several nominal endemic species have been described from isolated lakes but these are considered by many taxonomists to be different populations or subspecies of a single species. Most authors recognize a single autochthonous British species of *Salvelinus, S. alpinus.* The taxonomic validity of *S. inframundus* is further called into question as the species is represented by only two specimens from Hellyal Lake, Hoy Island, Orkneys, which could be morphological variants. Accepting *S. inframundus* as a valid species, then it is extinct based on Regan's (1909) report that attempts to relocate the species in years preceding 1909 had been unsuccessful. Regan (1911) stated that stocking of Hellyal Lake with trout may account for the disappearance of the char.

[b]Barbour (1974) noted that *Chirostoma compressum* is known only from two specimens (one now lost) collected from Lake Cuitzeo in 1940 and it apparently became extinct after drying of the lake in 1941 so as to fill a nearby reservoir (De Buen, 1943). Barbour (1974) indicated that the species is probably conspecific with *C. grandocule* from nearby Lake Pátzcuaro, although it shows slight morphological variation. Mayden *et al.* (1992) listed the species as extinct and warranting additional taxonomic investigation. Lyons *et al.* (1998) listed the species as "extinct, endangered" but did not refer to any additional material.

Table 3. Category C—Unresolved Extinctions

Family	Species	Common name	Range	EED	Hypodigm	Causes	References	Notes
Petromyzontidae	*Lampetra minima*	Miller Lake lamprey	Miller Lake, Oregon, US	1952–1959	‡	7	1	
Cyprinidae	*Cephalakompsus pachycheilus*		Lake Lanao, Philippines	1921–1982 [1963–1982]	H	1,3,4	2–4	
	Chondrostoma scodrensis		Lake Skadar, Montenegro	1881–1990	‡	2?	5,6	*a*
	Cyprinus yilongensis		Lake Yilong, China	1981–1993 [1981]	‡	1	7	
	Evarra bustamantei	Mexican chub	Valley of Mexico	1957–1983	‡	1,2	8,9	*b*
	Evarra eigenmanni	plateau chub	Valley of Mexico	1954–1983	‡	1,2	8,9	*b*
	Evarra tlahuacensis	endorheic chub	Valley of Mexico	1902–1954	H	1,2	8,9	*b*
	Gila crassicauda	thicktail chub	California, US	1957–1989	‡	1,3	8	
	Hybopsis amecae	Ameca shiner	Río Ameca, Mexico	1969	‡	1–3	8,10	
	Hybopsis aulidion	Durango shiner	Río Tunal, Mexico	1961–1985	‡	1–3	8,10	
	Lepidomeda altivelis	pahranagat spinedace	springs, Nevada, US	1938–1959	‡	3	8	
	Mandibularca resinus	bagangan	Lake Lanao, Philippines	1922–1982 [1963–1982]	‡	1,3,4	2–4	
	Notropis orca	phantom shiner	Río Grande, US and Mexico	1975–1981	‡	1–3, 6	8,11	*c*
	Ospatulus palaemophagus		Lake Lanao, Philippines	1924–1982 [1963–1982]	H	1,3,4	2–4	
	Ospatulus trunculatus	bitungu	Lake Lanao, Philippines	1921–1982 [1963–1982]	H	1,3,4	2–4	
	Pogonichthys ciscoides	Clear Lake splittail	Clear Lake and tributaries, California, US	1970–1976	‡	1–3	8	
	Rhinichthys deaconi	Las Vegas dace	Lake Vegas Creek, US	1940–1965 [1955–1965]	‡	1	8	
	Sprattlicypris palata	palata	Lake Lanao, Philippines	1921–1982 [1963–1982]	‡	1,3,4	2–4	*b*
	Stypodon signifer	stumptooth minnow	Parras Spring, Mexico	1903–1953 [1930s]	‡	1,2	8,16	

Characidae	*Brycon acuminatus*		São Paulo, Brazil	1900–1982	H+	2		*d*
	Henochilus wheatlandi		Minas Gerais, Brazil	1865–1990 [1880s]	H+	1		*e*
Schilbeidae	*Platytropius siamensis*		central Thailand	1966–1997	‡	1		*f*
Salmonidae	*Coregonus johannae*	deepwater ciscoe	Great Lakes, US	1952–1980	‡	3,4	8,12	*g*
Bedotiidae	*Rheocles sikorae*	Zona	River Manambola, Madagascar	1966–1990	‡	1	13	*h*
Cyprinodontidae	*Cyprinodon ceciliae*	cachorrito del la Presa	Ojo de Agua la Presa, Mexico	1988–1990	‡	1	14,15	
	Cyprinodon inmemoriam	cachorrito del la Presa	Ojo la Trinidad, Mexico	1985	H	1	14,15	
	Cyprinodon latifasciatus	Parras pupfish	Parras, Mexico	1903–1953 [1930s]	‡	1,2	8,16	
	Cyprinodon longidorsalis	cachorrito de Charco Palma	Charco La Palma, Mexico	1985–1994	‡	1	14,15	
	Leptolebias marmoratus	ginger pearlfish	Rio de Janeiro, Brazil	1944–1995	?	1,2	17,18	*i*
	Orestias cuvieri	Lake Titicaca orestias	Lake Titicaca, South America	1937–1962 [1941–1962]	‡	3	19,20	*j*
Fundulidae	*Fundulus albolineatus*	whiteline topminnow	Spring Creek, Alabama, US	1899–1966	‡	1,3	8	
Goodeidae	*Characodon garmani*	Parras characodon	Parras, Mexico	1903–1953 [1930s]	H	1,2	8,16,21	
	Empetrichthys merriami	Ash Meadows killifish	springs, Nevada, US	1948–1953	‡	1,3	8	
Poeciliidae	*Gambusia amistadensis*	Amistad gambusia	spring, Texas, US	1974–1987	‡	1	8,22	*k*
Gasterosteidae	*Pungitius kaibarae*	Kyoto ninespine stickleback	springs/streams, Hyogo Prefecture, Japan	1959–1993 [1960s]	‡	1,2	23	*l*

(*continued*)

Table 3. (*Continued*)

References

1, Bond and Kan (1973); 2, Kornfield and Carpenter (1984); 3, Kornfield (1982); 4, Kornfield (1985); 5, Kottelat (1997); 6, Maric (1995); 7 Chen and Cui (1993); 8, Miller *et al.* (1989); 9, Alvarez and del Villar Navarro (1957); 10, Chernoff and Miller (1986); 11, Chernoff *et al.* (1982); 12, Todd (1980); 13, Stiassny (1990); 14, Lozano-Vilano and Contreras-Balderas (1993); 15, Contreras-Balderas and Lozano-Vilano (1996); 16, Miller (1964); 17, Costa (1995); 18, Rosa and Menezes (1996); 19, Parenti (1984); 20, Lauzanne (1992); 21, Smith and Miller (1986); 22, Minckley *et al.* (1991); 23, Kawanabe and Watanabe (1993).

Notes

[a]Maric (1995) noted other species of *Chondrostoma* have declined in number in Lake Skadar because of overfishing (not applicable in the case of *C. scodrensis,* known only by the type series) and pollution. Rakaj and Flloko (1995) include this species in a list of Albanian freshwater fish, but make no judgments of the conservation status of the species.

[b]Alvarez del Villar and Navarro (1957) indicated *Evarra bustamantei* was extant at the time of their publication, *E. eigenmanni* had been collected in 1954 after a period when it had been thought extinct, and *E. tlahuacensis* was known only by the holotype collected in 1902 and perhaps extinct by 1957. Miller *et al.* (1989) listed all three species as presumed extinct by 1983; there is no published documentation of sampling programs between 1957 and 1983 but we understand the habitat is probably destroyed by Mexico City suburban development.

[c]The last specimen was collected from the lower Río Grande in 1975 and not seen in this region in 1981; extinction is presumed on this basis, although the former widespread distribution of this species increases the possibility that it still exists.

[d]This species is represented by two specimens: the holotype, collected from Río Tiete before 1900; and a second specimen collected in 1911 from Taubaté (Río Paraíba basin), São Paulo (R. Vari, personal communication). A third specimen, held in the Royal Ontario Museum (catalog number ROM 64026), was collected from Peru in 1991 and incorrectly identified as this species (I.J.H., personal observation). Extensive collecting from both Río Tiete and Río Tiete and Río Paraíba has produced no further specimens (Howes, 1982) and T. R. Roberts (personal communication) believes it to almost certainly be extinct. The type locality in São Paulo is now heavily polluted (Howes, 1982; R. Vari and M. Toledo-Piza, personal communication).

[e]*Henochilus wheatlandi* is represented only by three specimens collected in 1865. R. Vari (personal communication) reports that collecting in the type locality and discussion with local fishermen in the early 1990s failed to locate the species. The proposed extinction may be correlated with habitat modification; the type locality was originally heavily forested and is now open grassland (R. Vari, personal communication). Vari is cited by Swerdlow (1998) as stating the species became extinct in the 1880s.

[f]T. Roberts (personal communication) states that this species has not been collected after 1966. T. Roberts and M. Kottelat (personal communication) suggest it is extinct, with Roberts attributing its extinction to canalization and damming of the Chao Phraya.

[g]The last known record of this species is in 1952; Todd (1980) suggested conspecificity with *C. zenithicus* but this is not mentioned in subsequent accounts.

[h]Stiassny (1990) stated this species probably had an extremely circumscribed distribution which is now deforested, cultivated, and degraded. Stiassny was unable to locate the species in 1990 and concluded it was probably extinct.

[i]*Leptolebias marmoratus* inhabited roadside ditches near Río de Janeiro which have been bulldozed and treated with DDT for land reclamation and mosquito control (Miller, 1968; K. Lazara, personal communication).

[j]Loubens (1992) discussed introduction of salmonids to Lake Titicaca, including the rainbow trout, *Salmo gairdneri* (= *Oncorhynchus mykiss*), in 1941. Loubens believed that the commonly stated hypothesis that *O. cuvieri* were out-competed by the rainbow trout is unprovable.

[k]Populations of *Gambusia amistadensis* have been maintained in artificial refugia since 1969, after loss of natural habitat. This species was declared extinct in 1987 after the collapse of refugial populations through human accident and genetic contamination (Minckley *et al.*, 1991).

[l]Ida (1984) recognized *Pungitius kaibarae* as the subspecies, *P. sinensis kaibarae,* but Kawanabe and Watanabe (1993) and Yuma *et al.* (1998) considered it to be a distinct species that became extinct in the 1960s through land reclamation and/or overdoses of phosphoric insecticides. An attempt to maintain a refugial population in Kyoto Zoo in 1959 was unsuccessful.

Table 4. Category D1—Unresolved Extinctions

Family	Species	Common name	Range	EED	Hypodigm	Causes	References	Notes
Cyprinidae	*Barbus microbarbis*		Lake Luhondo, Rwanda	1934–1952	H	3	1	*a*
	Puntius amarus	pait	Lake Lanao, Philippines	1910–1982 [1963–1982]	‡	1,3,4	2–4	*b*
	Puntius baoulan	baolan	Lake Lanao, Philippines	1926–1982 [1963–1982]	‡	1,3,4	2–4	*b*
	Puntius clemensi	bagangan	Lake Lanao, Philippines	1921–1982 [1963–1982]	‡	1,3,4	2–4	*b*
	Puntius disa	disa	Lake Lanao, Philippines	1932–1982 [1963–1982]	H+	1,3,4	2–4	*b*
	Puntius flavifuscus	katapa-tapa	Lake Lanao, Philippines	1921–1982 [1963–1982]	H+	1,3,4	2–4	*b*
	Puntius herrei		Lake Lanao, Philippines	1908–1982 [1963–1982]	H	1,3,4	2–4	*b*
	Puntius lanaoensis	kandar	Lake Lanao, Philippines	1922–1982 [1963–1982]	‡	1,3,4	2–4	*b*
	Puntius manalak	manalak	Lake Lanao, Philippines	1924–1982 [1963–1982]	H+	1,3,4	2–4	*b*
	Puntius sirang	sirang	Lake Lanao, Philippines	1932–1982 [1963–1982]	‡	1,3,4	2–4	*b*
	Puntius tras	tras	Lake Lanao, Philippines	1925–1982 [1963–1982]	H	1,3,4	2–4	*b*
Salmonidae	*Coregonus alpenae*	longjaw ciscoe	Great Lakes, US	1975–1989	‡	1–4	5,6	*c*
	Coregonus confusus	pfärrig	Lake Morat, Switzerland	1885–1950	1?	?	7,8	*d*
	Coregonus fera	féra	Lake Geneva, Switzerland	1950–1958	?	3,4,6	7–11	*e*
	Coregonus gutturosus	kilch	Lake Constance, Switzerland	1972–1992	?	?	7	*f*
	Coregonus hiemalis	gravenche	Lake Geneva, Switzerland	1900–1950	?	4	7–9	*g*
	Coregonus restrictus	férit	Lake Morat, Switzerland	1885–1950	?	?	7,8	*h*
	Salvelinus scharffi	Scharff's charr	Lough Owel, Ireland	1908	H	?	7,12	*i*
Cyprinodontidae	*Cyprinodon* sp. ?	Monkey Spring pupfish	Monkey Spring, Arizona, US	1971	‡	1,3	13	*j*

(continued)

Table 4. *(Continued)*

References

1, De Vos *et al.* (1990); 2, Kornfield and Carpenter (1984); 3, Kornfield (1982); 4, Kornfield (1985); 5, Miller *et al.* (1989); 6, Wells and McLain (1972); 7, Kottelat (1997); 8, Dottrens (1950); 9, Kirchhofer (1996); 10, Spillman (1958); 11, Laurent (1972); 12, Quigley and Flannery (1996); 13, Minckley *et al.* (1991).

Notes

[a]De Vos *et al.* (1990) indicated that the only known specimen of *Barbus microbarbis,* collected from Lake Luhondo in 1934, may be an aberrant morphotype or a hybrid; thus, its taxonomic status remains uncertain. *B. microbarbis* and two other cyprinids were not collected from the lake in 1952 or during the 1980s; De Vos *et al.* (1990) attribute this to the introduction of *Tilapia* from 1935 to 1938.

[b]Reid (1980) commented that Herre's descriptions of the Lanao cyprinids indicate strong similarities between species that may be accounted for by ecophenotypic variation, and his descriptions are often based on very few specimens. Resolution of the taxonomic status of many of these species is now difficult if not impossible because of loss of type material.

[c]Todd *et al.* (1981) suggested that *Coregonus alpenae* was a morphological variant of *C. zenithicus,* which is "vulnerable" (IUCN, 1996) but not extinct. Authors subsequent to Todd *et al.* (1981) have variously accepted or ignored their suggested synonymy.

[d]Kottelat (1997:105) suggested *Coregonus confusus* is a valid species, and according to his "pragmatic species concept," it is endemic to Lake Morat, although he is unable to confirm this because the type specimens are lost. However, the validity of the species should be regarded tentatively; the taxonomy of coregonines is exceptionally complex and several nominal endemic species of coregonines have been described from isolated lakes but these are considered by many taxonomists to be different populations or subspecies of a single species. Dottrens (1950) noted coregonines had disappeared from Morat; the original hydrography of the lake has been extensively modified by the "Jura water corrections" in 1868–1878 and 1962–1970 (Kottelat, 1997). The loss of the species could be associated with these corrections, although there is no proof.

[e]*Coregonus fera* has been reported from Lake Geneva (Lac Léman), where it is called "féra," and from Lake Constance, where it has been called "sandfelchen" (Dottrens, 1959; Nümann, 1972). Kottelat (1997:111) described the sandfelchen as a new species, *Coregonus arenicolus,* and so restricts *C. fera* to the "Féra du Léma" (Dottrens, 1950). However, Kottelat considered these species required a more detailed taxonomic review. Accepting *Coregonus fera* as the Lake Geneva endemic "féra," this species is variously reported as extinct. Dottrens (1950) described féra as rare and close to disappearing, but not extinct. Spillman (1958) stated the féra seemed to have disappeared and Laurent (1972) confirmed this, blaming heavy fishing and hybridization with extensive introductions of coregonines from other lakes from 1881 to 1946 (see also Dottrens, 1950). Kirchhofer (1996) and Kottelat (1997:104) both confirmed this extinction.

[f]Kottelat (1997:112) noted *C. gutturosus* (the kilch) as extinct, citing Ruhlé and Kindle (1992) and personal communications. Dottrens (1959) referred to the kilch as *C. acronius,* which Kottelat assumes as a junior synonym of *C. gutturosus.* The taxonomy of these coregonids requires further review (see notes for other coregonines). Nümann (1972) referred to *C. acronius* as present in Lake Constance but showing a rapid decrease in numbers of fish caught. Nümann stated that fishing and eutrophication may be affecting coregonines of Lake Constance but did not believe this to be significant in the case of *C. acronius* and stated that the declining catches did not necessarily reflect a genuine decline in stock. Accepting Kottelat's definition of *C. gutturosus* (including *C. acronius* as a synonym), we still do not have direct evidence of the species' extinction (i.e., population decline unproven; environmental threat to the species unproven). However, we have not seen Ruhlé and Kindle's report and accept that they, and Kottelat, must have additional data to support the extinction as occurring sometime after 1972.

[g]Kottelat's (1997:104) checklist of European freshwater fishes recognizes *Coregonus hiemalis* (gravenche) as a valid species, endemic to Lake Geneva, according to his "pragmatic species concept." Previously, Dottrens (1959) recognized *C. hiemalis* as a subspecies of the more widespread, European species *C. lavaretus.* Accepting Kottelat's definition of *C. hiemalis* as a distinct species endemic to Lake Geneva, there is evidence that a combination of overfishing and introductions from the 1920s onwards resulted in the extinction of this species by 1950 (Dottrens, 1950; Kirchhofer, 1996).

[h]Kottelat (1997:107) accepted *Coregonus restrictus* as a valid species but indicated the possibility of conspecificity with *C. candidus,* although this cannot be proven because the types of *C. restrictus* are lost. Absence of coregonines from Lake Morat is discussed in note *d* above.

[i]Kottelat (1997:155) recognized *Salvelinus scharffi* as a valid species according to his "pragmatic species concept." However, the validity of the species should be accepted tentatively given the taxonomic complexity of *Salvelinus* (see Table 2, note *a*). The taxonomic validity of *S. scharffi* is called further into question because the species is represented by only one specimen from Lough Owel, Ireland, which could very easily be a morphological variant. Accepting *S. scharffi* as a valid species, it is extinct as Quigley and Flannery (1996) reported that charrs are extinct from Lough Owel.

[j]*Cyprinodon* sp. was extirpated in its natural habitat in 1969; refugial populations failed in 1971 (Minckley *et al.*, 1991).

Table 5. Category D2—Unresolved Extinctions

Family	Species	Common name	Range	EED	Hypodigm	Causes	References	Notes
Cyprinidae	*Acanthobrama hulensis*		Lake Huleh, Israel	1957–1996	‡	1	1	*a*
	Anabarilius alburnops		Lake Dianchi, China	1904–1993 [1950s–1993]	‡	1–3	2,3	*b*
	Anabarilius polylepis		Lake Dianchi, China	1904–1993 [1950s–1993]	H	1–3	2,3	*b*
	Phoxinellus egridiri	yag baligi	Lake Egridir, Turkey	1955–1995	‡	3	4,5	*c*
	Phoxinellus handlirschi	çiçek	Lake Egridir, Turkey	1955–1995	?	3	4,5	*c*
Trichomycteridae	*Rhizosomichthys totae*	greasefish	Lake Tota, Columbia	1957–1990	‡	1,3	6	*d*
Adrianichthyidae	*Adrianichthys kruyti*	duck-billed buntingi	Lake Poso, Sulawesi	1983–1987	‡	3,5	7–9	*e*
Cichlidae	*Tristramella magdalenae*		Damascus, Israel	1997	‡	1		*f*
Cottidae	*Cottus echinatus*	Utah Lake sculpin	Utah Lake, Utah, US	1928–1963 [1928–1936]	‡	1,2	10,11	*g*
Gobiidae	*Weberogobius amadi*	Poso bungu	Lake Poso, Sulawesi	1985–1987	‡	3,5	7–9	*e*

References

1, Krupp and Schneider (1989); 2, Kottelat and Chu (1988); 3, Chen and Cui (1993); 4, Balik (1995); 5, Crivelli (1995); 6, Arratia (1990); 7, Whitten *et al.* (1987); 8, Kottelat (1990); 9, Kottelat and Whitten (1996); 10, Bailey and Bond (1963); 11, Miller *et al.* (1989).

Notes

[a]D. Golani (personal communication) informs us that Lake Huleh was drained in 1957; the description by Goren *et al.* (1973) of *Mirogrex terraesanctae hulensis* (=*A. hulensis,* in part; see Krupp and Schneider [1989]) indicated the species existed in remaining pools for at least some of the time that the lake was drained. Some people claim they have collected *A. hulensis* after the reflooding of Lake Huleh, but Golani (personal communication) states that all specimens he has seen that were reported as *A. hulensis* were misidentifications. Krupp and Schneider (1989) noted *A. hulensis* was probably "on the verge of extinction" and the species was listed as extinct by IUCN (1996). The evidence above is not conclusive either way; the species may have survived in isolated pools and repopulated the lake, although we know of no confirmed reports of the species after the lake was reflooded and so assign the species to Category D2 with the suggestion that further sampling be conducted to confirm this hypothesis.

[b]Kottelat and Chu (1988) noted *Anabarilius alburnops* and *A. polylepis* as endemic to Lake Dianchi, and they listed *A. polylepis* as endangered or extinct. They have also recorded environmental threats to the lake (p. 71). Chen and Cui (1993) reported both species as extinct; Kottelat (personal communication) also listed *A. polylepsis* as extinct. Supporting evidence of unsuccessful sampling has not been available.

[c]Balik (1995) and Crivelli (1995) noted *Phoxinellus egridiri* and *P. handlirschi* are endemic to Lake Egridir and, following introduction of *Stizostedion lucioperca,* have disappeared and may be considered extinct.

[d]IUCN (1996) listed *Rhizosomichthys totae* as extinct and the presumed cause is the introduction of rainbow trout (Miles, 1942; J. Ivan Mojica, personal communication). The presumed extinction may have been accelerated by seismic events causing mass mortalities of the species in Lake Tota, the only known habitat for this species (Miles, 1942). *Rhizosomichthys totae* is known only from nine specimens (Miles, 1942, 1943). Four type specimens were collected in 1942; D. Catania (personal communication) states that paratype material I.C.N. 20a and one of the two paratypes sent to Stanford (SU 37074) are probably the head and posterior part of the same specimen. Four additional specimens (USNM 120130) were collected in 1942 and one (ICN-MHNH 354) in 1957. Two further specimens (USNM 120129), coming from the same collection as the other USNM material, do not appear to be the same species (IJH, personal observation). The species has probably been poorly collected because it is believed to inhabit deep waters of Lake Tota (Miles, 1942); it is possible that the species is extant in the more inaccessible deep waters of the lake (J. Ivan Mojica, personal communication). G. Arratia, M. de Pinna, and S. Weitzman (personal communication) have noted the possibility that *R. totae* is simply an unusually fat form of *Trichomycterus*.

[e]The last specimen of *Adrianichthys kruyti* and *Weberogobius amadi* were seen in 1983 and 1987, respectively (Kottelat, 1990). Kottelat (1990) questioned whether the species still exist but thought it premature to consider them extinct as sampling had been limited; he advocated further sampling in deeper waters. Kottelat and Whitten (1996) referred to the species as extinct. We have tentatively placed them in this category but feel they require further sampling.

[f]D. Golani (personal communication) states that this species has not been collected following drainage of the small lake and pools around Damascus where it was previously found.

[g]The last collection of *Cottus echinatus* was in 1928. There is no mention of detailed attempts to collect the species after this date, although we presume the area has been well surveyed. The species might not have survived the low lake levels in the early 1930s (Tanner, 1936; Bailey and Bond, 1963).

Table 6. Category D3—Unresolved Extinctions

Family	Species	Common name	Range	EED	Hypodigm	Causes	References	Notes
Catostomidae	*Chasmistes muriei*	Snake River sucker	Snake River (Jackson Lake?), Wyoming, US	1927–1970	H	?	1	*a*

References

1, Baxter and Simon (1970); Miller and Smith (1981); Miller *et al.* (1989).

Notes

[a]Very little documentation exists for the biology of this species and therefore the cause of extinction is speculative. First noted as possibly extinct by Baxter and Simon (1980). Miller and Smith (1981) and Miller *et al.* (1989) suggested that *C. muriei* became extinct through hybridization with *Chasmistes ardens.* This seems unprovable based on the single specimen. No evidence is given for subsequent attempts specifically to catch *C. muriei,* but the Snake River and associated lakes have probably been representatively sampled over the past 70 years by fisheries surveys.

Table 7. Unclassifiable Species

Family	Species	Common name	Range	Hypodigm	IUCN (1996) Status	Notes
Petromyzontidae	*Eudontomyzon sp.*		Rivers Prut, Dniester and Dnieper; Byelorussia and Ukraine	?	—	*a*
Carcharhinidae	*Glyphis gangeticus*	Ganges shark	Bengal; Houghly River; possibly Karachi; India	H+	CR	*b*
Acipenseridae	*Pseudoscaphirhynchus fedtschenkoi*	Syr-Darya shovelnose	Sry Darya drainage, Uzbekistan and Kazakhstan	?	CR	*c*
	Pseudoscaphirhynchus hermanni	little Amu-Darya shovelnose	Amu Darya drainage, Uzbekistan	?	CR	*c*
Clupeidae	*Alosa vistonica*		Lake Vistonis, Greece	‡	—	*d*
Balitoridae	*Sphaerophysa dianchiensis*		China	?	VU	*e*
Cyprinidae	*Acheliognathus elongatus*		Lake Dianchi, China	‡	EN	*e, f*
	Algansea barbata	Lerma chub	Río Lerma basin, Mexico	‡	—	*g*
	Gymnocypris biswasi		Ladakh, India	‡?	—	*h*
	Leuciscus turskyi		River Cikola, Croatia	‡	EX	*i*
	Leuciscus ukliva		River Cetina, Croatia	‡	CR	*j*
	Notropis saladonis	Salado shiner	Mexico	‡	EX	*k*
	Pararasbora moltrechtii		Mekong River, downstream of Nakorn Phanom	H+?	—	*l*
Characidae	*Hyphessobrycon parvellus*		Rio Catu, Brazil	‡	—	*m*
Amblycipitidae	*Liobagrus kingi*		China	H?	—	*e*
	Liobagrus nigricauda		Lake Dianchi, China	‡	EN	*f*
Bagridae	*Leiocassis medianalis*		Lake Dianchi, China	H+?	—	*f*
Ictaluridae	*Noturus trautmani*	Scioto madtom	tributaries of Scioto River, Ohio, US	‡	CR	*n*
Schilbeidae	*Irveneia voltae*		lower River Volta, Ghana	‡	—	*o*
Siluridae	*Silurus mento*		Lake Dianchi, China	‡	EN	*f*
Salmonidae	*Salmo montenegrinus*		tributaries to Lake Ohrid and Skadar, and River Radika, Montenegro	?	DD	*p*
	Salmo schiefermuelleri		Lake Attersee, Traunsee and Fuschlersee, Austria	H+?	—	*q*
	Salvelinus profundus	tiefseesaibling	Lake Constance, Switzerland	—	—	*r*
Atherinidae	*Chirostoma charari*	least silverside	Río Lerma basin, Mexico	?	—	*s*

(continued)

Table 7. *(Continued)*

	Teramulus waterloti		northwest Madagascar	‡	—	*t*
Adrianichthyidae	*Xenopoecilus poptae*	Popta's buntingi	Lake Poso, Sulawesi	‡	CR	*u*
	Xenopoecilus sarasinorum		Lake Lindu, Sulawesi	‡?	EN	*v*
Cyprinodontidae	*Cualac tessellatus*	tesselated pupfish	Media Luna Lake, San Luis Potosi, Mexico	‡	EN	*w*
	Cyprinodon sp.	Perritos de Sandia	Mexico	?	EX	*x*
	Orestias incae		Lake Titicaca, South America	‡	—	*y*
	Orestias taquiri		Lake Titicaca, South America	‡	—	*y*
	Orestias tutini		Lake Titicaca, South America	‡	—	*y*
	Orestias uruni		Lake Titicaca, South America	‡	—	*y*
Goodeidae	*Chapalichthys peraticus*	alien splitfin	San Juanico, Mexico	‡	—	*z*
Poeciliidae	*Gambusia georgii*	San Marcos gambusia	San Marcos River, Texas, US	‡	EX	*aa*
	Pantanodon madagascariensis		northeastern Madagascar	‡	EN	*t*
	Phalloptychus eigenmanni		Río Catu, Brazil	‡	—	*bb*
	Priapella bonita	graceful priapella	Mexico	‡	EX	*cc*
Gasterosteidae	*Gasterosteus crenobiontus*		River Techirghiol, Romania	H+?	—	*dd*
Cichlidae	*Stomatepia mongo*		Lake Barombi Mbo, Cameroon	‡	CR	*ee*
Belontiidae	*Betta* cf. *tomi*		Singapore	‡	VU	*ff*
Gobiidae	*Sicyopterus nigrescens*		Hawaiian Islands	‡	—	*gg*
Percidae	*Etheostoma sellare*	Maryland darter	creeks, Maryland, US	‡	EX	*hh*

Notes

[a]Extinct according to Kottelat (1997), who cited Holcík and Renaud (1986). The reason for extinction is unknown and there is no available evidence of sampling.

[b]*Glyphis gangeticus* is only known reliably from three specimens, collected in 1867 or earlier (Compagno, 1988, and personal communication; Paepke, 1988; Zorzi, 1995). *Carcharhinus murrayi,* known only by the holotype collected from Karachi prior to 1883, might be synonymous (Compagno, 1984, 1988, personal communication). Other Indo-Pacific reports of *G. gangeticus* are unconfirmed or based on misidentifications of *Carcharinus leucas* or *C. ambionensis.* The numerous shark attacks in the Ganges have traditionally and perhaps wrongly, been attributed to *G. gangeticus; C. leucas* has been confirmed as present in the river and may be responsible for the attacks (Compagno, 1988). T. Roberts (personal communication) saw fresh specimens of *G. gangeticus* in 1996. Further judgment on the conservation status of *G. gangeticus* should await publication of reviews currently in preparation by T. Roberts and L. Compagno, and assimilation of further information on the biology and distribution of the species.

[c]*Pseudoscaphirhychus fedtschenkoi* and *P. hermanni* have not been seen since 1970 or for the last 20 years, respectively. Both species might be extinct, although *P. fedtschenkoi* may still exist in the upper reaches of the Syr-Darya river and *P. hermanni,* which has always been rare, may still exist in the upper reaches of the Amu-Darya river (Pavlov *et al.,* 1985; Mukhina and Kreuzberg, personal communication). Poulation declines can be attributed to changes in hydrography through irrigation.

[d]Described in 1986 as a subspecies *Alosa caspia vistonica,* endemic to Lake Vistonis. A taxonomic review is necessary; Kottelat (1997:7, 41) noted this as a distinct species that had not been seen for 2 years. Kottelat did not list the species as extinct, suggesting further sampling is required. Economidis (1995) noted pollution and water extraction as environmental threats to the species.

[e]Kottelat (personal communication) suggests this species might be extinct; further evidence is required to support this.

[f]An endangered or extinct endemic species of Lake Dianchi, threatened by habitat alteration, pollution, and introduction of exotics (Kottelat and Chu, 1988).

[g]*Algansea barbata,* collected from the headwaters of the Río Lerma (Alvarez and Cortés, 1964), is endemic to this drainage and listed as "extinct?, endangered" by Lyons *et al.* (1998). No further information is available.

[h]Sehgal (1994) listed *G. biswasi* as extinct only because it has not been seen since its first capture (described in 1977); but there is no evidence of sampling or environmental threat. The hypothesized extinc-

tion has been questioned by other authors (Mahanta *et al.*, 1994). *Gymnocypris biswasi* is listed as extinct in the 1998 report of the Conservation Assessment and Managemnt Plan (C.A.M.P.) Workshops, compiled by the Zoo Outreach Organisation and Conservation Breeding Specialist Group of India, but without any further evidence.

[i]Mrakovcic *et al.* (1995) referred to two subspecies of *Leuciscus turskyi* (*tenellus* and *microlepis*) not listed as extinct but of unknown conservation status. Kottelat (1997) did not list *L. turskyi* as extinct.

[j]Mrakovcic *et al.* (1995) recorded this as an endangered/extinct endemic that was common but has not been found in the last few years.

[k]Listed as threatened according to Mayden *et al.* (1992), but extinct according to IUCN (1996). No further information is available.

[l]Tzeng (1986) listed this species as present in rivers Tadu and Choshui. Kwang-Tsao (personal communication) notes the species might be extinct. Further sampling is required.

[m]Rosa and Menezes (1996) state that confirmed reports for *Hyphessobrycon parvellus* are restricted to the type locality (Rio Catu, Alagoinhas, Brazil) and reports from Rio Tietê are possibly erroneous; Rosa and Menezes (1996) listed the species as possibly extinct.

[n]*Noturus trautmani* noted as extremely rare by Rohde (1980). Mayden *et al.* (1992) listed this species as extinct and Bondhus (1995) noted it as probably extinct according to U.S. Fish and Wildlife Service. No further information is available.

[o]*Irveneia voltae* not collected since the original collection before 1943. T. Roberts (personal communication) looked intensively but unsuccessfully for this species on collection trips from 1961 to 1988 and believes it may be extinct. Construction of the Volta Dam changed the downstream type locality; if the species is extinct, then the construction of the dam probably contributed to its extinction. De Vos (1995) stated that local fishermen in the area still knew the species, had a name for it, but would not verify having seen it.

[p]*Salmo montenegrinus* was described in 1933 with no subsequent data; the continued existence of this species needs to be confirmed (Maric, 1995). The taxonomic validity of the species should be accepted tentatively; the taxonomy of *Salmo* is exceptionally complex and several nominal endemic species have been described but are considered by many taxonomists to be no more than different populations or subspecies of a single species.

[q]Kottelat (1997:140) tentatively accepted *Salmo schiefermuelleri* as valid (see taxonomic comments for *S. montenegrinus*, above) and stated the original stock now seems to be extinct. No further supporting data are available.

[r]*Salvelinus profundus* was named by Schillinger in 1901 but a full description has never been published (Kottelat, 1997:154). Recognized as a distinct species by Behnke (1972, 1908a). *S. profundus* is only reported definitely from Lake Constance (deep waters); other nominal material from other subalpine lakes (Behnke, 1972, 1980a; Brenner, 1980) is not proven as conspecific according to Kottelat (1997). Kottelat also noted specimens currently collected from deep waters of Lake Constance do not appear to be *S. profundus;* he stated *S. profundus* "might be extinct," attributable to pollution, eutrophication, and hybridization. Kottelat accepted the species may still be present in deeper areas not usually fished and this requires investigation.

[s]*Chirostoma charari* is endemic to the Río Lerma drainage and listed as "extinct, endangered" by Lyons *et al.* (1998). No further information is available.

[t]Possibly extinct, on the basis of its absence from recent collections by one of us (MLJS). More sampling is required.

[u]See notes for *Adrianichthys kruyti* (Table 5). Kottelat and Whitten (1996) listed *A. kruyti* as extinct but referred to *X. poptae* as "very seriously threatened." Further sampling is required to confirm the conservation status of the two species.

[v]*Xenopoecilus sarasinorum* was noted as "virtually extinct" by Kottelat and Whitten (1996), presumably through predation by tilapia, introduced in the 1950s (Whitten *et al.*, 1987). Further sampling is required to confirm the conservation status of *X. sarasinorum*.

[w]Minckley *et al.* (1991) listed *C. tessellatus* as uncommon but not extinct. Lambert (1992) noted that the species was present in 1991 but could not be found during 2 days of sampling in 1992. More sampling is required through the lake.

[x]*Cyprinodon* sp. noted as endangered by Minckley *et al.* (1991). Listed as extinct by IUCN (1996) but no further data.

[y]Not collected since 1937; possibly extinct but this is a small species and may have been overlooked (Lauzanne, 1992).

[z]Mayden *et al.* (1992) list *C. peraticus* as extinct. No further information is available to confirm this.

[aa]Populations of *G. georgii* declined through disruption of the species' habitat and introduction of exotic species; genetically pure *G. georgii* have not been collected since 1983 (Miller *et al.*, 1989). The species is possibly extinct (Minckley *et al.*, 1991; WCMC, 1992).

[bb]*Phalloptychus eigenmanni* is known only from the type locality (Rio Catu, Alagoinhas, Brazil) and is listed as possibly extinct by Rosa and Menezes (1996).

[cc]Meyer and Pérez (1990) stated *P. bonita* "has not been collected for many years and perhaps is extinct or at least very rare." Mayden *et al.* (1992) noted the species as endangered. Although listed as extinct by WCMC (1992), the possibility of extant populations was recognized.

[dd]Kottelat (1997:165) stated *Gasterosteus crenobiontus* is extinct, citing Banarescu (1994:8). The species's taxonomy is unconfirmed and no further information is available.

[ee]The local fishermen around Lake Barombi Mbo report the absence of *S. mongo,* although this might be because the fishermen do not set their basket traps deep enough (Reid, 1991). Reid stated the *S. mongo* may be "the first rare West Afica cichlid to become extinct, although its status is at present indeterminate in Lake Barombi Mbo."

[ff]Ng and Kottelat (1994) described *B. tomi* from 29 specimens from Peninsula Malaysia. Seven other specimens, collected from Singapore, were tentatively assigned to this species. The Singapore specimens show some meristic differences from the Malaysian ones, perhaps indicating they are not conspecific. The collection site for the Singaporean specimens, a peaty swamp, has been destroyed and collection attempts from the only other comparable site in Singapore have been unsuccessful. If the Singaporean specimens are specifically distinct from the Malaysian ones, then they represent an apparently extinct species.

[gg]Watson (1997) reported *S. nigrescens* as extinct since the 1880s but provided no further information.

[hh]*Etheostoma sellare* is probably extinct according to Bondhus (1995), citing Wheeler (1991). No further information is available.

Table 8. Species Categorized as Disqualified Extinctions or Species Represented by Captive Populations (Excluding Lake Victoria Cichlids)

Family	Species	Common name	Range	IUCN (1996) Status	Alternative extinction reference	Notes on disqualification
Catostomidae	*Chasmistes liorus liorus*	june sucker	Utah, US	—	Miller *et al.* (1989)	*a, c*
Cyprinidae	*Cyprinus carpio chilia*		Lake Dianchi, China		Kottelat and Chu (1988)	*a, d*
	Cyprinus micristius		China	EN	Kottelat and Chu (1988)	*a, e*
	Epalzeorhynchos bicolor		Thailand	EW	—	*b*
	Gila bicolor insulator	Independence Valley tui chub	Nevada, US	—	Miller *et al.* (1989)	*a, f*
	Gnathopongon elongatus suwae	Suwa gudgeon	Lake Suwa, Japan	—	Yuma *et al.* (1998)	*a, g*
	Notropis lutrensis blairi	Maravillas red shiner	Texas, US	—	Miller *et al.* (1989)	*a, h*
	Notropis simus simus	Río Grande bluntnose shiner	Texas and New Mexico, US	—	Miller *et al.* (1989)	*a, i*
	Osteobrama belangeri		India	—	C.A.M.P. Workshop Report (1998)	*b, j*
	Rhinichthys cataractae smithi	Banff longnose dace	Alberta, Canada	—	Miller *et al.* (1989)	*a, k*
	Rhinichthys osculus reliquus	Grass valley speckled dace	Nevada, US	—	Miller *et al.* (1989)	*a, k*
	Sinocylocheilus grahami grahami		Lake Dianchi, China		Kottelat and Chu (1988)	*a, l*
Homalopteridae	*Yunnanilus nigromaculatus*		China	EN	Kottelat and Chu (1988)	*m*
Salmonidae	*Coregonus kiyi orientalis*	Lake Ontario kiyi	Lake Ontario	—	Miller *et al.* (1989)	*a, n*
	Coregonus nigripinnis nigripinnis	Lake Ontario kiyi	Lake Ontario	EX	Miller *et al.* (1989)	*a, o*
	Oncorhynchus clarki macdonaldi	yellowfin cutthroat trout	Colorado, US	—	Miller *et al.* (1989)	*a, p*
	Oncorhynchus clarki ssp.	Alvord cutthroat trout	Oregon/Nevada, US	—	Miller *et al.* (1989)	*a, p*
	Oncorhynchus nerka kawamurae	black kokanee	Lak Tazawa, Japan	—	Yuma *et al.* (1998)	*a, q*
Cyprinodontidae	*Cyprinodon alvarezi*	Potosí pupfish	Mexico	EW	—	*b*
	Cyprinodon nevadensis calidae	Tecopa pupfish	California, US	—	Miller *et al.* (1989)	*a, r*

	Cyprinodon sp.		Mexico	—	WCMC (1992)	*s*
	Cyprinodon sp.		Mexico	—	WCMC (1992)	*s*
	Megupsilon aporus	catarina pupfish	Mexico	EW	—	*b*
Goodeidae	*Allotoca maculata*	opal allotoca	Mexico	CR	Miller *et al.* (1989)	*t*
	Ameca splendens		Mexico	EW	—	*b, u*
	Empetrichthys latos concavus	Raycraft ranch killifish	Nevada, US	—	Miller *et al.* (1989)	*a, v*
	Empetrichthys latos pahrump	Pahrump ranch killifish	Nevada, US	—	Miller *et al.* (1989)	*a, v*
	Skiffia francesae	golden skiffia	Mexico	EW	Mayden *et al.* (1992)	*b*
Gobiidae	*Mistichthys luzonensis*	sinarapan	Philippines	LR (cd)	Kottelat and Whitten (1996)	*w*
Percidae	*Stizostedion vitreum glaucum*	blue pike	Lake Erie and Ontario	—	Miller *et al.* (1989)	*a, x*

Notes on disqualifications

[a]Subspecies are not evaluated; see text Section 4.
[b]"Extinct in the Wild" is not recognized as an extinction category according to the current scheme; see text Section 8.2.
[c]*C. liorus mictus* extant in Utah Lake, Utah, US (Miller and Smith, 1981).
[d]*Cyprinus c. chilia* also present in Fuxian Lake and most lakes on Yunnan plateau (Yang, 1991).
[e]*Cyprinus m. fuxianensis* present in Fuxian and Xingyun Lakes (Yang, 1991); hence, subspecies *C. micristius micritius* not evaluated.
[f]*Gila bicolor* and subspecies are widespread throughout western US (Smith, 1980).
[g]*Gnathopogon elongatus elongatus* is extant in Japan (Yuma *et al.*, 1998).
[h]*Notropis lutrensis* and nominate subspecies extant through Mississippi and Gulf drainages (Matthews, 1980).
[i]*N. simus pecosensis* is endangered but extant in New Mexico (Chernoff *et al.*, 1982).
[j]*Osteobrama belangeri* is noted as "extinct in the wild" according to the 1998 report of the Conservation Assessment and Management Plan (C.A.M.P.) Workshops, compiled by the Zoo Outreach Organisation and Conservation Breeding Specialist Group of India. This species is also known from China and Myanmar.
[k]*Rhinichthys cataractae* and *R. osculus* are widespread throughout North America (Gilbert and Shute, 1980; Wallace, 1980).
[l]*Sinocylocheilus g. tingi* present and endemic to Fuxian Lake (Yang, 1991).
[m]*Yunnanilus nigromaculatus* has been extinct from Lake Dianchi since ca. 1964; Zheng (1989) and Yang (1991) record it as extant in Lake Fuxian.
[n]*Coregonus kiyi* is regionally extirpated but exists as a "vulnerable" species in Lake Superior (McAllister *et al.*, 1985).
[o]*Coregonus n. regalis* is extant in Lake Nipigon (McAllister *et al.*, 1985).
[p]*Oncorhynchus clarki* and subspecies widespread throughout northwestern US (Behnke, 1980b).
[q]*Oncorhynchus nerka* and subspecies widespread around northern Pacific rim (Araga and Ida, 1984).
[r]*Cyprinodon nevadensis* subspecies extant in various springs (Minckley *et al.*, 1991).
[s]No further data; this species may have been subsequently named and listed as extinct under its scientific name.
[t]*Allotoca maculata* rediscovered in 1990 (Almada-Villeja, 1990).
[u]G. McG. Reid believes *Ameca splendens* may be extant in the wild (personal communication).
[v]*Empetrichthys latos latos* is "endangered," extant at Manse Ranch Spring, Nevada (Minckley *et al.*, 1991).
[w]*Mistichthys luzonensis* is currently found in Lakes Buhi, Bato, and Manapao, Luzon (Soliman, 1993; Aypa *et al.*, 1995; E. Capuli, personal communication).
[x]*Stizostedion vitreum* is widespread throughtout North America (McAllister *et al.*, 1985).

Table 9. Lake Victoria Cichlids Provisionally Placed in Extinction Category ?D1

Species	Possible EED	IUCN (1996) status	Notes
Haplochromis "bartoni-like"	1982–1987[1]	—	*a–c*
Haplochromis "bicolor"	1982–1987[1]	—	*a–c*
Haplochromis (?) "big teeth"	1982–1987[1]	—	*a, c, d*
Haplochromis (Lipochromis) "black cryptodon"	1982–1987[1]	CR	*a, c, e*
Haplochromis "black pectoral"	1982–1987[1]	—	*a–c*
Haplochromis "chlorocephalus"	1982–1987[1]	—	*a–c*
Haplochromis (?) "citrus"	1982–1987[1]	—	*a, c, d*
Haplochromis "coop"	1982–1987[1]	—	*a–c*
Haplochromis (Paralabidochromis) "elongate rockpicker"	1993–1995[4]	—	*c, f*
Haplochromis "filamentus"	1982–1987[1]	—	*a–c*
Haplochromis (Paralabidochromis) "fleshy lips"	1993–1996[3]	—	*g*
Haplochromis "grey pseudo-nigricans"	1979–1990[1]	—	*b, c, h, i*
Haplochromis (Harpagochromis) "large eye guiarti"	1982–1987[1]	—	*c, j*
Haplochromis "lividus-frels"	1982–1987[1]	—	*a–c*
Haplochromis (Prognathochromis) "longurius"	1979–1990[1]	—	*c, d, k*
Haplochromis "macrops-like"	1982–1987[1]	—	*a–c*
Haplochromis (Lipochromis) "micro-obesus"	1982–1987[1]	—	*a–c*
Haplochromis (Enterochromis) "morsei"	1990–1991[4]	—	*l, m*
Haplochromis (Astatotilapia) "orange cinereus"	1990–1991[4]	—	*c, l*
Haplochromis "orange macula"	1982–1987[1]	—	*a–c*
Haplochromis "orange yellow big teeth"	1982–1987[1]	—	*a–c*
Haplochromis "orange yellow small teeth"	1982–1987[1]	—	*a–c*
Haplochromis "paropius-like"	1982–1987[1]	—	*a–c*
Haplochromis "pink paedophage"	1982–1987[1]	—	*a–c*
Haplochromis "pseudo-morsei"	1982–1987[1]	—	*a–c*
Haplochromis (Enterochromis) "purple head"	1990–1991[4]	—	*l, m*
Haplochromis "purple miller"	1982–1987[1]	—	*a–c*
Haplochromis (?) "purple rocker"	1990–1991[4]	—	*c, n*
Haplochromis "red empodisma"	1982–1987[1]	—	*a–c, o*
Haplochromis "red eye scraper"	1982–1987[1]	—	*a–c*
Haplochromis "reginus"	1982–1987[1]	—	*a–c*
Haplochromis "regius"	1989[2]	—	*p*
Haplochromis "short supramacrops"	1982–1987[1]	—	*a–c*
Haplochromis (Astatotilapia) "small blue zebra"	1991–1993[4]	—	*c, q*
Haplochromis "small empodisma"	1982–1987[1]	—	*a–c, o*
Haplochromis "smoke"	1979–1990[1]	—	*b, c, h*
Haplochromis "soft grey"	1982–1987[1]	—	*a–c*
Haplochromis (?) "stripmac"	1990–1991[4]	—	*c, l*
Haplochromis "supramacrops"	1982–1987[1]	—	*a–c*
Haplochromis "theliodon-like"	1982–1987[1]	—	*a–c*
Haplochromis "tigrus"	1982–1987[1]	—	*a–c*
Haplochromis "too small"	1982–1987[1]	—	*a–c*
Haplochromis "twenty"	1982–1987[1]	—	*a–c*
Haplochromis "two stripe white lip"	1982–1987[1]	—	*a–c*
Haplochromis "wyber"	1982–1987[1]	—	*a–c*
Haplochromis (Ptyochromis) "xenognathus-like"	1982–1987[1]	—	*a, r*
Haplochromis "yellow"	1982–1987[1]	—	*a–c*
Haplochromis "yellow-blue"	1991–1996[3]	—	*c, s*

Table 9. *(Continued)*

Details concerning hypodigms and common names have not been published for most of these species and therefore are not included in this table.

Possible EED

The possible EED is based on data from: [1]Witte *et al.* (1992b); [2]Kaufman and Ochumba (1993); [3]Seehausen (1996); [4]Seehausen *et al.* (1997b). For those publications that do not refer to the year when the species was first absent from collections, we use the year of publication of the information for the EED.

Notes

[a]Species collected from the southern part of the lake during 1979–1982 but not 1987/1990 (Witte *et al.*, 1992b).
[b]No reference to this species in a recent surveys of southern Lake Victoria (Seehausen, 1996; Seehausen *et al.*, 1997b).
[c]No reference to this species in surveys of northern parts of the lake (Kaufman and Ochumba, 1993).
[d]Seehausen (1996) noted this species as possibly extinct.
[e]*Haplochromis* (*L.*) *cryptodon* was described from specimens from Uganda and Tanzanian waters (Greenwood, 1959) but the taxonomic identity is problematic and "cryptodon-like" fishes from Mwanza Gulf are usually referred to as "black cryptodon" (Seehausen, 1996).
[f]*Haplochromis* (*P.*) "elongate rockpicker" not recorded from sampled areas of southern Lake Victoria since 1993 (Seehausen *et al.*, 1997b); Seehausen (1996) noted a similar form that may be identical with this species.
[g]Collected in 1993 and not subsequently (Seehausen, 1996).
[h]Collected from the southern part of the lake during 1978–1979 but not 1990 (Witte *et al.*, 1992b).
[i]Witte *et al.* (1992b) noted that because of identification problems they could not be sure that no individuals of *H.* "grey pseudonigricans" had been caught.
[j]Collected from the southern part of the lake during 1979–1982 but not subsequently (Witte *et al.*, 1992b, Seehausen, 1996).
[k]Collected from the southern part of the lake during 1978–1979 but not subsequently (Witte *et al.*, 1992b, Seehausen, 1996).
[l]Not observed in sampled areas of the southern part of the lake after 1990 (Seehausen *et al.*, 1997b).
[m]Kaufman and Ochumba (1993) list *Enterochromis* spp. as absent from sampled northern waters but do not give precise taxonomic details.
[n]From Mwanza Gulf and not collected from that region since 1990, perhaps extinct (Witte *et al.*, 1992b; Seehausen, 1996; Seehausen *et al.*, 1997b).
[o]See comments for *H. empodisma* complex (Table 11).
[p]Not observed in sampled areas of northern lake during 1989–1992 (Kaufman and Ochumba, 1993); surveys of southern parts of the lake do not refer to this species (Witte *et al.*, 1992b; Seehausen *et al.*, 1997b).
[q]Rarely caught and last collected from southern lake in 1991 (Seehausen, 1996; Seehausen *et al.*, 1997b).
[r]*Haplochromis* (*Ptyo.*) *xenognathus* were collected from northern parts of the lake during 1989–1992 (Kaufman and Ochumba, 1983), from the southern parts of the lake during 1991–1995 (Seehausen *et al.*, 1997b), and are maintained as a captive breeding stock (Reid, 1990); but there is no other reference to *H.* "xenognathus-like," except Witte *et al.* (1992b).
[s]Collected in 1991 and not subsequently (Seehausen, 1996).

Table 10. Lake Victoria Cichlids Provisionally Placed in Extinction Category ?D2

Species	Hypodigm	Possible EED	IUCN (1996) status	Notes
Haplochromis (*Harpagochromis*) *altigenis*	‡	1989–1996[5]	—	*d–f*
Haplochromis apogonoides	‡	1982–1987[1]	—	*a, c, d, g*
Haplochromis (*Prognathochromis*) *arcanus*	‡	1969–1996[4]	EX	*b–d, h*
Haplochromis (*Prognathochromis*) *argenteus*	‡	1982–1987[1]	EX	*a, c, d, i*
Haplochromis (*Harpagochromis*) *artaxerxes*	H	1962–1996[4]	EX	*b–d, j, k*
Haplochromis (*Astatotilapia*) *barbarae*	‡	1967–1989[2]	EN	*e, l–n*
Haplochromis bareli	‡	1982–1987[1]	—	*a, c, d*
Haplochromis (*Prognathochromis*) *bartoni*	‡	1962–1996[4]	EX	*b–d, o*
Haplochromis (*Prognathochromis*) *bayoni*	‡	1962–1996[4]	EX	*b–d, p*
Haplochromis (*Harpagochromis*) *boops*	H+	1962–1996[4]	EX	*b–d, g, k*
Haplochromis (*Psammochromis*) *cassius*	‡	1982–1987[1]	EX	*a, c, d, q*
Haplochromis (*Enterochromis*) *cinctus*	‡	1982–1987[1]	—	*a, c, d, h*
Haplochromis cnester	‡	1982–1987[1]	—	*a, c, d*
Haplochromis (*Prognathochromis*) *decticostoma*	‡	1969–1996[4]	EX	*b–d, h*

(continued)

Table 10. (*Continued*)

Species	Hypodigm	Possible EED	IUCN (1996) status	Notes
Haplochromis (*Prognathochromis*) *dentex* group	‡	1982–1987[1]	EX	*c, l, o, r*
Haplochromis (? *Harpagochromis*) *diplotaenia*	H	1928–1996?[5]	—	*k, s*
Haplochromis (*Prognathochromis*) *estor*	‡	1962–1996[4]	EX	*b–d, o*
Haplochromis (*Prognathochromis*) *flavipinnis*	‡	1962–1989[2]	EX	*b, c, l, p*
Haplochromis (*Prognathochromis*) *gilberti*	‡	1978–1996[4]	EX	*b–d, t*
Haplochromis (*Prognathochromis*) *gowersi*	‡	1962–1996[4]	EX	*b–d, o*
Haplochromis (*Harpagochromis*) *guiarti*	?	1982–1987[1]	—	*d, u, v*
Haplochromis (*Yssichromis*) *heusinkveldi*	‡	1990–1991[6]	—	*d, w, x*
Haplochromis (*Gaurochromis*) *hiatus*	‡	1982–1987[1]	—	*a, c, d*
Haplochromis (*Gaurochromis*) *iris*	‡	1982–1987[1]	—	*a, c, d*
Haplochromis (*Prognathochromis*) *longirostris*	‡	1982–1987[1]	EX	*a, c, d, p*
Haplochromis (*Prognathochromis*) *macrognathus*	‡	1982–1987[1]	EX	*d, i, p, v*
Haplochromis (*Harpagochromis*) *maculipinna*	‡	1967–1996[4]	EX	*b–d, g*
Haplochromis (*Prognathochromis*) *mandibularis*	‡	1962–1996[4]	EX	*b–d, p*
Haplochromis (*Astatotilapia*) *martini*	‡	1960–1989[2]	EX	*b, c, l, y*
Haplochromis (*Astatotilapia*) *megalops*	‡	1982–1987[1]	EX	*a, d, n, o*
Haplochromis (*Harpagochromis*) *michaeli*	‡	1982–1987[1]	EX	*a, c, d, h*
Haplochromis (*Lipochromis*) *microdon*	‡	1982–1987[1]	EX	*d, z, aa*
Haplochromis (*Labrochromis*) *mylergates*	‡	1978–1995[3]	EX	*b–d, i, q*
Haplochromis (*Prognathochromis*) *nanoserranus*	‡	1978–1996[4]	EX	*b–d, q*
Haplochromis (*Prognathochromis*) *nigrescens*	?	1912?–1996[4]	EX	*b–d, bb*
Haplochromis (*Harpagochromis*) *nyanzae*	‡	1962–1996[4]	EX	*b–d, o*
Haplochromis (*Gaurochromis*) *obtusidens*	‡	1960–1996[4]	EX	*b–d, y*
Haplochromis (*Harpagochromis*) *pachycephalus*	‡	1967–1996[4]	EX	*b–e*
Haplochromis (*Prognathochromis*) *paraguiarti*	‡	1967–1996[4]	EX	*b–e*
Haplochromis (*Harpagochromis*) *paraplagiostoma*	‡	1969–1996[4]	EX	*b–d, h*
Haplochromis (*Pyxichromis*) *parorthostoma*	H+	1967–1996[4]	EX	*b–d, g*
Haplochromis (*Prognathochromis*) *percoides*	‡	1962–1996[4]	EX	*l, p, x*
Haplochromis (*Labrochromis*) *pharyngomylus*	‡	1960–1989[2]	EX	*b, c, l, y*
Haplochromis (*Prognathochromis*) *prognathus*	‡	1967–1996[4]	EX	*b–e*
Haplochromis (*Prognathochromis*) *pseudopellegrini*	‡	1967–1996[4]	EX	*b–d, g*
Haplochromis pyrrhopteryx	‡	1982–1987[1]	—	*a, c, d*
Haplochromis (*Harpagochromis*) *spekii*	‡	1967–1996[4]	EX	*b–d, g*
Haplochromis (*Labrochromis*) *teegelaari*	‡	1982–1987[1]	EX	*a, c, d, q*
Haplochromis (*Harpagochromis*) *thuragnathus*	H+	1967–1996[4]	EX	*b–d, g*
Haplochromis (*Prognathochromis*) *tridens*	‡	1967–1989[2]	—	*e, cc*
Haplochromis (*Harpagochromis*) *victorianus*	‡	1962–1996[4]	EX	*b–d, o*
Haplochromis (*Prognathochromis*) *xenostoma*	‡	1982–1987[1]	EX	*a, c, d*
Hoplotilapia retrodens	‡	1991–1996[4]	EX	*l, dd*
Psammochromis cryptogramma group	‡	1969–1989[2]	—	*ee*

Common names have not been published for most of these species and therefore are not included in this table.

Possible EED

The possible EED is based on data from: [1]Witte *et al.* (1992b); [2]Kaufman and Ochumba (1993); [3]Seehausen and Witte (1995); [4]IUCN (1996); [5]Seehausen (1996); [6]Seehausen *et al.* (1997b). For those publications that do not refer to the year when the species was first absent from collections, we use the year of publication of the information for the EED.

Notes

[a]Collected from southern parts of Lake Victoria during 1979–1982 but not 1987/1990 (Witte *et al.*, 1992b).

[b]No reference to this species in surveys of southern parts of Lake Victoria by Witte *et al.* (1992b).

[c]No reference to this species in surveys of southern parts of Lake Victoria by Seehausen (1996) and Seehausen *et al.* (1997b).

[d]No reference to this species in surveys of northern parts of Lake Victoria by Kaufman and Ochumba (1993).

[e]Formerly reported from northern and southern parts of lake (Greenwood, 1967).

Table 10. *(Continued)*

[f]Not recorded from Mwanza Gulf since 1989 (Seehausen, 1996).
[g]Formerly reported from northern parts of Lake Victoria (Greenwood, 1967).
[h]Formerly reported from northern parts of Lake Victoria (Greenwood and Gee, 1969).
[i]Extinct according to Seehausen and Witte (1995).
[j]Known only from the type specimen caught from northern part of Lake Victoria in 1962 (Oijen *et al.*, 1991).
[k]Presumably extinct because of the effects of the introduction of Nile perch (as for other haplochromines) but it is unusual the species was not collected again prior to the collapse of the haplochromine populations in Lake Victoria.
[l]Not observed in sampled northern areas of Lake Victoria during 1989–1992 (Kaufman and Ochumba, 1993).
[m]Not observed in sampled southern areas of Lake Victoria after 1989 (Witte *et al.*, 1992b; Seehausen, 1996).
[n]Noted as possibly extinct (Seehausen, 1996).
[o]Formerly reported from northern parts of Lake Victoria (Greenwood, 1962).
[p]Formerly reported from northern and southern parts of Lake Victoria (Greenwood, 1962).
[q]Formerly reported from Mwanza Gulf (Greenwood and Barel, 1978).
[r]*Prognathochromis dentex* listed as extinct (IUCN, 1996). *H.* "dentex-like" collected from the southern part of the lake during 1979–1982 but not 1987/1990 (Witte *et al.*, 1992b).
[s]Known from only a single specimen collected prior to 1928 (Seehausen, 1996). Seehausen (1996) states that fish similar to *H. diplotaenia* were exported to Sweden under the name *Prognathochromis* sp. aff. *longirostris*.
[t]Formerly reported from northern Lake Victoria (Greenwood and Barel, 1978).
[u]Formerly common in Mwanza Gulf (Seehausen, 1996).
[v]Collected from the southern part of Lake Victoria during 1979–1982 but not susequently (Witte *et al.*, 1992b; Seehausen, 1996).
[w]Formerly widespread through Lake Victoria (Greenwood, 1957).
[x]Not observed in sampled areas of southern part of Lake Victoria after 1990 (Seehausen *et al.*, 1997b).
[y]Formerly reported from northern and southern parts of Lake Victoria (Greenwood, 1960).
[z]Formerly reported from northern parts of Lake Victoria (Greenwood, 1959).
[aa]A similar species, *H.* (*L.*) "blue microdon," was recorded in 1995 from Speke Gulf (Seehausen, 1996; Seehausen *et al.*, 1997b).
[bb]Formerly reported from northern parts of Lake Victoria (Oijen *et al.*, 1991).
[cc]The *H.* (*P.*) *tridens* group was not observed in sampled areas of northern Lake Victoria during 1989–1992 (Kaufman and Ochumba, 1993); surveys of southern parts of the lake do not refer to this species (Witte *et al.*, 1992b; Seehausen *et al.*, 1997b) but note the absence of *H.* "red tridens" (see Table 11).
[dd]Formerly widespread through Lake Victoria (Greenwood, 1956); Seehausen *et al.* (1997b) list *Hoplotilapia retrodens* as disappeared from Mwanza Gulf after 1987/1990 but recorded from another locality in the southern part of the lake in 1991.
[ee]*Psamochromis cryptogramma* was perhaps formerly widespread (Greenwood and Gee, 1969); not observed in sampled areas of northern lake during 1989–1992 (Kaufman and Ochumba, 1993); *H.* "erythrocryptogramma" is an undescribed species reported from southern waters of the lake until 1982 but not during 1987/1990 (Witte *et al.*, 1992b) and considered extinct by Seehausen and Witte (1995).

Table 11. Lake Victoria Cichlids Provisionally Placed in Category ?U

Species	Hypodigm	IUCN (1996) status	Notes
Enterochromis spp.	?	—	*h*
Haplochromis (*H.*) *annectidens*	‡	CR	*i*
Haplochromis (*Astatotilapia*) *brownae*	‡	EN	*a, f, j*
Haplochromis (*Harpagochromis*) *cavifrons*	‡	EX	*b, k*
Haplochromis (*Paralabidochromis*) *chromogynos*	‡	VU	*c, g, l*
Haplochromis (*Lipochromis*) *cryptodon* complex	‡	—	*a, m*
Haplochromis (?) "curvicephalus"	?	—	*a, d, n*
Haplochromis (*Astatotilapia*) "diversicolor"	?	—	*o*
Haplochromis "double stripe"	?	—	*p*
Haplochromis (*Enterochromis*) "dusky wine-red fin"	?	—	*q*
Haplochromis (*Gaurochromis*) *empodisma* complex	‡	—	*a, r*
Haplochromis (*Ptyochromis*) *granti*	‡	EN	*c, s*
Haplochromis (?) "kribensis"	?	—	*b, t*

(continued)

Table 11. (*Continued*)

Species	Hypodigm	IUCN (1996) status	Notes
Haplochromis (*Lipochromis*) *maxillaris*	‡	CR	*c, d, u*
Haplochromis (*Lipochromis*) *melanopterus*	H	CR	*a, e, v*
Haplochromis (*Prognathochromis*) *mento*	‡	CR	*a, f, w*
Haplochromis (*Xystichromis*) *nuchisquamulatus*	‡	EN	*b, x*
Haplochromis (*Lipochromis*) *obesus*	‡	—	*b, y*
Haplochromis (*H.*) *obliquidens*	‡	EN	*b, z*
Haplochromis (*Lipochromis*) *parvidens*	‡	—	*b, d, g, aa*
Haplochromis (*Harpagochromis*) *plagiostoma*	‡	CR	*c, d*
Haplochromis (*Yssichromis*) "plumbus"	?	—	*a, d, bb*
Haplochromis (*Labrochromis*) *ptistes*	‡	—	*c, cc*
Haplochromis "red tridens"	?	—	*dd*
Haplochromis (*Paralabidochromis*) "rockmacula"	?	—	*ee, ff*
Haplochromis (*Paralabidochromis*) "rockpicker"	?	—	*ff, gg*
Haplochromis tanaos	?	—	*hh*
Haplochromis (?) "thick skin"	‡	—	*a, e, ii*
Haplochromis (*Astatotilapia*) "two stripe yellow green"	?	—	*a, jj*
Haplochromis "ursus"	?	—	*kk*

Common names have not been published for most of these species and therefore are not included in this table.

Notes

[a]No reference to this species in surveys of northern parts of Lake Victoria by Kaufman and Ochumba (1993).

[b]Not observed in sampled northern areas of Lake Victoria during 1989–1992 (Kaufman and Ochumba, 1993).

[c]Collected from northern parts of Lake Victoria during 1989–1992 (Kaufman and Ochumba, 1993).

[d]Collected from southern parts of Lake Victoria during 1979–1982 but not 1987/1990 (Witte *et al.*, 1992b).

[e]Collected from southern parts of Lake Victoria during 1979–1982 but not 1990 (Witte *et al.*, 1992b).

[f]Formerly reported from northern and southern parts of Lake Victoria (Greenwood, 1962).

[g]Formerly reported from northern and southern parts of Lake Victoria (Greenwood, 1959).

[h]Species of this genus not observed in sampled area of northern lake during 1989–1992 (Kaufman and Ochumba, 1993). No further information available for confirmation of extinction of this species.

[i]From Lake Nabugabo (satellite of Lake Victoria). Ogutu-Ohwayo (1993) recorded *Haplochromis* (*H.*) *annectidens* in Lake Nabugabo during 1991–1992; Kaufman and Ochumba (1993) list this species as present but very rare.

[j]Seehausen *et al.* (1997b) reported *Haplochromis* (*A.*) *brownae* as absent from Mwanza Gulf but *H.* cf. *brownae* was recorded from other southern parts of the lake in 1991 and 1993.

[k]Formerly widespread, *Haplochromis* (*H.*) *cavifrons* was not seen for many years after the Nile perch boom; Seehausen (1996) recently collected a single specimen from the southern part of the lake.

[l]*Haplochromis* (*P.*) *chromogynos* not observed in sampled areas of southern lake since 1982 (Witte *et al.*, 1992b; Seehausen, 1996). Seehausen *et al.* (1997b) collected *H.* cf. *chromogynos* from southern waters in 1993 and 1995.

[m]The *cryptodon* complex was not observed in sampled areas of southern Lake Victoria during 1987/1990 (Witte *et al.*, 1992b). Seehausen (1996) and Seehausen *et al.* (1997b) indicated species of this complex to be present but rare.

[n]Seehausen *et al.* (1997b) recorded *H.* cf. "curvicephalus" from Mwanza Gulf in 1993.

[o]*Haplochromis* (*A.*) "diversicolor" was absent from Mwanza Gulf but collected from other parts of the lake in 1991 and 1993 (Seehausen *et al.*, 1997b).

Table 11. *(Continued)*

[p]*Haplochromis* "double stripe" is conspecific with *H.* (?) *tanaos;* see text Section 11.3.1 and this table for notes on that species.
[q]*Haplochromis* (*E.*) "dusky wine-red fin" was absent from the southern part of Lake Victoria during 1987/1990 (Witte *et al.*, 1992b) but recorded in 1993 and 1995 (Seehausen *et al.*, 1997b).
[r]*Haplochromis* (*G.*) *empodisma* was formerly reported from northern and southern parts of Lake Victoria (Greenwood, 1960). *Haplochromis empodisma* complex was not observed in samples from southern parts of Lake Victoria during 1987/1990 (Witte *et al.*, 1992); Seehausen *et al.* (1997b) recorded *H. empodisma* group in Mwanza Gulf in 1993 and list it as rare.
[s]*Haplochromis* (*Pyto.*) *granti* was formerly widespread through Lake Victoria (Greenwood, 1957). Listed as absent from Mwanza and Speke Gulfs after 1990 (Seehausen *et al.*, 1997b).
[t]*Haplochromis* (?) "kribensis" was listed as almost extinct by Seehausen and Witte (1995); reported from Mwanza Gulf of southern part of lake in 1993 and from Speke Gulf in 1996, when it was abundant (Seehausen, 1996; Seehausen *et al.*, 1997b).
[u]No reference to *H.* (*L.*) *maxillaris* in recent surveys of southern Lake Victoria (Seehausen, 1996; Seehausen *et al.*, 1997b).
[v]*Haplochromis* (*L.*) *melanopterus* was described from a single specimen from Mwanza Gulf, southern Lake Victoria (Seehausen, 1996); *H.* cf. *melanopterus* has been collected from southern Lake Victoria from 1991 to 1995 (Seehausen, 1996; Seehausen *et al.*, 1997b).
[w]Seehausen *et al.* (1997b) referred to a single specimen of *H.* cf. *mento* collected from Mwanza/Speke Gulf region in 1995 and listed that species as rare.
[x]*Haplochromis* (*X.*) *nuchisquamulatus* was formerly reported from Victoria Nile and Lake, including islands in the central part of the lake. Surveys of southern parts of the lake have not referred to this species (Witte *et al.*, 1992b; Seehausen *et al.*, 1997b).
[y]*Haplochromis* (*L.*) *obesus* was formerly reported from Lakes Victoria and Kwania; collected from Mwanza/Speke Gulfs in southern parts of Lake Victoria in 1992 and 1995, and listed as rare (Seehausen *et al.*, 1997b). No recent data for Lake Kwania.
[z]*Haplochromis* (*H.*) *obliquidens* was formerly reported from Victoria Nile and Lake (Oijen *et al.*, 1991; also listed as present in Lake Bunyoni, Tanzania, by IUCN, 1996); collected from Mwanza and Speke Gulfs, southern Lake Victoria, in 1992 and 1995 (Seehausen *et al.*, 1997b).
[aa]Seehausen *et al.* (1997b) collected *H.* (*L.*) cf. *parvidens* in 1995 from the southern part of the lake and note the species as rare.
[bb]Seehausen *et al.* (1997b) recorded *H.* cf. "plumbus" from Mwanza Gulf in 1991 and list this as rare.
[cc]*Haplochromis* (*L.*) *ptistes* was described from Speke Gulf, southern Lake Victoria (Greenwood and Barel, 1978); absent from southern part of the lake during 1987/1990 (Witte *et al.*, 1992b) and no reference to this species in more recent surveys of southern Lake Victoria (Seehausen, 1996; Seehausen *et al.*, 1997b).
[dd]*Haplochromis* "red tridens" absent from southern part of the lake during 1987/1990 (Witte *et al.*, 1992b) but recorded in 1993 (Seehausen *et al.*, 1997b).
[ee]*Haplochromis* (*P.*) "rockmacula" not observed in sampled areas of southern part of the lake since 1993.
[ff]Seehausen (1996) stated the species may be extinct; Seehausen *et al.* (1997b) listed the species as rare.
[gg]*Haplochromis* (*P.*) "rockpicker" not recorded from sampled areas of southern Lake Victoria since 1991 (Seehausen *et al.*, 1997b). Kaufman and Ochumba (1993) listed a species, *H.* "rockpicker," as collected from northern waters of Lake Victoria during 1989–1992. However, unlike the southern species which is listed as an insectivore, the northern form is listed as an algal grazer; hence, conspecificity is unclear.
[hh]*Haplochromis tanaos* was described from southern Lake Victoria; numbers declined after introduction of Nile perch and was listed as absent in 1987/1990 (Witte *et al.*, 1992b; who refer to this species as *Haplochromis* "double stripe"). *Haplochromis tanaos* was recollected from the southern part of the lake in 1993 and 1995, having shifted its habitat and its diet (Seehausen, 1996; Seehausen *et al.*, 1997b).
[ii]Seehausen (1996) described *H.* (?) "thick skin" as abundant and Seehausen *et al.* (1997b) listed it as collected from southern parts of lake in 1991, 1993, and 1995.
[jj]*Haplochromis* (*A.*) "two stripe yellow green" last collected in 1991 as a single specimen from Mwanza Gulf (Seehausen, 1996); Seehausen *et al.* (1997b) list this as a rare species.
[kk]*Haplochromis* "ursus" not collected after 1983 (Witte *et al.*, 1992b). No further data available for this species in recent surveys of southern and northern parts of lake (Kaufman and Ochumba, 1993; Seehausen, 1996; Seehausen *et al.*, 1997b).

Table 12. Lake Victoria Cichlids Categorized as Disqualified Extinctions or Species Represented by Captive Populations

Species	Hypodigm	IUCN (1996) status	Notes
Haplochromis (?) "argens"	‡	EW	*a–c, e, f*
Haplochromis (Labrochromis) ishmaeli	‡	EW	*a, b, d, g*
Haplochromis (H.) lividus	‡	EW	*a, d, h*
Haplochromis (Harpagochromis) pectoralis	‡	EX	*i*
Haplochromis (Prognathochromis) perrieri	?	EW	*a, c, j*
Haplochromis (Astatotilapia) piceatus	‡	EN	*b, c, e, k*
Haplochromis (Yssichromis) pyrrhocephalus	‡	VU	*l*
Haplochromis "red-eye guiarti"	?	EW	*a, m*
Haplochromis (Ptyochromis) sauvagei	‡	VU	*b, d, n*
Haplochromis (Prognathochromis) vittatus	‡	EX	*o*
Haplochromis (Allochromis) welcommei	‡	CR	*p*
Macropleurodus bicolor	‡	CR	*b, d, q*
Platytaeniodus degeni	‡	EW	*a, b, d, r*
Xystichromis bayoni	‡	EX	*s*

Common names have not been published for most of these species and therefore are not included in this table.

Notes

[a]"Extinct in the Wild" is not recognized as an extinction category according to the current scheme; see text Section 8.2.

[b]Reid (1990) referred to captive breeding stocks.

[c]No reference to this species in surveys of northern parts of Lake Victoria by Kaufman and Ochumba (1993).

[d]Not observed in sampled northern areas of Lake Victoria during 1989–1992 (Kaufman and Ochumba, 1993).

[e]Not observed in sampled areas of southern parts of Lake Victoria after 1987 and noted as "possibly extinct" Seehausen, 1996; Seehausen *et al.*, 1997b).

[f]IUCN (1996) lists *Yssiichromis "argens"* as extinct in the wild.

[g]*Haplochromis (L.) ishmaeli* was formerly reported from northern and southern parts of Lake Victoria (Greenwood, 1960); last recorded from southern areas in 1991 and listed as rare (Seehausen *et al.*, 1997b).

[h]*Haplochromis (H.) lividus* was formerly reported from northern and southern parts of Lake Victoria (Greenwood, 1956b); the species is uncommon but present in southeastern parts of the lake (Seehausen 1996).

[i]Oijen *et al.* (1991) listed this species in synonymy with *H. squamulatus,* for which there are no data indicating an extinction.

[j]*Haplochromis (P.) perrieri* was collected from the southern part of Lake Victoria during 1979–1982 but not subsequently (Witte *et al.*, 1992b; Seehausen, 1996).

[k]*Haplochromis (A.) piceatus* was described from specimens from northern Lake Victoria (Greenwood and Gee, 1969).

[l]Reid (1990) noted *Haplochromis (Yssi.) pyrrhocephalus* as extinct in the wild, otherwise represented by captive breeding stocks. Seehausen and Witte (1995) reported it as a surviving species in Lake Victoria; recorded from southern parts of the lake during 1991–1995 (Seehausen *et al.*, 1997b).

[m]The conservation status of *H.* "red-eye guiarti" requires confirmation through further sampling. See comments for *H.* "large eye guiarti" (Table 9).

[n]*Haplochromis (Ptyo.) sauvagei* was formerly reported as widespread through Lake Victoria (Greenwood, 1957); present in southern parts of the lake from 1991 to 1995 (Seehausen *et al.*, 1997b).

[o]*Haplochromis (P.) vittatus* was incorrectly listed as an extinct Lake Victoria species (IUCN, 1996); it is extant and found in Lake Kivu (Snoeks, 1994).

[p]*Haplochromis (Allochromis) welcommei* was collected from the southern part of Lake Victoria during 1979–1982 but not subsequently (Witte *et al.*, 1992b) and was noted as entinct by Seehausen and Witte (1995); it has recently been rediscovered in the satellite Lake Nawampasa to the north of Lake Victoria (Kaufman *et al.*, 1997).

[q]*Macropleurodus bicolor* was formerly widespread through Lake Victoria but is now almost extirpated (Seehausen, 1996); absent from sampled areas of Mwanza Gulf and represented by only one known population, observed in Speke Gulf in 1996 (Seehausen, 1996; Seehausen *et al.*, 1997b).

[r]*Platytaeniodus degeni* was reported from several areas in Lake Victoria (Greenwood, 1956a). Seehausen *et al.* (1997b) listed this species as disappeared from Mwanza Gulf after 1987/1990 but recorded it from another locality in the southern part of the lake in 1991.

[s]Oijen *et al.* (1991) listed *X. bayoni* in synonymy with the extant species *Haplochromis niloticus,* described from the Victoria Nile.

Table 13. Species Assigned to Extinction Categories in This Chapter but Not Listed Extinct by IUCN (1996)

Family	Species	IUCN (1996) status	Extinction category in this chapter
Cyprinidae	*Anabarilius alburnops*	—	D2
	Anabarilius polylepis	EN	D2
	Barbus microbarbis	—	D1
	Cephalakompsus pachycheilus	CR	C
	Chondrostoma scodrensis	CR	C
	Mandibularca resinus	CR	C
	Ospatulus palaemophagus	EN	C
	Ospatulus trunculatus	CR	C
	Phoxinellus egridiri	CR	D2
	Phoxinellus handlirschi	CR	D2
	Puntius amarus	CR	D1
	Puntius baoulan	CR	D1
	Puntius clemensi	CR	D1
	Puntius disa	CR	D1
	Puntius flavifuscus	CR	D1
	Puntius herrei	CR	D1
	Puntius lanaoensis	CR	D1
	Puntius manalak	CR	D1
	Puntius sirang	VU	D1
	Puntius tras	CR	D1
	Sprattlicypris palata	CR	C
Characidae	*Brycon acuminatus*	DD	C
	Henochilus wheatlandi	—	C
Schilbeidae	*Platytropius siamensis*	—	C
Salmonidae	*Coregonus confusus*	—	D1
	Coregonus fera	—	D1
	Coregonus gutturosus	—	D1
	Coregonus hiemalis	—	B1
	Coregonus restrictus	—	D1
	Salvelinus inframundus	—	B1
	Salvelinus scharffi	—	D1
Atherinidae	*Chirostoma compressum*	—	B1
Adrianichthyidae	*Adrianichthys kruyti*	CR	D2
Cyprinodontidae	*Cyprinodon* sp. ?	—	D1
	Leptolebias marmoratus	VU	C
	Orestias cuvieri	DD	C
Gasterosteidae	*Pungitius kaibarae*	—	C
Cichlidae	*Tristramella magdalenae*	—	D2
Gobiidae	*Weberogobius amadi*	CR	D2

See Tables 9 and 10 for comparable data on Lake Victoria cichlids.

Table 14. Taxonomic Distribution of Possible Extinctions

Family	Extinction categories								Excluding Lake Victoria cichlids		Including Lake Victoria cichlids	
	A	B1	C	D1	?D1	D2	?D2	D3	Total number of extinctions for famiy	Percentage of all freshwater extinctions	Total number of extinctions for family	Percentage of all freshwater extinctions
Petromyzontidae			1						1	1	1	<1
Catostomidae	1							1	2	3	2	1
Cyprinidae			18	11		5			34	49	34	20
Characidae			2						2	3	2	1
Schilbeidae			1						1	1	1	<1
Trichomycteridae						1			1	1	1	<1
Retropinnidae	1								1	1	1	<1
Salmonidae	1	1	1	7					10	14	10	6
Bedotiidae			1						1	1	1	<1
Atherinidae		1							1	1	1	<1
Adrianichthyidae						1			1	1	1	<1
Cyprinodontidae			6	1					7	10	7	4
Fundulidae			1						1	1	1	<1
Goodeidae			2						2	3	2	1
Poeciliidae			1						1	1	1	<1
Gasterosteidae			1						1	1	1	<1
Cichlidae					48	1	54		1	1	103	60
Cottidae						1			1	1	1	<1
Gobiidae						1			1	1	1	<1
Totals for extinction categories	3	2	35	19	48	10	54	1	70		172	
Percentage of all freshwater extinctions excluding Lake Victoria cichlids	4	3	50	27		14		1				
Percentage of all freshwater extinctions including Lake Victoria cichlids	2	1	20	11	28	6	31	<1				

References

Allan, J. D., and Flecker, A. S. 1993. Biodiversity conservation in running waters. *BioScience* **43**:32–43.

Almada-Villela, P. C. 1990. Status of threatened fishes. *J. Fish Biol.* **37**(Suppl. A):197–199.

Alvarez, J., and Cortés, M. T. 1964. Una especie de Algansea capturada en el Alto Lerma (Pisc. Cyprin.). *Ciencia (Mexico City)* **23**(3):104–106.

Alvarez del Villar, J., and Navarro, G. L. 1957. Los peces del Valle de Mexico. *Secretaria de Marina, Direccion General de Pesca e Industrias Conexas* **1957**:1–62.

Araga, C., and Ida, H. 1984. Salmonidae, in: H. Masuda, K. Amoaka, C. Araga, T. Uyeno, and T. Yoshino (eds.), *The Fishes of the Japanese Archipelago,* pp. 34–40. Tokai University Press, Tokyo, Japan.

Arratia, G. 1990. The South American Trichomycterinae (Teleostei: Siluriformes), a problematic group, in: G. Peters and R. Hutterer (eds.), *Vertebrates in the Tropics,* pp. 395–403. Museum Alexander Koenig, Bonn.

Aypa, S. M., Galicia, Jr., A. M., and Penolio, L. L. 1995. The present status and ecology of sinarapan, (*Mistichthys luzonensis*) in Lake Buhi, Camarines Sur, in: R. B. Edra, E. V. Manalili, and R. D. Guerrero III (eds.), *Lake Fisheries management in the Philippines (Proceedings of the National Symposium-Workshop on Lake Fisheries Management, 28–39 October 1993, PCARRD, Los Baños, Laguna)*, pp. 27–74. PCAMRD Book Ser. No. 19/1995.

Bailey, J. R., and Oliver, J. A. 1939. The fishes of the Connecticut watershed, in: H. E. Warfel, *Biological Survey of the Connecticut Watershed,* pp. 150–189. New Hampshire Fish and Game Department.

Bailey, R. M., and Bond, C. E. 1963. Four new species of freshwater sculpins, genus *Cottus,* from western North America. *Occas. Pap. Mus. Zool. Univ. Mich.* **634**:1–27.

Balik, S. 1995. Freshwater fish in Anatolia, Turkey. *Biol. Conserv.* **72**:213–223.

Banarescu, P. 1994. The present-day conservation status of the fresh water fish fauna of Romania. *Ocrotirea Nat. Med. Inconjurator* **38**:5–20.

Barbour, C. D. 1974. Redescription and taxonomic status of *Chirostoma compressum,* a Mexican atherinid fish. *Copeia* **1974**:277–279.

Barel, C. D. N., Dorit, R., Greenwood, P. H., Fryer, G., Hughes, N., Jackson, P. B. N., Kanawabe, H., Lowe-McConnell, R. H., Witte, F., and Yamoaka, K. 1985. Destruction of fisheries in Africa's lakes. *Nature* **315**:19–20.

Barnard, K. H. 1938. Description of a new species of fresh-water fish from Natal. *Ann. Natal Mus.* **8**(3):525–528.

Baxter, G. T., and Simon, J. R. 1970. Wyoming fishes. *Wyo. Game Fish Dep. Bull.* **4**:1–168.

Behnke, R. J. 1972. The systematics of salmonid fishes of recently glaciated lakes. *J. Fish. Res. Board Can.* **29**:639–671.

Behnke, R. J. 1980a. A systematic review of the genus *Salvelinus,* in: E. K. Balon (ed.), *Charrs, Salmonid Fishes of the Genus Salvelinus,* pp. 441–480. Junk, The Hague.

Behnke, R. J. 1980b. *Salmo clarki* Richardson, in: D. S. Lee, C. R. Gilbert, C. H. Hocutt, R. E. Jenkins, D. E. McAllister, and J. R. Stauffer, Jr. (eds.), *Atlas of North American Freshwater Fishes,* p. 105. North Carolina Biological Survey.

Benke, A. C. 1990. A perspective on America's vanishing streams. *J. North Am. Bentholog. Soc.* **91**:77–88.

Beverton, R. J. H. 1992. Fish resources; threats and protection. *Neth. J. Zool.* **42**(2–3):139–175.

Bianco, P. G. 1995. Mediterranean endemic freshwater fishes of Italy. *Biol. Conserv.* **72**:159–170.

Böhlke, J. E., Weitzman, S. H., and Menezes, N. A. 1978. Estado atual da sistemática dos peixes de água doce de América do Sul. *Acta Amazonica* **8**(4):657–677.

Bond, C. E., and Kan, T. T. 1973. *Lampetra (Entosphenus) minima* n. sp., a dwarfed parasitic lamprey from Oregon. *Copeia* **1973**:568–574.

Bondhus, J. 1995. North American Native Fishes Association Endangered Species Conservation Program. *Aquat. Surviv.* **4**(1):9–12.

Brenner, T. 1980. The arctic charr, *Salvelinus alpinus salvelinus,* in the prealpine Attersee, Austria, in: E. K. Balon (ed.), *Charrs, Salmonid Fishes of the Genus Salvelinus,* pp. 765–772. Junk, The Hague.

Bruton, M. N. 1995. Have fish had their chips? The dilemma of threatened fishes. *Environ. Biol. Fish.* **43**:1–27.

Chen, Y., and Cui, Y. 1993. Some aspects of biological diversity in freshwater ecosystems. *Chin. Biodivers.* **1**:46–49.

Chernoff, B., and Miller, R. R. 1986. Fishes of the *Notropis calientis* complex with a key to the southern shiners of Mexico. *Copeia* **1986**:170–183.

Chernoff, B., Miller, R. R., and Gilbert, C. R. 1982. *Notropis orca* and *Notropis simus,* cyprinid fishes from the American southwest, with description of a new subspecies. *Occas. Pap. Mus. Zool. Univ. Mich.* **698**:1–49.

Collar, N. J. 1998. Extinction by assumption; or, the Romeo Error on Cebu. *Oryx* **32**(4):239–243.

Compagno, L. J. V. 1984. Sharks of the world. An annotated and illustrated catalogue of sharks known to date [Charcharhiniformes]. *FAO Species Catalogue* **4**(2):251–655.

Compagno, L. J. V. 1988. *Sharks of the Order Charcharhiniformes.* Princeton University Press, Princeton, N.J.

Contreras-Balderas, S., and de L. Lozano-Vilano, M. 1996. Extinction of most Sandia and Potosí valleys (Nuevo Léon, Mexico) endemic pupfishes, crayfishes and snails. *Ichthyol. Explor. Freshwaters* **7**:33–40.

Costa, W. J. E. M. 1995. Pearl killifishes. The Cynolebiatinae. Systematics and biogeography of a neotropical annual fish subfamily (Cyprinodontiformes: Rivulidae). T. F. H. Publications Inc., Neptune City, New Jersey. 128 pp.

Courtenay, J., W. R., and Moyle, P. B. 1992. Crimes against biodiversity: the lasting legacy of fish introductions. *Transactions of the 57th North American Wildlife and Nature Reserve Conference* **1992**:365–372.

Crivelli, A. J. 1995. Are fish introductions a threat to endemic freshwater fishes in the northern Mediterranean region? *Biol. Conserv.* **72**:311–319.

Day, D. 1981. *The Doomsday Book of Animals: A Natural History of Vanished Species.* Viking Press, New York.

De Buen, F. 1943. Los Lagos Michoacanos. I. Caracteres generales. El Lago de Zirahuen. *Rev. Soc. Mex. Hist. Nat.* **4**(3–4):211–232.

De Vos, L. 1995. A systematic revision of the African Schilbeidae (Teleostei, Siluriformes). *Ann. Mus. R. Afri. Cent. Sci. Zool.* **271**:1–449.

De Vos, L., Snoeks, J., and Thys van den Audenaerde, D. 1990. The effects of *Tilapia* introductions in Lake Luhondo, Rwanda. *Environ. Biol. Fish.* **27**:303–308.

Diamond, J. M. 1987. Extant unless proven extinct? Or, extinct unless proven extant? *Conserv. Biol.* **1**:77–79.

Dottrens, E. 1950. Le corégone actuel du Léman. *Rev. Suisse Zool.* **57**:789–813.

Dottrens, E. 1959. Systématique des corégones de l'Europe occidentale, basée sur une étude biométrique. *Rev. Suisse Zool.* **66**:1–66.

Dynesius, M., and Nilsson, C. 1994. Fragmentation and flow regulation of river systems in the northern third of the world. *Science* **266**:753–762.

Economidis, P. S. 1995. Endangered freshwater fishes of Greece. *Biol. Conserv.* **72**:201–211.

Eldredge, N., and Cracraft, J. 1980. *Phylogenetic Patterns and the Evolutionary Process.* Columbia University Press, New York.

Eschmeyer, P. H., and Bailey, R. M. 1955. The pygmy whitefish, *Coregonus coulteri,* in Lake Superior. *Trans. Am. Fish. Soc.* **84**:161–199.

Faragher, R. A., and Harris, J. H. 1994. The historical and current status of freshwater fish in New South Wales. *Aust. Zool.* **29**(3–4):166–176.

Gilbert, C. R., and Shute, J. R. 1980. *Rhinichthys cataractae* (Valenciennes), in: D. S. Lee, C. R. Gilbert, C. H. Hocutt, R. E. Jenkins, D. E. McAllister, and J. R. Stauffer, Jr. (eds.), *Atlas of North American Freshwater Fishes,* p. 353. North Carolina Biological Survey.

Goren, M., Fishelson, L., and Trewavas, E. 1973. The cyprinid fishes of *Acanthobrama* Heckel and related genera. *Bull. Br. Mus. Nat. Hist. Zool.* **24**(6):293–315.

Greenwood, P. H. 1956a. The monotypic genera of cichlid fishes in Lake Victoria. *Bull. Br. Mus. Nat. Hist. Zool.* **3**(7):295–333.

Greenwood, P. H. 1956b. A revision of the Lake Victoria *Haplochromis* species (Pisces, Cichlidae), Part I. *Bull. Br. Mus. Nat. Hist. Zool.* **4**(5):223–244.

Greenwood, P. H. 1957. A revision of the Lake Victoria *Haplochromis* species (Pisces Cichlidae), Part II. *bull. Br. Mus. Nat. Hist. Zool.* **5**(4):73–97.

Greenwood, P. H. 1959. A revision of the Lake Victoria *Haplochromis* species (Pisces, Cichlidae), Part III. *Bull. Br. Mus. Nat. Hist. Zool.* **5**(7):179–218.

Greenwood, P. H. 1960. A revision of the Lake Victoria *Haplochromis* species (Pisces, Cichlidae), Part IV. *Bull. Br. Mus. Nat. Hist. Zool.* **6**(4):227–281.

Greenwood, P. H. 1962. A revision of the Lake Victoria *Haplochromis* species (Pisces, Cichlidae), Part V. *Bull. Br. Mus. Nat. Hist. Zool.* **9**(4):139–214.

Greenwood, P. H. 1967. A revision of the Lake Victoria *Haplochromis* species (Pisces, Cichlidae), Part VI. *Bull. Br. Mus. Nat. Hist. Zool.* **15**(2):29–119.

Greenwood, P. H. 1992. Are the major fish faunas well-known? *Neth. J. Zool.* **42**(2–3):131–138.

Greenwood, P. H., and Barel, C. D. N. 1978. A revision of the Lake Victoria *Haplochromis* species (Pisces, Cichlidae), Part VIII. *Bull Br. Mus. Nat. Hist. Zool.* **33**(2):141–192.

Greenwood, P. H., and Gee, J. M. 1969. A revision of the Lake Victoria *Haplochromis* species (Pisces, Cichlidae), Part VII. *Bull. Br. Mus. Nat. Hist. Zool.* **18**(1):1–65.

Greenwood, P. H., and Jubb, R. A. 1967. The generic identity of *Labeo quathlambae* Barnard (Pisces, Cyprinidae). *Ann. Cape Prov. Mus. Nat. Hist.* **6**(2):17–37.

Groombridge, B. 1993. *1994 Red List of Threatened Animals.* IUCN, Gland, Switzerland and Cambridge, UK.

Harrison, I. J. 1993. The west African sicydiine fishes, with notes on the genus *Lentipes* (Teleostei: Gobiidae). *Ichthyol. Explor. Freshwaters* **4**:201–232.

Harrison, K., Crimmen, O., Travers, R., Maikweki, J., and Mutoro, D. 1989. Balancing the scales in Lake Victoria. *Biologist* **36**:189–191.

Holcík, J., and Renaud, C. B. 1986. *Eudontomyzon mariae* (Berg, 1931), in: J. Holcík (ed.), *The Freshwater Fishes of Europe.* Part 1. Petromyzontiformes, pp. 165–185. AULA-Verlag, Wiesbaden.

Howes, G. J. 1982. Review of the genus *Brycon* (Teleostei: Characoidei). *Bull. Br. Mus. Nat. Hist. Zool.* **43**(1):1–47.

Howes, G. J. 1991. Note on the holotype and unique specimen of *Barilius longirostris* Boulenger, 1902 (Teleostei: Cyprinidae). *J. Afr. Zool.* **105**:163–169.

Hughes, N. F. 1983. *A study of the Nile perch, an introduced predator in the Kavirondo Gulf Lake Victoria.* Oxford University Nile perch Project, Oxford. 75 pp.

Ida, H. 1984. Family Gasterosteidae, in: H. Masuda, K. Amaoka, C. Araga, T. Uyeno, and T. Yoshino (eds.), *The Fishes of the Japanese Archipelago,* pp. 83–84. Tokai University Press, Tokyo, Japan.

International Code of Zoological Nomenclature. 1985. W. D. L. Ride, C. W. Sabrowsky, G. Bernardi, and R. V. Melville (editorial committee). International Trust for Zoological Nomenclature, London.

IUCN. 1994. *IUCN Red List Categories.* IUCN, Gland, Switzerland.

IUCN. 1996. *1996 Red List of Threatened Animals.* IUCN, Gland, Switzerland.

IUCN. 1997. Review of IUCN Red List Categories and Criteria. *Species* **29**:9.

Kaufman, L. 1992. Catastrophic change in species-rich freshwater ecosystems. *BioScience* **42**(11):846–858.

Kaufman, L., and Ochumba, P. 1993. Evolutionary and conservation biology of cichlid fishes as revealed by faunal remnants in northern Lake Victoria. *Conserv. Biol.* **7**:719–730.

Kaufman, L. S., Chapman, L. J., and Chapman, C. A. 1997. Evolution in fast forward: haplochromine fishes of the Lake Victoria region. *Endeavour* **21**(1):23–30.

Kawanabe, H., and Watanabe, K. 1993. Kyoto ninespine stickleback, *Pungitius kaibarae,* in: S. Ueno (ed.), *Japanese Animals Going Extinct,* pp. 133–135. Tsukiji Shokan, Tokyo.

Keith, P., and Allardi, J. 1996. Endangered freshwater fish: the situation in France, in: A. Kirchhofer and D. Hefti (eds.), *Conservation of Endangered Freshwater Fish in Europe,* pp. 35–54. Birkhäuser Verlag, Berlin.

Kirchhofer, A. 1996. Fish conservation in Switzerland—three case-studies, in: A. Kirchhofer and D. Hefti (eds.), *Conservation of Endangered Freshwater Fish in Europe,* pp. 135–145. Birkhäuser Verlag, Berlin.

Kornfield, I. 1982. Report from Mindanao. *Copeia* **1982**:493–495.

Kornfield, I. 1985. The cyprinids of Lake Lanao. *Trop. Fish Hobbyist* **33**(1):40–57.

Kornfield, I., and Carpenter, K. E. 1984. Cyprinids of Lake Lanao, Philippines: taxonomic validity, evolutionary rates and speciation scenarios, in: A. A. Echelle and I. Kornfield (eds.), *Evolution of Species Flocks,* pp. 69–84. University of Maine at Orono Press.

Kottelat, M. 1990. Synopsis of the endangered Buntingi (Osteichthyes: Adrianichthyidae and Oryziidae) of Lake Poso, Central Sulawesi, Indonesia, with a new reproductive guild and descriptions of three new species. *Ichthyol. Explor. Freshwaters* **1**:49–67.

Kottelat, M. 1994. *Scriptor pretentiosus* ist keine bedrohte Art. *Datz Aquarien Terr.* **47**(3):172–173.

Kottelat, M. 1995. Systematic studies and biodiversity: the need for a pragmatic approach. *J. Nat. Hist.* **29**:565–569.

Kottelat, M. 1997. European freshwater fishes: an heuristic checklist of the freshwater fishes of Europe (exclusive of former USSR), with an introduction for non-systematists and comments on nomenclature and conservation. *Biologia (Bratislava)* **52**(Suppl. 5):1–271.

Kottelat, M., and Chu, X.-L. 1988. Revision of *Yunnanilus* with descriptions of a miniature species flock and six new species from China (Cypriniformes: Homalopteridae). *Environ. Biol. Fish.* **23**:65–93.

Kottelat, M., and Whitten, T. 1996. Freshwater biodiversity in Asia with special reference to fish. *World Bank Technical Paper 343.* World Bank, Washington, D.C. ix+59 pp.

Krupp, F., and Schneider, W. 1989. The fishes of the Jordan River Drainage basin and Azraq Oasis. *Fauna Saudi Arabia* **10**:347–416.

Lambert, D. 1992. How to save the blue-tailed goodeid. *Pract. Fishkeeping* **June**:52–53.

Laurent, P. J. 1972. Lac Léman: effects of exploitation, eutrophication, and introductions on the salmonid community. *J. Fish. Res. Board Can.* **29**:867–875.

Lauzanne, L. 1992. VI.5. Fish fauna. VI.5a. Native species. The *Orestias*, in: C. Dejoux and A. Iltis (eds.), *Lake Titicaca: A Synthesis of Limnological Knowledge*, pp. 405–419. *Monographiae Biologicae*, Volume 68. Kluwer, Dordrecht.

Lazara, K. J. 1995. History of the genera *Lebia* Oken 1817 and *Lebias* Goldfuss 1820 (Teleostei: Cyprinodontiformes: Cyprinodontidae) with designation of a type species for *Lebias*. *Copeia* **1995**(2):501–503.

Loubens, G. 1992. VI.5. Fish fauna. VI.5b. Introduced species. 1. *Salmo gairdneri* (rainbow trout), in: C. Dejoux and A. Iltis (eds.), *Lake Titicaca: A Synthesis of Limnological Knowledge*, pp. 420–426. *Monographiae Biologicae*, Volume 68. Kluwer, Dordrecht.

Lozano-Vilano, Ma de L., and Contreras-Balderas, S. 1993. Four new species of *Cyprinodon* from southern Nuevo Léon, Mexico, with a key to the *C. eximus* complex (Teleostei: Cyprinodontidae). *Ichthyol. Explor. Freshwaters* **4**:295–308.

Lyons, J., González-Hernández, G., Soto-Galera, E., and Guzmán-Arroyo, M. 1998. Decline of freshwater fishes and fisheries in selected drainages of west-central Mexico. *Fisheries* **23**(4):10–18.

Mace, G. M. 1994. Classifying threatened species: means and ends. *Philos. Trans. R. Soc. London Ser. B* **344**:91–97.

Magin, C. D., Johnson, T. H., Groombridge, B., Jenkins, M., and Smith, H. 1994. Species extinctions, endangerment and captive breeding, in: P. J. S. Olney, G. M. Mace, and A. T. C. Feistner (eds.), *Creative Conservation: Interactive Management of Wild and Captive Animals*, pp. 3–31. Chapman & Hall, London.

Mahanta, P. C., Kapoor, D., Dayal, R., and Ponniah, A. G. 1994. Prioritization of the Indian fish species for conservation, in: P. V. Dehadrai, P. Das, and S. R. Verma (eds.), *Threatened Fishes of India*. Natcon Publication 4, pp. 379–385. Nature Conservators, Muzaffarnagar, India.

Mainguet, M., and Létolle, R. 1997. The ecological crisis of the Aral Sea basin in the frame of a new time scale: the "anthropo-geological scale." *Naturwissenschaften* **84**:331–339.

Mann, C. C. 1991. Extinction: are ecologists crying wolf? *Science* **253**:736–738.

Maric, D. 1995. Endemic fish species of Montenegro. *Biol. Conserv.* **72**:187–194.

Martin, N. V., and Olver, C. H. 1980. The lake charr, *Salvelinus namaycush*, in: E. K. Balon (ed.), *Charrs, Salmonid Fishes of the Genus Salvelinus*, pp. 205–277. Junk, The Hague.

Matthews, W. J. 1980. *Notropis lutrensis* (Baird and Girard), in: D. S. Lee, C. R. Gilbert, C. H. Hocutt, R. E. Jenkins, D. E. McAllister, and J. R. Stauffer, Jr. (eds.), *Atlas of North American Freshwater Fishes*, p. 285. North Carolina Biological Survey.

Mayden, R. L., Burr, B. M., Page, L. M., and Miller, R. R. 1992. The native freshwater fishes of North America, in: R. L. Mayden (ed.), *Systematics, Historical Ecology, and North American Freshwater Fishes*, pp. 827–863. Stanford University Press, Stanford.

McAllister, D. E. 1995. Status of the World Ocean and its biodiversity. *Sea Wind* **9**(4):1–72.

McAllister, D. E., Parker, B. J., and McKee, P. M. 1985. Rare, endangered and extinct fishes in Canada. *Syllogeus* **54.** National Museums of Canada. 192 pp.

McAllister, D. E., Hamilton, A. L., and Harvey, B. 1997. Global freshwater biodiversity: striving for the integrity of freshwater ecosystems. *Sea Wind* **11**(2):1–106.

McDowall, R. M. 1989. Filling in the gaps—the introduction of exotic species into New Zealand, in: D. A. Pollard (ed.), Introduced and translocated fishes and their ecological effects, pp. 69–82. *Bureau of Rural Resources Proceedings* No. 8. Bureau of Rural Resources, Canberra.

McDowall, R. M. 1996. Volcanism and freshwater fish biogeography in the northeastern North Island of New Zealand. *J. Biogeogr.* **23**:139–148.

Menon, A. G. K. 1994. Criteria for determining the status of threatened categories of Indian freshwater fishes, in: P. V. Dehadrai, P. Das, and S. R. Verma (eds.), *Threatened Fishes of India*. Natcon Publication 4, pp. 1–5. Nature Conservators, Muzaffarnagar, India.

Meyer, M. K., and Pérez, H. E. 1990. *Priapella olmecae* sp. n., a new species from Veracruz (México) (Teleostei: Poecilidae). *Zool. Abh. Staatliches Mus. Tierkd. (Dresden)* **45**(12):121–126.

Miles, C. 1942. Descripcion sistematica del "pez grasso" del Lago de Tota (Boyacá). *Caldasia* **5**:55–58.

Miles, C. 1943. On three recently described species and a new genus of pygidiid fishes from Colombia. *Rev. Acad. Colomb. Cienc. Exactas Fis. Nat.* **5**(19):367–369.

Miller, P. J. 1988. Studies on *Sihouettea* Smith 1959 and an account of *Ebomegobius* Herre 1946 (Pisces: Gobiidae). *Senckenbergiana Biol.* **68**(4/6):241–273.

Miller, R. R. 1964. Redescription and illustration of *Cyprinodon latifasciatus*, an extinct cyprinodontid fish from Coahuila, Mexico. *Southwest. Nat.* **9**(2):62–67.

Miller, R. R. 1968. Pisces. Freshwater fishes, in: *IUCN Red Data Book.* Switzerland.

Miller, R. R. 1971. Pisces. Freshwater fishes, in: *IUCN Red Data Book.* Switzerland.

Miller, R. R., and Smith, G. R. 1981. Distribution and evolution of *Chasmistes* (Pisces: Catostomidae) in western North America. *Occas. Pap. Mus. Zool. Univ. Mich.* **696**:1–46.

Miller, R. R., Williams, J. D., and Williams, J. E. 1989. Extinctions of North American fishes during the past century. *Fisheries* **14**(6):22–38.

Minckley, W. L., Meffe, G. K., and Soltz, D. L. 1991. Conservation and management of short-lived fishes: the cyprinodontoids, in: W. L. Minckley and J. E. Deacon (eds.), *Battle Against Extinction: Native Fish Management in the America West,* pp. 247–282. University of Arizona Press, Tucson.

Moyle, P. B. 1997. The importance of an historical perspective: fish introductions. *Fisheries* **22**(10):14.

Mrakovcic, M., Misetic, S., and Povz, M. 1995. Status of freshwater fish in Croatian Adriatic river systems. *Biol. Conserv.* **72**:179–185.

Myers, N. 1996. Two key challenges for biodiversity: discontinuities and synergisms. *Biodivers. Conserv.* **5**:1025–1034.

Nelson, J. S. 1994. *Fishes of the World,* 3rd ed. Wiley, New York.

Ng, P. K. L., and Kottelat, M. 1994. Revision of the *Betta waseri* species group (Teleostei: Belontiidae). *Raffles Bull. Zool.* **42**(3):593–611.

Nümann, W. 1972. The Bodensee: effects of exploitation and eutrophication on the salmonid community. *J. Fish. Res. Board Can.* **29**:833–847.

Ogutu-Ohwayo, R. 1990. The decline of the native fishes of lakes Victoria and Kyoga (East Africa) and the impact of introduced species, especially the Nile perch, *Lates niloticus,* and the Nile tilapia, *Oreochromis niloticus. Environ. Biol. Fish.* **27**:81–96.

Ogutu-Ohwayo, R. 1993. The effects of predation by Nile perch, *Lates niloticus* L., on the fish of Lake Nabugabo, with suggestions for conservation of endangered endemic cichlids. *Conserv. Biol.* **7**:701–711.

Oijen, M. J. P. van. 1991. A systematic revision of the piscivorous haplochromine Cichlidae (Pisces: Teleostei) of Lake Victoria (East Africa). Part I. *Zool. Verh. Leiden* **272**:1–95.

Oijen, M. J. P. van, and Witte, F. 1996. Taxonomical and ecological description of a species complex of zooplanktivorous and insectivorous cichlids from Lake Victoria. *Zool. Verh. Leiden* **302**:1–56.

Oijen, M. J. P. van, Witte, F., and Witte-Maas, E. L. M. 1981. An introduction to ecological and taxonomic investigations on the haplochromine cichlids from the Mwanza Gulf of Lake Victoria. *Neth. J. Zool.* **31**:149–174.

Oijen, M. J. P. van, Snoeks, J., Skelton, P. H., Maréchal, C., and Teugels, G. G. 1991. Haplochromis, in: J. Daget, J.-P. Gosse, G. G. Teugels, and D. F. E. Thys van den Audenaerde (eds.), *Check-List of the Freshwater Fishes of Africa 4,* pp. 100–184. ISNB Brussels, MRAC Tervuren, and ORSTOM Paris.

Paepke, H.-J. 1988. Im Museum "wiederentdeckt": der Gangeshai. *Aquarien- und Terrarienfachgruppe* **35**(4): 129–130.

Parenti, L. R. 1984. A taxonomic revision of the Andean killifish genus *Orestias* (Cyprinodontiformes, Cyprinodontidae). *Bull. Am. Mus. Nat. Hist.* **178**(2):107–214.

Pavlov, D. S., Reshetnikov, Y. S., Shatunovskiy, M. I., and Shilin, N. I. 1985. Rare and disappearing fishes in the USSR and the principles of their inclusion in the "Red Book." *J. Ichthyol.* **25**(1):88–99.

Pike, T., and Tedder, A. J. 1973. Rediscovery of *Oreodaimon quathlambae* (Barnard). *Lammergeyer* **19**:9–15.

Pimental, D., Houser, J., Preiss, E., White, O., Fang, H., Mesnick, L., Barsky, T., Tariche, S., Schreck, J., and Alpert, A. 1997. Water resources: agriculture, the environment, and society. *BioScience* **47**(2):97–121.

Ponniah, A. G. 1994. Categorisation of India's threatened fishes, in: P. V. Dehadrai, P. Das, and S. R. Verma (eds.), *Threatened Fishes of India.* Natcon Publications 4, pp. 365–377. Nature Conservators, Muzaffarnagar, India.

Postel, S. L., Daily, G. C., and Ehrlich, P. R. 1996. Human appropriation of renewable fresh water. *Science* **271**:785–788.

Povz, M. 1995. Status of freshwater fishes in the Adriatic catchment of Slovenia. *Biol. Conserv.* **72**:171–177.

Quigley, D. T. G., and Flannery, K. 1996. Endangered freshwater fish in Ireland, in: A. Kirchhofer and D. Hefti (eds.), *Conservation of Endangered Freshwater Fish in Europe,* pp. 27–34. Birkhäuser Verlag, Berlin.

Rakaj, N. and Flloko, A. 1995. Conservation status of freshwater fish of Albania. *Biol. Conserv.* **72**:195–199.

Regan, C. T. 1909. The char (*Salvelinus*) of Great Britain. *Ann. Mag. Nat. Hist. Ser. 8* **3**:111–122.

Regan, C. T. 1911. *The Freshwater Fishes of the British Isles.* Methuen, London.

Reid, G. M. 1980. 'Explosive speciation' of carps in Lake Lanao (Philippines)—fact or fancy? *Syst. Zool.* **29**(3):314–316.

Reid, G. M. 1990. Captive breeding for the conservation of cichlid fishes. *J. Fish Biol.* **37**(Suppl. A):157–166.

Reid, G. M. 1991. Threatened rainforest cichlids of Lower Guinea, West Africa—a case for conservation, in M. J. H. Nelissen (ed.), Proceedings of the Fifth European Congress "Biology of Cichlids," State University of Antwerp, January, 1990, *Ann. Mus. Roy. l'Afrique Centrale, Tervuren* **263**:109–119.

Ribbink, A. J. 1991. Distribution and ecology of the cichlids of the African Great Lakes, in: M. H. A. Keenleyside (ed.), *Cichlid Fishes: Behaviour, Ecology and Evolution,* pp. 36–59. Chapman & Hall, London.

Richter, B. D., Braun, D. P., Mendelson, M. A., and Master, L. L. 1997. Threats to imperiled freshwater fauna. *Conserv. Biol.* **11**:1081–1093.

Roberts, T. R. 1993. Just another dammed river? Negative impacts of Pak Mun Dam on fishes of the Mekong basin. *Nat. Hist. Bull. Siam Soc.* **41**:105–133.

Rohde, F. L. 1980. *Noturus trautmani* Taylor, in: D. S. Lee, C. R. Gilbert, C. H. Hocutt, R. E. Jenkins, D. E. McAllister, and J. R. Stauffer, Jr. (eds.), *Atlas of North American Freshwater Fishes,* p. 471. North Carolina Biological Survey.

Rosa, R. S., and Menezes, N. A. 1996. Relação preliminar das espécies de peixas (Pisces, Elasmobranchii, Actinopterygii) ameaçadas no Brasil. *Rev. Bras. Zool.* **13**(3):647–667.

Ruhlé, C., and Kindle, T. 1992. Morphological comparison of river-spawning whitefish of the alpine Rhine with the whitefish of Lake Constance. *Pol. Arch. Hydrobiol.* **39**:403–408.

Schillinger, A. 1901. Der Teefseesaibling. *Allge. Fisch. Ztg. N. F.* **16**:149–151.

Seehausen, O. 1996. *Lake Victoria Rock Cichlids—Taxonomy, Ecology, and Distribution.* Verduyn Cichlids, Germany.

Seehausen, O., and Bouton, N. 1998. The community of rock-dwelling cichlids in Lake Victoria. *Bonn. Zool. Beitr.* **47**(3–4):301–311.

Seehausen, O., and Witte, F. 1994. Eine zweite Aussterbewelle im Victoriasee und die Notwendigkeit der Verknupfung von Forschung und Naturschutz. *Datz Aquarien Terrarien* **47**(9):586–591.

Seehausen, O., and Witte, F. 1995. Extinction of many, survival of some. *Trop. Fish Hobbyist* **43**(7):96–105.

Seehausen, O., van Alphen, J. J. M., and Witte, F. 1997a. Cichlid fish diversity threatened by eutrophication that curbs sexual selection. *Science* **277**:1808–1811.

Seehausen, O., Witte, F., Katunzi, E. F., Smits, J., and Bouton, N. 1997b. Patterns of the remnant cichlid fauna in southern Lake Victoria. *Conserv. Biol.* **11**:890–904.

Sehgal, K. L. 1994. State-of-art of endangered, vulnerable and rare coldwater fishes of India, in: P. V. Dehadrai, P. Das and S. R. Verma (eds.), *Threatened Fishes of India, Natcon Publication,* **4**:127–135. Nature Conservators, Muzaffarnagar, India.

Smith, F. D. M., May, R. M., Pellew, R., Johnston, T. H., and Walter, K. R. 1993. How much do we know about the current extinction rate? *Trends Ecol. Evol.* **8**(10):375–378.

Smith, M. L., and Miller, R. R., 1986. Mexican goodeid fishes of the genus *Characodon,* with description of a new species. *Am. Mus. Novit.* **2851**:1–14.

Smith, R. E. 1980. *Gila bicolor* (Girard), in: D. S. Lee, C. R. Gilbert, C. H. Hocutt, R. E. Jenkins, D. E. McAllister, and J. R. Stauffer, Jr. (eds.), *Atlas of North American Freshwater Fishes,* p. 161. North Caroline Biological Survey.

Smith, S. H. 1972. Factors of ecologic succession in oligotrophic fish communities of the Laurentian Great Lakes. *J. Fish. Res. Board Can.* **29**:717–730.

Snoeks, J. 1994. The haplochromines (Teleostei, Cichlidae) of Lake Kivu (East Africa). A taxonomic revision with notes on their ecology. *Ann. Mus. R. Afr. Cent. Sci. Zool.* **270**:1–221.

Soliman, V. S. 1993. Status of Lake Manapao, Buhi, Camarines Sur: sanctuary of the only viable stock of sinarapan, the world's smallest commercial fish. *Outreach* **6**(1):3p.

Spillman, J. 1958. Sur deux poissons des collections du Muséum, témoins de l'existence au XIXe siècle, dans le lac Léman, de corégones du type "*dispersus*." *Bull. Mus. Natl. Hist. Nat. Paris Ser. 2* **30**:144–145.

Stiassny, M. L. J. 1990. Notes on the anatomy and relationships of the bedotiid fishes of Madagascar, with a taxonomic revision of the genus *Rheocles* (Atherinomorpha: Bedotiidae). *Am. Mus. Novit.* **2979**:1–33.

Stiassny, M. L. J. 1994. Systematics and conservation, in: G. K. Meffe and C. R. Carroll (eds.), *Principles of Conservation Biology,* 1st edition, pp. 64–66. Sinauer, Sunderland, Mass.

Stiassny, M. L. J. 1996. An overview of freshwater biodiversity: with some lessons from African fishes. *Fisheries* **21**(9):7–13.

Stiassny, M. L. J. In press. The medium is the message: freshwater biodiversity in peril, in: J. Cracraft and F. Grifo (eds.), *The Living Planet in Crisis: Biodiversity Science and Policy,* Columbia University Press, New York.

Swerdlow, J. L. 1998. Making sense of the millenium. *Nat. Geogr.* **193**(1):8.

Tanner, V. M. 1936. A study of the fishes of Utah. *Utah Acad. Sci. Arts Lett.* **13**:155–184.

Todd, T. N. 1980. *Coregonus johannae* (Wagner), in: D. S. Lee, C. R. Gilbert, C. H. Hocutt, R. E. Jenkins, D. E. McAllister, and J. R. Stauffer, Jr. (eds.), *Atlas of North American Freshwater Fishes,* p. 82. North Carolina Biological Survey.

Todd, T. N., Smith, G. R., and Cable, L. E. 1981. Environmental and genetic contributions to morphological differentiation in ciscoes (Coregoninae) of the Great Lakes. *Can. J. Fish. Aquat. Sci.* **38**:59–67.

Tzeng, C.-S. 1986. Distribution of the freshwater fishes of Taiwan. *J. Taiwan Mus.* **39**(2):127–146.

Viederman, S., Meffe, G. K., and Carroll, C. R. 1994. The role of institutions and policymaking in conservation, in: G. K. Meffe and C. R. Carroll (eds.), *Principles of Conservation Biology,* First Edition, pp. 466–490. Sinauer Associates, Sunderland.

Vitousek, P. M., Mooney, H. A., Lubchenco, J., and Melillo, J. M. 1997. Human domination of the Earth's ecosystems. *Science* **277**:494–499.

Wallace, R. L. 1980. *Rhinichthys osculus* (Girard), in: D. S. Lee, C. R. Gilbert, C. H. Hocutt, R. E. Jenkins, D. E. McAllister, and J. R. Stauffer, Jr. (eds.), *Atlas of North American Freshwater Fishes,* p. 356. North Carolina Biological Survey.

Watson, R. E. 1997. Extinction of *Sicyopterus nigrescens* and threats to sicydiine gobies, in: 5th Indo-Pacific Fish Conference, *Abstracts,* p. 111. ORSTOM and SFI.

Wells, L., and McLain, A. L. 1972. Lake Michigan: effects of exploitation, introductions, and eutrophication on the salmonid community. *J. Fish. Res. Board Can.* **29**:889–898.

Wetzel, R. G. 1992. Clean water: a fading resource. *Hydrobiologia* **243/244**:21–30.

Wheeler, T. 1991. Maryland darter may have vanished. *Am. Curr.* **Summer**:27–28.

Whitfield, A. 1997. Red Data Book: estuarine pipefish. *Endangered Wild.* **27**:30–31.

Whitfield, A. K., and Bruton, M. N. 1996. Extinction of the river pipefish *Syngnathus watermeyeri* in the Eastern Cape Province, South Africa. *S. Afr. J. Sci.* **92**:59–60.

Whitten, A. J., Nash, S. V., Bishop, K. D., and Clayton, L. 1987. One or more extinctions from Sulawesi, Indonesia? *Conserv. Biol.* **1**:42–48.

Wiley, E. O. 1981. *Phylogenetics: The Theory and Practice of Phylogenetic Systematics.* Wiley, New York.

Williams, J. D., and Nowak, R. M. 1993. Vanishing species in our own backyard: extinct fish and wildlife of the United States and Canada, in: L. Kaufman and K. Mallory (eds.), *The Last Extinction,* 2nd ed., pp. 115–148. MIT Press, Cambridge, Mass.

Williams, J. E., Johnson, J. E., Hendrickson, D. A., Contreras-Balderas, S., Williams, J. D., Navarro-Mendoza, M., McAllister, D. E., and Deacon, J. E. 1989. Fishes of North America endangered, threatened, or of special concern: 1989. *Fisheries* **14**(6):2–20.

Witte, F. 1984. Ecological differentiation in Lake Victoria haplochromines: comparison of cichlid species flocks in African lakes, in: A. A. Echelle and I. Kornfield (eds.), *Evolution of Species Flocks,* pp. 155–167. University of Maine at Orono Press.

Witte, F., Goldschmidt, T., Goudswaard, P. C., Ligtvoet, W., van Oijen, M. J. P., and Wanink, J. H. 1992a. Species extinction and concomitant ecological changes in Lake Victoria. *Neth. J. Zool.* **42**(2–3):214–232.

Witte, F., Goldschmidt, T., Wanink, J., van Oijen, M., Goudswaard, K., Witte-Maas, E., and Bouton, N. 1992b. The destruction of an endemic species flock: quantitative data on the decline of the haplochromine cichlids of Lake Victoria. *Environ. Biol. Fish.* **34**:1–28.

World Conservation Monitoring Centre. 1992. Species extinction, in: WCMC, *Global Biodiversity: Status of the Earth's Living Resources,* pp. 192–233. Chapman & Hall, London.

Yang, J.-X. 1991. The fishes of Fuxian Lake, Yunnan, China, with full description of two new species. *Ichthyol. Explor. Freshwaters* **2**:193–202.

Yuma, M., Hososya, K., and Nagata, Y. 1998. Distribution of the freshwater fishes of Japan: an historical overview. *Environ. Biol. Fish.* **52**:97–124.

Zheng, C.-Y. 1989. [*Fishes of the Zhujiang River*]. Science Press, Beijing, 438 pp. [Chinese].

Zorzi, G. D. 1995. The biology of freshwater elasmobranchs: an historical perspective, in: M. I. Oetinger and G. D. Zorzi (eds.), The biology of freshwater elasmobranches. A symposium to honor Thomas B. Thorson, pp. 10–90. *J. Aquaricult. Aquat. Sci.* **7**:10–90.

CHAPTER 13

Requiem Æternam

The Last Five Hundred Years of Mammalian Species Extinctions

R. D. E. MacPHEE and CLARE FLEMMING

1. Introduction

This contribution provides an analytical accounting of mammal species that have disappeared during the past 500 years—the "modern era" of this chapter. Our choice of date was dictated by several considerations, but two are paramount. AD 1500 marks more or less precisely the beginning of Europe's expansion across the rest of the world, a portentous event in human history by any definition. It is an equally momentous date for natural history because it marks the point at which empirical knowledge of the planet began to burgeon exponentially. These two factors, linked for both good and ill for the past half-millennium, have affected every aspect of life on earth, and are thus fitting subjects to commemorate in a record such as this.

A number of modern-era extinction lists (hereafter, "compilations") for Mammalia have been published in recent years (e.g., Goodwin and Goodwin, 1973; Williams and Nowak, 1986, 1993; World Conservation Monitoring Centre [WCMC], 1992; International Union for the Conservation of Nature and Natural Resources [IUCN], 1993, 1996; Cole *et al.*, 1994). One may wonder why yet another recension is needed. Superficially, the empirical problem seems uncomplicated: A species either exists or it doesn't—what is there to disagree about? Regrettably, things are not so simple. A target taxon that has not been collected or seen for an appreciable interval may indeed be extinct—or it may simply not have been collected or seen. Unless someone has taken the trouble to search carefully and seriously for the target, mere lack of information on status should not be accepted blindly as evidence of extinction. Similarly, poor or outdated systematic studies are no more adequate for evaluating the status of a species than is incomplete inventorying or surveying. As we shall attempt to show, the quality of the evidence for evaluating whether a particular extinction has occurred varies widely from case to case, and even in the best circumstances a clear-cut deci-

---

R. D. E. MacPHEE and CLARE FLEMMING • Department of Mammalogy, American Museum of Natural History, New York, New York 10024.

Extinctions in Near Time, edited by MacPhee. Kluwer Academic/Plenum Publishers, New York, 1999.

sion is often not possible. Does any of this matter? We think so, because reliable databases are essential to any science. Extinction lists are likely to play an ever-increasing role in conservation policy debates, and it is important to ensure that such lists are not replete with erroneous data. As we shall proceed to show in the rest of this chapter, here we can offer only a preliminary database for studies of very recent mammalian extinctions.

If it were necessary to be absolutely certain that a loss had occurred before it could be counted, the list of modern-era extinctions would be very short indeed. This is the major reason that extinction rates are usually estimated by proxy measures, such as the one that relates annual loss of rain forest to some calculation of species richness (for an extended discussion, see May *et al.,* 1995). The trouble is that proxy estimates are difficult to verify and can lead to rate estimates of enormous magnitude (see Brooks *et al.,* 1997). For example, we regard as being beyond meaningful empirical verification the recent claim that "approximately 1800 populations per hour (16 million annually) are being destroyed in tropical forests alone" (Hughes *et al.,* 1997).

Epistemologically, a more reasonable approach is to establish whether there is adequate justification for one's belief that any particular extinction has taken place. The efficient way to do this is to arrange knowledge about specific extinctions according to levels of apparent reliability. We have tried to achieve this goal by constructing a series of tables (Tables 1–4), each defined by the degree to which available evidence relating to extinction appears to justify a specific conclusion.

An extinction (or, better, the evidence for an extinction) can be said to be *resolved* when it is found that it meets the following criteria: The name of the species thought to be extinct is valid systematically (e.g., the name is available and not preoccupied by another valid name); the species has been well characterized and, optimally, has been subjected to revisionary review; no individuals belonging to this species have been collected, inventoried, or observed reliably in the last 50 years; and there is evidence of a positive nature that places the loss date within the period spanning AD 1500–1950. Extinctions thus resolved are placed in Table 1. *Unresolved* extinctions are those for which a meaningful prima facie case can be made. Because they currently fail to meet one or more of the stipulated criteria, however, their status cannot be resolved fully and must be held in abeyance until appropriate information is forthcoming. These potential instances of extinction are apportioned among Tables 2–4 according to the nature of their deficiencies (see below for additional descriptions of tables). We are aware that several taxa might have been placed, with about equal justification, in a table other than the one we selected. Our allocations reflect the merits of individual cases as we see them.

Obviously, the current organization of this extinction database is not immutable. Unresolved extinctions are potentially resolvable if evidence can be gathered that clarifies whatever point is in doubt. Similarly, a case that was thought to be fully adjudicated may be reopened if new evidence warrants it. This last point is important to note because we have taken special pains to identify taxa that should not be regarded as modern-era species-level extinctions according to our criteria, even though they have been so listed in other compilations (see Table 7).

It is appropriate to end this introduction with a word about the cause(s) of modern-era extinctions. This is a compendium of modern-era extinctions, not an explanation thereof (but see Section 7, Coda: Modern-Era Extinctions in Quaternary Perspective). The choice of wording is exact: This work is an actuarial account, in which our concern is to establish what losses occurred, where, and when. *How* they occurred is a topic of undoubted importance, but in only a tiny fraction of cases is it reasonably clear why particular species disappeared (Caughley and Gunn, 1996; Flemming and MacPhee, 1997; MacPhee and Flemming, in press). Systematic research on species that happen to be extinct is no more likely to uncover evidence regarding causes of loss than is ordinary armchair speculation. On the other hand, any investigation of causation that intends to stay grounded in reality requires reliable databases that assess the systematic facts as rigorously as possible.

The companion chapter by Harrison and Stiassny (this volume) should be consulted for a parallel treatment of extinctions among freshwater fishes, using criteria similar to the ones defined here. It is of interest to compare the nature of the two databases, which differ notably in some respects (e.g., absence of fossil material of fishes, very large number of apparently extinct taxa based on limited hypodigms, number of extinctions in last half-century).

2. Abbreviations and Conventions

2.1. Abbreviations

CITES	Convention on International Trade in Endangered Species of Wild Fauna and Flora
EED	effective extinction date (last reliable record of collection or observation)
I	Island, Islands
IUCN	International Union for the Conservation of Nature and Natural Resources of the World Conservation Union
rcyrbp	radiocarbon years before present (i.e., before radiocarbon datum AD 1950)
WCMC	World Conservation Monitoring Centre
WI	West Indies
yrbp	years before present

2.2. Conventions Used in Tables 1–4

2.2.1. Effective Extinction Date (EED)

* Target taxon died out post-1500, but date of loss not otherwise constrained

2.2.2. Evidence Categories

1 Target taxon collected or seen in living state between AD 1500 and 1950

2 Target taxon associated with remains of exotics having a documented introduction date of AD 1500 (or later);

3 Target taxon dated radiometrically; calibrated ^{14}C date ± two sigma (standard errors) lying close to or within the period AD 1500–1950

2.2.3. Hypodigm Descriptions

H Hypodigm consists of holotype only

H+ Hypodigm consists of two or three assigned specimens (including holotype)

‡ Hypodigm consists of four or more specimens

F Hypodigm consists of fossil material only (no wild-caught specimens). Here, "fossil" is applied to bony remains of any age, such as mammal bones retrieved in modern-era deposits in caves.

2.2.4 Body Size Classes

BW Body weight (in kg). Unless otherwise indicated, weight estimates are derived from data reported by Silva and Downing (1995) for the species stipulated in the tables or close relatives in the same body size range.

I ≤0.1 kg

II 0.1 to <1.0 kg

III 1.0 to <10 kg

IV 10 to <100 kg

V 100 to <1000 kg

VI ≥1000 kg

3. Methods

One of the motivations for the present chapter comes from our own sense of frustration with previous lists of modern-era extinctions (Flemming and MacPhee, 1995). In general, they have not been accompanied by the kinds of discussions that enable the interested but nonspecialist reader to interpret authors' justifications for including (or excluding) particular species. We consider this to be important. Although for reasons of space we present justifications in summary form in this chapter, the detailed assessments on which they are based are presented elsewhere (Table 7; MacPhee and Flemming, in press). We expect that our tables will be in a state of ongoing revision for the next several years until consensus is achieved on still-doubtful cases (see also Section 5, Notes on Major Taxa).

3.1. Table 1: Resolved Extinctions, AD 1500–1950

Listings in Table 1 are limited to valid species of Recent mammals for which there is positive evidence of an EED within the period 1500–1950. Most of the species allocated to Table 1 have been reviewed systematically in recent revisionary works, although inevitably professional opinion on the status of some species is not uniform. Criteria for identifying EEDs are explained more fully in the next section, Organization of Tables.

3.2. Table 2: Unresolved Extinctions, AD 1500–1950—Missing Data

For a number of allegedly extinct taxa, the evidence for their systematic validity or EED (or both) is ambiguous, i.e., it neither confirms nor denies the possibility that they died out between 1500 and 1950. For example, the systematic validity of a taxon may be considered ambiguous when the species has never been reviewed adequately or when reasonable queries about its supposed distinctiveness have been raised in the professional literature. Until such queries are followed up in an appropriate systematic review, the status of a species cannot be considered as settled and is best regarded as unresolved. Similar reasoning applies to insecure dating. For example, we include some species of the Antillean insectivore genus *Nesophontes* (island-shrews) in Table 1 because remains attributed to these species have been found in contexts that also yielded the black rat, *Rattus rattus*, introduced into the West Indies from Europe about AD 1500. By contrast, several other island-shrews have to be excluded from Table 1 because no reliable associations with *Rattus* have been reported for them. Their EEDs are properly considered as unresolved, and will stay that way until evidence appropriate for a decision is available.

3.3. Table 3: Unresolved Extinctions, AD 1500–1950—Missing Names

The purpose of systematic research is to exhaustively organize the world's biota in a meaningful, hierarchical arrangement. The cornerstone of the systematic hierarchy is the species; and until a biotic entity conforming to a species has a valid name, it does not have a secure identity or place in the hierarchy. We note this point because there are a number of Recent biotic entities that have not been named or diagnosed formally that nonetheless appear frequently on lists of extinct species. Unnamed, undiagnosed species are not truly open to systematic review, and for this reason their status must be considered unresolved. On the other hand, for most of these entities there is positive evidence for EEDs between 1500 and 1950, which means that they should not be ignored if they are, in fact, good species. The obvious solution is to list them separately, pending a final determination of their status and review by other systematists. Although nomina nuda, these entities are best referred to as *taxa* because they have been assigned (with one exception) to valid genera, at least provisionally. It is to be hoped that most of these partially named taxa will have been formally described before publication of the next version of the extinction list.

3.4. Table 4: Unresolved Extinctions—Losses after AD 1950

How long after the last collection or sighting of a species should the taxon be declared extinct? At present, the recommendation of the IUCN and CITES is 50 years, although this period is no more inherently meaningful than half or twice this term. However, database reliability clearly requires that some substantial "waiting period" be set before a species is posted formally to an extinction list. For example, the swamp deer *Cervus schomburgki*, believed to have become extinct more than 60 years ago, has apparently managed to survive in a small area in Laos: Antlers referable to this species have turned up recently as hunting trophies (Schroering, 1995). The Vietnamese warty hog, *Sus bucculentus* (often considered a subspecies of *Sus verrucosus*), also thought to be extinct, has recently been seen alive (Groves *et al.*, 1997), as has the "extinct" woolly flying squirrel *Eupetaurus cinereus* of Pakistan (Zahler, 1996). Burbidge *et al.* (1988) have shown on the basis of interview evidence that a number of small marsupials thought to have disappeared more than 100 years ago may have survived until the 1960s or later. Although examples could be multiplied, the point here is that cases of *very* recent extinction should not be treated as resolved (in favor of complete loss) unless the evidence is overwhelming. We deal with this by separately tabulating the names of valid species whose extinction is thought to have occurred since 1950.

3.5. Other Tables

The remaining tables (Tables 5–7) are concerned with additional analyses. Tables 5 and 6 present information on taxa not included (or interpreted differently) in the IUCN (1996) *Red List.* Disqualified taxa are separately noted in Table 7, together with a brief statement of the reason for disqualification (e.g., species invalid, species still extant, species disappeared prior to our cutoff date). For additional information on disqualified taxa, see MacPhee and Flemming (in press).

4. Organization of Tables

For each taxon considered, relevant information is presented in tabular format under eight headings: valid species name, common name, range, effective extinction date (EED), evidence category for EED, hypodigm size, body weight, and notes. Symbols used in the tables are defined in Section 2.2.

4.1. Valid Species Name

A valid name is defined as the correct name for a taxon according to the rules of zoological nomenclature, as codified in the most recent (1985) edition of the *International Code of Zoological Nomenclature.* Our basic authority file for valid species of Mammalia is Wilson and Reeder's (1993) edited compendium, with additions and corrections based on expert information from colleagues. All names in Tables 1 and 4 are considered valid, while the validity of some names in Table 2 has been questioned as indicated. The taxa noted in Table 3 do not yet have formal binomina and have not been published properly. Table 7 lists names of taxa (whether valid or invalid) that cannot be shown to be modern-era, species-level extinctions.

Some compilers have listed extinctions at the subspecies as well as the species level (e.g., Goodwin and Goodwin, 1973; Williams and Nowak, 1986, 1993), but we avoid doing this here. Although any number of nominal mammalian subspecies may qualify as full species depending on the species concept used (see papers collected by Kimbel and Martin, 1993, and Claridge *et al.,* 1997), we prefer to see this determined in individual cases by original revisionary work.

Species that are "extinct in the wild" raise a related concern. The IUCN (1996) applied this designation to species lacking free-living populations, and, in the mammal section of the current *Red List,* provided such a rating for Saudi gazelle (*Gazella saudiya*), Przewalski's horse (*Equus przewalskii*), and black-footed ferret (*Mustela nigripes*). Although Extinct in the Wild may be an important distinction for conservation purposes, it lies outside our system of evaluation. Consistency requires that if a recognized species is represented by one or more local populations (or even individuals), whether wild or captive, that species cannot be considered extinct.

4.2. Common Name

Because many extinct species are known only from accounts in specialist literature, it is not surprising that, for many of them, common names either do not exist or are not well fixed. Whether a species that was rarely or perhaps never seen by humans should have a "common" name is partly a matter of taste, although we argue that, given the widening interest in recently extinct species, there is nothing objectionable in manufacturing names that are somewhat more accessible to the nonspecialist than the typical neo-Latin binomen. For the sake of completeness we have included common names in this chapter if well fixed, but have generally avoided creating any new ones for taxa currently lacking such names.

4.3. Range

Although the range of a species is vital biological information, for reasons of space we supply only minimal geographical data in the tables. This permits the reader to glean basic biogeographical associations and compare losses by both major taxon and geographical area. "Range" is defined as apparent distribution in 1500 or later. Thus, the range for *Thylacinus cynocephalus* is limited to Tasmania, even though its range earlier in the Quaternary included the Australian mainland as well as New Guinea (Rounsevell and Mooney, 1995). The location of islands within their respective water bodies is as defined by the *Times Atlas* (1994:xiv).

4.4. Effective Extinction Date (EED)

The information included in this column is the date or interval when authenticated specimens of target taxa were last collected or observed scientifically, whether in nature or in captivity. Scientifically, the last observation of a species is an important datum—regardless of whether or not it straggled on for a few more years—because it provides a minimum date for final loss, here termed the effective extinction date. This is practical, as well as justifiable empirically, because the EED is the point at which opportunity to study the target taxon as a living entity terminates.

When does a species become extinct? The obvious answer—extinction occurs when the last member of a species dies—is unambiguous in its meaning but not in its application. Unless taxa are regularly censused in an exacting manner, to follow the demise of "last" survivors is impractical—and, of course, only applicable to species that are still extant. For older extinctions especially, estimates of the time of loss of rare taxa may be highly inaccurate (cf. Signor and Lipps, 1982; McFarlane, this volume; Alroy, this volume). It may therefore strike the reader as odd that times of loss are often cited in calendar years in studies of modern-era extinctions. Yet the precision is not necessarily false, if it is properly understood that the date should correspond to the last authentic observation of living members of the taxon in question.

In the case of taxa known only from fossils, "last collected specimen" is not a meaningful concept and the EED is given as an interval. Because of dating uncertainties, this interval is usually indeterminate ("later than 1500") and signaled as such by an asterisk in the EED column.

For some taxa our EEDs will differ from "last observation" dates provided by other compilers, prompting the reader to wonder whether a lapsus has occurred. In this chapter, any date given in calendar years is the date of last collection of identified specimens, if this were possible to ascertain from museum records or reliable literature. Dates given to decade (e.g., "1960s") clearly imply that EEDs are less well constrained.

Other compilers have made recourse to the publication date of the species's name as equivalent to a last record, but we do this only when no other information is available. Thus, the date of "last observation" of the St. Lucia giant rice-rat (*Megalomys luciae*) is sometimes given as 1901, which is the date of Forsyth Major's (1901) publication of the species name. However, the two specimens that comprise the entire hypodigm of this species are known to have been collected before 1881 (G. M. Allen, 1942). Its EED is therefore more precisely given as "pre-1881" rather than as "before 1901."

4.5. Evidence for EED

Acceptable evidence that a given extinction occurred in the modern era can be of three kinds, or categories. In the case of category 1, there must be evidence that the target species was actually collected or observed in the living state, at a specified time or times during the past 500 years. Category 2 evidence refers to associations of remains of target species with those of exotic species (known to have been introduced into the relevant area after 1500). Category 3 evidence consists of radiometric dating results that confirm the target taxon's survival into the modern era. An additional category, for fossils with attached hair or other soft tissues but no exotic associations, was rejected because it is not a reliable proxy for "extant within the last 500 years" (cf. permafrost mummies from Alaska and Siberia, which may be tens of thousands of years old; see Haynes and Eiselt, this volume; MacPhee *et al.*, 1999).

Although category 1 evidence is potentially the most detailed, it is basically limited to times and places in which literate observers were collecting specimens and keeping records. For this reason, evidence in categories 2 and 3 is often the only basis for adjudication; if clear-cut, such evidence should be accepted as dispositive.

Parentheses are used if there is some question about the quality or completeness of the evidence of a specific EED. For example, it is possible but not definite that Schomburgk (1848) saw a living example of the Barbados rice-rat (*Oryzomys* sp. 1). This taxon is otherwise known only from fossil evidence (Ray, 1964). However, because of ambiguities in Schomburgk's observation, we rate this evidence as "(1)," implying that it is still unclear whether this species survived into the modern era (see Table 3, footnote *c*).

Categories are defined additionally as follows.

4.5.1. Category 1

This category comprises collection-based or reliable observational evidence regarding the last appearance of the target taxon, with subsequent unfruitful efforts to relocate the target in nature.

"Collection-based evidence" consists of reliably identified, wild-caught specimens (vouchers). Usually, the vouchers of interest are ones that can be regarded as the "last"-collected specimens of the species in question. The amount of ancillary information accompanying vouchers is sometimes incomplete, especially for taxa last collected in the nineteenth century or earlier. However, as this part of the database is unlikely to improve, we have not excluded any taxon from being given a category 1 rating solely because last-collection records are imperfect.

"Reliable observational evidence" is the detection, by a qualified witness, of living specimens of the target taxon at stated times and places and the recording thereof. Only sightings by persons with intimate knowledge of the taxon in question should be accepted as reliable. Although we recognize that this stipulation may be too narrow, its real purpose is to control for the "thylacine effect," which we define as the tendency to accept the continued existence of a species according to the frequency with which unsubstantiated reports of sightings of it are made.

Optimally, before a determination of extinction is made, it should be settled beyond a reasonable doubt that vouchers have been identified securely and that observational evidence relates truly to the target taxon. This is particularly important in cases in which the taxon might be confused with another that is still extant.

EXAMPLE: The extinction of the Falkland Islands dog, *Dusicyon australis,* was probably caused by overhunting in the nineteenth century. By the 1870s the Falkland Islands dog was very rare; the last confirmed killing took place in 1876 (Dollmann, 1937). Because a comparatively large, island-dwelling vertebrate will not be overlooked easily, we regard the last reliable record for *D. australis* as a very good indicator of its final extinction date.

4.5.2. Category 2

Category 2 evidence requires a demonstrable association of remains assignable to the target taxon with those of the introduced species whose own date of introduction can be inferred reliably.

Category 1 evidence requires the existence of a collected specimen or observational materials concerning the EED of the target taxon. However, a relatively high proportion of mammal species included on modern-era extinction lists are known only as fossils and were never the subject of a recorded observation in the living state. Their inclusion on a list of resolved extinctions can be justified by proxy evidence, such as the climacteric provided by the introduction of *Rattus rattus* and *R. norvegicus* into the New World during the late fifteenth and mid-eighteenth centuries, respectively. (The *Rattus* proxy is not valid globally: Some *Rattus* introductions into Mediterranean islands may have antedated 1500, and introduction of the kiore rat, *R. exulans,* onto South Pacific islands took place on a completely different schedule [Holdaway, 1996, this volume].) The proxy provides a minimum estimate of the EED, although in many cases its relationship to the true extinction date may not be very close.

Ideally, it should be ascertainable from the excavator's report that the bones of the target were found in unimpeachable association with the remains of introduced species. This is an important consideration in cave sites with very low sedimentation rates, in which specimens of very different ages may occur at the same apparent level (MacPhee *et al.,* 1999).

EXAMPLE: Evidence for the extinction of *Solenodon marcanoi* meets category 2 requirements because bones referable to this species have been found in association with *Rattus rattus* (Morgan and Woods, 1986), introduced into the New World at the start of the modern era.

4.5.3. Category 3

Evidence in this category derives from high-quality chronometric determinations, based optimally on bones or other tissues of the target taxon.

Within limits, radiocarbon determinations can be of assistance in determining EEDs (for excellent examples, see Steadman *et al.,* 1991). Optimally, the determination should be made on bone or other tissue of the target itself, although this may not be feasible if the target is represented by few specimens. High-quality dates for bones are ones based on the collagen fraction, with $^{13}C/^{12}C$ ratios and standard errors within acceptable limits. Obviously, it is important to base one's conclusions on remains believed to be among the most recent available for the target taxon.

Because of the expense involved in radiocarbon dating, this method has not been used routinely for dating modern-era extinctions. Radiocarbon determinations offer some promise for better constraining EEDs that are based presently on category 2 evidence. For example, possible survival of *Nesophontes* into this century, as alleged by Miller (1930), G. M. Allen (1942), and several other workers on the basis of specimen "freshness," is not supported by available radiocarbon dates (cf. entry under *N. micrus;* MacPhee *et al.,* 1999).

The EED provided by a radiocarbon date may be given as a range in calendar years equivalent to the calibrated mean $^{14}C \pm$ two standard errors (95% confidence level). Because the calibration curve is nonlinear (Stuiver and Reimer, 1993), uncalibrated ^{14}C dates may correspond to more than one calendrical date, which has the effect of increasing the error. Given the narrow time frame of interest, it would be an excess of refinement to cite ^{14}C dates in the form of EEDs.

EXAMPLE: Radiocarbon dating of the holotype (and only specimen) of the alleged endemic Cuban spider monkey *Ateles "anthropomorphus"* by MacPhee and Rivero de la Calle (1996) established that the specimen is less than three centuries old, not pre-Columbian as previously thought. However, because the species is not valid (specimen appears to be referable to the extant spider monkey, *Ateles fusciceps*), it therefore does not count as a modern-era extinction.

4.6. Hypodigm

The term *hypodigm* refers to the entire collected material of a species. Abundance of specimens in systematic collections is a measure of how well a particular taxon has been collected; it is not necessarily a measure of how common the species is or was in nature. This point affects the weighing of evidence in the construction of extinction lists in two ways. First, whether or not a taxon was originally rare in nature, it should be obvious that a comparative lack of specimens may complicate or limit revisionary work. For example, when a species in a poorly known but speciose group is represented by only one or two specimens, it may be exceedingly difficult to ascertain the species's validity if it differs little from close congeners. In such cases, the usual systematic practice is to give the target the benefit of the doubt and continue to recognize it. We follow this practice in most cases, except that extinct species whose validity has been questioned by appropriate authorities are placed on one of the unresolved lists. Second, for a species that may be truly rare in na-

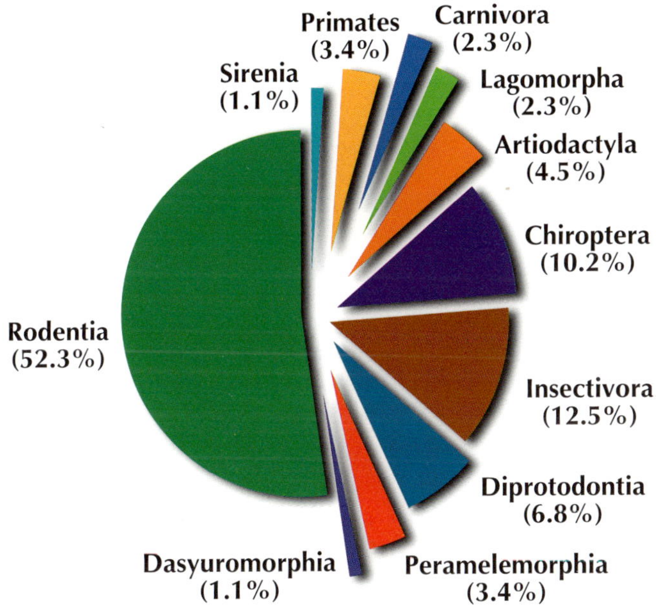

FIGURE 1. Mammalian species extinctions since AD 1500: losses per order as a percentage of all losses (N = 88).

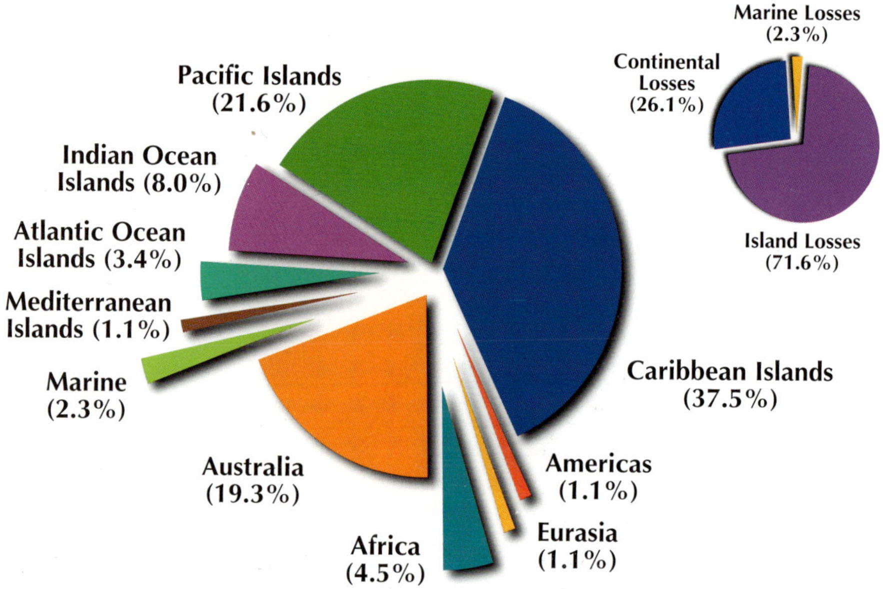

FIGURE 2. Mammalian species extinctions since AD 1500: distribution of losses—continental, island, and marine species ($N = 88$).

ture, it would seem important to know how efficient recent efforts to collect it have been, before pronouncing it extinct. Obviously, this point may be hard to evaluate in many cases. Here we rate specimen abundance in collections in order to invite reexamination of certain taxa through additional collecting or revisionary efforts, as appropriate (see Section 2.2.3). Abundance in collections is not rated for fossil taxa.

4.7. Notations

Although detailed notes on each taxon cannot be presented here, in each table we separately footnote points of special importance for interpretation of evidence (see MacPhee and Flemming, in press).

5. Notes on Major Taxa

For many reasons (see Stiassny, 1994) it is important to make a distinction between the loss of one or two species within a group that has many living representatives, and the complete obliteration of all members of a well-differentiated clade of substantial probable antiquity (cf. loss of the Guam flying fox, *Pteropus tokudae,* in a genus with almost five dozen extant species, versus loss of the entire insectivore family Nesophontidae). Some tragedies are, we believe, greater than others.

Distribution of modern-era losses by order is illustrated in Fig. 1. Of the 26 orders of Recent mammals recognized by Wilson and Reeder (1993), 11 have suffered species-level losses according to our criteria. The order most affected by modern-era extinctions is Rodentia (52.3% of all losses), followed distantly by Insectivora (12.5%), Chiroptera (10.2%), and three orders of Australian marsupials (accounting for another 11.3%).

5.1. Marsupialia

Of ten marsupial extinctions listed in the tables (all of which are Australian), four are placed in Table 1 and the rest in Table 4. Even if some marsupials listed in Table 4 still survive, this does not reduce the force of the argument that a number of marsupial species have been lost or brought to critically low numbers in very recent decades (Flannery, 1995a). Note that a high proportion of the Australian (and Meganesian) total reflects extinctions among endemic placentals, particularly conilurine rodents (Smith and Quin, 1996).

5.2. Insectivora

Although ten species of Nesophontidae (Antillean island-shrews) are accepted as valid for present purposes, this family is being revised by H. Whidden (personal communication) and colleagues, and some major changes in the taxonomy of the group can be expected. (An 11th species, Puerto Rican *N. edithae,* is valid and extinct; however, according to the available evidence it cannot be shown to have survived into the modern era and is therefore ex-

Table 1. Resolved Extinctions, AD 1500–1950 ($N = 40$)

Valid species name	Common name	Range	EED	Category	Hypodigm	BW	Notes
Hydrodamalis gigas	Steller's sea cow	Bering Sea	1768	1	‡	VI	
Prolagus sardus	Sardinian pika	Sardinia & Corsica	1777	(1)	F	II	
Hippotragus leucophaeus	bluebuck	South Africa	1800	1	‡	V	
Pteropus subniger	Réunion flying fox	Mascarenes	pre-1866	1	‡	II	
Pteropus pilosus	large Palau flying fox	Carolines: Palau	1874	1	H+	II	
Conilurus albipes	white-footed rabbit rat	Australia	1875	1	‡	II	
Potorous platyops	broad-faced potoroo	Australia	1875	1	‡	III	
Dusicyon australis	Falkland Islands dog	Falklands	1876	1	‡	IV	*a*
Oryzomys antillarum	Jamaican rice-rat	WI: Jamaica	1877	1	H+	I	*b*
Megalomys luciae	St. Lucia giant rice-rat	WI: St. Lucia	pre-1881	1	H+	III	*a*
Uromys porculus	little pig rat	Solomons: Guadalcanal	1887	1	H	II	*c*
Leimacomys buettneri	groove-toothed forest mouse	Ghana	1890	1	H+	I	*a*
Lagorchestes leporides	eastern hare-wallaby	Australia	1890	1	‡	III	
Pharotis imogene	large-eared nyctophilus	Papua New Guinea	1890	1	‡	I	*a*
Oligoryzomys victus	St. Vincent pygmy rice-rat	WI: Saint Vincent	1892	1	H	I	*a*
Nyctimene sanctacrucis	Nendö tube-nosed fruit bat	Solomons: Santa Cruz	pre-1892	1	H	I	*d*
Gazella rufina	red gazelle	Algeria	pre-1894	1	H+	IV	
Notomys amplus	short-tailed hopping-mouse	Australia	pre-1896	1	H+	I	
Notomys longicaudatus	long-tailed hopping-mouse	Australia	1901	1	‡	I	
Megalomys desmarestii	Martinique giant rice-rat	WI: Martinique	1902	1	‡	III	*a*
Rattus macleari	Maclear's rat	Christmas I. (Indian Ocean)	1903	1	‡	II	*a*
Rattus nativitatis	bulldog rat	Christmas I. (Indian Ocean)	1903	1	‡	II	*a*
Peromyscus pembertoni	Pemberton's deer mouse	Mexico: San Pedro Nolasco I.	1931	1	‡	I	*e*
Caloprymnus campestris	desert rat-kangaroo	Australia	1932	1	‡	II	
Thylacinus cynocephalus	thylacine	Tasmania	1936	1	‡	IV	*f*
Nesoryzomys darwini	Galápagos rice-rat	Galápagos: Santa Cruz	pre-1940	1	‡	I	*a*

Nesophontes hypomicrus	Atalaye island-shrew	WI: Hispaniola	*	2,3	F	I	*a*
Nesophontes micrus	western Cuban island-shrew	WI: Cuba	*	2,3	F	I	*a, g*
Nesophontes paramicrus	St. Michel island-shrew	WI: Hispaniola	*	2,3	F	I	*a*
Nesophontes zamicrus	Haitian island-shrew	WI: Hispaniola	*	(2)	F	I	*g*
Solenodon marcanoi	Marcano's solenodon	WI: Hispaniola	*	2	F	II	
Megaladapis, cf. *M. edwardsi*	tretretretre	Madagascar	*	(1),3	F	IV	*h*
Palaeopropithecus cf. *P. ingens*	sloth lemur	Madagascar	*	3	F	IV	*h*
Xenothrix mcgregori	Jamaican monkey	WI: Jamaica	*	(1),2	F	III	*a, i*
Megaoryzomys curioi	Curio's giant rat	Galápagos: Santa Cruz	*	2,3	F	II	*a, j*
Brotomys voratus	muhoy	WI: Hispaniola	*	(1),2	F	I	*a, k*
Geocapromys columbianus	Cuban coney	WI: Cuba	*	2	F	III	*a*
Isolobodon montanus	montane hutia	WI: Hispaniola	*	(1),2	F	III	*a*
Isolobodon portoricensis	Allen's hutia	WI: Hispaniola & Puerto Rico	*	(1),3	F	III	*a*
Plagiodontia ipnaeum	Johnson's hutia	WI: Hispaniola	*	2	F	III	*l*

[a]Weight estimate by authors.
[b]Morgan (1993) suggested that *O. antillarum* be revived as a species distinct from extant (and mainland) *O. couesi*, based on measurements and discrete traits.
[c]The related species *U. imperator* (Table 4) is as poorly known as *U. porculus*. We treat these taxa differently because Groves and Flannery (1994) inferred that *imperator* may have been extant until the 1960s (even though no specimens of either taxon have been recovered since the collection of the type series).
[d]Possibly seen as late as 1907 in Nendö (Santa Cruz group), according to K. F. Koopman (personal communication).
[e]Extensive exploration and trapping activities by Lawlor (1983) in the 1970s failed to yield any additional specimens of *P. pembertoni*.
[f]There are frequent but unsubstantiated reports of thylacines having been seen in Tasmania and even on mainland Australia. Weight estimate (15–35 kg) is taken from Flannery *et al.* (1990).
[g]*Nesophontes zamicrus* not stated by Miller (1930) to have been found in association with *Rattus*, but included in a larger collection for which such an association was reported. Specimens of *N. micrus* from Cueva Martin (Sierra de Escambray, Cuba) have been dated to 590 ± 50 rcyrbp (corrected; MacPhee *et al.*, 1999).
[h]Simons (1997) reported (?uncorrected) dates of 630 ± 50 rcyrbp for *Megaladapis* and 510 ± 80 rcyrbp for *Palaeopropithecus*, for samples from caves on the Manamby Plateau. Suggested species allocations are the present authors' and are based on known distributions of taxa concerned. Weight estimates are from Godfrey *et al.* (1997).
[i]Based on association with *Rattus* in caves in Jackson's Bay (MacPhee and Flemming, 1997) and an apparent reference by Sloane (1707–1725); size estimate is by MacPhee and Fleagle (1991).
[j]Most recent ^{14}C date for this taxon is 210 ± 55 rcyrbp (Steadman *et al.*, 1991).
[k]*Boromys* regarded as subgenus of *Brotomys*.
[l]Weight estimate based on Woods's (1984) estimate for another member of genus.

Table 2. Unresolved Extinctions, AD 1500–1950 ($N = 20$)

Valid species name	Common name	Range	EED	Category	Hypodigm	BW	Notes
Notomys macrotis	big-eared hopping-mouse	Australia	1843	1	H	I	*b1*
Notomys mordax	Darling Downs hopping-mouse	Australia	1846	1	H	I	*b2*
Pteropus brunneus	dusky flying fox	Percy I. (Coral Sea)	1874	1	H	II	*b3, c*
Kerivoula africana	Dobson's painted bat	Tanzania	pre-1878	1	H	I	*b4*
Oryzomys nelsoni	Nelson's rice-rat	Maria Madre I. (Gulf of California)	1897	1	‡	I	*b5, c*
Pseudomys gouldii	Gould's mouse	Australia	1930	1	‡	I	*b2*
Quemisia gravis	quemi	WI: Hispaniola	*	(1)	F	III	*a*
Malpaisomys insularis	volcano mouse	Canaries	*	(3)	F	II	*a, d*
Hippopotamus lemerlei	Lemerle's Malagasy hippo	Madagascar	*	(1),3	F	V	*a, e*
Hippopotamus madagascariensis	common Malagasy hippo	Madagascar	*	(1)	F	V	*a, e*
Nesophontes longirostris	long-nosed island shrew	WI: Cuba	*	(2)	F	I	*a, e*
Nesophontes major	—	WI: Cuba	*	(2)	F	I	*a*
Nesophontes submicrus	—	WI: Cuba	*	(2)	F	I	*a*
Nesophontes superstes	—	WI: Cuba	*	2	F	I	*a, b6*
Megalomys audreyae	Barbuda giant rice-rat	WI: Barbuda	*	2,3	F	II	*a, f*
Oryzomys hypenemus	Barbuda rice-rat	WI: Barbuda & Antigua	*	2,3	F	II	*b7, f*
Brotomys offella	Cuban esculent spiny rat	WI: Cuba	*	2	F	I	*a, g*
Brotomys torrei	de la Torre's esculent spiny rat	WI: Cuba	*	2	F	I	*a, g*
Hexolobodon phenax	—	WI: Hispaniola	*	(2)	F	III	*a*
Rhizoplagiodontia lemkei	—	WI: Hispaniola	*	2	F	III	*a, h*

[a]Weight estimate by authors.
[b]Species validity is challenged by: *1*, Mahoney (1975); *2*, Calaby and Lee (1989); *3*, Koopman (1984); *4*, Corbet and Hill (1991); *5*, Hershkovitz (1971); *6*, J. A. Ottenwalder (personal communication); *7*, C. Flemming and D. A. McFarlane (unpublished data).
[c]Percy Island and Maria Madre Islands are Pacific islands.
[d]Cultural associations with remains of *M. insularis* indicate that this species become extinct "between 800 years and now" (Hutterer, 1993). A ^{14}C date of ca. 1070 rcyrbp is available for the stratigraphic level that yielded this rodent (Hutterer *et al.*, 1988).
[e]Although oral traditions indicate hippos may have survived in Madagascar until ca. AD 1500, the latest published date is 980 ± 200 rcyrbp (Dewar, 1984).
[f]*Megalomys audreyae* and *Oryzomys hypenemus* may be identical, but we keep these taxa separated pending further analysis (Flemming and McFarlane, in preparation). Material assigned to *M. audreyae* from Darby Sink, Barbuda, has been AMS ^{14}C-dated to 750 ± 50 rcyrbp ($^{13}C/^{12}C$ corrected)—close enough to warrant inclusion in this table.
[g]*Boromys* regarded as subgenus of *Brotomys*.
[h]According to Woods (1989a:67, table 3), *R. lemkei* occurs in association with *Rattus* and *Mus*.

Table 3. Unnamed Taxa Presumed Extinct since AD 1500 ($N = 11$)

Taxon	Common name	Range	EED	Category	Hypodigm	BW	Notes
Oryzomys sp. 1	Barbados rice-rat	WI: Barbados	pre-1890	(1)	F	I	*b1, c*
Notomys sp. 1	great hopping-mouse	Australia	pre-1900	2	F	I	*b2, d*
Nesophontes sp. 1	Grand Cayman island-shrew	WI: Grand Cayman	*	2	F	I	*a, b3*
Nesophontes sp. 2	Cayman Brac island-shrew	WI: Cayman Brac	*	2	F	I	*a, b3*
Megaoryzomys sp. 1	Isabela giant rice-rat	Galápagos: Isabela	*	3	F	II	*a, b4*
Nesoryzomys sp. 2	Isabela island rice-rat "A"	Galápagos: Isabela	*	2	F	I	*a, b4*
Nesoryzomys sp. 3	Isabela island rice-rat "B"	Galápagos: Isabela	*	2	F	I	*a, b4*
Oryzomyin sp. 1	Vespucci's rice-rat	Brazil: Fernando da Noronha I.	*	(1)	F	I	*a, b5, e*
Capromys sp. 1	Cayman hutia	WI: Cayman Is.	*	2,3	F	III	*a, b3*
Geocapromys sp. 1	Great Cayman coney	WI: Grand Cayman	*	2	F	III	*a, b3*
Geocapromys sp. 2	Cayman Brac coney	WI: Cayman Brac	*	(2)	F	III	*a, b3*

[a]Weight estimate by authors.
[b]Relevant information may be found in: *1,* Ray (1964); *2,* Flannery (1995b); *3,* Morgan (1994); *4,* Steadman *et al.,* (1991); *5,* Olson (1981).
[c]Physically known only from fossils, but Schomburgk may have seen animal in the 1840s (see Ray, 1964). Because Fielden (1890) found no evidence of this rodent, Marsh (1985) raised the possibility that the rat died out between 1847 and 1890; for purposes of calculations, we have used the end of this time period (1890) as the EED. We consider Barbados to be a Caribbean rather than an Atlantic Ocean island.
[d]Flannery (1995b) suspected that this taxon became extinct in the last half of the nineteenth century; for purposes of calculations, we have used the end of this time period (1900) as the EED.
[e]Fernando da Noronha is an Atlantic Ocean island.

Table 4. Unresolved Extinctions after AD 1500 ($N = 17$)

Valid species name	Common name	Range	EED	Category	Hypodigm	BW	Notes
Monachus tropicalis	Caribbean monk seal	Caribbean Sea	1950s	1	‡	V	*a*
Geocapromys thoracatus	Little Swan Island coney	Little Swan I. (Caribbean Sea)	1950s	1	‡	II	*b*
Crateromys paulus	small Ilin cloud rat	Philippines: Ilin I.	1953	1	H	II	
Onychogalea lunata	crescent nailtail wallaby	Australia	1956	1	‡	III	*c*
Chaeropus ecaudatus	pig-footed bandicoot	Australia	1960s	1	‡	II	*a*
Macrotis leucura	lesser bilby	Australia	1960s	1	‡	II	
Perameles eremiana	desert bandicoot	Australia	1960s	1	‡	II	
Lagorchestes asomatus	central hare-wallaby	Australia	1960s	1	H	III	
Uromys imperator	giant naked-tailed rat	Solomons: Guadalcanal	1960s	1	H+	II	*d*
Microtus bavaricus	Bavarian pine vole	Germany	1962	1	‡	I	
Nesoryzomys swarthi	San Salvador rice-rat	Galápagos: San Salvador	1965	1	‡	I	*a, e*
Mystacina robusta	greater short-tailed bat	New Zealand: Big Cape I.	1965	1	‡	I	
Pteropus tokudae	Guam flying fox	Guam	1968	1	H+	II	*f*
Dobsonia chapmani	Dobson's fruit bat	Philippines: Negros I.	1970s	1	‡	II	*g*
Leporillus apicalis	lesser stick-nest rat	Australia	1970	1	‡	II	
Macropus greyi	toolache wallaby	Australia	1972	1	‡	IV	
Sylvilagus insonus	Omilteme cottontail	Mexico: Omilteme	1991	1	H+	III	

[a]Weight estimate by authors.
[b]Weight estimate reported by Morgan (1989).
[c]Flannery *et al.* (1990) cited 1956 as the "last unequivocal record" of the existence of *O. lunata*.
[d]Groves and Flannery (1994) reported "few or no sightings over last 40 years."
[e]*N. swarthi* is possibly part of *N. indefessus* according to Patton and Hafner (1983), although Musser and Carleton (1993) disagreed with this assessment.
[f]Possible sighting in 1984, but no further details (Flannery, 1995c).
[g]L. Heaney (personal communication) was unable to secure further *D. chapmani* in Negros despite major collecting efforts in 1980s. Local informants said that a large bat, evidently this species, had not been seen in 20 years.

Table 5. Comparison of Mismatches of Taxa Considered Extinct: Tables 1–4 versus IUCN (1996) *Red List*

Taxon	IUCN (1996) classification	This chapter[a]
Acerodon lucifer	extinct	disqualified
Pteropus loochooensis	extinct	disqualified
Phyllonycteris major	extinct	disqualified
Myotis milleri	extinct	disqualified
Myotis planiceps	extinct	disqualified
Nyctophilus howensis	extinct	disqualified
Procyon gloveralleni	extinct	disqualified
Equus quagga	extinct	disqualified
Sus bucculentus	extinct	disqualified
Cervus schomburgki	extinct	disqualified
Gazella arabica	extinct	disqualified
Canariomys tamarani	extinct	disqualified
Coryphomys buhleri	extinct	disqualified
Papagomys theodorverhoeveni	extinct	disqualified
Paulomys naso	extinct	disqualified
Spelaeomys florensis	extinct	disqualified
Nesoryzomys fernandinae	extinct	disqualified
Sphiggurus pallidus	extinct	disqualified
Brotomys contractus	extinct	disqualified
Heteropsomys antillensis	extinct	disqualified
Heteropsomys insulans	extinct	disqualified
Puertoricomys corozalus	extinct	disqualified
Plagiodontia velozi	extinct	disqualified
Plagiodontia araeum	extinct	disqualified
Amblyrhiza inundata	extinct	disqualified
Clidomys osborni	extinct	disqualified
Clidomys parvus	extinct	disqualified
Elasmodontomys obliquus	extinct	disqualified
Pharotis imogene	critically endangered	Table 1
Leimacomys buettneri	critically endangered	Table 1
Crateromys paulus	critically endangered	Table 4
Isolobodon portoricensis	critically endangered	Table 1
Sylvilagus insonus	critically endangered	Table 4
Uromys imperator	endangered	Table 4
Oligoryzomys victus	endangered	Table 1
Kerivoula africana	data deficient	Table 2

[a]See Table 7 and MacPhee and Flemming (in press). IUCN (1996) listed the following mammal species as Extinct in the Wild: *Mustela nigripes, Equus przewalskii, Gazella saudiya* (category not utilized in this chapter).

cluded; see Table 7). The quality of information respecting reported associations between island-shrews and exotics is quite variable, and this is reflected in table assignments.

5.3. Chiroptera

Many bats are island dwellers, and one might imagine that chiropteran losses on islands should have been substantial in the modern era, as that is the pattern for terrestrial mammals. However, it has proven difficult to verify how many species-level bat extinctions

Table 6. Entities in Tables 1–3 Not Included in IUCN (1996) *Red List*

Table 1	*Megaladapis*, cf. *M. edwardsi*
	Palaeopropithecus, cf. *P. ingens*
	Xenothrix mcgregori
	Uromys porculus
	Oryzomys antillarum
	Megaoryzomys curioi
Table 2	*Nesophontes longirostris*
	Nesophontes major
	Nesophontes submicrus
	Nesophontes superstes
	Hippopotamus madagascariensis
	Hippopotamus lemerlei
	Malpaisomys insularis
	Megalomys audreyae
	Oryzomys hypenemus
Table 3	*Nesophontes* sp. 2[a]
	Notomys sp. 1
	Oryzomys sp. 1
	Oryzomyin sp. 1
	Megaoryzomys sp. 1
	Nesoryzomys sp. 2
	Nesoryzomys sp. 3
	Geocapromys sp. 2[a]

[a]Morgan (1994) identified two unnamed species each of *Nesophontes* and *Geocapromys* from the Cayman Islands; IUCN (1996) listed only one of each.

(as opposed to local extirpations and range reductions) have actually occurred, either on mainlands or on islands. Nine species meet our minimum criteria for inclusion in the tables. Of these, three are very recent losses (Table 4), and at least one other has a validity problem (*Pteropus brunneus,* Table 2). Although many bat species are undoubtedly threatened at present, documented losses of Chiroptera in the modern era are few.

5.4. Xenarthra

Xenarthra suffered major losses in the late Pleistocene (Martin, 1984; Martin and Steadman, this volume), but bona fide Holocene extinctions within this order have been negligible except among the tardigrades (sloths). It appears likely that all Quaternary Antillean megalonychids—probably amounting to some dozen species—disappeared during the Holocene. However, reliable dates are too few to place EEDs more precisely. At least some megalonychids were still extant when humans first came to the Greater Antilles, 5000–7000 years ago, as a date of 6330 ± 50 rcyrbp (corrected) is available for *Megalocnus* sp. from the site of Cueva Beruvides in Matanzas province, Cuba (MacPhee *et al.*, 1999; see also Pino and Castellanos, 1985; MacPhee, 1997a). Despite the asseverations of Miller (1929a) and others, there are no conclusive observational records or exotic associations that suggest post-Columbian survival of any megalonychid species.

Table 7. Disqualified Taxa

Name	Common name	Range[a]	Reason for rejection[b]
Antechinus apicalis	dibbler	Australia	extant as *Parantechinus apicalis*
Bettongia gaimardi	Gaimard's rat-kangaroo	Australia	extant (Tasmania only)
Potorous gilbertii	long-nosed potoroo	Australia	extant as *P. tridactylus*
Acratocnus odontrigonus	Puerto Rican sloth	WI: Puerto Rico	date of loss not established as within modern era[c]
Tolypeutes tricinctus	northern three-banded armadillo	Brazil	extant
Solenodon arredondoi	Arredondo's solenodon	WI: Cuba	date of loss not established as within modern era
Solenodon cubanus	almiquí	WI: Cuba	extant
Nesophontes edithae	Edith's island-shrew	WI: Puerto Rico	date of loss not established as within modern era
Cryptogale australis	large-eared tenrec	Madagascar	extant as *Geogale aurita*
Crocidura goliath	goliath white-toothed shrew	C. Africa	extant
Soriculus corsicanus	Corsican shrew	Corsica & Sardinia	date of loss not established as within modern era
Acerodon lucifer	Philippine fruit bat	Philippines: Panay I.	extant as *A. jubatus*[d] (Luzon, Negros, & nearby islands)
Pteropus loochooensis	Marianas flying fox	Ryukyu Is.: Okinawa	extant as *P. mariannus*[e] (Marianas, Carolines, & Ryukyu Is.)
Mormoops magna	—	WI: Cuba	date of loss not established as within modern era
Pteronotus pristinus	—	WI: Cuba	date of loss not established as within modern era
Reithronycteris aphylla	Jamaican pallid flower bat	WI: Jamaica	extant as *Phyllonycteris aphylla*
Phyllonycteris obtusa	Poey's pallid flower bat	WI: Cuba & Hispaniola	extant as *P. poeyi*
Phyllonycteris major	Puerto Rican flower bat	WI: Puerto Rico	date of loss not established as within modern era
Stenoderma rufum	red fruit bat	WI: Puerto Rico & Virgin Is.	extant
Phyllops vetus	—	WI: Cuba	date of loss not established as within modern era
Natalus primus	large funnel-eared bat	WI: Cuba, Hispaniola & Jamaica	extant as *N. stramineus*
Myotis milleri	Miller's myotis	Mexico: Baja California	extant
Myotis planiceps	flat-headed myotis	Mexico: Coahuila	extant, probably extremely rare in nature
Nyctophilus howensis	Lord Howe Island bat	Australia: Lord Howe I.	date of loss not established as within modern era
Pipistrellus sturdeei	Sturdee's Bonin pipistrelle	Japan: Bonin Is.	extant[e]
Scotophilus borbonicus	Réunion vesper bat	Mascarenes: Réunion	extant[f]
Cheirogaleus trichotis	hairy-eared dwarf lemur	Madagascar	extant as *Allocebus trichotis*
Ateles anthropomorphus	Cuban spider monkey	(WI: Cuba)	extant as *A. ?fusiceps* (NW South America)
Antillothrix bernensis	Hispaniolan monkey	WI: Hispaniola	date of loss not established as within modern era[g]
Mustela macrodon	sea mink	coastal NE North America	extant as *M. vison*
Mustela nigripes	black-footed ferret	USA: Midwest	extant

(continued)

Table 7. (*Continued*)

Name	Common name	Range[a]	Reason for rejection[b]
Procyon gloveralleni	Barbados raccoon	(WI: Barbados)	extant as *P. lotor*
Eschrichtius gibbosus	Atlantic gray whale	Atlantic Ocean	extant as *E. robustus*[a] (Pacific Ocean)
Equus przewalskii	Przewalski's horse	China & Mongolia	extant
Equus quagga	quagga	South Africa	extant as *E. burchellii*[h]
Sus bucculentus	Vietnamese warty pig	Vietnam	extant
Cervus schomburgki	Schomburgk's deer	Thailand	extant (Laos)
Rangifer dawsoni	Queen Charlotte Islands dwarf caribou	Queen Charlotte Is.	extant as *R. tarandus*[b] (Holarctic)
Gazella arabica	Arabian gazelle	Saudi Arabia: Farasan Is.	systematic status uncertain
Gazella bilkis	Queen of Sheba gazelle	Saudi Arabia	systematic status uncertain[i]
Gazella saudiya	Saudi gazelle	Saudi Arabia	extant
Bos primigenius	aurochs (cattle)	Eurasia	extant as *Bos taurus* (worldwide)
Bubalus mephistopheles	—	China	extinction predates modern era
Epixerus ebii	African palm squirrel	W. Africa	extant
Geomys fontanelus	Sherman's pocket gopher	SE North America	extant as *G. pinetis*
Microtus nesophilus	Gull Island vole	USA: New York	extant as *M. pennsylvanicus* (E North America)
Tyrrhenicola henseli	large Corsican field vole	Sardinia & Corsica	date of loss not established as within modern era
Canariomys tamarani	—	Canaries: Grand Canary I.	date of loss not established as within modern era
Coryphomys buhleri	Timor giant rat	Indonesia: Timor I.	date of loss not established as within modern era
Papagomys theodorverhoeveni	Flores giant rat	Indonesia: Flores I.	date of loss not established as within modern era
Paulamys naso	—	Indonesia: Flores I.	extant
Pseudomys fieldii	Shark Bay mouse	Australia	extant[j]
Pseudomys sp. 1	basalt plains mouse	Australia	date of loss not established as within modern era[k]
Rhagamys orthodon	Hensel's field mouse	Sardinia & Corsica	date of loss not established as within modern era
Solomys salamonis	Florida rat	Solomons: Florida Is.	extant
Spelaeomys florensis	—	Indonesia: Flores I.	date of loss not established as within modern era
Zyzomys pedunculatus	central rock-rat	Australia	extant[l]
Megalomys curazensis	—	Netherlands Antilles: Curaçao I.	date of loss not established as within modern era
Nesoryzomys fernandinae	Fernandina island rice-rat	Galápagos: Fernandina I.	extant[m]
Nesoryzomys indefessus	Santa Cruz island rice-rat	Galápagos: Santa Cruz & Baltra	extant[n]
Nesoryzomys sp. 1	Rabida island rice-rat	Galápagos: Rabida	date of loss not established as within modern era[o]
Peromyscus nesodytes	giant deer mouse	USA: Channel Is.	date of loss not established as within modern era
Sphiggurus pallidus	West Indian prehensile-tailed porcupine	West Indies	genus and species indet.[p]

Brotomys contractus	small Hispaniolan esculent spiny rat	WI: Hispaniola	date of loss not established as within modern era
Heteropsomys antillensis	large Puerto Rican spiny rat	WI: Puerto Rico	date of loss not established as within modern era[q]
Heteropsomys insulans	hutia-like Puerto Rican spiny rat	WI: Puerto Rico & Vieques	date of loss not established as within modern era[q]
Puertoricomys corozalus	old Puerto Rican spiny rat	WI: Puerto Rico	extinction predates modern era
Plagiodontia araeum	—	WI: Hispaniola	date of loss not established as within modern era
Plagiodontia velozi	—	WI: Hispaniola	extinct as *P. ipnaeum*[r]
Plagiodontia hylaeum	Dominican hutia	WI: Hispaniola	extinct as *P. aedium*
Alterodon major	—	WI: Jamaica	extinct as *Clidomys osborni*
Clidomys osborni	Osborn's Jamaican plate-tooth	WI: Jamaica	extinction predates modern era
Clidomys parvus	small Jamaican plate-tooth	WI: Jamaica	extinction predates modern era
Amblyrhiza inundata	Anguillan giant plate-tooth	WI: Anguilla & St. Martin	extinction predates modern era
Elasmodontomys obliquus	Puerto Rican plate-tooth	WI: Puerto Rico	date of loss not established as within modern era
Plesiorycteropus madagascariensis	greater bibymalagasy	Madagascar	date of loss not established as within modern era

[a]"Range" refers to distribution (known or alleged) of entity named in left-hand column; if distribution of current name combination is different, appropriate range is given in right-hand column (in parentheses).

[b]See Methods. "Extant as . . . " means that the nominal taxon listed under "Name" is considered to be part of the valid and accepted species named in the right-hand column, "Reason for rejection." This interpretation applies whether or not the nominal taxon is a named subspecies (e.g., *Rangifer tarandus dawsoni*) and whether or not it has been extirpated (e.g., *Eschrichtius robustus gibbosus*, the Atlantic gray whale).

[c]Williams and Nowak (1993) listed *A. odontrigonus* as having become extinct around 1500, mentioning that it "survived on Puerto Rico until around the time Columbus arrived." So far as we know, there are no published associations with *Rattus* or *Mus*, nor any radiocarbon dates suggesting survival into modern era. Woods (1989a:67, table 3) listed remains of "ground sloths (several species)" as having been found in association with *Rattus* and *Mus* in Trouing Jeremie #5, a Haitian cave locality. However, later in his text (p. 68) Woods stated that sloths were found in "deeper layers" in which introduced murids do not occur.

[d]Utzurrum (1992); L. Heaney (personal communication).

[e]Yoshiyuki (1989); Koopman (1993).

[f]Includes *S. leucogaster; S. borbonicus* has priority and is the valid name for this living species (see Cheke and Dahl, 1981).

[g]Woods's (1989a) evaluation of remains of *Antillothrix* (= *Saimiri*) *bernensis* is ambiguous. The only published radiocarbon dates for *Antillothrix* are 3850 and 9550 rcyrbp (MacPhee *et al.*, 1995), well beyond the cutoff for modern taxa.

[h]*E. quagga* has priority and is the valid name for this living species if this combination is accepted.

[i]May be part of *G. arabica.*

[j]Sometimes included in "*P. praeconis,*" but *P. fieldii* has priority and is the valid name for this living species if this combination is accepted.

[k]Never recorded as a living animal, but Flannery (1995d) believed that it may have survived late into nineteenth century on the basis of fossil associations with other (still extant) endemic murids no longer found in the basalt plains mouse's known range.

[l]Extant according to T. Flannery (personal communication).

[m]*Nesoryzomys fernandinae* was originally described on the basis of owl-pellet material (Hutterer and Hirsch, 1979); it has since been recorded as living (D. Steadman, personal communication).

[n]"*N. indefessus*" is only an extirpated subspecies, *N. indefessus indefessus* (Musser and Carleton, 1993).

[o]Dated by Steadman *et al.* (1991) at 8540 ± 100 to 5700 ± 70 rcyrbp.

[p]Whether nominal taxon belongs in *Sphiggurus* or *Coendu* (and in which species thereof) is not settled (cf. Handley and Pine, 1992); we see no reason to accept it as a separate, valid species (contra Woods, 1993).

[q]Anthony (1916) reported that remains of *H. insulans* at the type locality were found in an intermediate sedimentary level, between older *Acratocnus* beds and younger *Isolobodon* beds. If *Acratocnus* died out well before the modern era, this information does not constrain EED of *H. insulans.* No dating information is available for related species *H. antillensis.*

[r]See *P. ipnaeum* in Table 1.

5.5. Primates

It has long been suspected that at least some of the so-called Malagasy "subfossil" taxa were still extant at the time of European discovery of Madagascar in 1502 (e.g., Grandidier, 1905; Standing, 1908; Godfrey, 1986). This inference was based primarily on the evident freshness of bone from certain paleontological localities as well as oral traditions that seem to refer to hippos, elephantbirds, and giant lemurs (Flacourt, 1661; Tattersall, 1982). However, reliable evidence for EED assessment is still almost entirely lacking: Most published ^{14}C dates on identified remains are considerably earlier than 1500, which may indicate that current search images for very late sites are inadequate (D. A. Burney, personal communication; this volume). It is therefore of considerable interest that Simons (1997) has obtained dates close to 1500 on unspecified samples from cave sites containing skeletal material of the giant lemurs *Megaladapis* and *Palaeopropithecus.* (Attribution of this material to species in Table 1 remains to be confirmed.) If these large-bodied lemurs managed to survived this late, then perhaps many other now-extinct Malagasy primates managed to do so as well (cf. Tattersall, 1982).

Interestingly, a passage in Sloane's (1707–1725) natural history of Jamaica makes reference to monkeys living in the interior of that island. Recently, we recovered cranial and postcranial material of *Xenothrix mcgregori* in stratified contexts in southern Jamaica that also yielded *Rattus* (MacPhee, 1997b; MacPhee and Flemming, 1997), indicating that this monkey also survived later than 1500.

5.6. Carnivora

Carnivore losses in the modern era at the species level have been exceptionally light—a dog (*Dusicyon australis*) and a seal (*Monachus tropicalis*). Losses at the populational level, however, have been woefully serious in many groups, which is why threatened and extinct carnivore subspecies dominate compilations such as the IUCN *Red List.*

5.7. Cetacea

It is astounding that no whale species has disappeared in the past 500 years—this, despite the unchallengeable fact that many cetacean species have suffered incredible population depletions. At least one distinct population, the Atlantic gray whale, has died out in recent times (see Table 7). Interestingly, the Atlantic gray whale is very poorly known from contemporary records (Ellis, 1991); it may have died out before the seventeenth century (Bryant, 1996), well before the era of maximum overexploitation in the nineteenth and twentieth centuries.

5.8. Sirenia

The sole sirenian extinction of the modern era, *Hydrodamalis gigas,* occurred off the Commander Islands in the subarctic Bering Sea. The last accepted record of Steller's sea cow dates from 1768 (Forsten and Youngman, 1982). Because this species would not have been confused with any other, we accept the apparent failure to find any individuals after 1768 as category 1 evidence. However, Nordenskiold (1881, vol. 2:276–278) recorded anecdotal information to the effect that Steller's sea cows were still in existence as late as 1780 or even 1854. This last date seems unlikely, given that whaling and seal hunting in the North Pacific were highly developed industries by the mid-nineteenth century (Ellis, 1991).

5.9. Artiodactyla

Few artiodactyls have become extinct in the modern era, despite the fact that this group suffered copious losses earlier in the Quaternary (Martin, 1984; Alroy, this volume; Owen-Smith, this volume). Our very brief roster includes only the red gazelle, *Gazella rufina,* about which virtually nothing is known (Groves, 1985); the bluebuck, *Hippotragus leucophaeus,* usually mentioned on extinction lists; and the two species of Malagasy hippos, usually ignored.

Anecdotal information summarized by Godfrey (1986) strongly supports the notion that at least one kind of Malagasy hippo survived until the modern era. However, the youngest ^{14}C date, on material of *Hippopotamus lemerlei* from Itampolo, is 980 ± 200 rcyrbp (Mahé and Sourdat, 1972; Dewar, 1984). Nevertheless, the ubiquity of hippo remains in sites of apparently recent vintage and the widespread oral tradition of the hippo-like *lalomena* (Flacourt, 1661) justify our inclusion of this species and its undated relative *H. madagascariensis* (*nec Hexaprotodon,* contra Harris [1991]) in Table 2.

5.10. Rodentia

Rodentia is the mammalian order that has lost the most species in the modern era—a total of 43 losses. The greatest loss by family amounts to 31 extinctions in Muridae, although no murid subfamily has lost more than a few percent of its Recent complement. According to our tables, 14.7% of all mammalian extinctions in the past 500 years have occurred within tribe Oryzomyini (13 species), but validity problems affecting several species in this group have not been satisfactorily resolved. According to Musser and Carleton (1993), almost two dozen nominal murid species are known from one or a few specimens. The status of these species is best described as utterly unknown. They are not included here because no reliable observer has claimed that they are actually extinct, although this inference has certainly been made in other cases where the evidence is just as poor.

To illustrate the slipperiness of anecdotal information, it is worth briefly noting the history of *Quemisia gravis.* Miller (1929b) believed that the distinctive fossil rodent to which he gave this name was the same as the *quemi* described by Oviedo y Valdez (1535), which therefore must have survived into post-Contact times. However, Woods (1989b) has argued that the term may refer instead to *Plagiodontia ipnaeum* (= *P. velozi*), already known to have survived into the modern era (Table 1). If Woods is correct, then *Quemisia gravis* lacks a definite EED.

Within Rodentia, island endemics have suffered proportionately more losses than continental species (36 insular extinctions, or 78.3% of all rodent losses in the modern era). The number of island losses in the modern era may be underestimated, if the poorly known efflorescences on various islands in the Lesser Antilles (Pregill *et al.,* 1994) and Galápagos Islands (Steadman *et al.,* 1991) prove to include additional very recent extinctions. On the other hand, fossil taxa in some groups (e.g., Capromyidae) seem to have been seriously oversplit (cf. Camacho *et al.,* 1995).

5.11. Lagomorpha

The Sardinian pika, *Prolagus sardus,* is known apparently only from fossil material; the basis for the assumption that it was still alive in the last quarter of the eighteenth century is a single, rather vague observation by Cetti (1777; see also Vigne, 1988).

The status of the Omilteme (or Omiltemi) cottontail *Sylvilagus insonus,* cited by MacPhee and Flemming (1997) as continental North America's only modern-era mammalian extinction, is difficult to evaluate. One "possible sighting" was reported in 1991 (Ceballos and Navarro, 1991), but this species has escaped detection in several recent surveys of the mammals of Parque Omiltemi (Chapman and Ceballos, 1990; Jiménez Almaraz *et al.,* 1993). It is clearly a judgment call as to whether it should be recognized as extinct (as opposed to severely endangered, which is the current view of the Mexican government [Cervantes and Lorenzo, 1997]). This example underlines the difficulty of rating putative extinctions that have occurred extremely recently.

6. Discussion

6.1. How Many Mammal Extinctions at the Species Level Have Occurred in the Modern Era?

Depending on criteria used, the question posed in this heading may have more than one answer. The least problematic group of losses consists of the 40 species listed in Table 1. These taxa meet the basic criteria for recognition as resolved extinctions, meaning that nothing has been placed in evidence to contradict either their status as valid species or the inference that they were lost between 1500 and 1950. Nevertheless, in some cases the evidence for inclusion is not of especially high quality, and it is possible that future revision or renewed collecting may result in some deletions. The next group consists of the 31 unresolved extinctions listed in Tables 2 and 3. Their credentials as examples of modern-era extinctions are often nearly as good as those of species in Table 1, but they have been listed separately because of outstanding problems (date of loss, species validity, or nomenclature). The 17 unresolved losses in Table 4 are also kept apart, because these species may—or may not—have died out since 1950.

If these results are combined to achieve a single estimate, then the number of species-level losses during the past 500 years for which there is at least some positive evidence comes to 88. This amount, by no means insignificant, could be pushed to 100 or more if one were to accept (as we do not) a wide variety of equivocal cases listed in some other compilations, and higher still if one were to make due allowance for lost species that never were, and may never be, detected empirically. However, for purposes of this discussion we shall stick with 88 as being the number of losses derived empirically from the accepted evidence.

6.1.1. Systematic and Temporal Considerations

A large number of nominal species listed in the tables—18 taxa, or 20.5%—are based on extremely limited hypodigms (≤3 specimens, H or H+). (The count would be even higher if fossil or "F" taxa, known from only one or a few specimens, were included.) Obviously, small hypodigms need not automatically raise suspicions concerning the validity of species. But the fact that rarely collected, thinly documented taxa make up nearly one-fifth of suspected modern-era extinctions is worth worrying about, for roughly the same reason that the possibility of a type 2 error (incorrect acceptance of a false null hypothesis) in a statistical analysis is worth worrying about. We believe that the odds are against all such taxa being valid, but this is something to be worked out systematically by students of the groups in question.

We emphasize that it is dangerous to assume that an extinction has occurred merely because a species has not been recently seen or collected. Certainly, species not seen for some time could be extinct—or they could be rare in nature, or unusually cryptic, or behaviorally adept at escaping detection or trapping. For species that are collected rarely, an inference of extinction should be avoided until appropriate collecting methods to increase sample sizes have been tried repeatedly, without success. As noted above, we do not include some two dozen murids catalogued by Musser and Carleton (1993) as having very small hypodigms, because no reliable observer has claimed that they have become extinct. It would seem reasonable to keep this moratorium in place until appropriate surveys are made.

It is practically inevitable that poorly documented taxa will have highly suspect EEDs. For example, the groove-toothed forest mouse *Leimacomys buettneri,* collected only once in 1890, fulfills the EED criterion for entry in Table 1, but the cloud rat *Crateromys paulus,* collected only once in 1953, has to be posted to Table 4 because of the 50-year rule. Although Table 1 is said to include only "resolved" extinctions, there is really nothing other than the spread of years that justifies the assumption that *L. buettneri* is more likely to be extinct than is *C. paulus.* We can only emphasize, once again, that the road to resolution is through collection and inventory.

6.1.2. Geographical Considerations

It has long been recognized that the vast majority of vertebrate losses during the modern era have occurred on islands (e.g., Diamond, 1984; Olson, 1989). Our data reveal that extinctions among island mammals are even higher than previously reported. No fewer than 71.6% (63/88) of all suspected species-level extinctions among mammals in the past 500 years have occurred on islands, according to Tables 1–4. If Table 4 data (unconfirmed post-1950 losses) are omitted from the calculation, this figure rises to 78.9%, which compares closely with figures published for losses of birds (>90%) and molluscans (80%) on islands (WCMC, 1992; Steadman, 1995).

It is often stated that the great majority of modern-era mammalian extinctions that have occurred on the planet took place in Australia, mostly within the past 150 years (Flannery, 1995a; Smith and Quin, 1996). However, we point out that it is the Caribbean region that has suffered the greatest concentration of losses (Figs. 2, 3). According to the data in Tables 1–4, 37.5% (33/88) of all modern-era extinctions occurred on West Indian islands. This is nearly equal to the combined total of losses on the Australian continent (19.3%, 17/88) plus the entire Pacific area (21.6%, 19/88).

The small number of losses on continents other than Australia is noteworthy (Figs. 2, 3). In the Americas (excluding the West Indies), two of three documented modern-era extinctions occurred on nearshore islands. The single mainland loss (*Sylvilagus insonus*) for the Americas is not fully resolvable in our framework because of its recency (ca. 1991). Similarly, according to our data, Eurasia has lost only one continental species (a vole), while mainland Africa has witnessed only four species-level extinctions in the past 500 years (two antelopes, a mouse, and a bat). Despite an immense amount of habitat degradation in this century, at least at the empirical level, continental biomes have suffered a mere handful of documented losses of mammalian species.

6.2. What Is the Modern-Era Extinction Rate for Mammals?

Figure 4 distributes the 88 losses in Tables 1–4 along a time line divided into 25-year increments. Of this number, only 51 (58.0%) are dated sufficiently well to assign their EEDs

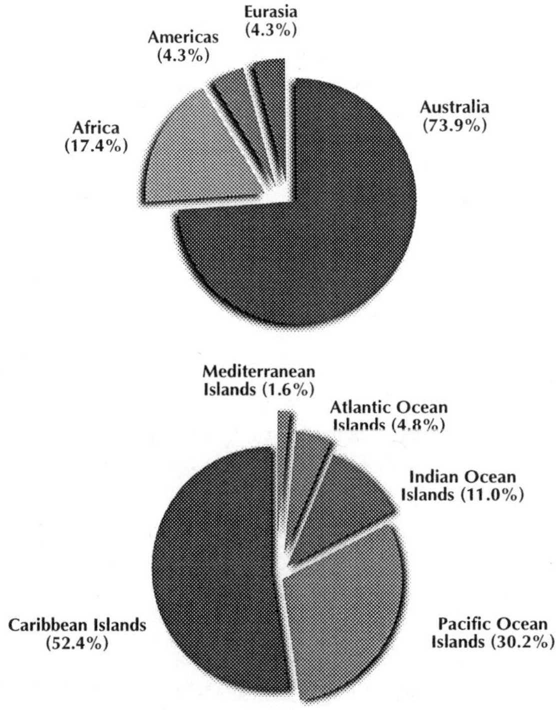

FIGURE 3. (Top) Continental extinctions since AD 1500: percentage by area ($N = 23$). (Bottom) Insular extinctions since AD 1500: percentage by area ($N = 63$).

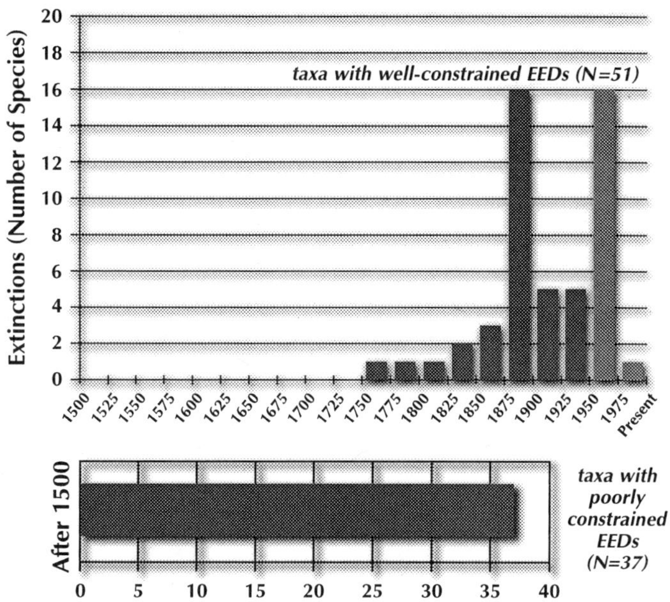

FIGURE 4. Mammalian species extinctions plotted at quarter-century intervals between AD 1500 and present (N = 51). Lower bar graph depicts known taxa that died out sometime after AD 1500 (N = 37).

to narrow intervals. (Quarter-century intervals were chosen for depiction, but the reader should be aware that some interval assignments are based on convention only.) Of these 51 species, only two (3.9%) apparently disappeared before the nineteenth century. Twenty-two species (43.1%) were lost in the nineteenth century (3 before 1850), whereas 27 species (52.9%) disappeared in the twentieth century (17 since 1950). It is noteworthy that the loss figure for the period 1850–1899 (19 species) is substantially higher than that for 1900–1949 (10 species). The cumulative curve for mammalian extinctions in the modern era (Fig. 5) thus has a distinctive appearance, rising quickly through the nineteenth century and flattening notably during the past half-century.

The figures just quoted cannot be accepted at face value, because they do not include the large number of "fossil" taxa (N = 37) whose EEDs usually cannot be approximated even to the nearest century (separate bar in Fig. 4, "After 1500"). Large numbers of temporally unallocated losses may severely complicate the estimation of extinction rates from empirical data, as the following example illustrates. In Fig. 6A we portray 51 extinctions by quarter-century intervals from the beginning of the sixteenth century through the third quarter of the twentieth century. In examining the sharp slope of the regression line in the inset, it might be concluded that the extinction rate has shot upward exponentially in the past two centuries, from few losses before 1750 to dozens in the last several 25-year intervals. However, it is of course unlikely that there was only one mammalian extinction between ca. 1500 and 1775, for example. The graph merely reflects the fact that we cannot narrowly date more of the "fossil" extinctions, which is a different matter. To model the possibility that many imprecisely dated "fossil" species disappeared within this early period, we added 37 "fossil" taxa (Fig. 6B) randomly to the time line. Interestingly, these unbiased

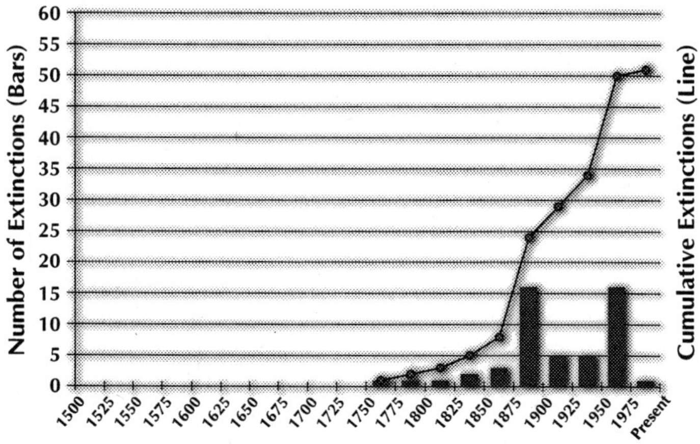

FIGURE 5. Mammalian species extinctions since AD 1500: losses with well-constrained EEDs ($N = 51$) by quarter-century (bars) and cumulatively (line).

additions have a negligible effect on the regression expression (inset), which is driven by high values on the right side of the graph (although the line's intercept is shifted slightly, to the mid-sixteenth century). Nonrandom additions can have a quite different effect, however: If we allot the 37 poorly dated extinctions in the tables to the period between 1500 and 1599, a distinctly bimodal graph results (Fig. 6C). Bimodality, combined with the much lower slope of the regression expression (inset), is inconsistent with the usual view that the modern-era extinction rate for mammals was very low until the mid-nineteenth century and then rose steeply thereafter. This scenario cannot be verified either, but that is not the point. The important thing is that extinction rate estimates and perceived patterns of loss are bound to be affected seriously by the quality of the data utilized. Given so many temporally unallocated losses (Fig. 4), the few facts in play can be made consistent with practically any extinction-rate scenario (Caughley and Gunn, 1996).

6.3. How Does This Compilation Differ from Others?

We shall answer this question by comparing our results with the most recent major compilation, the IUCN (1996) *Red List of Threatened Animals.* This is the first *Red List* issued using the new, more elaborate set of criteria for ranking threat status (Mace, 1993). Although the IUCN cutoff date is AD 1600 rather than 1500, the difference is unimportant inasmuch as there are no mammalian extinctions that can be dated explicitly to the sixteenth century according to our criteria.

The 88 losses counted in Tables 1–4 amount to 1.9% of the 4629 Recent species catalogued in Wilson and Reeder's (1993) *Mammal Species of the World* (which, however, excludes a number of the extinct species tabulated here). Our total differs trivially from the IUCN (1996) *Red List* estimate of 85 losses. At first glance, it might appear that such apparently close similarity in final totals indicates that basic data sets and methodologies must be highly compatible. Closer inspection reveals, however, that the similarity is almost accidental, because there is nearly a 2:1 difference in taxic content and status ratings in the two papers.

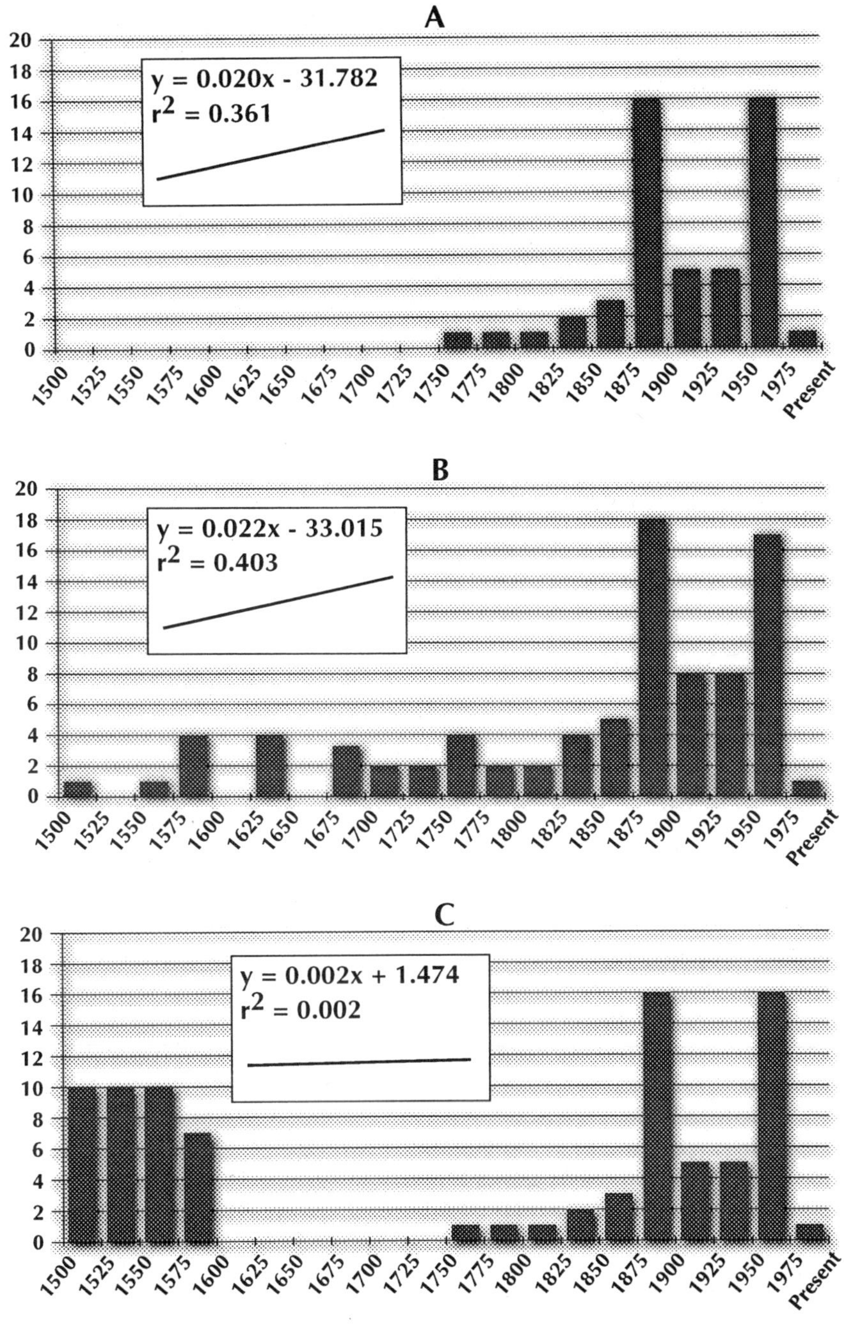

FIGURE 6. Mammalian species-level extinctions per century since AD 1500: (A) modern-era extinctions with well-constrained EEDs ($N = 51$); (B) $N = 88$ (as in A, with the addition of 37 fossil species placed at random in the interval 1500–1999); (C) $N = 88$ (as in A, with the addition of 37 fossil species placed in the interval 1500–1599).

This striking difference results from several factors (see Tables 5 and 6). The IUCN (1996) *Red List* includes 28 instances of alleged extinction that we exclude for systematic or other reasons, whereas we include 23 species extinct in the modern era that the *Red List* does not mention at all. There are 8 additional species regarded by us as extinct that the IUCN classifies differently (as Critically Endangered, Endangered, or Data Deficient). Overall, the two lists classify only 57 species in the same way, that is, as instances of modern-era extinction. This yields a compatibility factor of only 49% (i.e., 57/116, number of status matches in the two papers divided by matches + mismatches).

Two questions spring immediately to mind. First, if the study of modern-era extinctions is truly amenable to scientific inquiry, how can assessments of the status of individual taxa differ as substantially as they do between our lists and the *Red List*? Second, where do we go from here?

It should be evident from remarks peppered throughout this chapter that we do not, in fact, have much of a "science" of modern-era extinctions of mammals at this time. It is therefore no wonder that differing criteria for assessing extinction status exist, and that this produces incompatible results. The biggest single source of disagreement is that the IUCN (1996) *Red List* and compilations published previously accept many instances of extinction that we reject because they do not meet minimal objective criteria (see Table 7). A possible reason for this discrepancy is that the burden of assembling databases like the *Red List* falls disproportionately to researchers whose interest and training in systematics may be limited. Scientists who are not systematists themselves, or are not adequately advised by systematists, face substantial hurdles in dealing with and interpreting basic taxonomic information. For example, it seems to us that the category "Extinct" was applied rather uncritically in the compilation of current and past *Red Lists*. Possibly, compilers relied too much on offhand, unsubstantiated guesses in the literature about the extinction status of specific taxa (see discussion by Harrison and Stiassny, this volume). It is also possible that it was difficult to get qualified systematists to scrutinize such claims.

The second question concerns the future of lists like the ones offered here. Dramatic differences among compilations indicate that systematic and evidentiary criteria are being applied unevenly. For future compilations, we recommend better use of the revisionary literature than is ordinarily done, because this literature is the only arbiter of what is valid and what is not. We think it should also be mandatory for compilers to search paleontological sources for relevant "fossil" taxa, although we recognize that this point has pertinency only for certain groups (e.g., vertebrates, molluscans). In this way it should be possible to reach a consensus among conservation biologists, systematists, and paleontologists working to the same end.

In this vein it is worth remarking that much better use of radiometric dating should be made for sorting out doubtful cases. Because of the paucity of published dates, category 3 information played only a very small role in the construction of our lists (Fig. 7). In the future, a substantial effort should be undertaken to resolve the status of the large number of extinct species with poorly constrained EEDs (currently 37 taxa).

We would like to end this section by answering an objection raised by some of our colleagues, that extinction lists like the ones presented here do an actual disservice to the cause of conservation biology. Such lists, it is claimed, cannot be comprehensive because they capture only a fraction of all species extinctions that have occurred in the modern era. In the wrong hands, the little knowledge that we do provide can become a truly dangerous

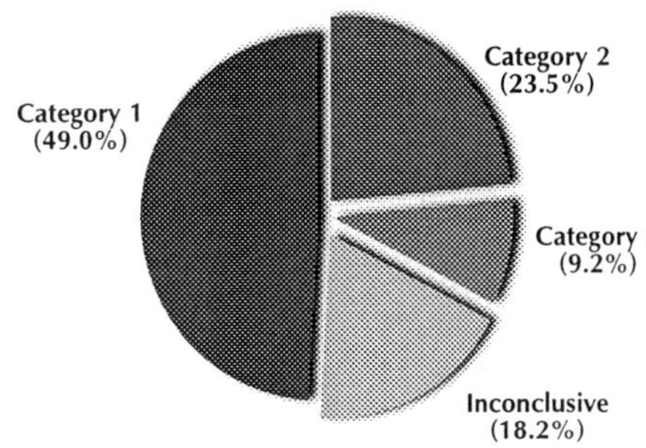

FIGURE 7. Mammalian species extinctions since AD 1500: availability of evidence of extinction, by category.

thing. Although we acknowledge the force of the sentiments that prompt this view, they are ultimately beside the point. We cannot control for varying tastes in the taxonomic recognition of species, and we cannot control for species extinctions that have left no accessible record. Nor can we control for the use others may make of extinction lists. What we can do is present and discuss the evidence, and identify where the existing database is probably adequate and where it is defective. Only in this way will extinction studies attain a scientific footing (see recommendations outlined by Harrison and Stiassny, this volume).

7. Coda: Modern-Era Extinctions in Quaternary Perspective

Compared with losses that occurred earlier in the Quaternary, such as the ones discussed elsewhere in this book, there are many things that modern mammalian extinctions are not.

7.1. Not Just Another Case of Overkill

The familiar mantra concerning human-induced extinctions intones that it is only logical to assume that currently recognized causes of species endangerment will ineluctably lead to extinction—unless, of course, remedial measures are undertaken. As is well known, habitat loss is usually rated as the leading cause of endangerment, followed by overexploitation and introduction of exotic species (cf. WCMC, 1992; Hanski *et al.*, 1995; Baillie, 1996; Czech and Krausman, 1997). "Other causes," such as incidental taking and disease, are often not rated or rated as having effects on isolated taxa only (cf. WCMC, 1992). It is also accepted that many, if not most, endangered taxa are threatened by more than one cause (Czech and Krausman, 1997). Such frameworks for threat analyses are usually self-limited to anthropogenic influences and therefore do not include "natural" factors like climate change (except "global warming," which is viewed as having an anthropogenic origin).

Interestingly, when Late Quaternary extinctions are under discussion, the relative ranking of causes is essentially inverted. Overkill is often identified as the chief, if not the only, cause of these extinctions, especially those that took place on continents and large islands. Introduced species are allowed a strong role in provoking extinctions, but only on islands. Habitat destruction is much more rarely invoked as a cause of prehistoric losses, but lack of evidence pointing to it in the earlier Quaternary has not prevented the assertion that humans have always been destroyers of nature: "on all continents, there is an extensive record of prehistoric human alteration and destruction of habitat coinciding with a high rate of species extinctions" (Primack, 1995:59). In actuality, there is no such record for any period before the modern era. Today, even in continental settings in which habitat modification has been enormous, the number of documented losses remains remarkably low, at least among mammals and birds (e.g., Brazilian Atlantic forest; Brooks and Balmford, 1996).

What of the "intermediate period" of extinction, the modern era that we focus on here that lies between the first major European voyages of discovery and the advent of the postindustrial biodiversity crisis? No more than a handful of recent mammalian extinctions can be plausibly attributed to overexploitation in any of its guises (here including overhunting and persecution). The best-supported cases of overexploitation culminating in extinction are *Hydrodamalis gigas* (hunted for meat), *Dusicyon australis* (for fur), and *Thylacinus cynocephalus* (persecuted as a pest). *Monachus tropicalis* was hunted intensively for its fat, which was rendered into lamp oil (cf. G. M. Allen, 1942). Demand for seal oil collapsed in the mid-nineteenth century with the appearance of better alternatives, but by this time the Caribbean monk seal was already so extremely rare that J. A. Allen (1887) described it as being nearly "mythical." Whether its population collapse was related solely to earlier overhunting is, however, completely unexplored. Likewise, the wallaby *Macropus greyi* may have been hunted incidentally for its fur, but there is no good evidence to indicate that furtrapping (or any other specific pressure) caused its extinction (Smith, 1995). Why rarely seen *Hippotragus leucophaeus* disappeared is similarly obscure. Overhunting may have had some effect if, as often claimed, the bluebuck's range was limited to a single district in Cape Colony (Swellendam). But there is no evidence on this point.

Some small mammals—especially ones living on islands—may also have been brought to the point of extinction through overexploitation. However, as with large mammals, the evidence is limited and tenuous. *Pteropus subniger* was evidently hunted for its meat, but so were (and are) many other pteropodid species (Heaney and Heideman, 1987). Yet among bats, such exploitation has led at most to local extirpations, not complete species extinctions (e.g., bat losses on 'Eua, Tonga [Koopman and Steadman, 1995]). In reference to extinctions in the West Indies, it is reasonable to assume that many of the endemic rodents were large enough to make the hunting of them worthwhile, although the archaeological evidence for direct impacts of this sort is negligible. The capromyid *Isolobodon portoricensis*—frequently encountered in archaeological sites—may even have been brought into semidomestication on some islands (Miller, 1929b; Flemming and MacPhee, in press). Yet, if a domesticant, why did it become extinct (cf. Alcover *et al.*, this volume, in reference to the Balearic goat *Myotragus*)?

7.2. Not A Climate Change Event

"Climate change," a grab-bag category, has come to mean any kind of environmental change that some author or another believes could be offered seriously as a cause of extinction. Often, it has an essentially nonspecific, negative meaning: It is whatever human-induced change is not (e.g., habitat destruction, overkill). To find it, one looks wherever people are not, which is not an easy thing.

It is of special interest that none of the serious climatic disruptions of recent centuries is known to have induced a documented mammal or bird extinction. Thus, the Little Ice Age, which resulted in average temperature well below normal for the better part of four centuries, has not been implicated in any extinctions, despite clear effects on species distributions (Luckman, 1986; Pielou, 1991). There is ample documentation that the El Niño/Southern Oscillation causes severe short-term depressions in sizes of Pacific seabird populations, but this phenomenon does not seem to have induced any recent species losses (Martin and Steadman, this volume). Extreme pulses, in the form of record highs/lows in temperature, snowfall, precipitation, and a multitude of other phenomena have been recorded for all parts of the planet, but in no case have these effects been implicated in a correlated vertebrate extinction. How much more severe must such occurrences be, and on what time scale, to cause outright extinction?

7.3. Not a Continental Event

As noted above, nearly three-quarters of all mammal extinctions in the past 500 years occurred on islands (Figs. 2, 3). Of the remainder, most occurred on just one continent, Australia. In the Americas 11,000 rcyrbp, by contrast, virtually all losses were mainland. No extinctions are known to have taken place at this time on the Greater and Lesser Antilles or the Galápagos, the New World's principal groups of oceanic islands.

7.4. Not a Megafaunal Event

As may be seen by examining the Body Weight (BW) column in Tables 1–4, the vast majority of species that have disappeared since 1500 were rather small (Fig. 8). Only 12.5% (11/88) of modern-era extinctions involved "large" species (defined here as body size ≥10 kg). Although empirical weights based on substantial samples are unavailable for many of the species concerned, it is unlikely that *Dusicyon australis, Gazella rufina, Thylacinus cynocephalus,* or *Palaeopro-pithecus,* cf. *P. ingens* met Martin's Limit for megafauna (i.e., body mass ≥44 kg). If these species are excluded, only 8% (7/88) of modern-era extinctions can be considered to lie within the "megafaunal range" as defined traditionally (in rough rank-order by size: *Hydrodamalis gigas, Hippopotamus lemerlei* and *H. madagascariensis, Monachus tropicalis, Hippotragus leucophaeus, Macropus greyi,* and *Megaladapis,* cf. *M. edwardsi*). By contrast, during the end-Pleistocene extinctions in North America, almost three-quarters (70.1%) of all losses occurred among mammal species having body sizes this large (Graham and Lundelius, 1984).

7.5. Not the Last Word

In end-Pleistocene extinctions in North America, artiodactyl and perissodactyl ungulates were highly affected (28 losses, for 41.2% of 68 species-level extinctions; data of Martin, 1984; Graham and Lundelius, 1984). By contrast, to the degree that they can be

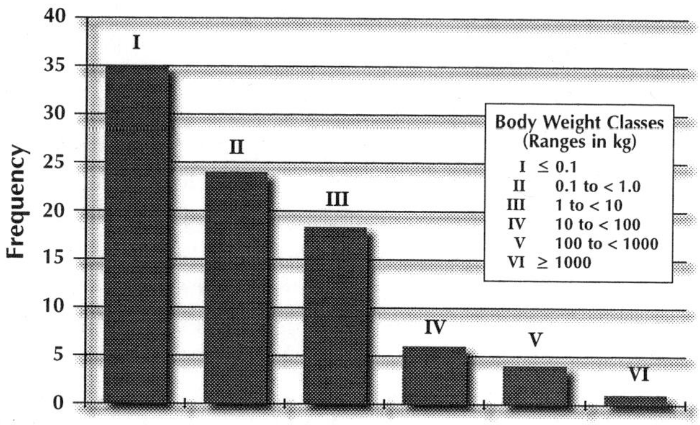

FIGURE 8. Mammalian species extinctions since AD 1500: body weights, by class (N = 88).

considered to have been fully detected by paleontological methods, losses among rodents (14.7%) and bats (5.7%) contributed relatively little to total wastage.

This situation has been effectively reversed in the past 500 years. Only 4.5% (4/88) of all documented extinctions in this period have occurred among ungulates (exclusively within Artiodactyla), whereas losses among bats and rodents contribute, respectively, 10.2% and fully 54.5% of the total.

It is important to recognize that our numbers would not change much even if certain taxa (e.g., several unlisted "giant" lemurs and ungulates rejected here because of insufficient information) were counted in the modern-era tally. Proboscidea and Perissodactyla, which have (or recently had) many members in the megafaunal size range, lost no species in the modern era despite suffering substantial wastage during the late Pleistocene (Martin and Steadman, this volume; Alroy, this volume). The low proportion of megafaunal extinctions in post-Pleistocene times may be partly artifactual, as, compared with earlier epochs, we live in a world largely bereft of big mammals (Owen-Smith, 1988, this volume). Yet the fact that so few megafaunal species have become extinct in recent times, despite egregious and persistent levels of high endangerment, remains a puzzle of the first order and demands a meaningful explanation (see discussion by MacPhee and Marx, 1997).

Additional information on mammalian extinctions in the modern era may be found at www.crea.org., the website of the Committee on Recently Extinct Organisms.

Acknowledgments

We thank Tim Flannery, Sara Gilson, Colin Groves, Larry Heaney, the late Karl Koopman, Don McFarlane, Veronica Mahanger, Guy Musser, Ronald Nowak, David Steadman, and Don Wilson for reading, criticizing, and improving the manuscript version of this chapter. Our colleagues Ian Harrison and Melanie Stiassny took some of the ideas developed here several steps further—a delightful result for us. We are especially grateful to Steven Shoji Kometani for the creativity he demonstrated in drafting the figures.

References

Allen, G. M. 1942. *Extinct and Vanishing Mammals of the Western Hemisphere with the Marine Species of all the Oceans.* Cooper Square, New York.

Allen, J. A. 1887. The West Indian seal (*Monachus tropicalus*). *Bull. Am. Mus. Nat. Hist.* **2**(1), 34 pp.

Anthony, H. E. 1916. Preliminary report on fossil materials from Porto Rico, with descriptions of a new genus of ground sloth and two new genera of hystricomorph rodents. *Ann. N. Y. Acad. Sci.* **27**:193–203.

Baillie, J. 1996. Analysis, in: J. Baillie and B. Groombridge (eds.), *1996 IUCN Red List of Threatened Animals,* pp. 24–43 [of introduction]. IUCN, Gland, Switzerland.

Brooks, T., and Balmford, A. 1996. Atlantic forest extinctions. *Science* **380**:115.

Brooks, T. M., Pimm, S. L., and Collar, N. J. 1997. Deforestation predicts the number of threatened birds in insular Southeast Asia. *Conserv. Biol.* **11**(2):382–394.

Bryant, P. J. 1996. Dating remains of gray whales from the eastern North Atlantic. *J. Mammal.* **76**(3):857–861.

Burbidge, A. A., Johnson, K. A., Fuller, P. J., and Southgate, R. 1988. Aboriginal knowledge of the mammals of the central deserts of Australia. *Aust. Wildl. Res.* **15**:9–39.

Calaby, J. H., and Lee, A. K. 1989. The rare and endangered rodents of the Australasian region, in: W. Z. Lidicker (ed.), Rodents: A World Survey of Species of Special Concern. *Occas. Pap. IUCN (SSC)* **4**:53–57.

Camacho, A., Borroto, R., and Ramos Garcia, I. 1995. Los capromidos de Cuba: estado actual y perspectivas de las investigaciones sobre su sistematica. *Marmosiana* **1**:43–56.

Caughley, G., and Gunn, A. 1996. *Conservation Biology in Theory and Practice.* Blackwell Science, Cambridge, Mass.

Ceballos, G., and Navarro, D. 1991. Diversity and conservation of Mexican mammals, in: M. A. Mares and D. J. Schmidly (eds.), *Latin American Mammalogy: History, Biodiversity and Conservation,* pp. 167–198. University of Oklahoma Press, Norman.

Cervantes, F. A., and Lorenzo, C. 1997. *Sylvilagus insonus. Mamm. Species* **568**, 4 pp. American Society of Mammalogists.

Cetti, F. 1777. Appendice alla Storia Naturale dei Quadrupedi di Sardegna, Sassari. 63 pp.

Chapman, J. A., and Ceballos, G. 1990. The cottontails, in: J. A. Chapman and J. E. C. Flux (eds.), *Rabbits, Hares and Pikas: Status Survey and Conservation Action Plan,* pp. 95–110. IUCN, Gland, Switzerland.

Cheke, A. S., and Dahl, J. F. 1981. The status of bats on western Indian Ocean islands, with special reference to *Pteropus. Mammalia* **45**:205–238.

Claridge, M. F., Dawah, H. A., and Wilson, M. R. 1997. *Species: The Units of Biodiversity.* Chapman & Hall, London.

Cole, F. R., Reeder, D. M., and Wilson, D. E. 1994. A synopsis of distribution patterns and the conservation of mammal species. *J. Mammal.* **75**:266–276.

Corbet, G. B., and Hill, H. E. 1991. *A World List of Mammalian Species,* 3rd ed. British Museum and Oxford University Press, London.

Czech, B., and Krausman, P. R. 1997. Distribution and causation of species endangerment in the United States. *Science* **277**:1116.

Dewar, R. E. 1984. Extinctions in Madagascar: The loss of the subfossil fauna, in: P. S. Martin and R. G. Klein (eds.), *Quaternary Extinctions: A Prehistoric Revolution,* pp. 574–593. University of Arizona Press, Tucson.

Diamond, J. M. 1984. Historic extinction: A Rosetta Stone for understanding prehistoric extinctions, in: P. S. Martin and R. G. Klein (eds.), *Quaternary Extinctions: A Prehistoric Revolution,* pp. 354–403. University of Arizona Press, Tucson.

Dollmann, J. G. 1937. Mammals which have recently become extinct and those on the verge of extinction. *J. Soc. Preserv. Fauna Empire* n. s. **30**:67–74.

Ellis, R. 1991. *Men and Whales.* Knopf, New York.

Fielden, H. W. 1890. Notes on the terrestrial mammals of Barbados. *Zoologist* Ser. 3 **14**:52–55.

Flacourt, E. 1661. *Histoire de la Grande Isle Madagascar.* Pierre L'Amy, Paris.

Flannery, T. F. 1995a. *The Future Eaters.* George Braziller, New York.

Flannery, T. F. 1995b. Great hopping-mouse, in: R. Strahan (ed.), *Mammals of Australia,* 2nd ed., p. 582. Smithsonian Institution Press, Washington, D.C.

Flannery, T. F. 1995c. *Mammals of the South-West Pacific and Moluccan Islands.* Cornell University Press (Comstock), Ithaca, N.Y.

Flannery, T. F. 1995d. Basalt plains mouse, in: R. Strahan (ed.), *Mammals of Australia,* 2nd ed., p. 619. Smithsonian Institution Press, Washington, D.C.

Flannery, T. F., Kendall, P., and Wynn-Moylan, K. 1990. *Australia's Vanishing Mammals, Endangered and Extinct Native Species.* RD Press, New South Wales.

Flemming, C., and MacPhee, R. D. E. 1995. The quick, the dead, and the uncertain: how many recently extinct mammals are there? Abstract with program, 75th Annual Meeting of the American Society of Mammalogists.

Flemming, C., and MacPhee, R. D. E. 1997. Then and now: 20,000 years of mammalian extinctions in the New World. *J. Vertebr. Paleontol.* **17**(3, Suppl.):46A.

Forsten, A., and Youngman, P. 1982. *Hydrodamalis gigas. Mamm. Species* **165**, 3 pp. American Society of Mammalogists.

Forsyth Major, C. I. 1901. The musk-rat of Saint Lucia (Antilles). *Ann. Mag. Nat. Hist. (*Ser. 7) **7**:204–206.

Godfrey, L. R. 1986. The tale of the tsy-aomby-aomby. *The Sciences (NY)* (Jan/Feb) pp. 49–51.

Godfrey, L. R., Jungers, W. L., Reed, K. E., Simons, E. L., and Chatrath, P. S. 1997. Subfossil lemurs: inferences about past and present primate communities in Madagascar, in: S. Goodman and B. Patterson (eds.), *Natural Change and Human Impact in Madagascar,* pp. 218–256. Smithsonian Institution Press, Washington, D.C.

Goodwin, H. A., and Goodwin, J. M. 1973. List of mammals which have become extinct or are possibly extinct since 1600. *IUCN Occas. Pap. No. 8,* 21 pp.

Graham, R. W., and Lundelius, E. L. 1984. Coevolutionary disequilibrium and Pleistocene extinctions, in: P. S. Martin and R. G. Klein (eds.), *Quaternary Extinctions: A Prehistoric Revolution,* pp. 223–249. University of Arizona Press, Tucson.

Grandidier, A. 1905. Recherches sur les lémuriens disparus et en particulier sur ceux qui vivaient á Madagascar. *Nouv. Arch. Mus.* (Ser. 4) **7**:1–140.

Groves, C. P. 1985. An introduction to the gazelles. *Chinkara* **1**:4–16.

Groves, C. P., and Flannery, T. F. 1994. A revision of the genus *Uromys* Peters, 1867 (Muridae: Mammalia) with descriptions of two new species. *Rec. Aust. Mus.* **46**:145–169.

Groves, C. P., Schaller, G. B., Amato, G., and Khounboline, K. 1997. Rediscovery of the wild pig *Sus bucculentus. Nature* **386**:335.

Handley, C. O., and Pine, R. H. 1992. A new species of *Coendou* Lacépède, from Brazil. *Mammalia* **56**:237–244.

Hanski, I., Clobert, J., and Reid, W. 1995. Ecology of extinction, in: V. H. Heywood (ed.), *Global Biodiversity Assessment.* Cambridge University Press, London.

Harris, J. M. 1991. Family Hippopotamidae, in: J. M. Harris (ed.), *Koobi Fora Research Project,* Volume 3, pp. 31–85. University Press (Clarendon), London.

Heaney, L. R., and Heideman, P. D. 1987. Philippine fruit bats: endangered and extinct. *Bats* **5**(1):3–5.

Hershkovitz, P. 1971. A new rice rat of the *Oryzomys palustris* group (Cricetinae, Muridae) from northwestern Colombia, with remarks on distribution. *J. Mammal.* **52**:700–709.

Holdaway, R. N. 1996. The arrival of rats in New Zealand. *Nature* **384**:225–226.

Hughes, J. B., Daily, G. C., and Erlich, P. R. 1997. Population diversity: Its extent and extinction. *Science* **278**:689–692.

Hutterer, R. 1993. Order Insectivora, in: D. E. Wilson and D. M. Reeder (eds.), *Mammal Species of the World: A Taxonomic and Geographic Reference,* 2nd ed., pp. 69–130. Smithsonian Institution Press, Washington, D.C.

Hutterer, R., and Hirsch, U. 1979. Ein neuer *Nesoryzomys* von der Insel Fernandina, Galápagos (Mammalia, Rodentia). *Bonn. Zool. Beitr.* **3**:276–283.

Hutterer, R., López-Martínez, N., and Michaux, J. 1988. A new rodent from Quaternary deposits of the Canary Islands and its relationships with Neogene and Recent murids of Europe and Africa. *Palaeovertebrata* **18**:241–262.

International Code of Zoological Nomenclature, 3rd ed. 1985. International Trust for Zoological Nomenclature, University of California Press, Berkeley.

IUCN. 1993. *1994 IUCN Red List of Threatened Animals* (B. Groombridge, ed.). IUCN, Gland, Switzerland.

IUCN. 1996. *1996 IUCN Red List of Threatened Animals* (J. Baillie and B. Groombridge, eds.). IUCN, Gland, Switzerland.

Jiménez Almaraz, T., Juarez Gomez, J., and Leon Paniagua, L. 1993. Mamíferos, in: I. Luna Vega and J. Llorente Bousquets (eds.), *Historía Natural del Parque Ecológico Estatal Omiltemi, Chilpancingo Guerrero, México,* pp. 503–549. Universidad Nacional Autonóma de México, C. de México, D. F.

Kimbel, W. H., and Martin, L. B. (eds.), 1993. *Species, Species Concepts, and Primate Evolution.* Plenum Press, New York.

Koopman, K. F. 1984. Taxonomic and distributional notes on tropical Australian bats. *Am. Mus. Novit.* **2778**, 48 pp.

Koopman, K. F. 1993. Order Chiroptera, in: D. E. Wilson and D. M. Reeder (eds.), *Mammal Species of the World: A Taxonomic and Geographic Reference,* 2nd ed., pp. 137–232. Smithsonian Institution Press, Washington, D.C.

Koopman, K. F., and Steadman, D. W. 1995. Extinction and biogeography of bats on 'Eua, Kingdom of Tonga. *Am. Mus. Novit.* **3125**, 13 pp.

Lawlor, T. 1983. The mammals, in: T. J. Case and M. L. Cody (eds.), *Island Biogeography in the Sea of Cortez,* pp. 265–289. University of California Press, Berkeley.

Luckman, B. H. 1986. Reconstruction of Little Ice Age events in the Canadian Rocky Mountains. *Geogr. Phys. Quat.* **40**:17–28.

Mace, G. M. 1993. The status of proposals to redefine the IUCN threatened species categories, in: B. Groombridge (ed.), *The 1994 IUCN Red List of Threatened Animals,* pp. xlviii–lv. IUCN, Gland, Switzerland.

MacPhee, R. D. E. 1997a. Digging Cuba: The lessons of the bones. *Nat. Hist.* **106**(11):50–54.

MacPhee, R. D. E. 1997b. Vertebrate paleontology of Jamaican caves, in: A. Fincham (ed.), *Jamaica Underground,* 2nd ed., pp. 47–56. The Press University of the West Indies, Kingston, Jamaica.

MacPhee, R. D. E., and Fleagle, J. G. 1991. Postcranial remains of *Xenothrix mcgregori* (Primates, Xenotrichidae) and other late Quaternary mammals from Long Mile Cave, Jamaica. *Bull. Am. Mus. Nat. Hist.* **206**:287–321.

MacPhee, R. D. E., and Flemming, C. 1997. Brown-eyed, milk-giving, and extinct: Losing mammals since AD 1500. *Nat. Hist.* **106**(3):84–88.

MacPhee, R. D. E., and Flemming, C. In press. Crossing the irremeable stream: An annotated catalogue of mammalian species extinctions since AD 1500. *Am. Mus. Novit.*

MacPhee, R. D. E., and Marx, P. A. 1997. The 40,000-year plague: humans, hyperdisease, and first-contact extinctions, in: S. Goodman and B. Patterson (eds.), *Natural Change and Human Impact in Madagascar,* pp. 169–217. Smithsonian Institution Press, Washington, D.C.

MacPhee, R. D. E., and Rivero de la Calle, M. 1996. Accelerator mass spectrometry ^{14}C age determination for the alleged "Cuban spider monkey", *Ateles (= Montaneia) anthropomorphus. J. Hum. Evol.* **30**:89–94.

MacPhee, R. D. E., Flemming, C. and Lunde, D. P. 1999. "Last Occurrence" of the Antillean insectivoran *Nesophontes:* New radiometric dates and their interpretation. *Am. Mus. Novit.*

MacPhee, R. D. E., Horovitz, I., Arredondo, O., and Jiménez Vasquez, O. 1995. A new genus for the extinct Hispaniolan monkey *Saimiri bernensis* Rímoli, 1977, with notes on its systematic position. *Am. Mus. Novit.* **3134**, 21 pp.

Mahé, J., and Sourdat, M. 1972. Sur l'extinction des vertébrés subfossiles et l'aridification du climat dans Sud-Ouest de Madagascar. *Bull. Soc. Geol. Fr.* **14**:295–309.

Mahoney, J. A. 1975. *Notomys macrotis* Thomas, 1921, a poorly known Australian hopping mouse (Rodentia: Muridae). *J. Aust. Mamm. Soc.* **1**:367–374.

Marsh, R. E. 1985. More about the Bajan mouse. *J. Barbados Mus. Hist. Soc.* **37**(3):310.

Martin, P. S. 1984. Prehistoric overkill: the global model, in: P. S. Martin and R. G. Klein (eds.), *Quaternary Extinctions: A Prehistoric Revolution,* pp. 354–403. University of Arizona Press, Tucson.

May, R. M., Lawton, J. H., and Stork, N. E. 1995. Assessing extinction rates, in: J. H. Lawton and R. M. May (eds.), *Extinction Rates,* pp. 1–24. Oxford University Press, London.

Miller, G. S. 1929a. A second collection of mammals from caves near St. Michel, Haiti. *Smithson. Misc. Collect.* **81**(9), 30 pp.

Miller, G. S. 1929b. Mammals eaten by Indians, owls, and Spaniards in the Coast Region of the Dominican Republic. *Smithson. Misc. Collect.* **82**(5), 16 pp.

Miller, G. S. 1930. Three small collections of mammals from Hispaniola. *Smithson. Misc. Collect.* **82**(15), 10 pp.

Morgan, G. S. 1989. *Geocapromys thoracatus. Mamm. Species* **341,** 5 pp. American Society of Mammalogists.

Morgan, G. S. 1993. Quaternary land vertebrates of Jamaica. *Mem. Geol. Soc. Am.* **182**:417–442.

Morgan, G. S. 1994. Late Quaternary fossil vertebrates from the Cayman Islands, in: M. A. Brunt and J. E. Davies (eds.), *The Cayman Islands: Natural History and Biogeography,* pp. 465–508. Kluwer Academic, Boston.

Morgan, G. S., and Woods, C. A. 1986. Extinction and the zoogeography of West Indian land mammals. *Biol. J. Linn. Soc.* **28**:167–203.

Musser, G. G., and Carleton, M. D. 1993. Family Muridae, in: D. E. Wilson and D. M. Reeder (eds.), *Mammal Species of the World: A Taxonomic and Geographic Reference,* 2nd ed., pp. 501–756. Smithsonian Institution Press, Washington, D.C.

Nordenskiold, A. E. 1881. *The Voyage of the Vega Round Asia and Europe,* 2 volumes (trans. by A. Leslie). Macmillan & Co., London.

Olson, S. L. 1981. Natural history of vertebrates on the Brazilian islands of the mid South Atlantic. *Nat. Geogr. Soc. Res. Rep.* **13**:481–489.

Olson, S. L. 1989. Extinction on islands: Man as a catastrophe, in: D. Western and M. Pearl (eds.), *Conservation for the Twenty-First Century,* pp. 50–53. Oxford University Press, London.

Oviedo y Valdez, G. F. de. 1535. *Natural History of the West Indies* [trans. and ed. by S. A. Stroudemire, 1959, *Univ. North Carolina Stud. Romance Lang. Lit.* **32**:1–140].

Owen-Smith, R. N. 1988. *Megaherbivores: The Influence of Very Large Body Size on Ecology.* Cambridge University Press, London.

Patton, J. L., and Hafner, M. S. 1983. Biosystematics of the native rodents of the Galapagos archipelago, Ecuador, in: R. L. Bowman, M. Berson, and A. E. Leviton (eds.), *Patterns of Evolution in Galapagos Organisms,* pp. 539–568. AAAS, San Francisco.

Pielou, E. C. 1991. *After the Ice Age: Return of Life to Glaciated North America.* University of Chicago Press, Chicago.

Pino, M., and Castellanos, N. 1985. Acerca de la asociación de perezosos cubanos extinguidos con evidencias culturales de aborígenes cubanos. *Reporte de Investigación del Instituto Ciencias Socials (La Habana),* No. 4, 29 pp.

Pregill, G. K., Steadman, D. W., and Watters, D. R. 1994. Late Quaternary vertebrate faunas of the Lesser Antilles: historical components of Caribbean biogeography. *Bull. Carnegie Mus. Nat. Hist.* **30**:1–51.

Primack, R. B. 1995. *A Primer of Conservation Biology.* Sinauer, Sunderland, Mass.

Ray, C. E. 1964. A small assemblage of vertebrate fossils from Spring Bay, Barbados. *J. Barbados Mus. Hist. Soc.* **31**(1):11–22.

Rounsevell, D. E., and Mooney, N. 1995. Thylacine, in: R. Strahan (ed.), *Mammals of Australia,* 2nd ed., pp. 164–165. Smithsonian Institution Press, Washington, D.C.

Schomburgk, R. H. 1848. *The History of Barbados.* Longmans, London.

Schroering, G. B. 1995. Swamp deer resurfaces. *Wildl. Conserv.* **98**(Dec.):22.

Signor, P. W., and Lipps, J. H. 1982. Sampling bias, gradual extinction patterns and catastrophes in the fossil record. *Spec. Pap. Geol. Soc. Am.* **190**:291–296.

Silva, M., and Downing, J. A. 1995. *CRC Handbook of Mammalian Body Masses.* CRC Press, Boca Raton.

Simons, E. L. 1997. Lemurs: old and new, in: S. M. Goodman and B. D. Patterson (eds.), *Natural Change and Human Impact in Madagascar,* pp. 142–168. Smithsonian Institution Press, Washington, D.C.

Sloane, H. 1707–1725. *A Voyage to the Islands Madera, Barbados, Nieves, S. Christophers and Jamaica . . . ,* 2 volumes [Volume 1, 1707; Volume 2, 1725]. British Museum, London.

Smith, A. P., and Quin, D. G. 1996. Patterns and causes of extinction and decline in Australian conilurine rodents. *Biol. Conserv.* **77**:243–267.

Smith, M. J. 1995. Toolache wallaby, in: R. Strahan (ed.), *Mammals in Australia,* 2nd ed., pp. 339–340. Smithsonian Institution Press, Washington, D.C.

Standing, H. 1908. On recently discovered subfossil primates from Madagascar. *Trans. Zool. Soc. London* **18**:69–162.

Steadman, D. W. 1995. Prehistoric extinctions of Pacific island birds: biodiversity meets zooarchaeology. *Science* **267**:1123–1131.

Steadman, D. W., Stafford, T. W., Donahue, D. J., and Jull, A. J. T. 1991. Chronology of Holocene vertebrate extinctions in the Galápagos Islands. *Quat. Res.* **36**:126–133.

Stiassny, M. L. J. 1994. Systematics and conservation, in: G. K. Meffe and C. R. Carroll (eds.), *Principles of Conservation Biology,* pp. 64–66. Sinauer, Sunderland, Mass.

Stuiver, M., and Reimer, P. J. 1993. Extended ^{14}C data base and revised CALIB 3.0 ^{14}C age calibration. *Radiocarbon* **35**:215–230.

Tattersall, I. 1982. *The Primates of Madagascar.* Columbia University Press, New York.

Times Atlas of the World. 1994. 9th ed. Random House of Canada, Toronto.

Utzurrum, R. C. B. 1992. Conservation status of Philippine fruit bats (Pteropodidae). *Silliman J.* **36**:27–45.

Vigne, J.-D. 1988. *Les Mammifères post-glaciaires de Corse: Etude archéozoologique.* Centre National de la Recherche Scientifique, Paris.

Williams, J. D., and Nowak, R. M. 1986. Vanishing species in our own backyard: extinct fish and wildlife of the

United States and Canada, in: L. Kaufman and K. Mallory (eds.), *The Last Extinction,* pp. 107–140. MIT Press, Boston.

Williams, J. D., and Nowak, R. M. 1993. Vanishing species in our own backyard: extinct fish and wildlife of the United States and Canada, in: L. Kaufman and K. Mallory (eds.), *The Last Extinction,* 2nd ed., pp. 115–148. MIT Press, Boston.

Wilson, D. E., and Reeder, D. M. (eds.). 1993. *Mammal Species of the World: A Taxonomic and Geographic Reference,* 2nd ed. Smithsonian Institution Press, Washington, D.C.

Woods, C. A. 1984. Hystricognath rodents, in: S. Anderson and J. K. Jones (eds.), *Orders and Families of Recent Mammals of the World,* pp. 389–446. Wiley, New York.

Woods, C. A. 1989a. A new capromyid rodent from Haiti: the origin, evolution, and extinction of West Indian rodents and their bearing on the origin of New World hystricognaths, in: C. C. Black and M. R. Dawson (eds.), *Papers on Fossil Rodents in Honor of Albert Elmer Wood,* pp. 59–89. Natural History Museum of Los Angeles County.

Woods, C. A. 1989b. Endemic rodents of the West Indies: the end of a splendid isolation, in: W. Z. Lidicker (ed.), Rodents: A World Survey of Species of Special Concern. *Occas. Pap. IUCN (SSC)* **4**:11–19.

Woods, C. A. 1993. Suborder Hystricognathi, in: D. E. Wilson and D. M. Reeder (eds.), *Mammal Species of the World: A Taxonomic and Geographic Reference,* 2nd ed., pp. 771–806. Smithsonian Institution Press, Washington, D.C.

World Conservation Monitoring Centre. 1992. *Global Biodiversity: Status of the Earth's Living Resources.* Chapman & Hall, London.

Yoshiyuki, M. 1989. *A Systematic Study of the Japanese Chiroptera.* National Science Museum, Tokyo.

Zahler, P. 1996. Rediscovery of the woolly flying squirrel (*Eupetaurus cinereus*). *J. Mammal.* **77**:54–57.

Systematic Index

An "*n*" following a page number indicates that the term may be found in a note on the page indicated.

General Index

An "*n*" following a page number indicates that the term may be found in a note on the page indicated.